3rd International Conference on Maritime Autonomous Surface Ships (ICMASS 2020)

Held online due to COVID-19

IOP Conference Series: Materials Science and Engineering
Volume 929

Ulsan, South Korea
11 - 12 November 2020

ISBN: 978-1-7138-3192-1
ISSN: 1757-8981

Printed from e-media with permission by:

Curran Associates, Inc.
57 Morehouse Lane
Red Hook, NY 12571

Some format issues inherent in the e-media version may also appear in this print version.

Institute of Physics
Dirac House, Temple Back
Bristol BS1 6BE UK

Phone: 44 1 17 929 7481
Fax: 44 1 17 920 0979

techtracking@iop.org

Additional copies of this publication are available from:

Curran Associates, Inc.
57 Morehouse Lane
Red Hook, NY 12571 USA
Phone: 845-758-0400
Fax: 845-758-2633
Email: curran@proceedings.com
Web: www.proceedings.com

TABLE OF CONTENTS

IOP Conf. Series: Materials Science and Engineering **929** (2020) 011001 doi:10.1088/1757-899X/929/1/011001

The International Conference on Maritime Autonomous Surface Ships (ICMASS 2020)

Preface

This special Issue presents peer-reviewed papers presented the 3rd International Conference on Maritime Autonomous Surface Ships (ICMASS) held in Ulsan, R. O. Korea, 11 – 12 November 2020. The ICMASS is the conference which presents cutting edge works on autonomous ships to both academic and industry researchers. The first International Conference on Maritime Autonomous Surface Ships (ICMASS) was arranged in 2018 in Busan, Korea and in 2019 in Trondheim in Norway. Due to COVID-19 pandemic, the conference program this year is provided with online conference. ICMASS 2020 is organized by UIPA (Ulsan ICT Promotion Agency), KMOU (Korea Maritime and Ocean University) and KAUS (Korea Autonomous and Unmanned Ship Forum) and, and co-organized with the International Network for Autonomous Ships (INAS).

The main tracks of ICMASS 2020 include:

1- Maritime Digitalization:

• Digitalization of shore-side decisions and manual processes, and digital integration with government/regulator portals or other relevant stakeholder for licenses, certifications etc.

• Research gap analysis for future needs such as integrated system or development of digital twin

• Digitalized and standardized data for the digital twin and intelligent navigation system

• Safe and secure ship-to-shore communication

2- Autonomous Systems and Vessels:

• Benefits and requirements for the safe operation of the autonomous ship

• Intelligent navigation system for the control and monitoring of the autonomous ship

• Design and operability for autonomous vessels – definition and the framework for the autonomous ship

• Human control and interactions for the safe operation of autonomous ship

• Cyber security and risk issues for the autonomous and automated operations including electronic failures

• Situational awareness and collision avoidance system with intelligent object detection and environment recognition system

• Ship-board enablers and systems for remote operations and management

• Analysis and evaluation methods of the autonomous ships, equipment and services

IOP Conf. Series: Materials Science and Engineering **929** (2020) 011001 doi:10.1088/1757-899X/929/1/011001

based on simulation and sea-trial

• Developing new performance standards, skill requirements and assessments for the remote fleet operators

3- Green Shipping:

• Emerging technologies to support smart shipping with remote control

• Improving energy efficiency through technical operations and optimized logistics planning

• Vessel traffic service with the autonomous ship

• Test-beds and research activities for the efficient shipping

We wish to thank all that have contributed their time to making this event successful. In all, 62 technical presentations have been scheduled as well as 2 keynotes and 2 panel discussion. This proceedings include 35 full length papers. We would like to thank the reviewers and the different committees and boards that have been responsible for the overall management. We also extend thanks to those who have helped with the practical arrangements for the conference. All submissions were subjected to a peer review process, which included the review of an abstract followed by the review of the full paper.

IOP Conf. Series: Materials Science and Engineering **929** (2020) 011001 doi:10.1088/1757-899X/929/1/011001

- Xiaofeng Wu, SICC
- Ando Hideyuki, MTI
- Kay Fjørtoft, SINTEF Ocean
- Ann-Sofie Pauwelyn , De Vlaamse Waterweg

ICMASS 2020 Steering Committee:

- Kwangil Lee, Korea Maritime and Ocean University
- Jungsik Jeong, Mokpo Maritime University
- Seung-Gi Gug, Korea Maritime and Ocean University
- Byungchul Kim, Ulsan ICT Promotion Agency

ICMASS 2020 Program Committee:

- Svein Peder Berge, SINTEF Ocean
- Ørnulf Jan Rødseth, SINTEF Ocean
- Lars Andreas Lien Wennersberg, SINTEF Ocean
- Thomas Porathe, NTNU
- Håvard Nordahl, SINTEF Ocean
- Roger Skjetne, NTNU
- Sashidharan Komandur, IFE
- Hans-Christoph Burmeister, Fraunfhofer CML
- Ann-Sofie Pauwelyn , De Vlaamse Waterweg
- Ando Hideyuki, MTI
- Xiaofeng Wu, SICC
- Chua Kie Hian, TCOMS
- Karvonen Hannu, VTT
- Robin Berglund, VTT

IOP Conf. Series: Materials Science and Engineering **929** (2020) 011001 doi:10.1088/1757-899X/929/1/011001

- Jukka Merenluoto, OneSea
- Gerben peeters, KU Luven
- J. van den Broek, University of Applied Sciences Rotterdam
- Jophannes Huffmeier, RISE University
- Deokje Cho, KRISO
- Hanseon Park, KMI
- Jongju Chae, KIMFT
- Seung-Gi Gug, Korea Maritime and Ocean University
- Jaehoon Kim, Korea Maritime and Ocean University
- Jinkyu Choi, Korea Maritime and Ocean University
- Hansol Kim, Korea Maritime and Ocean University
- Seokje Cho, Korea Maritime and Ocean University
- Yunsu Ha, Korea Maritime and Ocean University
- Seungwook Hwang, Korea Maritime and Ocean University
- Jonghwa Kim, Korea Maritime and Ocean University

ICMASS 2020 Local Organizing Committee:

- Kwangil Lee, Korea Maritime and Ocean University
- Hana, Lee, Ulsan ICT Promotion Agency
- Seoungmi Mun, Seanus
- Sewon Kim, Sejong University
- Mincheol Jo, KOMSA
- Yongsik Oh, Korea Maritime and Ocean University
- Jonghyun Park, Korea Maritime and Ocean University
- Joohyun Woo, Korea Maritime and Ocean University

IOP Conf. Series: Materials Science and Engineering **929** (2020) 011001 doi:10.1088/1757-899X/929/1/011001

UIPA 울산정보산업진흥원

국립 한국해양대학교
NATIONAL KOREA MARITIME & OCEAN UNIVERSITY

KAUS Korea Autonomos and Unmanned Ship Forum

INAS
International Network for Autonomous Ships

과학기술정보통신부
Ministry of Science and ICT

울산광역시
ULSAN METROPOLITAN CITY

nipa 정보통신산업진흥원
National IT Industry Promotion Agency

KMI 한국해양수산개발원
KOREA MARITIME INSTITUTE

KOMSA
한국해양교통안전공단

ETRI 한국전자통신연구원
Electronics and Telecommunications Research Institute

IOP Conf. Series: Materials Science and Engineering **929** (2020) 011002 doi:10.1088/1757-899X/929/1/011002

Peer review declaration

All papers published in this volume *of IOP Conference Series: Materials Science and Engineering* have been peer reviewed through processes administered by the Editors. Reviews were conducted by expert referees to the professional and scientific standards expected of a proceedings journal published by IOP Publishing.

- **Type of peer review:** Single-blind
- **Conference submission management system:** Yes
- **Number of submissions received:** 72
- **Number of submissions sent for review:** 38
- **Number of submissions accepted:** 35
- **Acceptance Rate (Number of Submissions Accepted / Number of Submissions Received X 100):** 48.61
- **Average number of reviews per paper:** 2
- **Total number of reviewers involved:** 30
- **Any additional info on review process:** N/A
- **Contact person for queries:**

Kwangil Lee

Korea Maritime and Ocean University

leeki@kmou.ac.kr

IOP Conf. Series: Materials Science and Engineering **929** (2020) 012001 doi:10.1088/1757-899X/929/1/012001

"PREParE SHIPS" for Automated Ship Passages by Modern Decision Support Tools by Exchanging Future Positions

J Hüffmeier[1], L Sanchez-Heres[1], R Rylander[1], S Nord[1], S Alissa[2], A Bagge[3], P Bergljung[3], P Henkel[4], A Rydlinger[5] and P Hansen[6]

[1] RISE Research Institutes of Sweden, Gothenburg, Sweden
[2] Lantmäteriet, Swedish Mapping, Cadastral and Land Registration Authority, Gävle, Sweden
[3] SAAB Transpondertech, Linköping, Sweden
[4] ANavS GmbH, Munich, Germany
[5] TELKO International, Gothenburg, Sweden
[6] TELKO A/S, Tjøme, Norway

E-mail: johannes.huffmeier@ri.se,

Abstract. Misunderstanding the intentions of other vessels is one of the most common causes of ship collisions today. Such misunderstandings are particularly dangerous during navigation in restricted areas, such as fairways, port areas, and inland water ways, and may be exacerbated with the introduction of unmanned vessels. To ensure safe operations, the understanding of the intentions and future positions of manned and unmanned vessels, the PREParE SHIPS project is creating a system capable of determining a ship's position with high accuracy, predicting its future positions, and communicating the current and future positions. The positioning subsystem combines the signals of Galileo and EGNOS, EGNSS and SBAS respectively, with land based RTK corrections, providing a positioning accuracy of centimetres. Furthermore, by taking advantage of Galileo's authentication features, the positioning system provides enhanced security measures against spoofing attacks. The prediction sub-system will use a ship dynamic model and machine learning to predict the future positions of the ship and improve said predictions with time, while the communication sub-system will broadcast the prediction through VDES ship2ship and ship2shore. PREParE SHIPS will improve the navigation and handling of ships, through high accuracy positioning and dynamic predictions, as well as reduce the risk of ship collisions, through the communication of present and predicted future positions. PREParE SHIPS is expected to increase safety and efficiency significantly and will be the base of future autonomous operations and standardisations.

1. Background

One of the main topics in research regarding autonomous shipping and navigation is collision avoidance and the challenges related especially to the navigational parts of COLREG (collision avoidance regulation), including Rule 2, 8 and 17. Most of the research topics here are focused on sensing and interpretation of sensor information, partly even perception, but little is done to ensure that ship2ship and ship2shore communication is enhanced to solve many of the challenges [1]. One of the exceptions is possibly the H2H project, which looks mainly at communication of relative position. By getting to know the intentions of manned or unmanned ships in the upcoming minutes and seconds,

IOP Conf. Series: Materials Science and Engineering **929** (2020) 012001 doi:10.1088/1757-899X/929/1/012001

allows for better decision making on all vessels and will solve one of the main challenges of today´s seafarers when it comes to ship collisions, a common situational awareness [2]. Accident statistics by a.o. EMSA are describing that collisions and loss of control are the dominating event types reported to European authorities. [3] The specific international rules on Prevention of Collision expressly or implicitly refer to the importance of situation awareness. Rule 5 (Look out) and Rule 6 (Safe Speed) are very good examples emphasizing on "Prevailing Circumstances and making a full appraisal of the situation" or rule 7 (Risk of Collision) requires the OOW "not making assumptions based on scanty information" and also paying attention to "Prevailing Circumstances" or Rule 19 (Conduct of Vessels in Restricted Visibility) which requires "Every vessel to proceed at a safe speed adapted to the prevailing circumstances and conditions". [4]

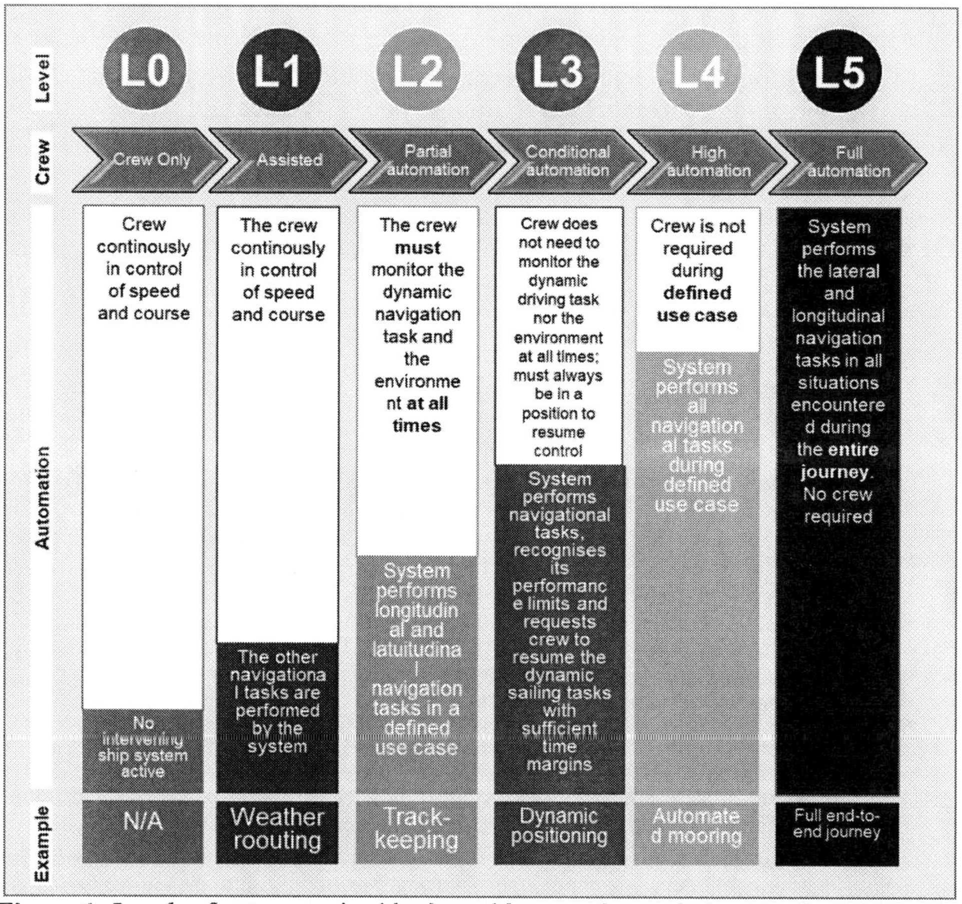

Figure 1: Levels of autonomy in shipping with examples as derived by the PREParE SHIPS project (partly adapted from OICA's Levels of Automated Navigation, based on SAE J3016)

Splitting the situational awareness into logical steps, the crew or any algorithm need to perceive and understand what is happening and then use this to think ahead. Providing decision support to ease these various steps will allow crews to make higher quality decisions and avoid certain biases.

Various studies and techniques pinpoint advanced decision support tools as the next steps towards automation [5] [6] [7], so no disruptive development can be expected but increased levels of automation will be realized through advanced decision support. A systematic review provided in [8] presents an analysis aiming at (1) increasing the understanding of the structure and contents of the academic field concerned with this topic; (2) determining and mapping scientific networks in this domain; (3) analyzing and visualizing Technology Readiness Level (TRL) of analyzed systems. Bibliometric methods are utilized to depict the domain of onboard DSSs for operations focused on safety ensurance and accident prevention. The results indicate that there are relatively many developments in selected DSS categories, such as collision avoidance and ship routing. However, even in these categories some issues and gaps still remain, so further improvements are needed. The analysis indicates a relatively low level of technology readiness of tools and concepts presented in academic literature. This signifies a need to move beyond the conceptual stages toward demonstration and validation in realistic, operating environments.

There are various approaches to describe the development towards more autonomy in navigation, where decision support tools will make higher automation feasible in the short and long term, one example derived in the PREParE SHIPS project context is shown in **Figure *1***.

Dynamic Predictors have been around for some time. Basic predictors would only use dead-reckoning, i.e. ship position, speed, COG (Course over Ground) and heading as sources of information, while more advanced can be based on environmental as well as ship-internal factors such as wind, rudder set and is values, bow-thrusters, etc. While most of the big suppliers provide basic predictors, there is a need of considering more sources of data to give more accurate predictions.

Machine learning is an application of artificial intelligence (AI) that provides systems the ability to automatically learn and improve from experience without being explicitly programmed. Although machine learning has already been used in numerous branches of the digital world, it is adaptation to maritime industry has remained limited so far. This could be partly explained to the low level of integration of ship system and connectivity of ships. Standardisation of data collection techniques is another barrier. As the maritime transportation requires smart tools, application of the machine learning provide utmost benefits for sustainable transportation. [8]

Black models have been challenging in certain applications in the maritime domain for applications such as fuel savings. Simulations models have been used for a long time but have limits in capturing all operating conditions. Grey models seem to be a good trade off combining the advantages of both sides.

2. Making use of current industry developments

There are game changing developments ongoing within the shipping industry and the positioning industry. Examples are EGNSS services for positioning with high accuracy and integrity, VDES as the next generation AIS, and the S100 suit of standards for the accurate information of water depth, currents and Under Keel Clearance (UKC). The project builds its navigation decision support system within shipping on this development which contains the following:

1) EGNSS resilience positioning:
The possibility to dynamically predict future positions of other vessels requires of precise positioning solutions by Galileo receiver/transmitter PPP and HAS enhanced with RTK correction.

2) Real-time dynamic predictor:
Use data learning in order to obtain information of earlier ship behavior to exchange near-future positions with vessels in the vicinity and VTS centers (Vessel Traffic Services) to increase safety and improve decision making.

3) Ship-to-ship / ship-to-shore interaction:
In order to define the correct requirements for PREParE SHIPS combined positioning solution, a collaborative automated vessel application is defined and developed. The vessel application will rely on the high availability positioning solution and use it to couple its various navigational systems with ship2ship/ ship2shore and aggregate information received from other connected vessels by using the next generation AIS - VDES.

4) Geo-fencing:
Implement and demonstrate a fairway geo-fencing with high precision positioning, utilizing various data sources (e.g. wind and current) as well as a traffic monitoring and predicted positions

IOP Conf. Series: Materials Science and Engineering **929** (2020) 012001 doi:10.1088/1757-899X/929/1/012001

so it can allow for safe decisions based on robust data. This means that 'PREParE SHIPS' also will implement perception layer sensor fusion that uses information collected historically in similar conditions based on machine learning-hybrid models.

The picture below shows the flow of information that will be the enabler for PrePARE Ship's Decision Support System.

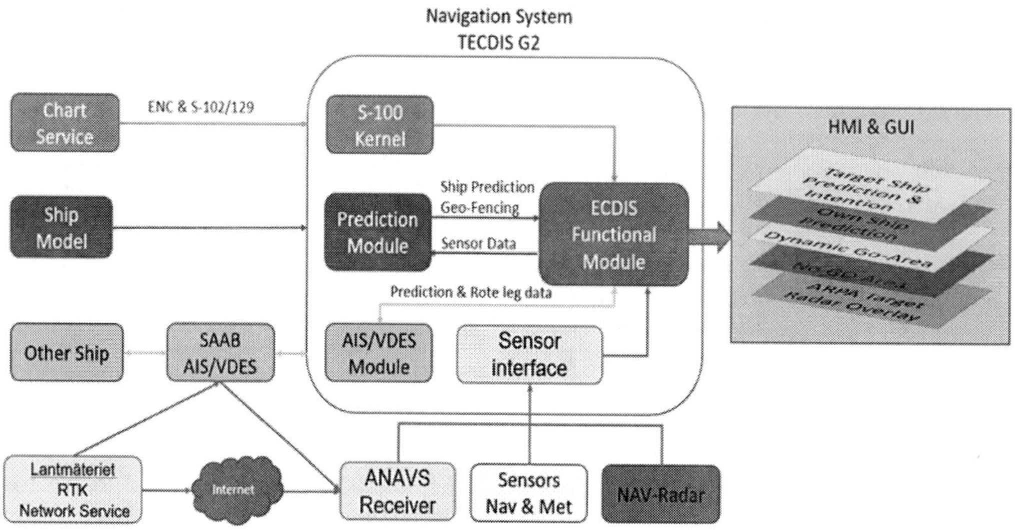

Figure 2: The PREParE SHIPS scope, modules and total integration into ship system

The objective of the project 'PREParE SHIPS' is the development and demonstration of a collaborative resilience navigation solution. It aims to develop and enhance existing software solutions by exploiting the distinguished features of the Galileo signals as well as combining it with other nautical information on internal as well as external parameters and sensor technologies.

2.1. EGNSS development

There are several techniques for augmented GNSS that can be used in order to achieve a higher precision positioning like e.g. DGPS, PPP and RTK. In the PREParE SHIPS project we will develop methods in order to use N-RTK for positioning of ships. As this technology requires land-based reference stations it will focus on positioning in fairways and ports. The reference stations are used to calculate Virtual Reference Stations (VRS) that are placed at suitable locations in the fairway and port. Lantmäteriet, the Swedish mapping, cadastral and land registration authority, is responsible for the N-RTK infrastructure in Sweden called SWEPOS. Based on the physical reference stations (PRS), SWEPOS provides reference data for the locations of the VRS and distributes the reference data to GNSS receivers typically via an Internet connection. In PREParE SHIPS, the N-RTK reference data will be transmitted to ships in fairways via VDES base stations. RTK service area is limited to the locations of PRS:es, therefore PEPaRE SHIPS will use PPP when N-RTK is not available. The N-RTK corrections shall have an accuracy (in positioning domain) of at least 2 cm in at least 95 % of the time if the ship located inside the coverage area for N-RTK, and the PPP corrections (which in the future will be available through the Galileo HAS service) shall have an accuracy of at least 20 cm in at least 95 % of the time. In the Prepare Ship project, PPP (Precise Point Positioning) are developed and tested increasing the understanding of how one can fuse/ integrate RTK services with it. Also, the utilization of correction integrity data in the project is beneficial for increased automation of shipping.

ANavS will provide the Multi-Sensor, Multi-Frequency and Multi-Constellation GNSS receiver system for PREParE SHIPS . The system will differ in three aspects from a standard Multi-Frequency and Multi-Constellation GNSS receiver: First, it will implement an innovative VRS RTK algorithm, that processes Virtual Reference Station (VRS) RTK corrections from a grid of VRSs concurrently instead of using RTK corrections from a single VRS. Second, the ANavS positioning system will determine also a PPP solution being initialized by the RTK solution and will perform an automatic handover between the RTK and PPP solution. The third contribution of ANavS will be the integrity analysis of the RTK/ PPP solution. The main objective of these contributions is to enhance the accuracy, robustness and integrity of the positioning system, which is the basis for the predictor.

2.2. VDES development

The dynamic ship predictor requires substantial communication bandwidth in order to sustain accurate and timely predictions ship to ship. AIS cannot deliver this bandwidth so the research will be performed using VDES as communication channel. VDES is the next generation AIS with up to 32 times the bandwidth compared to AIS. The VDES terrestrial link denoted VDE-TER link ID 19 can deliver 702 bytes per slot after FEC (Forward Error Correction) and the project has selected this link as the primary link for the dynamic ship predictor research. An alternative, but discarded solution, could be VDE-ASM with link ID 5 that can deliver 36 bytes after FEC per slot. There are 2250 slots/minute in the AIS/VDES communication system with 37.5 slots per second.

With VDE-TER link ID 19 the dynamic ship predictors "payload" limit has been set to 650 bytes/slot as an initial assumption and the rest of the bytes (52) have been reserved for link management.

VDE-TER link ID 19 is the link in VDES with the highest bandwidth using 16 QAM modulation scheme. Since VDE-TER link ID 19 will be used at comparably short ranges ship to ship, a few nautical miles, the physical communication link will be very reliable since it also includes FEC.

If the dynamic ship predictor is broadcast every second as an example this will utilize 1/37.5= 2.67 % of the total VDES bandwidth in that cell. The work with the VDES predictor exchange follows the standard development at IALA level [5].

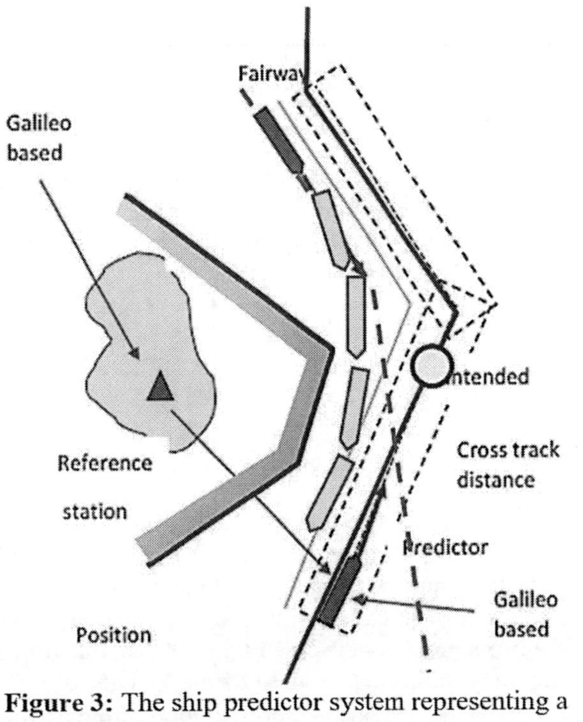

Figure 3: The ship predictor system representing a fairway with draught restricted waters

2.3. Integration and display of S-100 Products

The ECDIS kernel i.e. the platform on which ECDIS functionality is built upon will be able to support the following new data products according to the S-100 product standard.

2.3.1. S-101. The ECDIS kernel and ECDIS GUI will support the use and presentation of nautical charts following the IHO data product specification S-101 Electronic Navigational Chart (ENC).

S-101 is the new product specification for Electronic Navigational Charts following the S-100 product standard that will replace the current ENC specification S-57. S-101 follows the same general structure and principles as S-57 but makes use of the expanded opportunities of the S-100 product standard to add new capabilities to the specification.

The major new capabilities include Machine readable catalogues (allowing dynamic changes to chart features and portrayal over time), Complex attributes (allowing feature attributes to be composed of a structure of sub-attributes), Update features (allowing chart producers to clearly indicate important updates to the chart), Cartographic features (allowing chart producers better control over chart text placement) and Production system attributes (allowing chart producers to provide ECDIS systems with precalculated conditional symbology procedure information, speeding up the chart portrayal).

The S-101 data product will be used for chart portrayal in the PrePARE Ships project as shown in Figure 2.

2.3.2. S-102 Bathymetric Surface product. The ECDIS kernel and ECDIS GUI will support the use and presentation of nautical charts following the IHO data product specification S-101 Electronic Navigational Chart (ENC). The primary purpose of the bathymetric surface product is to provide high resolution bathymetry with special resolution down to meter level, in gridded form, as a supplementary layer to an S-101 ENC within an ECDIS. Portrayal of S-102 bathymetry with other S-100 compliant products is intended to support safe passage, precise berthing and mooring, as well as route planning of marine vessels. The secondary purpose of a bathymetric surface product is to provide high resolution bathymetric data for other maritime applications.

The S-102 bathymetric surface product is a digital elevation model which represents the seafloor in a regular grid structure. It will be used as an additional overlay in the ECDIS. The product specification is based on the IHO S-100 framework specification and the ISO 19100 series of standards. It comprises the content model (spatial structure and metadata), encoding structure, portrayal and exchange file format for a bathymetric surface product.
S-102 data product will be used for two main purposes in the PrePARE Ships project:

1. To determine GO and NO-GO areas with better accuracy than currently possible with current ENCs.
2. To improve the output of the Predictor System by providing more detailed information of underkeel-clearance, i.e. the distance between the hull and the seafloor, and thus improving the estimation of bank, squat, and channel effects.

2.3.3. S-129 Under Keel Clearance Management (UKCM). The primary purpose of the S-129 UKCM Product Specification is to provide a standard for encoding of information a ship needs to determine the time periods when there are suitable tidal conditions for it to transit a UKCM operational area. The UKCM data will be offered by service providers as a UKCM service. Based on the maturity of the standard, the data that can be derived from the S129 could be used in the system for more precise Underkeel Clearance Management based on the dynamic predictor.

3. The PREParE SHIPS Predictor

The Predictor System predicts the future positions and attitudes of the vessel in at least 1 second timesteps and with an indication of prediction accuracy at a minimum frequency of 1Hz. The predictions are based on the current position and attitude, accelerations, ship propulsion settings and environmental conditions. Figure below illustrates the content of the prediction. The predictions and accuracy indicators are used by the ECDIS to:

- Create a visual representation of the predictions and display them in the GUI.
- Create a message to be sent through VDES.

The contents of the datastore are initially defined during the setup up of the predictor system. The setup requires the vessel parameters and the initial settings for the prediction system. Once both are defined, the initial coefficients for the Dynamic Predictor are generated by Dynamic Predictor and the Coefficient Tuner and saved in the datastore.

IOP Conf. Series: Materials Science and Engineering **929** (2020) 012001 doi:10.1088/1757-899X/929/1/012001

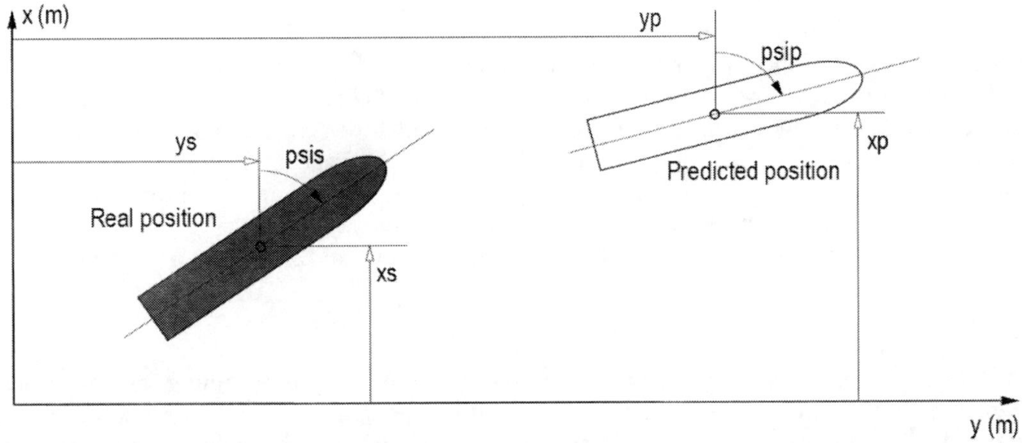

Figure 4: Real and predicted position of a ship.

The Dynamic Predictor module takes the sensor data and voyage data as inputs and returns a prediction of the future position and attitudes of the vessel as defined by the system settings. The predictions are calculated using the coefficients in the data store.

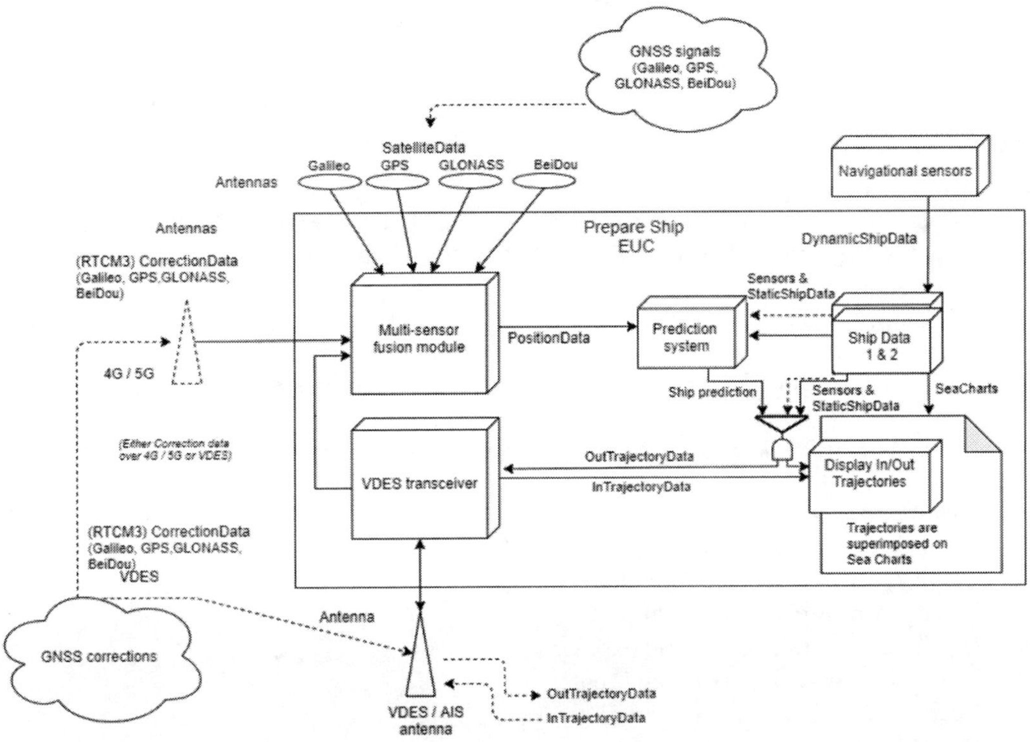

Figure 5: The PREParE SHIPS system topology with the Prediction System in the center

The Prediction Monitor module takes the predictions as input and returns them with a measure of quality unless they are below an absolute minimum safety value, if so, the Prediction Monitor outputs an error. The measures of quality are the result of:

- a learned assessment of the accuracy of the Dynamic Predictor module under the current and future conditions (e.g. deep or shallow waters) and,
- a continuous comparison of the recent predictions against the corresponding sensor data.

IOP Conf. Series: Materials Science and Engineering **929** (2020) 012001 doi:10.1088/1757-899X/929/1/012001

The learned assessment of the accuracy of the Dynamic Predictor is the result of training a machine learning algorithm to predict the accuracy based on previous performance.
The Coefficient Tuner, as mentioned, generates the initial coefficients for the Dynamic Predictor during the setup of the system. After the setup, the Coefficient Tuner continuously tunes the coefficients to improve the predictions of the Dynamic Predictor. The tuning of the coefficients is done through a machine learning algorithm. Class approval of self learning, improving or automated systems are a matter of current research and have a direct impact on the functional safety requirements. As the predictor system and its continously improvements are limited y physical limits of the grey model, these barriers are still able to be overcome.
Further detailed descriptions of some of the essential elements of the Predictor System are given in the following subsections.

3.1. Adaptive Learning
The Adaptive Learning Module improves the prediction over time by tuning the Dynamic Predictor parameters. It compares previous ship positions with its predictions and then minimizes the deviation. This is done by performing tests using new parameters. If new predictions are closer to the actual positions, the new parameters will be used.
The new set of parameters are created by add/ remove a change factor to each parameter element. Assuming linear independence of the elements, each set needs only to change one element at a time. Thus, the total number of "interesting" permutations is only two times the number of elements.
To assure that the new parameters gives better predictions, the position/prediction subset on which the test is performed, needs to represent all possible environmental conditions. To assure stability, the new predictions shall not differ too much from previous. The adaptive learning is based on suitable machine learning algorithms.

The figure on the next page is the use case diagram for the Prepare System. The diagram shows five types of actors:

- Officer On Watch onboard own ship (OOW): The target user of the Prepare System.
- Prepared target ship: A ship with a fully functional Prepare System that is within the VDES range of the own ship.
- STM target ship: A ship with an ECDIS capable of sending and receiving monitored routes through an AIS or ASM message according to the STM's S2SREX specification.
- Multi-GNSS: Multiple constellations of Global Navigation System Satellites (GNSS) sending signals for position fixing.
- Lantmäteriet RTK/PPP: Lantmäteriet's service of RTK and PPP corrections for GNSS signals.

As a list, the use cases in the PREParE Ships project, ensuring safe bybrid navigation, are:

- Determine own ship's present position.
- Display own ship's present position.
- Predict own ship's future positions.
- Display own ship's predicted positions.
- Broadcast own ship's monitored route.
- Receive target ships' monitored routes.
- Display target ship's monitored routes.
- Broadcast own ship's predicted positions.
- Receive target ships' predicted positions.
- Display target ships' predicted positions.
- Display warnings of possible conflicts between own ships and target ships.
- Receive target ship's monitored routes

IOP Conf. Series: Materials Science and Engineering **929** (2020) 012001 doi:10.1088/1757-899X/929/1/012001

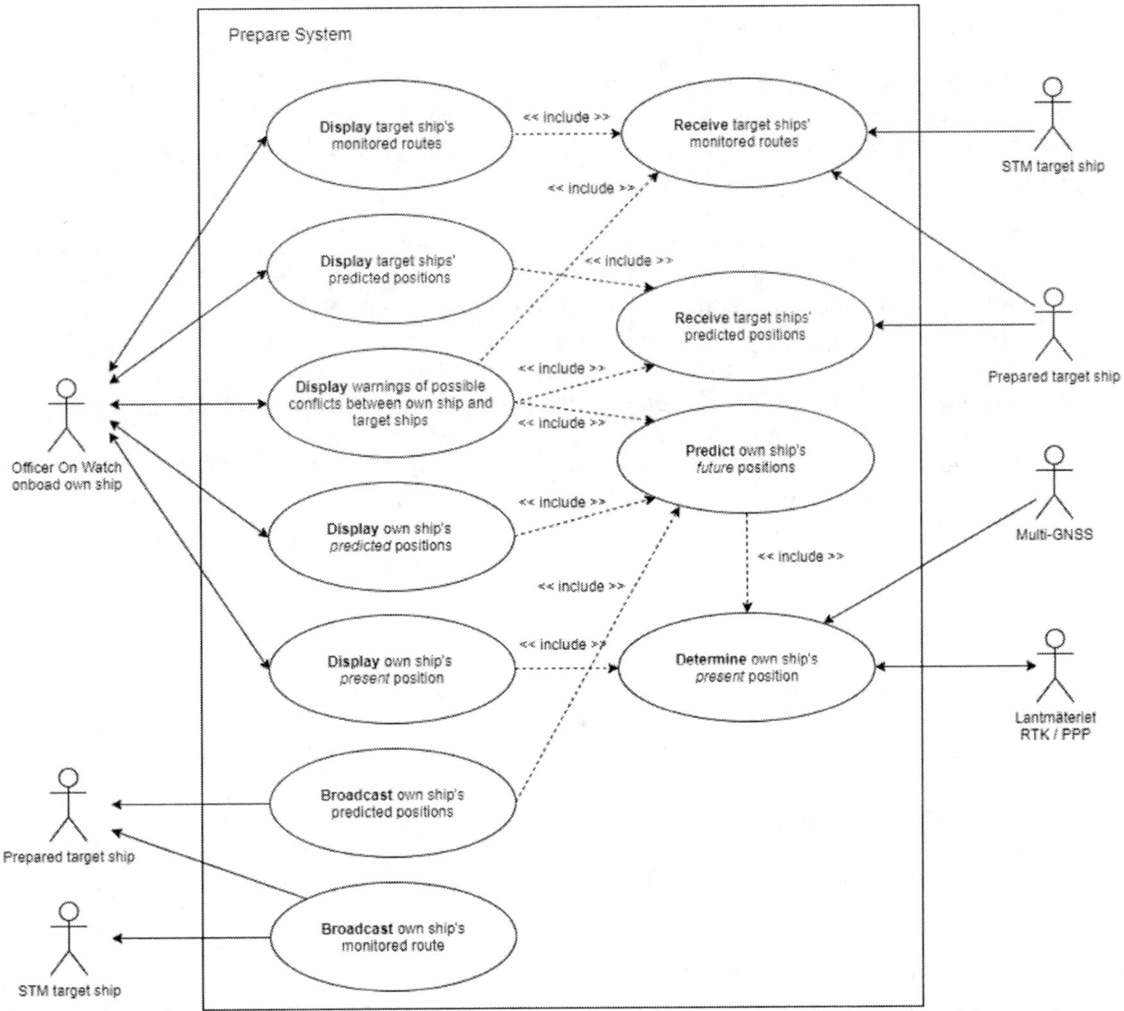

Figure 6: Use Case diagram for the Prepare System including present/ current position and future postiions. Officer on watch placed on «own» ship

3.2. Based on Functional Safety Analysis

The project has made use of established methods to define suitable SIL levels for the PREParE SHIPS concept as a basis for the design of all components and software. Functional safety as such, is a concept applicable across all industry sectors and applied more and more in the shipping domain. The trend towards more complex technology used for safety-related systems is enabled by the functional safety approach ensuring that the safety-related systems will provide the necessary risk reduction measures which are required to achieve safety for the equipment. The dynamic predictor module and all other related systems required to make the PREParE SHIPS system working, are computer based and are performing safety functions (hardware / software / other measures).

To enable the design-, development-, verification- and validation of the system, the functional safety was performed by support of the use cases and application scenarios.

The first step was to conduct a Hazard Analysis and Risk Assessment (HARA) for the PREParE SHIPS concept deriving high level safety requirements. Different scenarios (regarding travelling ships) were identified and the dimensioning (worst-case) scenario was selected as input for the risk assessment. Two safety functions - as described below - were also identified based on the Prepare Ship system architecture and assigned with the safety requirements derived from the HARA. The standard used as reference for this analysis was IEC 61508:2010 The result of the analysis showed that the two identified safety functions became assigned with strict safety requirements (SIL 3). A safety related electric/electronic or programmable electronic system that shall implement safety functions which must fulfil SIL 3 implies that each module included in the system, which plays part in implementing the safety functions, must fulfil SIL 3 themselves. It should be noted that IEC 61508 is an extensive standard in seven parts. However, IEC 61508 provides some flexibility for the developer on how to achieve SIL 3 by selection of different design patterns, but some summarized examples of common measures for SIL 3 are:

- **Management and development processes**. Usage of semi-formal notation and methods for system design specification. Strict following of the safety life cycle, including verification/validation and functional safety assessment with a high degree of independence.
- **Hardware**. Usage of highly reliable electric/electronic components only; Usage of two-channeled architectures (hardware redundancy) for all sensors, logic resolvers and final elements; No usage of commercial off the shelf products unless developed in accordance with IEC 61508.
- **Software**. All software is developed and tested by strictly following the V-model applying all applicable methods listed for SIL 3 in IEC 61508-3. No use of pre-existing software elements unless developed in accordance with IEC 61508 or supported by extensive statistical background data from use in an identical application (such as commercially available operating systems, communication stacks, etc.).

This activity can only be performed by the developers of the different modules composing the PREParE SHIPS system and who have knowledge of the modules detailed design.

The security analysis identified 10 threats, where two of those were of a criticality such that they should be further analysed for their safety impact – :

1. Spoofing of trajectory data. If an attacker sends a fabricated trajectory data and lie about a ship's intended route, this may lead to a collision as the OOW may take wrong decisions based on this.
2. Replay of trajectory data. If an attacker re-sends previous trajectory data, the data displayed on the display does not reflect the current position and intended route, thus, may lead to a collision as the OOW take wrong decisions based on this.

3.3. Providing an advanced decision support system

The decision support system will be built on a Type Approved ECDIS that combines precise positioning from the Galileo based Global Navigation Satellite System including Real Time Correction Data with a Dynamic Ship prediction model supported with high fidelity S-100 Chart Data products. The system will allow interaction with other ships through the VDES communication module allowing sharing of Ships routes and ship predictions.

The ship prediction model will use available sensor and chart data to predict the future movement of the vessel, and machine learning will be used to adapt the prediction model to the characteristics of the vessel. The decision support system will portray predicted future movement of both own ship and received predictions from other ships.

Effective navigation, ship handling, and conflict resolution require the prediction of the future events to come, as well the possible results of different courses of action. Predicting the future movements of a ship can be challenging. A navigation officer may find it difficult to consider the effects of wind, currents, shallow water, trim, and several other factors on its mental model of the ship's response. If so, the control of the ship is reactive instead of predictive, leading to less precise movements. Clearly then, a prediction of the ship's future movements that included transient factors would be of great help. The system developed by the PREParE SHIPS project will include a sub-system capable of predicting the future positions of the vessel through a dynamic vessel motions model (see Figure 6). Because all ships respond differently, the dynamic predictor will leverage machine learning models to tune and improve itself continuously. With time, the dynamic predictor will provide accurate predictions of the ship movements in the most challenging conditions. Furthermore, for conflict resolution, the predictions of future movements will be broadcasted through VDES and displayed on the ECDIS of compatible vessels. Doing so will provide a clear picture of the conflict situation and how its development continues if no action is taken, supporting the selection of actions that will avoid a collision. Overall, the dynamic predictor and the communication module will provide an advanced decision support system.

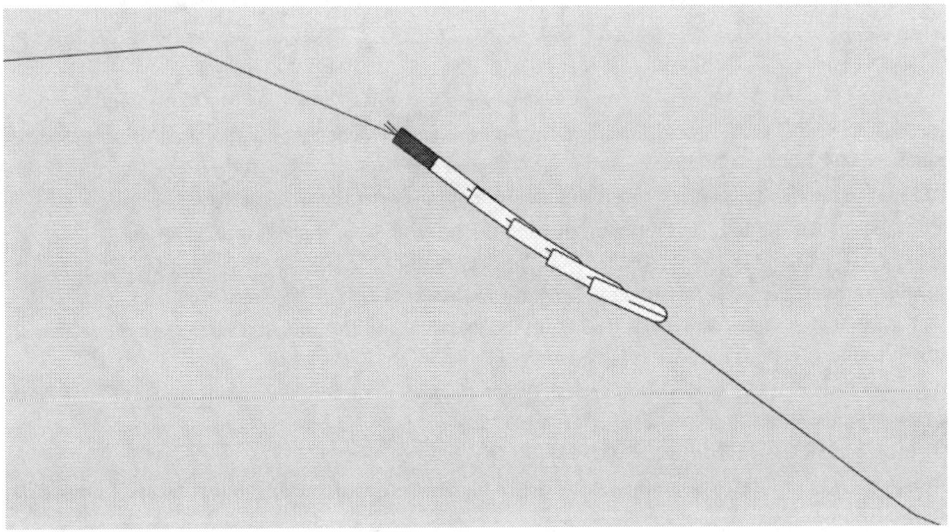

Figure 7: The Dynamic Ship Predictor

4. Discussions and Conclusions

The Ship Decision Support System combines standard information in ECDIS such as Electronic Navigation Charts (ENC), sensor data, AIS and ARPA overlay with state-of-the-art technology developed in the framework of PrePARE Ship project. Position and attitude data determined through the Galileo Global Navigation Satellite Service, Ships next route leg and predicted path both for own ship as well as close ships where route and prediction is transferred via VDES, Bathymetric Surface data and up to date Under Keel Clearance data to give the officer of the watch a better understanding of where she or he can navigate. This all together will provide a situation display and decision support system allowing officer of the watch to look ahead and understand how the navigation situation will develop in the nearest future.

Besides a decreased risk for collisions, this also means additional benefits in the form of a more energy effective maneuvering of the vessels, something which can also reduce the environmental impact of shipping in line with IMO's targets. The project will present and validate a positioning solution. This will be done through development of existing software using Galileo-signals and combining it with nautical information regarding internal and external parameters and sensor technology.

The project has finalized the technical requirement specification and is currently in the state of technical development which will be followed by component testing and integration testing during early 2021 as well as demonstration on commercial ships by the end of 2021. The project aims therefore to reach TRL levels corresponding to prototype level, while components will reach market maturity. Challenges the project tries to solve relate to the project market implementation are related to the standardization and rulemaking in the maritime industry.

Acknowledgement and Disclaimer
The research leading to these results has received funding from the European GNSS Agency under the European Union's Horizon 2020 research and innovation programme under grant agreement No 870239.

The content reflects only the author's view. The Union is not liable for any use that may be made of the information contained therein."

References

[1] S. Øie, E. Norstein och H. J. Johnsrud 2020, "SAFEMASS - Study of the risks and regulatory issues of specific cases of MASS," DNV GL Consulting, Hövik,.

[2] J. Hüffmeier och L. Sanchez-Heres, 2020 "Prepare Ships: D1.1 - Use Cases and Application Scenarios," RISE Research Institutes of Sweden, Gothenburg,.

[3] EMSA, 2019, "Annual Overview of Marine Casualties and Incidents," European Maritime Safety Agency, Lisbon.

[4] M. Safahani och L. S. Tuttle, 2007, *Situation Awareness and its Practical Application in Maritime Domain,* Marine Simulation, BC Ferries Inc,.

[5] ROLLS ROYCE, "Ship intelligence -Transforming future marine operations," ROLLS ROYCE.

[6] Rolls-Royce, , 2016 "Remote and Autonomous ships - The Next Step," Rolls-Royce, London.

[7] J. Walker, "Autonomous Ships Timeline – Comparing Rolls-Royce, Kongsberg, Yara and More," 2019. [Online]. Available: https://emerj.com/ai-adoption-timelines/autonomous-ships-timeline/.

[8] M. Gil, K. Wróbel, J. Montewka och F. Goerlandt, 2020, *A bibliometric analysis and systematic review of shipboard Decision Support Systems for accident prevention,* Elsvier Safety Science 128.

[9] E. Akyuz, K. Cicek och M. Celik, *A Comparative Research of Machine Learning Impact to Future of Maritime Transportation,* Elsevier, p. 275–280.

[10] IALA, *IALA, G1139 THE TECHNICAL SPECIFICATION OF VDES,* IALA.

[11] IEEE, 493-2007, 2007 - IEEE Recommended Practice for the Design of Reliable Industrial and Commercial Power Systems, IEEE.

[12] A. Rydlinger, P. Hansen, P. Ottosson, J. Hüffmeier och L. Sanchez-Heres, 2020, "Prepare Ships: D1.2 - Report on Ship Application System Requirements," RISE Research Institutes of Sweden, Gothenburg.

[13] IEC, "IEC 61508 2010:2010 Functional safety of electrical/electronic/programmable electronic safety-related systems," International Electrotechnical Commission (IEC),.

[14] G. Macher, H. Sporer, R. Berlach, E. Armengaud och C. Kreiner, 2015, *SAHARA: A security-aware hazard and risk analysis method,* Design, Automation Test in Europe Conference Exhibition, p. 621–624.

[15] G. Macher, E. Armengaud, E. Brenner och C. Kreiner, 2016, *Threat and Risk Assessment Methodologies in the Automotive Domain,* vol. 83, Procedia Computer Science, p. 1288–1294.

[16] IEC, , 2010 *IEC 61508-1: Functional safety of electrical/electronic/programmable electronic safety-related systems – Part 1: General requirements,* International Electrotechnical Commission (IEC),.

[17] IEC, 2010, *IEC 61508-5: Functional safety of electrical/electronic/programmable electronic safety-related systems – Part 5: Examples of methods for the determination of safety integrity levels,* International Electrotechnical Commission (IEC),.

IOP Conf. Series: Materials Science and Engineering **929** (2020) 012002 doi:10.1088/1757-899X/929/1/012002

Adaptive Neural Tracking Control for an Autonomous Azimuthing Stern Drive (ASD) Tug

Bernard Voon Ee HOW, Yuting JIN, Lucas J YIEW,
Daniel LIU and Allan R MAGEE

Technology Centre for Offshore and Marine Singapore (TCOMS), 12 Prince George's Park, Singapore 118411

E-mail: `bernard_how@tcoms.sg`

Abstract. The ASD tug is considered the modern habour tug due to its versatility in performing a wide variety of tugging/towing operations. With the advent of autonomous systems and desire to improve operations, there are benefits to automating the tug operations. Under different operating environments, there could be uncertainties in the model and the disturbance from the environment significant enough to affect the performance and control of the tug. In this paper, we consider the problem of tracking a desired trajectory for an autonomous ASD tug in the presence of uncertainties and unknown disturbances. The numerical modelling of an ASD tug based on a modified 4-DoF MMG model is first presented. The modified Maneuvering Modeling Group (MMG) model is then used to design a model based backstepping control. Thereafter, a adaptive neural network approximator is introduced which has the capability to account for uncertainties and unknown disturbance. The combination of approximation-based and backstepping design techniques allows us to handle time-varying model uncertainties. Stability analysis is carried out for the control design via Lyapunov analysis. Simulation are carried out for the tracking of maneuvering motion paths to demonstrate the performance of proposed approach.

1. INTRODUCTION

For future maritime autonomous surface ships (MASS), the on-board navigation and collision detection system require close to real-time projection of environmental influence on vessel behaviour in order to fill the role played by experience. This promotes the system-based manoeuvring models considering their rapid computational time despite the compromise on accuracy that sometimes must be made for long term trajectory forecasting. The most widely adopted models that are able to fulfil such requirements are, the Abkowitz model [1] based on truncated Taylor series, the Norrbin second order modulus model [2], and the modular-type model (MMG) which was recently standardised by the Japan Society of Naval Architects and Ocean Engineers [3]. When studying ship manoeuvrability in waves, the MMG model gains popularity since wave effect can be isolated from the hydrodynamic damping terms.

To allow the autonomous surface ships, to track a desired trajectory, the control system needs to be designed. Ship manoeuvring control systems need to cater for operating conditions and environment, such as inertia due to the massive structure, the hydrodynamics, wind, wave, current, and within the control design the parametric and functional uncertainties, un-modeled dynamics [4, 5]. On these challenges, a number of control approaches have been proposed with significant results achieved, such as classical Proportional-Integral-Derivative (PID) control [6, 7], Linear Quadratic Gaussian (LQG) control [8], H_∞ control [9, 10], Model Reference Adaptive Control [11], Feedback Linearization Control [12], Sliding Mode Control [13, 14]. For example, a direct feedback linearization based adaptive control was proposed for automatic steering of ships [12] in which both coursekeeping and course-changing were considered.

In the control domain, the backstepping control design technique is a useful approach for for a range of nonlinear systems [15]. For example, for ship steering, a nonlinear backstepping controller was designed in strict-feedback form [16] and further optimized using genetic algorithm [17]. To construct and apply the nonlinear backstepping control, accurate parameters and nonlinear functions of ship manoeuvring model are needed. By incorporating the Nussbaum-type gain into adaptive backstepping design, a robust controller was proposed for ship heading nonlinear system in the presence of unknown sign of uncertain control coefficients [18]. Later, this approach was extended to ship heading nonlinear system in the presence of uncertain control coefficients [19] as well as input saturation [20, 21]

For work with parametric uncertainties, adaptive neural network (ANN) control design becomes an advanced approach in dealing with the highly uncertain, nonlinear and complex systems [22]. Then, the minimal learning parameter (MLP) techniques was applied to control scheme [23] to reduce the number of design control parameters update online [24]. This elegant adaptive control scheme was extended to ship steering uncertain nonlinear system with rudder actuator dynamic [25]. In [26], a class of feedforward approximators was designed for planar control of surface vessels. Recently, a modified DSC approach and NNs based adaptive control scheme for steering system of a robotic unmanned surface vehicle (USV) with model uncertainties and measurement noises was proposed [27]. In addition, several adaptive NN control approaches were employed to carry out the simulation studies for ship steering systems [28, 29] recently.

Inspired by the above works, in this paper, the numerical modelling uses a modified 4-DoF MMG model to describe the manoeuvrability of a generic Azimuthing Stern Drive (ASD) tug developed at TCOMS. To extend beyond planar and steering type adaptive neural network control, we focus the control design on the 4-DoF ASD MMG type model developed. The contributions of this paper are:

- the 4DoF modeling of an ASD tugboat based on a modified MMG model.
- the design of a 4DoF model based control for the using a backstepping approach and
- the design of the 4DoF adaptive neural control for ASD type vessel which has the capabilities to account for uncertainties in the model and time varying disturbances

2. ASD TUG MODEL & COORDINATES

An example of a ASD tug is shown in Figure 1. Two right-handed coordinate systems are used in this study as shown in Figure 2. The local reference frame is represented as (x, y, z) which moves along with the manoeuvring motion, and the origin is located at the centre of gravity (CoG). The global reference frame is denoted as (x_o, y_o, z_o) and is fixed in the earth-fixed-frame. The two coordinate systems align at the initial instant and the global velocities of the vessel are related to the local velocities as

$$\begin{aligned} \dot{x} &= u\cos\psi(t) - v\sin\psi(t) \\ \dot{y} &= u\sin\psi(t) + v\cos\psi(t) \\ \dot{\phi} &= p \\ \dot{\psi} &= r \end{aligned} \tag{1}$$

where u, v, p and r are the surge, sway, roll angular and yaw angular velocity in the body-fixed coordinates of the vessel respectively. ϕ and ψ as the roll and yaw angle in the earth-fixed coordinates of the vessel and $(\dot{*})$ is the derivative of $(*)$ with respect to time.

3. FORMULATION OF 4-DOF MANOEUVRING MODEL

Manoeuvring motions in this paper are represented by surge, sway, roll and yaw, where only low frequency disturbances are considered. In addition, the hydrodynamic forces acting on the ship are treated quasi-steadily. The 4-DoF manoeuvring model is written by the following equations of motion.

Figure 1. OVERVIEW OF TCOMS G-TUG MODEL

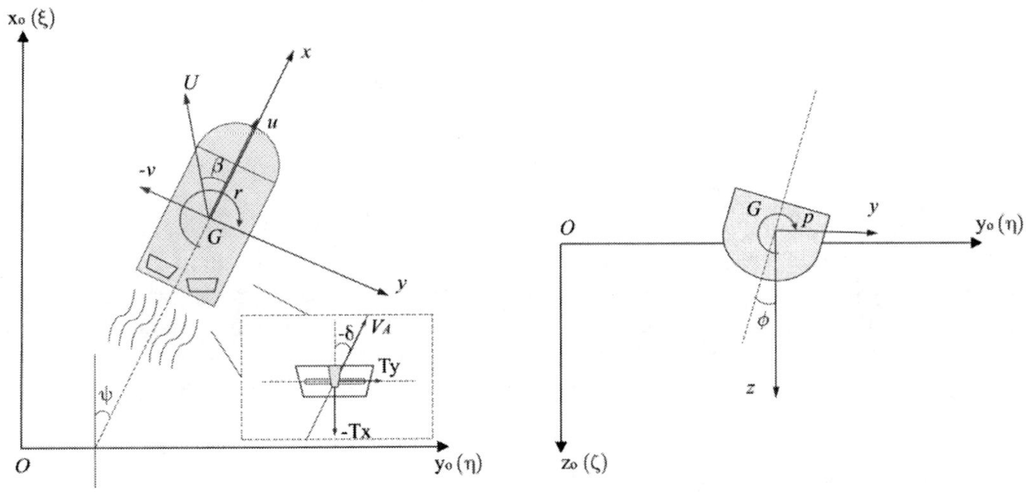

Figure 2. COORDINATE SYSTEM OF 4-DOF MMG MODEL

The subscripts H and T represent the hydrodynamic loads exerted on or generated from the bare hull and thrusters.

$$\begin{aligned}(m+m_x)\dot{u}-mvr &= X_H+X_{T1}+X_{T2}+d_u\\(m+m_y)\dot{v}+mur &= Y_H+Y_{T1}+Y_{T2}+d_v\\(I_{xx}+J_{xx})\dot{p} &= m_x z_H ur - mg\overline{GZ}(\phi)+K_H+K_{T1}+K_{T2}+d_p\\(I_{zz}+J_{zz})\dot{r} &= N_H+N_{T1}+N_{T2}+d_r\end{aligned} \tag{2}$$

where m is the displacement of the vessel, m_x, m_y are the added mass of the vessel, I_{xx}, I_{zz} are the moment of inertia of the vessel, J_{xx}, J_{zz} the added moment of inertia of the vessel, $\overline{GZ}$ the hydro-static righting arm of the vessel, d_u, d_v, d_p, d_r are exogenous disturbance due to the environment in surge, sway, roll and yaw DoF respectively.

3.1. HYDRODYNAMICS OF HULL

In Equation (2), the hull hydrodynamic loads (X_H, Y_H, K_H, N_H) are non-dimensionlised by the Prime-system II [30] with the following expressions:

$$
\begin{aligned}
X_H &= \frac{\rho L T u^2}{2} R'(u) + \frac{\rho L T U^2}{2} X'_H(u', v') \\
Y_H &= \frac{\rho L T U^2}{2} Y'_H(v', r') \\
K_H &= \frac{\rho L^2 T U^2}{2} K'_H(v', r', p') \\
N_H &= \frac{\rho L^2 T U^2}{2} N'_H(v', r')
\end{aligned} \tag{3}
$$

where $R'(u)$ is the Calm water resistance coefficient, X'_H, Y'_H, K'_H and N'_H are the non-dimensionalised surge, sway force and roll and yaw moment respectively. Expanding the non-dimensionlised hull damping (X'_H, Y'_H, K'_H, N'_H) are expanded as Taylor series of (u', v', r', p'):

$$
\begin{aligned}
R' &= X_0 + X_{uu}uu + X_{uuuu}uuuu \\
X'_H &= X_{vv}v'v' + X_{rr}r'r' + X_{vr}v'r' \\
Y'_H &= Y_v v' + Y_r r' + Y_{vvv}v'v'v' + Y_{vvr}v'v'r' + Y_{vrr}v'r'r' + Y_{rrr}r'r'r' \\
K'_H &= K_p p' + K_{ppp}p'p' + K_v v' + K_r r' + K_{vvv}v'v'v' + K_{vvr}v'v'r' + K_{vrr}v'r'r' + K_{rrr}r'r'r' \\
N'_H &= N_v v' + N_r r' + N_{vvv}v'v'v' + N_{vvr}v'v'r' + N_{vrr}v'r'r' + N_{rrr}r'r'r'
\end{aligned} \tag{4}
$$

r' is the on-dimensionalised yaw rate $r' = r(L_{pp}/U)$, p' the non-dimensionalised roll rate $p' = p(L_{pp}/U)$, v' the non-dimensionalised rate rate $v' = v/U$ and U the resultant ship over water speed (m/s). Terms such as ($X_v v$, $X_v r$, Y_v, N_r and etc) are the so called hydrodynamic derivatives or manoeuvring coefficients. In the surge equation, even terms are dominant whereas for the sway and yaw functions, the odd terms are superior to other components. These parameters are initially generated from a series of quasi-steady manoeuvres with CFD virtual Planar Motion Mechanism computations.

Assumption 1 *Referring to [31], the relation between the roll moment K and the sway force Y is assumed to be linear and is expressed as,*

$$
\begin{bmatrix} K_v & K_r & K_{vvv} & K_{vvr} & K_{vrr} & K_{rrr} \end{bmatrix} = z_H \begin{bmatrix} Y_v & Y_r & Y_{vvv} & Y_{vvr} & Y_{vrr} & Y_{rrr} \end{bmatrix} \tag{5}
$$

Assumption 2 *The exogenous disturbances d_u, d_v, d_p, and d_r are smooth functions, dependent on the states of the system x, y, ψ, ϕ, u, v, p, r and time t, and are bounded.*

Remark 1 *Assumption 2 is reasonable since the disturbances from the environment acting on the tug changes with the states of the tug and the time-dependent component of the disturbance can be largely attributed to the exogenous effects of the environment, which have finite energy and, hence, are bounded.*

4. MMG MODEL BASED AND ADAPTIVE NEURAL CONTROL DESIGN

Consider the multiple-input-multiple-output (MIMO) dynamics of the 4 DoF ASD tug defined in (1) and (2). Expressing the model as a concise state space model, we have

$$
\begin{aligned}
\dot{\eta} &= J(\eta)v \\
M\dot{v} &= \Delta(\eta, v, t) + \tau + d(\eta, v, t)
\end{aligned} \tag{6}
$$

where

$$J = \begin{bmatrix} \cos\psi & -\sin\psi & 0 & 0 \\ \sin\psi & \cos\psi & 0 & 0 \\ 0 & 0 & 1 & 0 \\ 0 & 0 & 0 & 1 \end{bmatrix}, \quad \Delta(\eta, v, t) = \begin{bmatrix} X_H + mvr \\ Y_H - mvr \\ m_x z_H ur - mg\overline{GZ}(\phi) + K_H \\ N_H \end{bmatrix} \tag{7}$$

where $J(\eta)$ is the Jacobian transformation matrix, $M = diag[m+m_x, m+m_y, I_{xx}+J_{xx}, I_{zz}+J_{zz}]^T \in R^{4\times4}$ the output $\eta = [x, y, \phi, \psi]^T \in R^4$ are the Earth-frame positions, roll and heading, respectively; $v = [v_x, v_y, v_\phi, v_\psi]^T \in R^4$ are the vessel-frame surge, sway, roll and yaw velocities, respectively; $\Delta(\eta, v, t) \in R^4$ the hull hydrodynamics and coupling terms, $d(\eta, v, t) = [d_u, d_v, d_p, d_r]^T \in R^4$ represents the unknown disturbance from the environment and can also be used to include un-modeled and uncertainties among others, $\tau = [X_{T1} + X_{T2}, Y_{T1} + Y_{T2}, K_{T1} + K_{T2}, N_{T1} + N_{T2}]^T \in R^4$ is the vector of thruster signals.

Lemma 1 *[32] For the continuous functions $d_i(\eta, v, t) : R^4 \times R^4 \times R \to R, i = 1, 2, 3, 4$, there exist positive, smooth nondecreasing functions $p_i(\eta, v) : R^4 \times R^4 \to R^+$ and $q_i(t) : R \to R^+$ such that*

$$|d_i(\eta, v, t)| \leq p_i(\eta, v) + q_i(t) \tag{8}$$

Lemma 1 allows one to separate the multivariable disturbance term $d_i(\eta, v, t)$, for $i = 1, 2, 3, 4$ into a bounding function in terms of v, η, the internal states of the tug, and a bounding function in terms of t, which generally includes exogenous effects and uncertainties.

The control objective is to ensure that all signals are bounded, while the output follows a desired trajectory $\eta_d = [x_d, y_d, \phi_d, \psi_d]^T \in \Omega_{\eta_d}$, such that the tracking errors converge to a small neighborhood of the origin, i.e., $\lim_{t\to\infty} \|\eta - \eta_d\| < \delta_e$ for some $\delta_e > 0$.

Assumption 3 *For the time-dependent functions $q_i(t), i = 1, 2, 3$, there exist constants $\bar{q}_i \in R^+, \forall t > t_0$, such that*

$$\|q_i(t)\| \leq \bar{q}_i \tag{9}$$

Assumption 4 *For all $t > 0$, there exists constants $N_1, N_2 > 0$ such that $\|\dot{\eta}_d(t)\| \leq N_1$ and $\|\ddot{\eta}_d(t)\| \leq N_2$*

Remark 2 *Assumption 4 requires that the desired trajectory be sufficiently smooth to avoid actuator saturation induced by sudden jumps of tracking error due to discontinuous command inputs.*

Lemma 2 *[33] For bounded initial conditions, if there exists a C^1 continuous and positive definite Lyapunov function $V(x)$ satisfying $\gamma_1(\|x\|) \leq V(x) \leq \gamma_2(\|x\|)$, such that $\dot{V}(x) \leq -\rho V(x) + c$, where $\gamma_1, \gamma_2 : R^n \to R$ are class K functions and c is a positive constant, then the solution $x(t)$ is uniformly bounded.*

4.1. ADAPTIVE NEURAL APPROXIMATORS

Function approximators can be represented as multilayer feedforward networks which may be nonlinearly- or linearly-parameterized. Examples of feedforward approximators include adaptive neural networks [22, 33, 34, 35] and adaptive fuzzy systems [36]. A class of linearly parametrized feedforward approximators used to approximate the continuous function $f(Z) : R^q \to R$ may be represented as follows:

$$f(Z) = W^T S(Z) + \varepsilon(Z)$$

where the vector $Z = [z_1, z_2, \ldots, z_q]^T \in R^q$ are the input variables to the approximator, $S(Z) \in R^l$ is a vector of known continuous (linear or nonlinear) basis functions, $W \in R^l$ is a vector of adaptable weights, and ε is the approximation error which is bounded over the compact set, i.e., $|\varepsilon(Z)| \leq \bar{\varepsilon}, \forall Z \in \Omega_Z$, where $\bar{\varepsilon} > 0$ is an unknown constant.

We consider a class of linearly parameterized feedforward approximators, which, according to the universal approximation property [15], can smoothly approximate any continuous function $f(Z)$ over a compact set $\Omega_Z \subset R^q$ to arbitrary any degree of accuracy as

$$f(Z) = W^{*T} S(Z) + \varepsilon^*(Z), \quad \forall Z \in \Omega_Z \subset R^q \tag{10}$$

where W^* are the ideal constant weights in the output layer, and $\varepsilon^*(Z)$ is the approximation error for the special case where $W = W^*$. The ideal weight vector W^*, an artificial quantity required for analytical purposes, is defined as the value of W that minimizes $|\varepsilon(Z)|$ for all $Z \in \Omega_Z \subset R^q$, i.e.

$$W^* := \arg \min_{W \in R^l} \left\{ \sup_{Z \in \Omega_Z} \left| f(Z) - W^T S(Z) \right| \right\} \tag{11}$$

Motivated by [36] the basis functions are defined by

$$s_i(Z) := \frac{\mu_i(Z)}{\sum_{j=1}^{N} \mu_j(Z)}, \quad \mu_i(Z) := \prod_{j=1}^{N} \nu_j \tag{12}$$

where ν_j is one of the three functions from the j th set:

$$\nu_j = \begin{pmatrix} 1/\left(1 + e^{a_{1j}(Z_j - b_{1j})}\right) \\ e^{-a_{2j}|Z_j - b_{2j}|^2} \\ 1/\left(1 + e^{-a_{3j}(Z_j - b_{3j})}\right) \end{pmatrix} \tag{13}$$

such that each μ_i is composed of a unique combination of the functions from the N sets. The constant parameters a_{kj} and b_{kj} are user-defined.

4.2. CONTROL DESIGN

Define error variables $z_1 = \eta - \eta_d$ and $z_2 = v - \alpha_1$ and consider Lyapunov function candidate $V_1 = (1/2) z_1^T z_1$. Differentiating z_1 with respect to time yields

$$\dot{z}_1 = J(\eta)\left(z_2 + \alpha_1\right) - \dot{\eta}_d \tag{14}$$

Noting the property $JJ^T = I$, and choosing the virtual control as

$$\alpha_1 = J^T(\eta)\left(\dot{\eta}_d - K_1 z_1\right) \tag{15}$$

where $K_1 = K_1^T > 0$, the time derivative of V_1 along (14) is given by

$$\dot{V}_1 = -z_1^T K_1 z_1 + z_1^T J(\eta) z_2 \tag{16}$$

The first term on the right-hand side (RHS) is stabilizing, and the second term will be handled in the next step. next, differentiating z_2 with respect to time yields

$$\dot{z}_2 = M^{-1}(\Delta(\eta, \nu, t) + \tau + d(\eta, \nu, t)) - \dot{\alpha}_1 \tag{17}$$

where $\dot{\alpha}_1 = (\partial\alpha_1/\partial\eta)\,\dot{\eta} + (\partial\alpha_1/\partial\dot{\eta}_d)\,\ddot{\eta}_d + (\partial\alpha_1/\partial z_1)\,\dot{z}_1$. Consider the Lyapunov function candidate $V_2^* = V_1 + (1/2)z_2^T M z_2$. From Lemma 1, Assumption 2 and 3 we have the following:

$$\begin{aligned}\dot{V}_2^* \;\le\; & -z_1^T K_1 z_1 + z_1^T J(\eta) z_2 \\ & + z_2^T\left(\Delta(\eta,\nu,t) - M\dot{\alpha}_1 + \tau\right) \\ & + \sum_{i=1}^{3} |z_{2,i}|\left(p_i(\eta,v) + \bar{q}_i\right)\end{aligned} \tag{18}$$

Consider the following desired control law

$$\begin{aligned}\tau^* = & -J^T(\eta)z_1 - K_2 z_2 - \Delta(\eta,\nu,t) \\ & + M\dot{\alpha}_1 - \mathrm{Sgn}\,(z_2)\,(p(\eta,v) + \bar{q})\end{aligned} \tag{19}$$

where $\mathrm{Sgn}\,(z_2) := \mathrm{diag}\,[\mathrm{sgn}\,(z_{2,i})]$, for $i = 1,2,3,4$ with $\mathrm{sgn}\,(\cdot)$ as the signum function. Substituting (18) into (19) the latter can be rewritten as $\dot{V}_2^* \le -z_1^T K_1 z_1 - z_2^T K_2 z_2$.

In the case where $\Delta(\eta,\nu,t)$,$p(\eta,v)$, and $q(t)$ are all unknown, the model-based control law (19) is not feasible. To overcome this problem, we utilize adaptive neural control to estimate the unknown entities in the control law as follows:

$$\tau = -J^T(\eta)z_1 - K_2 z_2 + \hat{W}^T S(Z) \tag{20}$$

$$\dot{\hat{W}}_i = -\Gamma_i\left(S_i(Z) z_{2i} + \sigma_i \hat{W}_i\right) \tag{21}$$

where $\hat{W} := \mathrm{diag}\left[\hat{W}_1^T, \hat{W}_2^T, \hat{W}_3^T, \hat{W}_4^T\right]$ contains the approximation weights, $S(Z) = \left[S_1^T(Z),\ S_2^T(Z),\ S_3^T(Z),\ S_4^T(Z)\right]^T$ are the basis functions and σ_i is a positive constant. The neural network $\hat{W}^T S(Z)$ approximates $W^{*T} S(Z)$ defined by

$$W^{*T} S(Z) = \Delta(\eta,\nu,t) + M\dot{\alpha}_1 - \mathrm{Sgn}\,(z_2)\,(p(\eta,v) + \bar{q}) - \varepsilon(Z) \tag{22}$$

where $\varepsilon(Z) \in R^4$ is the approximation error, $\mathrm{Sgn}\,(z_2) := \mathrm{diag}\,[\mathrm{sgn}\,(z_{2,i})]$, for $i = 1,2,3,4$ with sgn(.) as the signum function, $Z = \left[\eta^T, v^T, \alpha_1^T, \dot{\alpha}_1^T\right]^T$ are the input variables to the feedforward approximators, and $z_{2,i} \in R$ for $i = 1,2,3,4$ are the elements of z_2

Consider the augmented Lyapunov function candidate

$$V_2 \;=\; V_1 + \frac{1}{2} z_2^T M z_2 + \frac{1}{2}\sum_{i=1}^{3} \tilde{W}_i^T \Gamma_i^{-1} \tilde{W}_i \tag{23}$$

where $\tilde{W}_i := \hat{W}_i - W_i^*$. Differentiating V_2 along (14), (17) and (21) yields

$$\dot{V}_2 \;\le\; -\rho_l V_2 + C \tag{24}$$

$$\rho_l \;:=\; \min\left(2\lambda_{\min}(K_1),\ \frac{2\lambda_{\min}\left(K_2 - \frac{1}{2}I_{3\times3}\right)}{\lambda_{\max}(M)}\ \min_{i=1,2,3,4}\left(\frac{\sigma_i}{\lambda_{\max}\left(\Gamma_i^{-1}\right)}\right)\right) \tag{25}$$

$$C \;:=\; \sum_{i=1}^{3} \frac{\sigma_i}{2}\|W_i^*\|^2 + \frac{1}{2}\|\bar{\varepsilon}\|^2 \tag{26}$$

where $\lambda_{\min}(\bullet)$ and $\lambda_{\max}(\bullet)$ denote the minimum and maximum eigenvalues of $\bullet$, respectively.

Theorem 1 *Consider the tug dynamics* (6) *with Assumptions 3 and 4 under the action of control law* (20) *and adaptation law* (21). *Then, for each compact set* Ω_0, *where* $\left(\eta(0), v(0), \hat{W}_1(0), \hat{W}_2(0), \hat{W}_3(0), \hat{W}_4(0)\right) \in \Omega_0$, *the trajectories of the closed-loop system are semiglobally uniformly bounded. The tracking error* z_1 *converges to a compact set* $\Omega_{zs} := \left\{z_1 \in R^4 \mid \|z_1\| \leq \sqrt{(2C/\rho_l)}\right\}$ *asymptotically, where* C *and* ρ_l *are defined in* (26) *and* (25) *respectively.*

Proof: From (24) and Lemma 1, it is clear that the signals $z_1, z_2, \tilde{W}_1, \tilde{W}_2, \hat{W}_3, \hat{W}_4$ are semiglobally uniformly bounded. From the boundedness of η_d in Assumption 2, we know that η is bounded. since $\dot{\eta}_d$ is also bounded, it follows that α_1 is bounded, and in turn v is bounded. With W_i^* as a constant, we know that $\hat{W}_i$ is also bounded, for $i = 1, 2, 3, 4$. Therefore, all signals are bounded. To show that z_1 converges to Ω_{zs} is straightforward using a similar approach as that found in [22].

5. SIMULATION RESULT

The physical geometry of the investigated generic ASD tug is developed as shown in Figure 1 and parameters in Table 1. For the hydrodynamic damping, the Taylor series of hydrodynamic damping forces and moments are ideal for small oblique flow angles $0^o < \beta < 30^o$ or $330^o < \beta < 360^o$ when the ASD tug advances forward. For operational conditions where β exceeds these ranges, the hydrodynamic damping can no longer be expressed in a concise mathematical formula. Therefore a database of (X, Y, K, N) was generated based on the quasi-static drift CFD simulations as given in Figure 3.

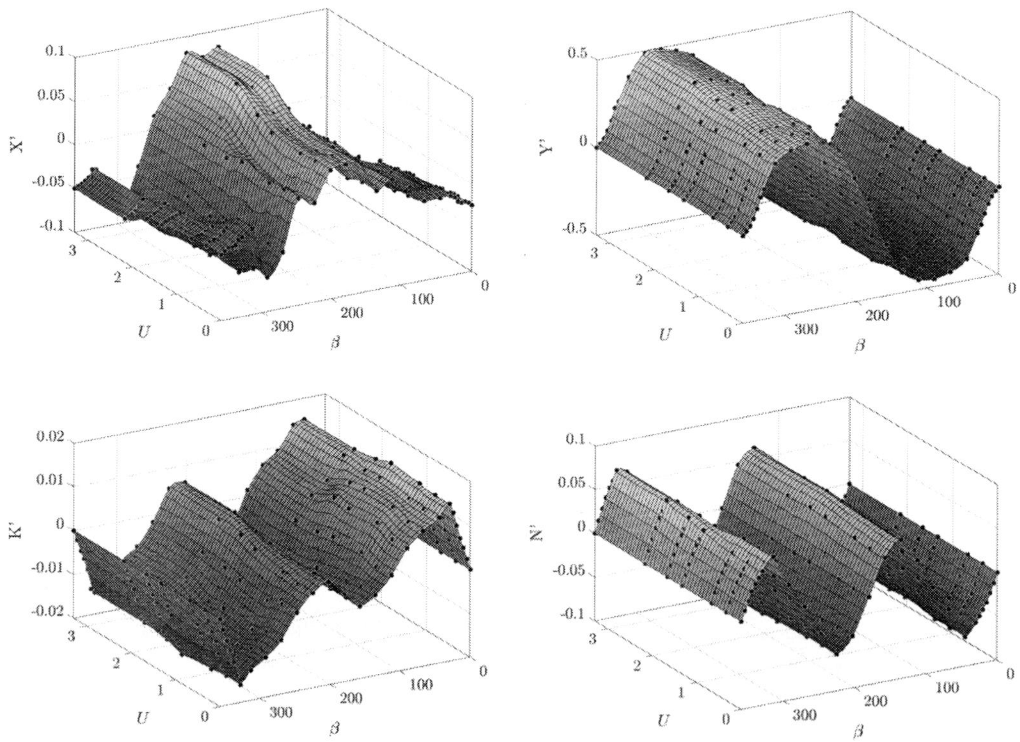

Figure 3. HYDRODYNAMIC DAMPING AT DIFFERENT OBLIQUE FLOW ANGLES

For control of the autonomous ASD tug, the control objective is to track the desired trajectory $\eta_d(t) = [x_d, y_d, \rho_d, \psi_d]^T$ where $x_d(t) = (1)\sqrt{2}t$ $y_d(t) = \cos(xd(t))$, $\rho_d = 0$ and $\psi_d(t) =$

Table 1. PHYSICAL PARTICULARS OF THE TCOMS G-TUG MODEL

Scale	λ=1.0	λ=9.78	**Scale**	λ=1.0	λ=9.78
Lpp (m)	33.0	3.37	Thruster model	Ka 4-70	Ka 4-70
Beam (m)	11.7	1.19	Thrusters to keel (m)	5.00	0.51
Draft (m)	4.80	0.49	Thrusters lateral (m)	5.20	0.53
Disp. (m^3)	825.6	0.88	Blade dia. (m)	2.20	0.22
Ixx (kgm^2)	18020089	201.4	Pitch ratio	0	0
Izz (kgm^2)	56190018	628.0	**Service cond.**		
LCG (m)	16.1	1.64	Speed (m/s)	6.68	2.14
KG (m)	4.80	0.49	Blade rps	4.8	15

$\tan^{-1}(dy_d/dx_d)$. The adaptive neural control (20) and adaptive law (21) are simulated with linearly parameterized basis function (12) in which $\hat{W} = \text{diag}\left[\hat{W}_1^T, \hat{W}_2^T, \hat{W}_3^T, \hat{W}_4^T\right]$ are adaptable parameters, and $Z = \left[\eta^T, \hat{v}^T, \alpha_1^T, \dot{\alpha}_1^T\right]^T \in R^{16}$ are the inputs to the approximator. We choose $\Gamma_1 = \Gamma_2 = \Gamma_3 =$ diag[5] and $\Gamma_3 =$ diag[50], $\sigma_i = 1 \times 10^{-6}$,, $K_1 =$ diag[10, 10, 5, 20] and $K_2 =$ diag[30, 30, 5, 50]. To evaluate the ability of the adaptive neural control to approximate the hull hydrodynamics, we set $d_i(\eta, \nu, t) = 0, i = 1, 2, 3, 4$. For comparison, we simulate the non-adaptive model-based control (19) wihtout the signum function with $K_1 =$ diag[5, 5, 3, 5] and $K_2 =$ diag[5, 5, 5, 8]. The initial conditions are $\eta(0) = [0, 0, 0, 5]^T$ and $\nu 0) = [0.1, 0, 0, 0]^T$.

5.1. CLOSED LOOP PERFORMANCE

Figure 4 shows the tracking performance of both the non-adaptive model-based and adaptive neural control. It can be observed that the tracking performance of the tugboat is satisfactory for both controls. The states of the system for both control methods as compared to the desired are also shown in Figure 4. It is noted that the roll control command can be keep the ASD tug roll within certain range if roll parameter is included into the 4-DoF control design.

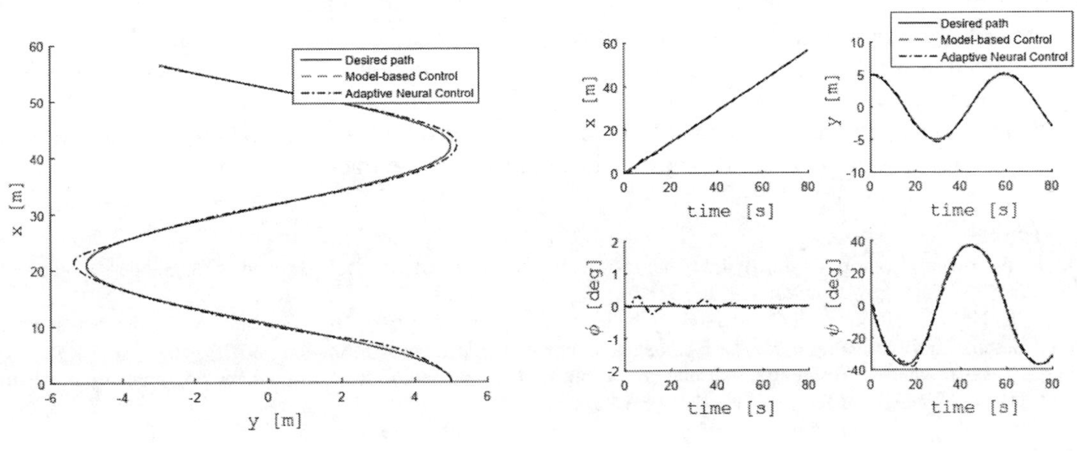

Figure 4. TRACKING PERFORMANCE AND SYSTEM STATES COMPARISON

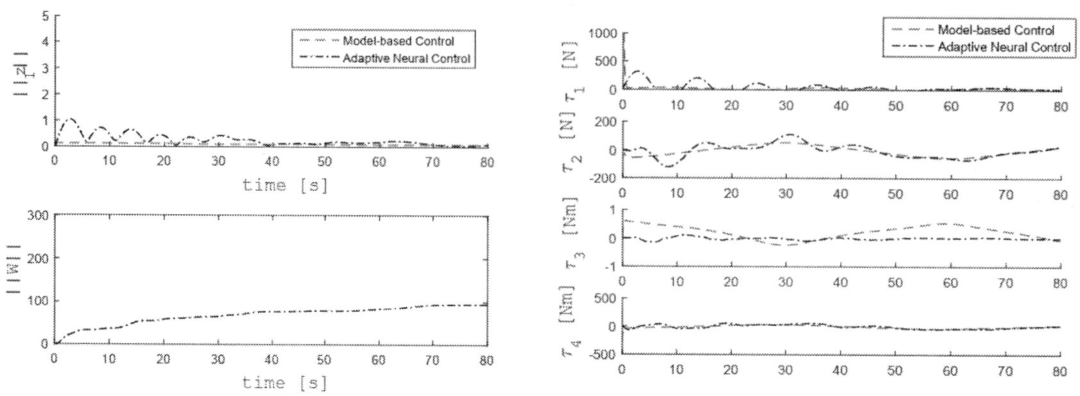

Figure 5. ERROR SIGNAL $||z_1||$, NORM OF ADAPTIVE WEIGHTS $||W||$ AND CONTROL τ

From Figure 5, error signal $||z_1||$, it is observed that the model-based controller τ_{mb}, performs better than the adaptive neural control. This is due to the complete knowledge of the hydrodynamics parameters which is utilized in the control strategy, with faster decay of tracking error and lower steady-state value. With time-varying disturbance, the adaptive neural is expected to perform better as the approximations can learn and compensate for the external disturbance $d(\eta, v, t)$ and will be treated in future work. The control commands τ_i, $i = 1, 2, 3, 4$ are shown with the adaptive neural control commands having slight oscillations. The norm of all adaptive neural weights $||W||$ are observed to be bounded and stable. Hence, we can conclude that the adaptive neural design is stable and able to compensate for the dynamics of the 4-DoF ASD tug.

6. Conclusion

In this paper, the 4DoF model of an ASD tugboat based on a modified MMG model has been presented. The model-based tracking control has been designed using a backstepping approach. A stable adaptive neural feedforward network which has the capabilities to account for uncertainties in the model and time varying disturbances has been derived. It has been shown that the closed-loop signals under the proposed control are semiglobally uniformly bounded. Simulation results have demonstrated that the surface vessel is able to track a desired trajectory satisfactorily. This control method is applicable as one of the basic building block on tracking control for autonomous tugboats. As future work, the tug parameters generated from quasi-steady manoeuvres with CFD virtual Planar Motion Mechanism computations can be refined via model tests. This paper also builds the foundation for future extension of the work for azimuth thruster allocation with coupled roll and tracking control.

References

[1] M. Abkowitz, "Lectures on ship hydrodynamics-steering and manoeuvrability, hydro-and aerodynamics laboratory, report no," *Hy-5, Lyngby, Denmark*, 1964.

[2] N. H. Norrbin, "Theory and observation on the use of a mathematical model for ship maneuvering in deep and confined waters," in *Proceedings of the 8th symposium on naval hydrodynamics, Pasadena, 1970*, 1970.

[3] H. Yasukawa and Y. Yoshimura, "Introduction of mmg standard method for ship maneuvering predictions," *Journal of Marine Science and Technology*, vol. 20, no. 1, pp. 37–52, 2015.

[4] T. I. Fossen, *Handbook of marine craft hydrodynamics and motion control*. John Wiley & Sons, 2011.

[5] T. I. Fossen, "A survey on nonlinear ship control: From theory to practice," *IFAC Proceedings Volumes*, vol. 33, no. 21, pp. 1–16, 2000.

[6] L. P. Perera and C. G. Soares, "Lyapunov and hurwitz based controls for input–output linearisation applied to nonlinear vessel steering," *Ocean Engineering*, vol. 66, pp. 58–68, 2013.

[7] J. M. Larrazabal and M. S. Peñas, "Intelligent rudder control of an unmanned surface vessel," *Expert Systems with Applications*, vol. 55, pp. 106–117, 2016.
[8] N. E. Kahveci and P. A. Ioannou, "Adaptive steering control for uncertain ship dynamics and stability analysis," *Automatica*, vol. 49, no. 3, pp. 685–697, 2013.
[9] G. Rigatos and S. Tzafestas, "Adaptive fuzzy control for the ship steering problem," *Mechatronics*, vol. 16, no. 8, pp. 479–489, 2006.
[10] L. Moreira and C. G. Soares, "h_2 and h_∞ designs for diving and course control of an autonomous underwater vehicle in presence of waves," *IEEE Journal of Oceanic engineering*, vol. 33, no. 2, pp. 69–88, 2008.
[11] Y. Yang, C. Zhou, and J. Ren, "Model reference adaptive robust fuzzy control for ship steering autopilot with uncertain nonlinear systems," *Applied Soft Computing*, vol. 3, no. 4, pp. 305–316, 2003.
[12] T. I. Fossen and M. Paulsen, "Adaptive feedback linearization applied to steering of ships," 1993.
[13] L. P. Perera and C. G. Soares, "Pre-filtered sliding mode control for nonlinear ship steering associated with disturbances," *Ocean Engineering*, vol. 51, pp. 49–62, 2012.
[14] J. Cheng, J. Yi, and D. Zhao, "Design of a sliding mode controller for trajectory tracking problem of marine vessels," *IET control theory & applications*, vol. 1, no. 1, pp. 233–237, 2007.
[15] M. Krstic, P. V. Kokotovic, and I. Kanellakopoulos, *Nonlinear and adaptive control design*. John Wiley & Sons, Inc., 1995.
[16] T. I. Fossen and J. P. Strand, "Tutorial on nonlinear backstepping: Applications to ship control," 1999.
[17] A. Witkowska, M. Tomera, and R. ŚMierzchalski, "A backstepping approach to ship course control," *International Journal of Applied Mathematics and Computer Science*, vol. 17, no. 1, pp. 73–85, 2007.
[18] J. Du, C. Guo, and S. Yu, "Adaptive robust nonlinear ship course control based on backstepping and nussbaum gain," *Intelligent Automation & Soft Computing*, vol. 13, no. 3, pp. 263–272, 2007.
[19] J. Du, C. Guo, S. Yu, and Y. Zhao, "Adaptive autopilot design of time-varying uncertain ships with completely unknown control coefficient," *IEEE Journal of Oceanic Engineering*, vol. 32, no. 2, pp. 346–352, 2007.
[20] Q. Zhang and X. Zhang, "Nonlinear improved concise backstepping control of course keeping for ships," *IEEE Access*, vol. 7, pp. 19258–19265, 2019.
[21] Y. Bai, Y. Cao, and T. Li, "Optimized backstepping design for ship course following control based on actor-critic architecture with input saturation," *IEEE Access*, vol. 7, pp. 73516–73528, 2019.
[22] S. S. Ge and C. Wang, "Adaptive neural control of uncertain MIMO nonlinear systems," *IEEE Transactions Neural Networks*, vol. 15, no. 3, pp. 674–692, 2004.
[23] D. Wang and J. Huang, "Neural network-based adaptive dynamic surface control for a class of uncertain nonlinear systems in strict-feedback form," *IEEE transactions on neural networks*, vol. 16, no. 1, pp. 195–202, 2005.
[24] T.-S. Li, D. Wang, G. Feng, and S.-C. Tong, "A dsc approach to robust adaptive nn tracking control for strict-feedback nonlinear systems," *IEEE transactions on systems, man, and cybernetics, part b (cybernetics)*, vol. 40, no. 3, pp. 915–927, 2009.
[25] C. Liu, T. Li, and N. Chen, "Dynamic surface control and minimal learning parameter (dsc-mlp) design of a ship's autopilot with rudder dynamics," *Journal of Harbin Engineering University*, vol. 33, no. 1, pp. 9–14, 2012.
[26] K. P. Tee and S. S. Ge, "Control of fully actuated ocean surface vessels using a class of feedforward approximators," *IEEE Transactions on Control Systems Technologies*, vol. 14, pp. 750–756, 2006.
[27] Z. Peng, D. Wang, W. Wang, and L. Liu, "Neural adaptive steering of an unmanned surface vehicle with measurement noises," *Neurocomputing*, vol. 186, pp. 228–234, 2016.
[28] X. Zheng, X. Zhao, R. Li, and Y. Yin, "Adaptive neural tracking control for a class of switched uncertain nonlinear systems," *Neurocomputing*, vol. 168, pp. 320–326, 2015.
[29] X. Wang, H. Li, and X. Zhao, "Adaptive neural tracking control for a class of uncertain switched nonlinear systems with unknown backlash-like hysteresis control input," *Neurocomputing*, vol. 219, pp. 50–58, 2017.
[30] T. I. Fossen *et al.*, *Guidance and control of ocean vehicles*, vol. 199. Wiley New York, 1994.
[31] H. Guo and Z. Zou, "System-based investigation on 4-DOF ship maneuvering with hydrodynamic derivatives determined by RANS simulation of captive model tests," *Applied Ocean Research*, vol. 68, pp. 11–25, 2017.
[32] W. Lin and C. Qian, "Adaptive control of nonlinearly parameterized systems: the smooth feedback case," *IEEE Transactions Automatic Control*, vol. 47, no. 8, pp. 1249 – 1266, 2002.
[33] S. S. Ge, T. H. Lee, and C. J. Harris, *Adaptive neural network control of robotic manipulators*, vol. 19. World Scientific, 1998.
[34] S. S. Ge, C. C. Hang, T. Lee, and T. Zhang, *Stable Adaptive Neural Network Control*. Boston: MA: Kluwer Academic, 2001.
[35] F. L. Lewis and T. Parisini, "Guest editorial: Neural network feedback control with guaranteed stability," 1998.
[36] L. X. Wang and J. M. Mendel, "Fuzzy basis functions, universal approximation and orthogonal least squares learning," *IEEE Transactions on Neural Network*, vol. 3, no. 5, pp. 807–814, 1992.

A taxonomy for autonomy in industrial autonomous mobile robots including autonomous merchant ships

Ørnulf Jan Rødseth[1*] and Marialena Vagia[2]

[1]SINTEF Ocean, Trondheim, Norway
[2]SINTEF Digital, Trondheim, Norway

*E-mail: OrnulfJan.Rodseth@sintef.no

Abstract. The concept of autonomous mobile robots (AMR) has gained much popularity in recent years, particularly in commercial settings where the name industrial autonomous mobile robot (IAMR) is proposed. In addition to automatic guided vehicles and automated mining trucks, IAMR also includes autonomous merchant ships. AMR is an old concept which was first introduced in the 1980s. Although the concept of AMRs is old and broadly used, there is still no common definition of autonomy when mobile robots are concerned. This paper will review some of the most known definitions and develop a taxonomy for autonomy in mobile autonomous robots. This will be used to compare the different definitions of robotic autonomy. This paper will mainly look at industrial autonomous mobile robots, i.e. systems that are designed to operate with a clear commercial objective in mind and which are normally supported by a remote control centre. This means that the robot is not fully autonomous, but to varying degrees dependent on humans in some control and monitoring functions.

1. Introduction

1.1. Autonomous mobile robots in an industrial perspective

The term "autonomous mobile robot" (AMR) has been around at least since 1984 [1] and should be obvious and self-explanatory. The first autonomous mobile robots were developed in the 1980's for research and education and later for scientific mission, e.g. in space or far under ice where it is necessary to have a certain degree of autonomy to handle fully or partly missing communication facilities. Around 1995 some companies started to test the technology in commercial applications, e.g. for supermarket floor cleaning [2]. This could be said to be the dawn of the *industrial* autonomous mobile robot (IAMR). We define an industrial autonomous system as *an autonomous unit, or a collection of such, that can operate safely and efficiently in a real world environment while doing operations of direct commercial value and which can be manufactured, maintained, deployed, operated and retrieved at an acceptable cost relative to the value it provides* [3]. This distinguishes industrial autonomous systems from many academic, space and military projects in that industrial autonomy is directly linked to commercial value creation as well as significant operational hazards.

A special version of the IAMR is the Maritime Autonomous Surface Ship (MASS) [4]. Although it is called an autonomous ship, it does in reality consist of the autonomous ship, normally a remote control centre (RCC), and other support systems, such as automatic mooring systems, position reference systems and so on. Thus, MASS could more correctly be used as an abbreviation for "Merchant Autonomous Ship System," also to highlights that MASS apply to merchant ships and not to, e.g. military or small research type vessels.

1.2. The availability of a Remote Control Centre (RCC)

The use of remote control centres is a common feature for almost all IAMR. High asset value, important commercial interests and often a high damage potential related to accidents will make most operators require that there is a remote monitoring and control function to safeguard their investments. This means that there will be humans in the control and monitoring loops for IAMR. This has two important consequences:

1. If these systems are to be defined as autonomous, we will need a definition of autonomy that goes beyond the idea of being fully independent of humans.
2. The quality and design principles for the Human-Automation Interface (HAI) will be an essential property of the IAMR.

The availability of human operators in the RCC also has an important function in managing the safety and cost of IAMR: The automation system does not have to be able to handle absolutely all situations the IAMR can encounter. If an RCC is available, one can use the human operators to assist in unusual or complex situation. This reduces the cost of implementation of the sensor and automation systems and simplifies testing and approval of the IAMR. However, this requires a high focus on the quality and design of the HAI as timely and correct human intervention will be essential for safe operation of the system.

1.3. IAMR needs a license to operate

Most IAMR will need some form of license to operate. For MASS, this involves approval from its flag state. This approval has to be in accordance with international rules set by the International Maritime Organization (IMO) to allow operation in international or other nations' territorial waters. For other types of IAMR, one will at least see that national Health, Safety and Environmental (HSE) regulations will set strict safety goals for the IAMR.

This means that an IAMR must be associated with an "operational envelope" [5] that defines exactly what the IAMR is able to do and what it is not able to do. If this is not defined, it may be impossible to test the system against safety performance criteria and, thus to issue the necessary license to operate. The implication of this is that an IAMR needs a certain level of determinism in its automation to be able to define its operational envelope.

1.4. Automatic or autonomous

AMR represent a high-growth sector in robotics and automation industry. Broadly speaking, AMR is any robot that can understand and move through its environment without being continuously overseen or controlled by an operator. It is assumed that the environment is not completely known and mapped, and that unmapped obstacles may appear during the mission. Thus, AMR need a set of sensors that enable them to understand and interpret their environment, which helps them to perform their task in the most efficient manner, navigating around fixed obstructions (building, racks, work stations, etc.) and variable obstructions (such as people, lift trucks, and debris).

The automated guided vehicles (AGV) have been moving things around on behalf of humans for more than half a century. They have become a familiar fixture in factories, warehouses, and anywhere there is a need for repetitive material delivery. The traditional difference between AGVs and AMRs is that the AGVs follow fixed routes, usually along wires or magnets embedded in the ground, while AMRs employ sensor systems to find its position and to avoid obstacles. However, when the guidewire is exchanged with a positioning system and sensors are added to avoid collisions, the AGV becomes an AMR, although the exact border is blurred. An example of this confusion are the transport systems that are used in in the St Olav Hospital in Norway that are commonly referred to as "autonomous", while it may also be called an AGV system [6].

Section 2 of this paper will come back to this issue and attempt to provide a definition of automatic and autonomous that can resolve some of these issues.

1.5. Dimensions of automation and autonomy
A well-known framework for defining autonomy levels for unmanned systems is ALFUS [7]. This defines levels of autonomy along three axes, *Mission Complexity*, *Environmental Difficulty* and *Human Independence*. This illustrates that the level of autonomy has to do with more than just human independence, at least the complexity of the mission and environment should be considered. This should also have an impact on the definition of when a system can be seen as autonomous versus only automatic. Several other variants of this "multi-dimensional" approach to autonomy will be discussed in section 4.

1.6. Autonomy for different processes and different mission phases
Large IAMR, like merchant ships complicate the definition of autonomy by having different requirements in different mission phases and for the different system processes. An example of the latter is the "periodically unmanned machinery spaces" that already is allowed on merchant ships. This automates the engine room, but leaves supervisory monitoring to the bridge officers. When discussing autonomy, one tends to focus on navigation. On a ship there are numerous other systems that need automation, such as energy production, fire safety, ballast, and hull integrity. In addition there will be different requirements for autonomy in different phases of the mission. Sailing on deep sea routes with few or no other ships in the vicinity is very different from sailing in narrow channels or in port areas. Thus, a definition of autonomy for MASS needs to take these factors into account.

1.7. Purpose of this paper
There is already a high and growing interest in MASS. Unmanned and autonomous ships have many properties that make them interesting in commercial applications, e.g. lower cost and lower energy consumption, ability to scale down ship sizes to realize flexible and more user oriented transport services and more [8]. On the other hand, merchant ships are internationally regulated through the International Maritime Organization (IMO) and must follow a strict approval regime, involving flag state and classification societies. This makes it important to provide good definitions of what an autonomous ship is and how it is differentiated from more conventional ship types, which also can have a high degree of automation. To provide some guidance in this process, this paper will investigate the following issues:

1. Is there a need to differentiate between autonomy and automation?
2. If so, how to define the difference?

As has been pointed out before, there are several issues that have an impact on these questions, in particular the number of dimensions in the space where autonomy appears, being it related to operation complexity, phases or ship processes. This will be discussed with the help of a simple taxonomy for IAMR autonomy. This taxonomy will be developed in section 6 and compared to a set of example IAMR described in section 4. Section 5 will give a brief overview of some official definitions of autonomy in the context of MASS.

2. Defining automatic and autonomous

2.1. Do we need the concept of autonomy?
The word "autonomy" is derived from the Greek word "autonomia" which means independent, or more literally, living by one's own laws [9]. The word represents a concept found in different kinds of sciences, like engineering, philosophy, biology and medicine. As far as the engineering etymology is concerned, the word autonomy has been proposed used in order to describe the ability of an engineering system to make its own decisions about its actions while performing different

tasks, without the need for the involvement of an exogenous system or operator [10].

Several sources suggest that autonomy as a term should be depreciated. The main arguments are that autonomy has already been given the meaning of "independent of external control", it should not be given a new definition in the context of IAMR and, thus the term should be avoided in professional literature. The exception may be systems that are truly and fully independent of any operator, including the need for any human fallback solutions. The American Society of Automotive Engineers (SAE) is one of them, and proposes to use the term "automated driving" [11]. The Central Commission for the Navigation of the Rhine (CCNR), which is the most important European organization for inland waterways, has adopted a strategy similar to SAE and currently uses "automated navigation" rather than autonomous [12]. Some classification societies have adopted a similar position (see section 5).

On the other hand, automation in the area of IAMR and ships in particular, is already far advanced and fully incorporated into existing international legal instruments. Traditionally, when new forms of automation emerges, e.g. as the automatic radar plotting aid (ARPA) in the 1960's, they are usually deployed with no need for change in regulations and used to assist the operators. Although, as use and experience evolve, some of these systems are eventually used to reduce crew sizes, and regulations are updated accordingly. The situation with autonomy in IAMR is very different. The current projects, at least in the MASS sector, aims at controlling the ship without operators directly attending to many critical tasks, although they still have a role in handling difficult situations and in high level monitoring of the system. This is clearly not a fully autonomous operation, but one can also argue that this new state of matters may require a new adjective other than automatic. It will in any case require a new approach to regulations in the maritime sector, something that the regulatory scoping exercise currently under way in IMO is trying to find an answer to.

It is also very difficult to find a good definition that clearly distinguishes automatic from autonomous. No nation, organism nor engineering system is completely free from any type of external control. There is virtually always an external set-point, relationship or other factor that guides the system's actions. Particularly for IAMR, there is absolutely no point in having an autonomous system that does not follow orders or mission plans. Thus, reserving autonomous for this type of "fully independent" operation is meaningless. Thus, it may be better to use the term *autonomous* for the major change in regulatory approaches that fully or partly unmanned operations require. Additionally, autonomous as a concept, has already been taken up in professional texts and it may be better to assign a clear meaning to it, rather than trying to remove it from the vocabulary.

2.2. A definition of automation

A common understanding is that automation includes the execution by a machine agent of a function that was previously carried out by a human [13]. This can concern different control systems for operating equipment in numerous applications such as production, medical equipment, traffic control or buildings to name only a few examples. However, the concept of automation has changed through time. It can mean a simple reallocation of a function from human to machine which is complete and permanent. In that case, the function will tend to be seen simply as machine operation not as automation [13], i.e. as can be seen with automatic elevators. Today, they are just elevators.

The roots of the word automatic are Greek, from *autos* "self" + *matos* "thinking, animated" [9]. It is normally used in several different meanings, e.g. 1) acting or operating in a manner essentially independent of external influence or control, or 2) Acting or done without volition or conscious control; involuntary [14]. *Automatic* was introduced around 1810 in the meaning of "self-acting" [9]. The word *automation* was introduced in the mid-19th century in conjunction with automatic production processes in car production [9].

When discussing the difference between automation and autonomy, it is tempting to use definition 2) above as basis and say that automation refers to a system that will do exactly what it is programmed by the programmer to do, without having any choice or possibility to act in any different way, and stay true to the operators instructions. Its actions are predefined from the beginning and it has no ability to change them into the future. However, this has at least two problems:

1. An adaptive controller, e.g. an autopilot that adjusts its control laws to adjust for variance in the weather or the ship's loading condition, must it be defined as autonomous, as it certainly seems to be more than automatic?
2. Most IAMR is deterministic in the sense that they need to limit its capabilities to an operational envelope and provide testable responses to all critical situations it can encounter, as discussed in section 1.3. Must all these systems be defined as automatic, independent of how "intelligent" they are within the operational envelope?

In general, most computer system must follow its programmed instruction, independent on how adaptable it is. Adaptability can be programmed into a system, e.g. by statistical estimators or by machine learning, that allows the computer program to give responses that seem more variable, but these responses are still limited by the control frameworks that they are programmed into.

A more pragmatic definition of automatic can be found in IEC 60050-351 [15] and in ISO/TR 11065 [16]. With some slight modifications this can be stated as:

Pertaining to a process, device or equipment that, under specified conditions, [can] function without human intervention."

Here we have added "can" before *function* to emphasize that some form of human instruction or supervision is necessary also in fully automated processes and changed to "*device or equipment*" as the two standards use the respective single word in this part of the definition. Automation is defined in
[16] as "*The implementation of processes by automatic means*." The above is also the definition that ISO has proposed to IMO as part of a new terminology for MASS [17].

2.3. A "weak" definition of autonomy

The input document from ISO to IMO [17] also proposes a definition of autonomy that, when generalized to IAMR instead of ships, reads as follows:

Autonomous, autonomy: In the context of IAMR, autonomy means that the IAMR uses automation to operate without human intervention, related to one or more processes, for the full duration or in limited periods of the IAMR's mission.

Note 1: This definition requires that the operation can be performed without human intervention during the specified time period. This differentiates autonomous functions from conventional automation by requiring that automation that facilitates autonomous operation must be designed and verified to allow it to be unattended during the whole of the relevant time period.

This definition also addresses the problem of including the multiple internal processes and the different voyage phases that ships and some other IAMR have to consider (see section 1.6).

A similar definition of autonomy can be found in [18] and is in that publication called a "weak" definition of autonomy.

"*the extent to which a system can carry out its own processes and operations without external control.*"

Both these definitions differ from the definition of automatic by requiring that the system *shall* be able to function without external control for a specified time and process. Beer et al. [18] also proposes a "strong" definition of autonomy which require a certain degree of "machine cognition," that will be returned to in section 4.

This differentiation between automation and autonomy allows a clear distinction to be made between the two concepts, although it has some side effects, e.g. that a simple thermostat now could be considered autonomous instead of just automatic!

3. Some examples of autonomous or automatic systems

Section 4 will go through some of the different factors that have been used to differentiate autonomy from automation in the literature. To provide a reference for the use of these factors, this section will provide four examples of IAMR systems that represent some typical classes of applications.

3.1. A ship autopilot

An autopilot is similar to a car cruise control, but is able to keep a constant heading as well as a constant speed. A modern ship autopilot will have adaptive control algorithms to better control the ship with different under-keel clearance, in different loading conditions and in different weather conditions. Autopilots may also make use of fuzzy control and other typed of "artificial intelligence" type technology [19] to provide parts of this adaptability. More advanced control criteria, in addition to minimizing speed and course variance, may be to minimize fuel use and rudder movements.

According to the weak definition of autonomy, an autopilot can be said to be autonomous as it can perform the relevant control processes independently of human supervision. The crew will attend to other processes, such as lookout, when the autopilot is engaged. However, the crew is still required to stay on the bridge to intervene if something goes wrong.

3.2. Automatic guided vehicle

In this paper we will use the AGV system at St. Olav's Hospital in Trondheim as example [6]. This is operating based on visual navigation and has anti-collision detectors as well as facilities for controlling doors and elevators. It can be operated without human control, although there is an operations centre that intervenes when needed. The robot is relatively simple-minded and will not try to avoid obstacles but will either automatically ask persons to remove themselves or call the operations centre when the obstacle remains. Then a service engineer needs to assist the vehicle to resolve the problem. Also, this AGV can be classified as autonomous according to the weak definition of autonomy.

3.3. Autonomous mining trucks

Autonomous mining trucks have been in use since 2006 and there is a code of practice for safe mobile autonomous mining in Australia [20]. These trucks are operating in controlled environment and mostly follow pre-planned routes. However, there are sensor systems that can detect other objects and either change the truck speed to avoid collisions or stop the truck. The trucks are normally uncrewed. Also this system will satisfy the weak requirements to being autonomous.

3.4. The MUNIN bulk carrier

The MUNIN project [21] made a concept study of a bulk carrier that was unmanned, but under control of a remote control centre for the deep sea passage between Europe and South America. The ship was manned during port approach and departure. The RCC operators would not normally operate the ship during the sea passage, but could be alerted in case of problems such as bad weather, unknown obstacles or technical faults. The automation system would use radar and visual sensors to detect and avoid obstacles, but it was assumed that this was relatively simple and that avoidance manoeuvres were limited to one or two other ships. Also object classification was mostly limited to identifying other ships, e.g. with the use of automatic identification system (AIS) messages. If this failed, one would rely on the RCC crew if identification was necessary. The MUNIN bulk carrier can also be defined as autonomous according to the weak definition of autonomy.

4. Automation and autonomy have more dimensions than just human supervision

Section 2.3 suggested a weak definition of automation and autonomy that allows us to more clearly differentiate between the two. One should note that from the regulatory side, this distinction may be the most important: Determining how certain human control processes can be taken over by machines and how to allow the humans to absent themselves during the operation.

However, the example of the "autonomous" thermostat in section 2.3 may indicate that we need a stronger definition of autonomy [18], and this section will go through some of the different dimensions of automation and autonomy that have been proposed as additional qualifying factors.

Table 1 shows how the weak definition of autonomy applies to the four example systems. All system can operate without direct operator attendance. However, for the autopilot this is only to a limited degree as the operator needs to be in the vicinity of the bridge, even when the autopilot is engaged.

Table 1: Environment as qualifying factor for autonomy

	Autopilot	AGV	Mining truck	MUNIN
Attendance	(Yes)	Yes	Yes	Yes

For larger IAMR it is also relevant to talk about different levels of autonomy in different processes and/or for different parts of the mission (see section 1.6). This has different relevance for different types of IAMR, but Table 2 shows how these attributes could be linked to the example systems.

Table 2: Different autonomy levels for processes and mission phases

	Autopilot	AGV	Mining truck	MUNIN
Functions	(All)	Some	Some	Some
Phases	Part	Part	Part	Part

Except for the autopilot, all systems will have different levels of autonomy in different processes. As an example, propulsion system maintenance on most systems will not be automatic, but will rely on periodic maintenance when the vehicle is inactive. This also means that all systems will only be automated in certain parts of their activities.

4.1. Environment sensing and complexity

A common way to define autonomy is to refer to the requirement of being able to sense real-world environment and act appropriately to hindrances that it may experience. One example is the following:

> *"Autonomy refers to systems capable of operating in the real-world environment without any form of external control for extended periods of time" [22].*

When applied to the four examples in section 3, we get the results listed in Table 3.

Table 3: Environment as qualifying factor for autonomy

	Autopilot	AGV	Mining truck	MUNIN
Env. sense	Yes	(Yes)	Yes	Yes

Here, one can argue that the inclusion of operation in a real-world environment does not really add any significant differentiating factor. Any automatic robot, including an AGV, will at least need to detect its position and will normally include some form of anti-collision function that allows it to act, perhaps just stop, based on the environment information.

A certain complexity of the environment could be a more relevant factor and that has been proposed as one of the axes defining autonomy in the ALFUS framework [23]. Here, both environmental difficulty and mission complexity are assigned metrics that are used to determine the "contextual autonomous capability". This allows the users of the framework to assign a "level of autonomy" to the system, but is difficult to use as a binary qualifying factor. Doing the comparison with the examples again, gives us Table 4. Here, the "level of complexity" has been indicated. Note that the mining truck and MUNIN environment are classified as medium complex. They are operating in a relatively controlled environment or in open sea, with limited unexpected obstacles.

Table 4: Environment complexity as qualifying factor for autonomy

	Autopilot	AGV	Mining truck	MUNIN
Env. complexity	Very low	Low	Medium	Medium

4.2. Cognition and planning

The following definition is a further development of the one presented in section 2.3. While the latter definition was described as "weak", the following is described as a "strong" definition [18]:

> *"The extent to which a robot can sense its environment, plan based on that environment, and act upon that environment with the intent of reaching some task-specific goal (either given to or created by the robot) without external control."*

A somewhat weaker variant of this is the following:

> *"autonomous system: system that, perceiving its environment and determining if this affects its goals, takes action to ensure as far as practicable that its goals will be safely achieved" [25].*

The second definition is more implicit in its requirements to a certain level of cognition. The inclusion of "cognitive functions" is obviously a clear difference from automation and could be relevant for differentiating between automation and autonomy. However, the term *plan* in the first definition poses a problem in that it is not well defined. For an IAMR, it will probably not be allowed to change the provided mission plans, although the IAMR will be expected to execute certain types of limited replanning, e.g. safe avoidance manoeuvres. While this in most cases qualifies as a cognitive function, it is not clear that it represents planning.

Both the truck and MUNIN allow a limited degree of avoidance plans to be made by the system, but it is questionable if this is sufficient to call it planning abilities. The AGV can stop if it encounters problems, but the autopilot does not replan at all.

Applying the idea of planning and cognition to the examples provides Table 5.

Table 5: Planning as qualifying factor for autonomy

	Autopilot	AGV	Mining truck	MUNIN
Planning	No	No	No (Very Low)	No (Very Low)
Cognition	(Yes)	No	Yes	Yes

The argument here is that the AGV only stops if it detects an obstacle and does not attempt to find a solution. The autopilot is to some degree able to adapt to the environment and ship conditions.

4.3. Use of artificial intelligence (AI) and learning

Definitions of advanced forms of autonomy will also commonly refer to the system's ability to learn over time [26]. This is also often associated with the use of artificial intelligence (AI) [27] which is also commonly referred to in conjunction with autonomous systems in general. However, none of the main sources refer to any need to implement learning or AI in today's IAMR.

Currently, it is not trivial to use AI in IAMR due to the general need for a license to operate (see section 1.3), which in turn requires strict test and approval regimes. Many AI technologies, in particular neural networks and deep learning, represent a form of decision making where performance may be difficult to quantify, being reliant on extensive testing and statistical methods. For certain operations, e.g. collision avoidance at sea [28], it can be argued that the possible number of situations that need to be tested for is too large to get sufficient significance levels from testing. This means that it may be impossible to prove that the system works sufficiently well in all situations [29]. Object detection and classification may be an easier task, as there probably are much fewer critical cases to classify and test. It is expected that at least object classification may use, e.g. machine learning in its implementation also in current projects [30]. The criticality of these tasks will obviously vary, dependent on the operational domain and type of IAMR.

In a supervised IAMR, it will be easier to deploy AI technology as one may rely on a human operator to intervene if the AI system fails to find a good solution to a problem. However, this also has implications on the human-automation interface as it introduces uncertainty in when the operator may be required to intervene. This could rule out the use of "constrained autonomy" [31], where it is possible to alert the operator to an intervention, in time for the operator to get sufficient situational awareness before actions are required.

Learning is also a complicating issue for IAMR as the IAMR generally needs an approval to operate based on certified technical performance (see section 1.3). If performance changes during operation, will then the original approval be valid? If it is possible to prove that the overall performance will improve and no previous capability is lost, this should be possible, but this is not trivial. Some recent advances in deep learning point to some possibilities for improved retention of knowledge [32], but even here there is degradation in knowledge retention, dependent on the number of patterns the system is required to recognize. For collision avoidance, this may be an issue due to a high number of distinct situations. It may be less of problem for object detection and classification.

If we apply the concept of learning and AI to the examples, we can tentative get a result as shown in Table 6.

Table 6: AI and learning as qualifying factor for autonomy

	Autopilot	AGV	Mining truck	MUNIN
AI	Yes	No	(Yes)	(Yes)
Learning	Yes	No	No	No

This is a somewhat arbitrary classification, but some autopilots are already using AI techniques, and they also include learning in their very limited scope of "expertise", i.e. they can "learn" a new loading condition for the vessel. The AGV will probably use neither due to its relatively simplistic operation. The mining truck may use AI in certain functions, e.g. object detection, which has also been suggested for MUNIN. It is not likely that either of the latter will use learning in their most critical functions, i.e. object detection, classification and collision avoidance.

4.4. Assigning liability to the IAMR

An interesting approach to the definition of robot autonomy is the following:

> *A system is considered autonomous if it can legally accept accountability for an operation, thereby assuming the accountability that was previously held by either a human operator or another autonomous system [33].*

This is a very distinct definition that captures some of the original concept of autonomy by trying to assign legal accountability to the system. This could be said to be in line with the idea of "living by one's own laws". In practical terms and today, this would not work, unless the autonomous robot can be defined to be a juridical person, which is not likely to happen in the foreseeable future. Generally, legal accountability can only be assigned to a real person or certain types of organizations, and not to machines. For IAMR, the legally responsible is expected to be the owner or the de facto operator of the IAMR.

5. Some official definitions of autonomous ship

As international shipping is based on internationally accepted legislation and rules, there has been a significant development in suggested definitions of autonomous ship is in the different contexts of authorities, standards organizations and class societies. This section will briefly go through some of the best known definitions.

IMO is the main organization for international regulation and has for the time being adopted the following definition [34]:

Maritime Autonomous Surface Ship (MASS) is defined as a ship which, to a varying degree, can operate independent of human interaction.

ISO [17] has proposed a similar, but somewhat more formalized version which was quoted in section 2.3.

The Norwegian Forum for Autonomous Ships [35] suggested the following definition:

Autonomy is the result of applying "advanced" automation to a ship so that it implements some form of self-governance, i.e. that it can select between alternative strategies without consulting the human.

NFAS has later adopted the ISO version of the definition.

Bureau Veritas [36] adopts the strategy of reserving autonomous only for ships that are fully automated.

Lloyd's Register does not define autonomy as such but uses a 7-stage scale to describe autonomy levels [37].

The first level where autonomous appears in the text is at level 4:

Decisions and actions at the ship level are performed autonomously with human supervision. High impact decisions are implemented in a way to give human operators the opportunity to intercede and over-ride them. Data may be provided by systems on or off the ship.

DNV GL proposes to use the term "autoremote" for any mix of decision making between automation systems and operators, except for fully autonomous ships [38]. An autoremote vessel is defined as:

vessel for which one or more key functions are remotely controlled from a remote control centre, possibly by assistance from personnel on board. To support safe and efficient operation of the vessel, the remotely controlled key function(s) is arranged with a defined level of automation ranging from simple decision support to complete automatic control.

Class NK adopts the principle of full automation to allow calling a *function* autonomous. Furthermore, it does not define autonomous in the context of ships but refer to automated/autonomous operations or tasks [39].

American Bureau of Shipping (ABS) defines autonomy as different from automation by:

(An autonomous system) requires a quality of self-governance and freedom from external control or influence. Rather than only predefined actions, the system can choose from the best or most appropriate option and not just follow a predetermined algorithm or script.

Most of the above definitions allow only parts of the ship functionality to be termed autonomous if it otherwise satisfies the definition, i.e. some operations or tasks may be autonomous, although the ship as a whole may not be.

6. Summary and conclusions

As can be seen from section 5, many of the sources prefer to describe systems as automatic rather than autonomous, defining a number of levels of automation up to full autonomy, which is only used when the system is truly independent of human operators. However, both IMO and ISO propose to use the broader and weak definition of autonomy more or less as it was presented in section 2.3. As was argued in section 2.1, the authors also believe it may be better to accept the term autonomy – also for "semi-autonomous" IAMR, but give autonomy a clearer meaning compared to automatic. This can be achieved by using the definition of automation that was proposed in section 2.2.

As for the definition of autonomy, we would prefer the weak definition presented in section 2.3, but there are also possibilities for qualifying this with other attributes as discussed in section 4. A summary of how the example systems satisfy these attributes is shown in Table 7.

Table 7: Environment as qualifying factor for autonomy

	Autopilot	**AGV**	**Mining truck**	**MUNIN**
Attendance	(Yes)	Yes	Yes	Yes
Function	(All)	Some	Some	Some
Phase	Part	Part	Part	Part
Env. sense	Yes	(Yes)	Yes	Yes
Env. complexity	Very low	Low	Medium	Medium
Planning	No	No	No (Very Low)	No (Very Low)
Cognition	(Yes)	No	Yes	Yes
AI	Yes	No	(Yes)	(Yes)
Learning	Yes	No	No	No

If one assumes that autopilot and perhaps the AGV should be called automatic rather than autonomous, it is difficult to pinpoint one qualifier that clearly adds to the weak definition. The reason for this is mainly that terms like, e.g. cognition and environment complexity are not well defined and do not lend themselves to "binary" definitions like autonomy versus automation. Other attributes may be forthcoming in the future or we may get more accurate definitions of the existing attributes. Until that happens, it may be better to stick with the weak definition of autonomy, combined with the corresponding definition of automation.

Acknowledgments
The work presented in this text has received funding from the European Union's Horizon 2020 research and innovation programme under grant agreement No 815012 (AUTOSHIP) and grant agreement No 859992 (AEGIS).

References

[1] Podnar, G., Dowling, K. and Blackwell, M., 1984. A functional vehicle for autonomous mobile robot research. Carnegie-Mellon University, Robotics Institute.

[2] Endres, H., Feiten, W. and Lawitzky, G., 1998, May. Field test of a navigation system: Autonomous cleaning in supermarkets. In *Proceedings. 1998 IEEE International Conference on Robotics and Automation (Cat. No. 98CH36146)* (Vol. 2, pp. 1779-1781). IEEE.

[3] Grøtli, E.I., Reinen, T.A., Grythe, K., Transeth, A.A., Vagia, M., Bjerkeng, M.C., Rundtop, P., Svendsen, E., Rødseth, Ø.J. and Eidnes, G., 2015. SEATONOMY Design, development and validation of marine autonomous systems and operations.

[4] IMO, 2017, Report of the Maritime Safety Committee on its Ninety-Eighth Session, MSC 98/23, 28 June 2017.

[5] Rødseth, Ø.J., 2019. Defining Ship Autonomy by Characteristic Factors. In *Proceedings of the 1st International Conference on Maritime Autonomous Surface Ships*. SINTEF Academic Press.

[6] Johnsen, S.O., Hoem, Å.S., Stålhane, T., Jenssen, G. and Moen, T., 2018. Risk based regulation and certification of autonomous transport systems. In *Proceedings of the 28th International European Safety and Reliability Conference (ESREL 2018), Trondheim, Norway, 17–21 June 2018*.

[7] Huang, H.M., Pavek, K., Novak, B., Albus, J. and Messin, E., 2005. A framework for autonomy levels for unmanned systems (ALFUS). *Proceedings of the AUVSI's Unmanned Systems North America*, pp.849-863.

[8] Rødseth Ø J 2018 Assessing Business Cases for Autonomous and Unmanned Ships. In: *Technology & Science for the Ships of the Future. Proceedings of NAV 2018*: 19th Int. Conference on Ship & Maritime Research. IOS Press 2018 ISBN 978-1-61499-870-9.

[9] The Online Etymology Dictionary, Retrieved July 2020 from: https://www.etymonline.com.

[10] Albus, J., Antsaklis, P., Meystel, A., Passino, K., Samad, T., 1998. Setting the stage: some autonomous thoughts on autonomy. In: IEEE ISIC/CIRA/ISAS Joint Conference, pp. 520e521.

[11] SAE J3016 (2016): "Taxonomy and definitions for terms related to on-road motor vehicle automated driving systems", Revision September 2016, SAE International.

[12] CCNR, Automated Navigation – Definitions of levels of automation in inland navigation, Resolution 2018-II-16. Retrieved June 2020 from https://ccr- zkr.org/files/documents/cpresse/cp20181219en.pdf

[13] Parasuraman, R. and Riley, V., 1997. Humans and automation: Use, misuse, disuse, abuse. *Human factors*, *39*(2), pp.230-253.

[14] American Heritage Dictionary of the English Language, Fifth Edition, 2016, Houghton Mifflin Harcourt Publishing Company.

[15] IEC, 2013, IEC 60050-351, International Electrotechnical Vocabulary - Part 351: Control Technology.

[16] ISO, 1992, ISO/TR 11065 Industrial automation glossary.

[17] ISO, 2020. Input document to IMO Maritime Safety Committee, Session 102, agenda item 5: Regulatory Scoping Exercise for the Use of Maritime Autonomous Surface Ships (MASS), Proposed terminology for MASS, Submitted by International Organization for Standardization (ISO), February 2020.

[18] Beer, J.M., Fisk, A.D. and Rogers, W.A., 2014. Toward a framework for levels of robot autonomy in human-robot interaction. *Journal of human-robot interaction*, *3*(2), pp.74-99.

[19] Velagic, J., Vukic, Z. and Omerdic, E., 2003. Adaptive fuzzy ship autopilot for track-keeping. *Control engineering practice*, *11*(4), pp.433-443.

[20] Department of Mines and Petroleum, 2015, *Safe mobile autonomous mining in Western Australia–code of practice*: Resources Safety, Department of Mines and Petroleum, Western Australia, 30 pp. ISBN 978 1 922149 41 1.

[21] Rødseth, Ø.J. and Burmeister, H.C., 2012. Developments toward the unmanned ship. In *Proceedings of International Symposium Information on Ships–ISIS* (Vol. 201, pp. 30-31).

[22] Bekey, G.A., 2005. Autonomous robots: from biological inspiration to implementation and control. MIT press.

[23] Huang, H.M., Pavek, K., Novak, B., Albus, J. and Messin, E., 2005. A framework for autonomy levels for unmanned systems (ALFUS). *Proceedings of the AUVSI's Unmanned Systems North America*, pp.849-863.

[24] Shattuck, L.G., 2015. Transitioning to autonomy: a human systems integration perspective. Transitioning to Autonomy: Changes in the Role of Humans in Air Transportation.

[25] ISO, 2020, ISO 21384-4:2020(en) Unmanned aircraft systems — Part 4: Vocabulary.

[26] Williams, A.P.; Scharre, P.D., 2015, *Autonomous Systems: Issues for Defence Policymakers*; NATO Communications and Information Agency: The Hague, The Netherlands.

[27] Luckcuck, M., Farrell, M., Dennis, L.A., Dixon, C. and Fisher, M., 2019. Formal specification and verification of autonomous robotic systems: A survey. *ACM Computing Surveys (CSUR)*,

52(5), pp.1-41.
[28] Perera, L.P., 2018, June. Autonomous Ship Navigation Under Deep Learning and the Challenges in COLREGs. In *International Conference on Offshore Mechanics and Arctic Engineering* (Vol. 51333, p. V11BT12A005). American Society of Mechanical Engineers.
[29] Teoh, E.R. and Kidd, D.G., 2017. Rage against the machine? Google's self-driving cars versus human drivers. *Journal of safety research*, *63*, pp.57-60.
[30] Prasad, D.K., Rajan, D., Rachmawati, L., Rajabally, E. and Quek, C., 2017. Video processing from electro-optical sensors for object detection and tracking in a maritime environment: a survey. *IEEE Transactions on Intelligent Transportation Systems*, *18*(8), pp.1993-2016.
[31] Rødseth, Ø.J., 2019. Defining Ship Autonomy by Characteristic Factors. In *Proceedings of the 1st International Conference on Maritime Autonomous Surface Ships*. SINTEF Academic Press.
[32] Kirkpatrick, J., Pascanu, R., Rabinowitz, N., Veness, J., Desjardins, G., Rusu, A.A., Milan, K., Quan, J., Ramalho, T., Grabska-Barwinska, A. and Hassabis, D., 2017. Overcoming catastrophic forgetting in neural networks. *Proceedings of the national academy of sciences*, *114*(13), pp.3521-3526.
[33] Myhre, B., Hellandsvik, A. and Petersen, S., 2019, October. A responsibility-centered approach to defining levels of automation. In *Journal of Physics: Conference Series* (Vol. 1357, No. 1, p. 012027). IOP Publishing.
[34] IMO, 2018, *Report of the Maritime Safety Committee on its One Hundredth Session*, Annex 2, MSC 100/20/Add.1, December 2018.
[35] Rødseth, Ø.J. and Nordahl, H. (Ed.), 2017. *Definition for autonomous merchant ships*. Version 1.0, October 10. 2017. Norwegian Forum for Autonomous Ships.
[36] Bureau Veritas (2019), *Guidelines for Autonomous Shipping*, October 2019, Guidance Note NI 641 DT R01 E.
[37] Lloyd's Register, 2016, *Cyber-enabled ships ShipRight procedure – autonomous ships*, First edition, July 2016.
[38] DNV GL (2018) Class Guideline DNVGL-CG-0264, *Autonomous and remotely operated ships*.
[39] Class NK (2020). *Guidelines for Automated/autonomous Operation on Ships* (Ver. 1.0), January 2020.
[40] ABS (2019) *Advisory on Autonomous Functionality*, American Bureau of Shipping.

IOP Conf. Series: Materials Science and Engineering **929** (2020) 012004 doi:10.1088/1757-899X/929/1/012004

A framework for description of autonomous ship systems and operations

Lars Andreas Lien Wennersberg[1,2], Håvard Nordahl[1], Ørnulf Jan Rødseth[1], Kay Fjørtoft[1] and Even Ambros Holte[1]

[1]SINTEF Ocean, Otto Nielsens veg 10, 7052 Trondheim, Norway
[2]Corresponding author, e-mail: Lars.Andreas.Wennersberg@sintef.no

Abstract. The Concept of Operations, or ConOps, has become a central document for the specification, design and approval of autonomous ship systems and operations in the absence of prescriptive rules and regulations. The flexible structure of the ConOps and the fact that it is written in prose text makes it very accessible for all involved stakeholders, but also prone to discrepancies between the descriptions and the actual design. This paper proposes a description framework, for autonomous ship systems and operations, that covers the information items requested through the ConOps. The proposed framework has the potential to facilitate development of a formalized ConOps, which in turn could lead to a standardization of the current approval procedures for autonomous ship systems and operations.

1. Introduction

The development of autonomous ship systems is to a large degree driven by the need to improve the sustainability and cost-effectiveness of zero-carbon shipping. The shift from conventional to autonomous shipping will have a significant impact on how ships interact with their surroundings. We expect that the autonomous ships will be designed to fit much more effectively into supply chains and logistics systems [1]. We also expect more automation and increased digital communication in ports and fairways. New interactions with onshore personnel that take over functions from the traditional ship crew will also be introduced, and the interaction between ships with and without crew onboard will need careful considerations. This and other elements in the automated ship's surroundings will require new communication principles and mechanisms. We expect an extensive shift from voice communication to digital information exchanges between automated entities.

The transition from conventional to autonomous shipping is challenging. The enabling technologies for autonomous ship systems are largely untested and technology development runs in parallel with implementation projects [2]. Issues with integration of technical systems has always been a challenge in the maritime industry, and these issues are not expected to diminish for autonomous ships. A tighter integration with shore-based systems adds on to the complexity of both the systems and operations. We expect that few, if any, large autonomous ships will operate completely without human control in the foreseeable future, but they will have personnel placed onshore both for supervision and control. Lack of crew onboard the ship will also require us to re-think how daily maintenance activities shall be carried out [3]. There are several regulatory gaps both on a national and international level. There are no prescriptive rules and regulations neither for ships nor for equipment, and the definitions and responsibilities of the master, crew and responsible person used in conventional shipping are not defined for autonomous shipping [4].

The purpose of this paper is to provide a systematic framework for description of autonomous ship systems and operations. The target group of the framework are the stakeholders involved in the design and specification, the approval, and the implementation of autonomous ship systems. The proposed framework is a result of ongoing work with development of a design standard and a holistic design methodology for autonomous ship systems and operations in the *AUTOSHIP project* [5]. The objective of the framework is to support a structured way of organizing information throughout the design process and the corresponding approval activities, as well as to lay the foundation for detailed design activities, system implementation, testing and final approvals.

The framework is needed for several reasons. Development of new technology that supports crew reductions onboard ships need clear and consistent descriptions of the relationships between the physical components and the roles of the human actors in the autonomous ship system. The system context is a critical part of the system design as it will define the system boundary and the most important communication paths. The definition of the components that are inside or outside of an autonomous ship system will have a major impact on the context.

The difference between the autonomous ship system and its context needs to be well understood, and higher degrees of automation and integration of the system into specific supply chains and logistics systems will require more context awareness during the design process compared to conventional ship design processes.

In the following sections, we will discuss the expected approval process for autonomous ships and analyse the information items that the authorities request through the Concept of Operations. We will explain how the description framework is built up and discuss the advantages of adopting the framework for design and approval activities.

2. The expected approval process

The approval process for conventional ships, seen from an industry perspective, is relatively straight forward today. Much of the components and equipment is already type approved and the final approval of the ship by the flag state or its recognized organisations is normally based on prescriptive rules that are well suited to the industrial production processes. In addition, the class societies will be involved, governed by the chosen class notations. All stakeholders involved in the design, the build and the operation of the ship must ensure that the systems comply with known requirements through the well-defined acceptance criteria and test regimes.

This is not the case for autonomous ship systems and operations as of today. The required technology is unproven, and the development of this technology runs in parallel with ongoing implementation projects. One could also say that the application domain for autonomous ships can be very different than the domain for conventional ships: The desire to integrate an autonomous ship into specific supply chains and logistics systems to optimize the cost-effectiveness could potentially reduce the system flexibility and put constraints on the allowable operating area. The lack of experience with the necessary technology, the operations and to some extent the application domain is reflected by the lack of well-defined acceptance criteria and consequently prescriptive rules.

There are however several initiatives that together work on closing the gaps. IMO has issued an interim guideline for MASS trials to guide the authorities and stakeholders in the process of planning and executing trials [6]. ISO has provided input to IMO on definitions of automation and autonomy with the purpose of aligning the understanding of the terminology for autonomous ships among involved stakeholders [7]. Several classification societies have issued guidelines for the design and operation of autonomous ship systems to facilitate development and implementation initiatives [8], [9], [10], [11]. We expect that the classification societies will gain experience through their involvement in the processes, and then contribute to the transition towards a more streamlined regulation framework. Until that happens, the IMO *Guidelines for the Approval of*

Alternatives and Equivalents [12] is thus the current norm for the design, the implementation and the approval of autonomous ship systems. The Norwegian Maritime Authority (NMA) has as a response issued a more specific circular [13] that is aligned with the IMO guidelines.

We do not expect that rules and regulations will be developed and made available for the industry within short time, as the scoping and drafting exercises within IMO are time-consuming processes. This will lead to many challenges and complications in the approval process of autonomous ship systems that will be time consuming and costly for the involved stakeholders if not managed in a structured and systematic way. Type approvals for new technology and equipment are not possible within the current regime. We expect more testing of both individual systems and integration testing prior to final approvals at commissioning and sea-trials. It is also likely that the ship needs to be approved together with the system, at least for those components that are tightly linked with the control and monitoring strategy. This leads to a case-by-case approval regime, where the autonomous ships cannot be easily relocated to another operating area without reapprovals. It also indicates that it probably not will be allowed to generalize the approval process of a ship, but perhaps parts of it.

Whereas the approval process primarily is concerned with the safety and the security of autonomous ship systems and operations, we expect that cost-effectiveness and sustainability aspects will be more closely linked compared to today's conventional ship regime. As cost-effective and sustainable operations are the primary motivation of the development initiatives, these aspects will play an increasingly significant role when designing out and mitigating safety and security hazards. We do not want autonomous ship systems that reduces the cost-effectiveness or worsens the environmental impact of shipping operations, unless there are special circumstances where a "higher than normal" safety and security level is desirable.

A key challenge is to precisely communicate all aspects of the autonomous ship system and operations that are relevant for the approval activities, and understandable for a wide range of involved stakeholders. For this purpose, the Concept of Operations has become a central document for the specification of autonomous ship operations in the absence of both international and national rules for the design, building and operation of autonomous ships.

3. The Concept of Operations (ConOps)

A number of maritime industry guidelines explicitly refers to the ConOps as a document that is to be used to communicate the design of autonomous ship systems and operations [9], [10], [11]. This also applies to the NMA circular that explicitly lists the ConOps as one of the documents that is to be submitted as part of the autonomous ship system design study to initiate the preliminary design approval process [13].

There are various definitions of the ConOps in international standards and it is not within the scope of this paper to discuss these. The framework for description of autonomous ship systems and operations that will be proposed in this paper should capture the information items that is requested through the ConOps. The previous mentioned maritime guidelines has for that reason been analyzed, and a summary of the main information items that these guidelines request as part of the ConOps is listed in Table 1. The summary includes some non-exhaustive examples of specific details, and also reflects information requests from a maritime guideline that implicitly requests similar type of information [8].

Categories one to five list information items intended to describe the ship and the components that it shall interact with, and this can be viewed as a description of the autonomous ship system and the system context. Categories six and seven are a mix of functional and operational requirements, as well as descriptions of the operating conditions. Category nine is concerned with the safety management and operational management of the autonomous ship system.

The amount of information requested through the ConOps is extensive and should together be considered as a collection of requirements that will need to be verified against the relevant authorities' or class societies' specific needs. The ConOps should be looked upon as a living

IOP Conf. Series: Materials Science and Engineering **929** (2020) 012004 doi:10.1088/1757-899X/929/1/012004

document that is maintained throughout the life cycle of the autonomous ship system. It is to be expected that the detail level will increase as the available information base is improved and the uncertainties reduced.

Table 1. Summary of information elements requested in the Concept of Operations.

1	Physical characteristics of ship - major system elements and how they are connected and interact, performance characteristics.
	1.1 Size, speed, cargo type and capacity, cargo handling. 1.2 Steering, propulsion, energy storage and capacity, endurance. 1.3 Navigation system, sensors, position fixing systems and accuracy, detection capabilities. 1.4 Hull integrity, stability, hull strength. 1.5 Fire protection, cargo monitoring. 1.6 Communication systems.
2	External supporting systems during normal operations. Performance characteristics.
	2.1 External sensor or positioning systems. 2.2 External automation in port for cargo, cold ironing, berthing. 2.3 Planned response services, tugs, escort.
3	Crew, passenger and others on board
	3.1 How persons can enter ship. 3.2 How safety of persons is catered for. 3.3 Life support systems, if any.
4	Remote Control Centre.
	4.1 Features of RCC if in use, manning levels, location, redundancy. 4.2 Communication systems for safe control. 4.3 Forms of human-machine interfaces where applicable. 4.4 Mechanisms for coordination of ship automation, ship crew and RCC personnel.
5	Communication with other ships, VTS, MRS.
	5.1 Responsibilities for communication (RCC, automation, other). 5.2 Procedures and preparedness.
6	Functions and operations.
	6.1 Intended area of operation, significant phases in operation and voyage, operational environment and characteristics, limitations and restrictions. 6.2 Functions to be performed during operations and voyage. 6.3 Operational risk factors, including e.g. traffic density, environmental conditions, geography. 6.4 Division of responsibility between human and automation in the different functions/phases of operation ("degree of automation/control/autonomy"). 6.5 Additional support where applicable, e.g. incidence response, planned response. 6.6 Fallback solutions, minimum risk conditions.
7	Recovery, incident and emergency preparedness.
	7.1 Incident or emergency preparedness systems and plans. 7.2 Towage, on board recovery teams.
8	Safety management system.
	8.1 Logistics management. 8.2 Operations planning. 8.3 Operational procedures and responsibilities. 8.4 System health monitoring and system maintenance plans and facilities.

IOP Conf. Series: Materials Science and Engineering **929** (2020) 012004 doi:10.1088/1757-899X/929/1/012004

The advantage of the ConOps is that the format is very accessible for all stakeholders. It is normally written in prose text and the structure is flexible. However, as the objective of the document also is to facilitate precise communication of a specific design between the involved stakeholders, the format has some shortcomings.

The use of prose text can be subject to individual interpretations. Information in the document will be transferred from prose text into other engineering systems and tools, and back again. Discrepancies between the ConOps description and the actual design could easily occur, causing a mismatch between the design intention, the approval basis and subsequently the acceptance criteria. The question whether this can be avoided has led to the question whether it is possible to formalize the ConOps, and subsequently the identification of the need to have a framework for description of autonomous ship systems and operations.

4. The AUTOSHIP description framework

The objective of the AUTOSHIP description framework is to support a structured way of organizing information throughout the design process and the corresponding approval activities for autonomous ship systems and operations, as well as to lay the foundation for detailed design activities, system implementation, testing and final approvals. To achieve this, we have identified that the framework should facilitate:

- a precise description and clear understanding of what an autonomous ship is,
- a precise description and clear understanding of the roles and relationships in the autonomous ship system,
- systematic analysis of the autonomous ship systems ability to carry out the intended voyages and operations, including safety, security and cost-effectiveness, and
- it should capture all information items that will be requested in the ConOps, as a minimum, to satisfy the authorities.

The relationship between the four description documents that together constitutes the description framework for autonomous ship systems and operations proposed in this paper, the scope of the AUTOSHIP design methodology and the scope of the ConOps is illustrated in Figure 1.

<table>
<tr><th colspan="5">AUTOSHIP design method</th></tr>
<tr><td rowspan="2">Autonomous ship system description</td><td rowspan="2">System context description</td><td rowspan="2">Operational envelope description</td><td colspan="2">Scenario description</td></tr>
<tr><td>General operations</td><td>Supply chain and logistics</td></tr>
<tr><td colspan="4">ConOps</td><td></td></tr>
</table>

Figure 1. Relationship between the four design description documents and the ConOps.

The objective of the *autonomous ship system description*, the *system context description* and the *operational envelope description* is to provide a general overview of what the proposed design can do. These three documents have a one-to-one relationship with the ConOps.

The objective of the *scenario description* is to provide an exact description of what the proposed design shall do, and this must match the abilities and constraints defined in the descriptions of the autonomous ship system, the system context, and the operational envelope. Note that the scenario description is divided into two parts: *General operations* and *supply chain and logistics*. The primary concern for the approval of a system design is related to the general operations part of the scenario description, which is important for the safety and security analysis.

The supply chain and logistics part will contain specific details of the intended operation, including more details on cargo flows and the overall logistics model. If the ship is moved to another operation, it will in many cases be possible to change only the supply chain and logistics part, if this is within the boundaries of the general operation. This would allow reapproval for a new operation by only checking that these two documents are consistent. It should not be necessary to go through a full approval process again. The AUTOSHIP design methodology will also evaluate the sustainability and cost-effectiveness of the autonomous ship system, and for this purpose an extensive supply chain and logistics description is needed.

The relationship between the description framework, the system objectives and the resulting design is illustrated in Figure 2. The system objectives are the highest-level objectives for the design of the autonomous ship system and its operations, and these form the input needed to start the design and specification process. They are typically linked to the business and societal effects of the resulting outcome of implementing the design [14]. One example of a system objective is to reduce the cost per transported unit of goods, and another is to increase the frequency of goods deliveries to customers from a biweekly to a weekly basis. The collection of the four description documents is the design that will be subject to approval activities, and it will, if deemed economically viable, safe and sustainable, be subject to further detailed design activities, implementation, testing and final approvals. The following subsections will outline the structure of each of the four description documents.

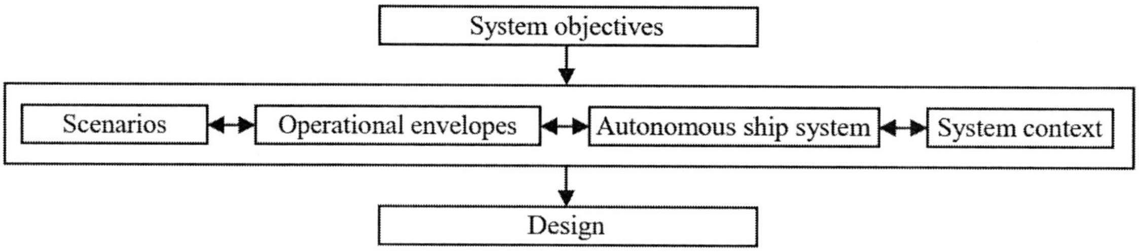

Figure 2. High-level structure of the framework.

4.1. Scenario description

The objective of the scenario description is to provide a description of what the proposed system design shall do. The first part of this document includes a high-level description of the operations that must be carried out given the specific system objectives. It also includes the properties of the specific voyage. The second part of the document contains a detailed description of the supply chain and logistics system that the autonomous ship system is part of. In the following, only the first part will be discussed as this is most relevant for the approval process.

The scenario description provides an overview of the general operations in a framework that divides the use cases into generic voyage phases [3]. This is illustrated in Figure 3, where each of the generic voyage phases contain one or more operations.

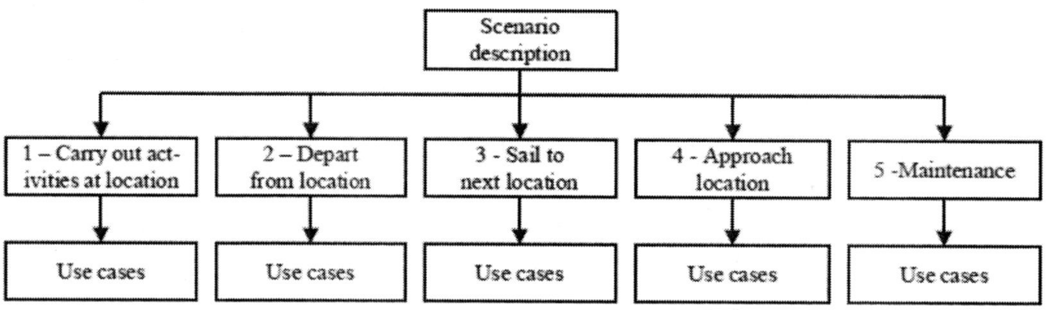

Figure 3. Structure of the general operations in the scenario description.

The process of identifying the general operations is supported by a systematic supply chain

IOP Conf. Series: Materials Science and Engineering **929** (2020) 012004 doi:10.1088/1757-899X/929/1/012004

analysis and a proposal for process subdivision. The systematic supply chain analysis supports identification of use cases based on semi-structured interviews of stakeholders as well as analysis of available quantitative data for the specific supply chain and logistics system in question [3]. The process subdivision provides guidance on relevant operations, functions, and tasks in an autonomous ship system divided into ship operations, ship management operations, inland waterways operations, port operations, coastal operations and on-site operations [5].

Identification and categorization of general operations and use-cases into generic voyage phases is not sufficient to describe the scenarios that the autonomous ship system shall be designed for. The properties of each operation must also be linked to the actual voyage and the specific sailing route which is described in the supply chain and logistics document. The general operations will have to describe each class of voyage phases, e.g. characterized by maximum wave height, visibility, duration etc.

In the supply chain and logistics document, the voyage is described using directed graphs with nodes and edges [15]. Each node corresponds to a geographical position and each edge is a voyage leg between two positions. This is illustrated in Figure 4. In the general operations these can be described as general use cases with parameters as suggested above.

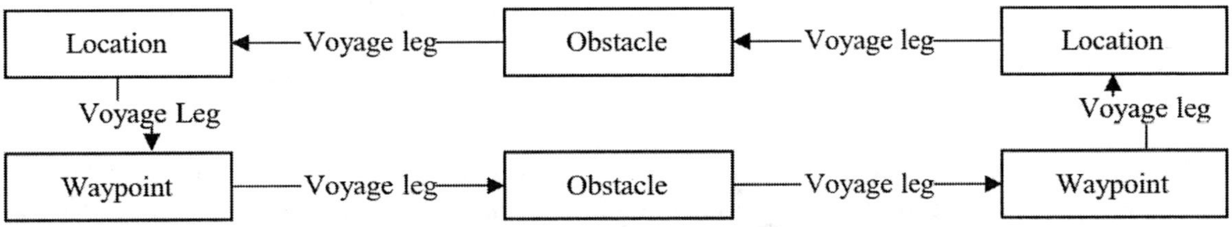

Figure 4. Structure of the voyage in the scenario description.

The voyage contains three main types of nodes. A location describes a factory, quay, or any equivalence where the ship carries out activities such as loading and offloading of goods. An obstacle is a location on the route where the ships needs to pass e.g. a water lock or bridge but does not offload or load any goods. A waypoint separates two voyage legs that have different characteristics. The objective of this description is to provide an accurate overview of the operational area and the characteristics of all locations and voyage legs that the ship will be subject to during operations.

The systematic supply chain analysis is also used to establish the voyage description, and much of the information that is obtained in this analysis will also be leveraged to create the operational envelope description.

4.2. Operational envelope description

The Operational Envelope (OE) is defined as "the specific conditions and scenarios under which a given autonomous ship system is designed to function" in [5]. An important step in the design process of autonomous ship systems is to do an operability study where the capabilities of the autonomous ship system is assessed. The capability assessment considers the tasks the autonomous ship shall perform, and under which conditions the tasks must be performed. The result is a set of operational limits for the autonomous ship system. The objective of the operational envelope is to provide an accurate overview of all operating conditions that the autonomous ship system will encounter, and translate this to boundaries for the identified operations and voyages, and a strategy for distribution of and transfer of control between automation and humans. The structure of the operational envelope description is illustrated in Figure 5.

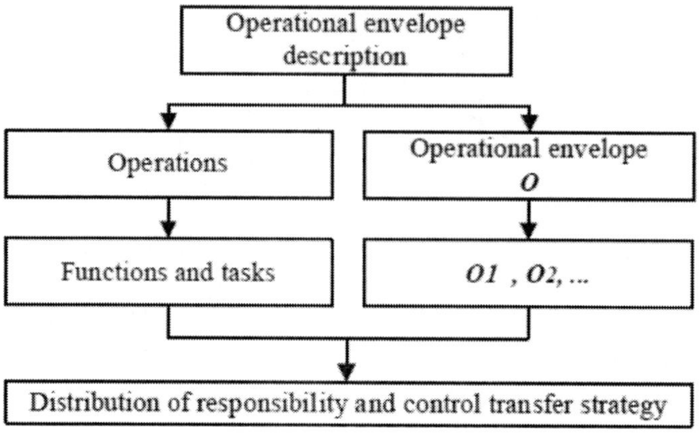

Figure 5. Structure of the operational envelope description.

The operational envelope description is based on the high-level overview of the operations and conditions identified in the general operations. Each scenario and operation consist of a set of functions that can be further subdivided into specific tasks until an adequate level of details is reached. All functions and tasks are subject to one or more operating conditions.

The conditions of the OE include weather complexity, traffic complexity and geographical complexity in the temporal and spatial dimension of the operating area. The autonomous ship system needs to be aware of changes in these conditions, and to be able to act in an appropriate manner to ensure that it stays within the limits of the operational envelope boundary, or alternatively fall back into a safe state or minimum risk condition outside the OE. Development of the operational envelope description will require iterations over the autonomous ship system description as it includes a description of the distribution of responsibility and transfer of control between the components and roles defined in the autonomous ship system.

4.3. Autonomous ship system description

The objective of the autonomous ship system description is to provide a description of all physical components and roles that together ensure effective monitoring and control of the autonomous ship processes for the ship's intended operation or voyage. The autonomous ship system must be viewed as a cyber-physical system of systems that consists of many physical components both onshore and onboard the ship. These physical components interact with each other and they are connected and communicate.

The structure of the autonomous ship system description is illustrated in Figure 6. The physical components and roles are either allocated to the ship or to shore. The ship will have a combination of various equipment, automation systems and crew onboard, and various equipment, automation systems and personnel will support the ship operation from one or several shore-based locations.

The automation systems, ship crew and shore personnel will together have the overall responsibility of operating both the ship and shore equipment, and this distribution of responsibility is reflected in the operational envelope description. The main components of the communication architecture, and the implementation of procedures, information models and protocols will affect how control can be transferred between humans and automation and explicitly contribute to the operability.

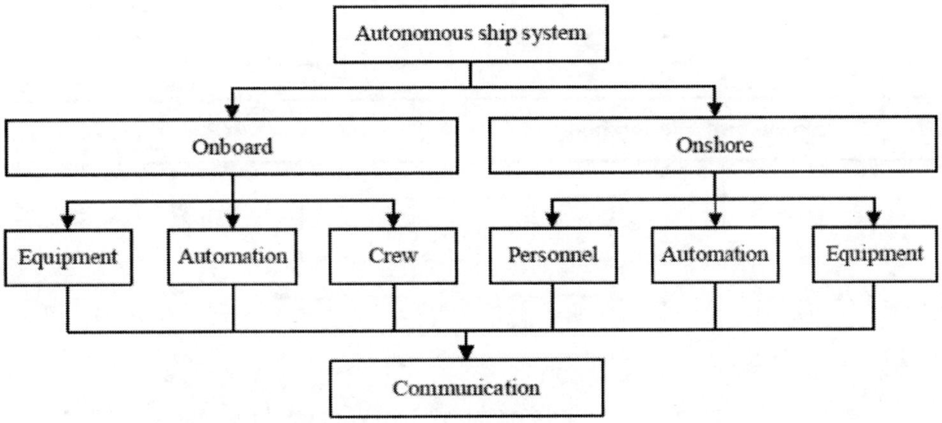

Figure 6. Structure of the autonomous ship system description.

The definition of the autonomous ship system proposed in the AUTOSHIP project is illustrated in Figure 7 and a detailed description of the components and roles is given in [5]. The physical components and roles within this definition has been selected based on the criteria of including the components that are tightly integrated with the control and monitoring strategy of the autonomous ship.

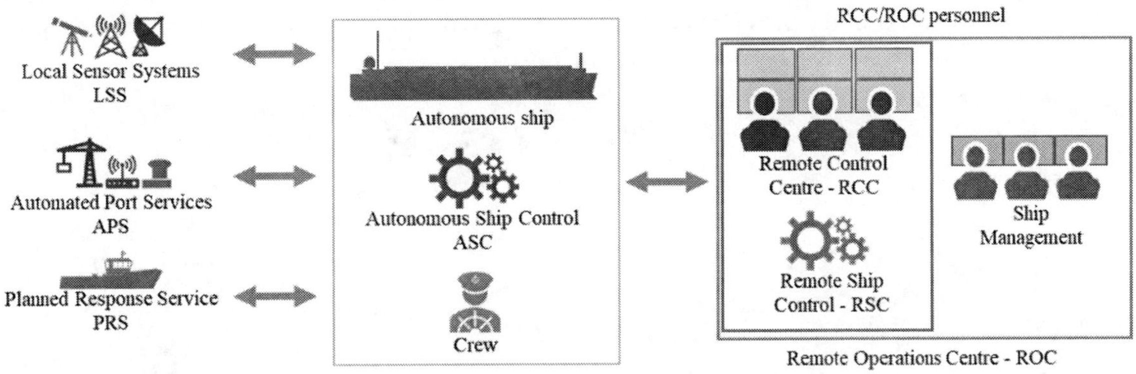

Figure 7. Examples of components and roles in an autonomous ship system [5].

It should be noted that it is possible that both the number and types of components, as well as the number of ships in the autonomous ship system could vary depending on the specific system objectives.

4.4. System context description

The objective of the system context description is to provide an accurate description of the boundaries between the autonomous ship system and its environment that includes the components, the roles and the most important communication paths. The structure of the system context description is illustrated in Figure 8.

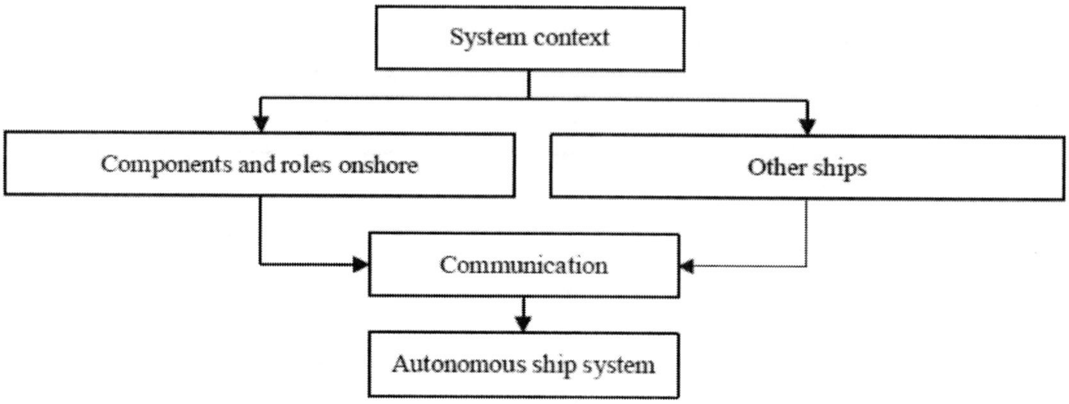

Figure 8. Structure of the system context description.

The description is divided into two parts. The first part contains a description of the shore-based components and roles that the autonomous ship system interacts with. The second part of the document contains a description of other ships that operate within the defined operating area of the autonomous ship system. The essence of the system context description is to capture the communication requirements between the autonomous ship system and the context components.

The definition of what is inside or outside the boundaries of the autonomous ship system is subjective and likely to be influenced by the individual designer or design team. The selection of the components that belong in the autonomous ship system creates the necessary boundary to define the system context of the autonomous ship system. The definition of the system context proposed in the AUTOSHIP project is illustrated in Figure 9.

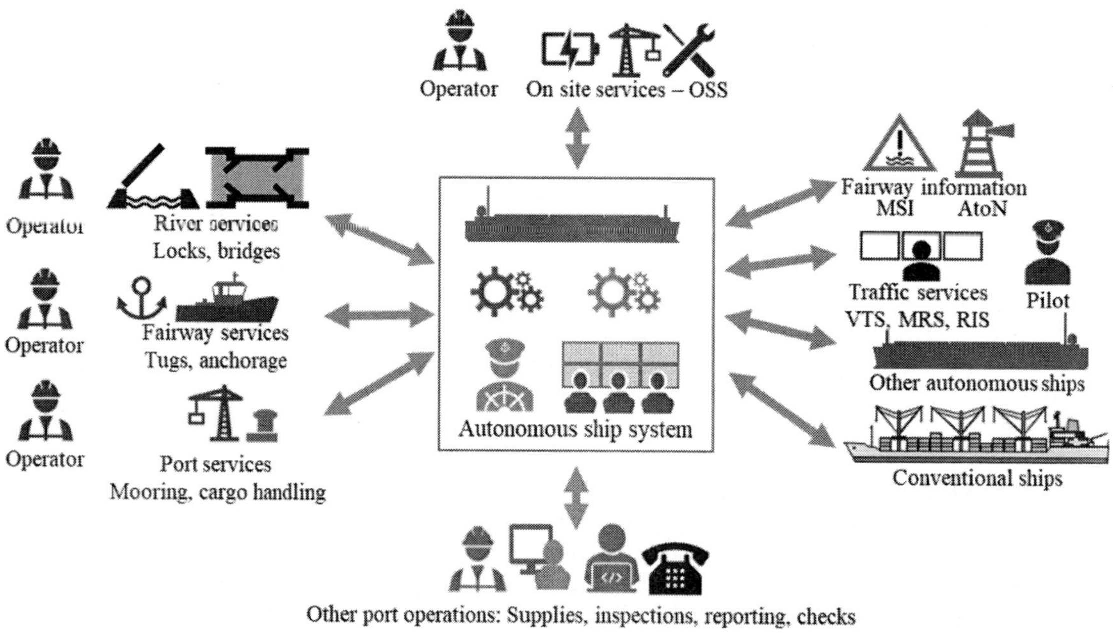

Figure 9. Examples of components and roles in the context of the autonomous ship system [5].

A detailed description of these components and roles is given in [5], and it should be noted that the system context only includes components that exchange information with the autonomous ship.

Unknown objects in the water are not included as they are mapped through the traffic complexity conditions in the operational envelope description.

5. Discussion

We believe that it is possible to formalize the ConOps, and the AUTOSHIP framework for description of autonomous ship systems and operations proposed in this paper is the first step in this process. The framework facilitates the use of a formal modeling approach, and even though it might be impossible to avoid prose text descriptions completely, the shift towards a formal description will most likely reduce discrepancies between the actual design and the design descriptions. A formal description of the autonomous ship system and operations would also facilitate effective reuse of information between different designs and allow us to create strong links between the design description and the approval basis, the acceptance criteria and test results.

The structure of the proposed framework also has some features that lead us to believe that it is possible to standardize parts of the approval processes. The high-level structure of the autonomous ship system description, the system context description and the operational envelope description will probably not change much from design to design. We anticipate that it is possible to create a maximum template that only needs to be configured according to the contents of the scenario description. This would allow for a more efficient process. This also has the implication that should one decide to change the operating area of an already built autonomous ship in the future, then reapproval would require that an updated scenario description is assessed and analyzed with the already existing autonomous ship description, the system context description and the operational envelope description.

The description of the operations in terms of the general operations description would also allow a much simpler reapproval process, should the ship be moved to another operating area.

The use of a common description framework will allow us to compare designs in cost and benefit analysis using the same basic structure. This also applies to safety, security, and cybersecurity analysis that requires a well-defined description of roles and relationships between the autonomous ship system components and its environment.

In a wider scope, a common framework for description of autonomous ship systems will ease work on standardisation activities, which are considered as essential for successful implementation of autonomous shipping. This particularly applies to the communication links between the autonomous ship system and its context. Furthermore, a common description framework complemented by a corresponding common terminology that is being developed for autonomous and automated ships, will enable more consistent communication in the technical and scientific community researching this area.

6. Conclusions

This paper has introduced a framework for description of autonomous ship systems and operations that consists of four description documents: The scenario description, the operational environment description, the autonomous ship system description and the system context description. The contents of the description framework cover the information items that the authorities request through the ConOps. The framework has the potential to facilitate development of a formalized ConOps. This could in turn lead to a standardization of the current approval procedures for autonomous ship systems and operations. Further research is required to validate the framework through implementation and testing on specific use-cases.

Acknowledgements

The work presented in this paper is a result of a collaboration between the AUTOSHIP project funded by EU Horizon 2020 with grant number 815012 and the AEGIS project funded by EU Horizon 2020 with grant number 859992

IOP Conf. Series: Materials Science and Engineering **929** (2020) 012004 doi:10.1088/1757-899X/929/1/012004

References

[1] Rødseth, Ø. J. (2018). Assessing business cases for autonomous and unmanned ships. Technology and Science for the Ships of the Future. Proceedings of NAV, 2018, 19th

[2] Heffner, K., & Rødseth, Ø. J. (2019, October). Enabling Technologies for Maritime Autonomous Surface Ships. *In Journal of Physics: Conference Series* (Vol. 1357, No. 1, p. 012021). IOP Publishing

[3] Wennersberg, L. A. L., Nordahl, H., Bolbot, V., and Theotokatos,G. "AUTOSHIP deliverable D2.1: Complete supply chain mapping & identification of interactions between SSS and IWW demonstrators" , November 2019

[4] Faivre J., Nzengu W., Bolbot V. "AUTOSHIP deliverable D2.3: Regulatory framework mapping & Identification of gaps and requirements for autonomous ships compliance", Revision 3.0, May 2020

[5] Rødseth Ø.J., Faivre J., Hjørungnes S.R., Andersen P., Bolbot V., Pauwelyn A.S., Wennersberg L.A.L. "AUTOSHIP deliverable D3.1: Autonomous ship design standards", Revision 1.0, June 2020

[6] IMO (2019). Interim Guidelines for MASS Trials, MSC.1/Circ.1604, 14 June 2019

[7] ISO (2020). Input document to Maritime Safety Committee, Session 102, agenda item 5: Regulatory Scoping Exercise for the Use of Maritime Autonomous Surface Ships (MASS), Proposed terminology for MASS, Submitted by International Organization for Standardization (ISO), February 2020

[8] Bureau Veritas (2019), Guidelines for Autonomous Shipping, October 2019, Guidance Note NI 641 DT R01 E

[9] LR (2017), ShipRight - Design and Construction, Additional Design Procedures, LR Code for Unmanned Marine Systems. February 2017

[10] ABS (2019) Advisory on Autonomous Functionality, American Bureau of Shipping

[11] DNV GL (2018) Class Guideline DNVGL-CG-0264, Autonomous and remotely operated ships

[12] IMO (2013). Guidelines for the Approval of Alternatives and Equivalents as provided for in Various IMO Instruments, MSC.1/Circ.1455, 24 June 2013

[13] Norwegian flag state authorities (2019): Rundskriv - Serie V (RSV 12-2020), "Requirements to documentation in conjunction with building autonomous, unmanned and/or remotely controlled vessels" – 04.06.2020 (In Norwegian)

[14] Samset, K. (2010). "Early project appraisal: making the initial choices". Springer, pp 12-18

[15] Cormen, T.H., Leiserson, C.E., Rivest, R.L., Stein, C., 2009. "Introduction to Algorithms", Second Edition, MIT Press, pp. 528

IOP Conf. Series: Materials Science and Engineering **929** (2020) 012005 doi:10.1088/1757-899X/929/1/012005

Autonomous vessels as ships – the definition conundrum

Mayank Suri[1]

[1]Assistant Professor, Jindal Global Law School,
India E-mail: msuri@jgu.edu.in

Abstract. The development of unmanned and autonomous vessels has accelerated, resulting in the requirement to study the law surrounding shipping. This article studies recent literature and case laws to determine the various interpretations of the word 'ship'. It, thereafter, seeks to determine which of these will have a bearing on autonomous vessels and whether the foreseeable technological incorporation is a challenge. Primarily, the focus is on showcasing that definitions across jurisdictions vary. The author points out that in the domain of private international law, this will result in disputes regarding applicability of admiralty laws.

1. Introduction

The operations of autonomous vessels are in the offing as proposed by several projects.[1] There is a steady momentum with respect to awareness of autonomous operations in the shipping industry.[2] Some leading technology manufacturers have started to market products that could tomorrow be one of the essential features of a commercial autonomous vessel.[3] Added to this, the commercial side of shipping operations is seeing a parallel jump in maritime intelligence products that offer efficient fleet management.[4] There are supporting views for international solidarity in regulating unmanned '*ships*'.[5] The Maritime Safety Committee of the International Maritime Organisation has initiated a regulatory scoping exercise to this extent. The Maritime Safety Committee has defined 'Maritime Autonomous Surface Ship' as "*a ship which, to a varying degree, can operate independent of human interaction*".[6] This definition does not reveal any characteristics that would constitute or at least assist in determining what is an autonomous '*ship*'. Some relief is sought in construing '*operate*' as meaning the entire operation of a vessel's movement and not just of one particular machinery because that would entail covering all modern ships as MASS. One can also reason that "independent of human interaction" is a reference to a lack of crew because if that is not the case then again ambiguity shrouds over the coverage of unmanned vessels.

It has also been pointed out that the development of autonomous '*ships*' will surpass the development of the regulatory framework surrounding them.[7] Taking, for example, the AUTOSHIP project which has received European Union funding to the tune of 20 million Euros (approx.), one can foresee a multi-country agreement on the movement of autonomous ships aided by appropriate regulatory frameworks.[8] Studies to investigate the economic viability of such a step are being conducted at an increasing pace.[9] Indeed there are conventional shipping routes between European nations that may be exploited to test the operations of autonomous ships depending on countries agreements.[10]

Therefore, it is important to investigate relevance of the existing jurisprudence in international maritime law with respect to adoption of autonomous vessels. It is also crucial to acknowledge that literature surrounding autonomous 'ships' has been on the rise and this is reflective of the broader outlook of the shipping industry. The question this paper seeks to answer is whether the current regulatory framework, allows for admission of the autonomous vessel as a 'ship' in the private law domain. Ergo, whether the definition of a '*ship*' in its current form will be accepting of an autonomous vessel. However, as we move further in this analysis it must be understood that this paper is written on the understanding of a clear difference between an unmanned vessel and an autonomous vessel. The former is a reference to a vessel without any onboard crew and can include an autonomous vessel. It will become clear as to why the current discussion does not need to demarcate this difference at each assertion.

2. Why the need for a study of the definition of a *ship*?

The primary premise on which this paper has been written is that there will be intercountry movement of autonomous vessels. As brought to light above, it has become evident that at least European countries will allow their movement.[11] However, it must not be lost that each country has its own definition of what a '*ship*' is, therefore at least for legal purposes one should see whether one country's '*ship'* will be accepted in another. Generally, the law governing what claims can be effected by way of detaining a '*ship*' is known as the Admiralty law of the country, a specialism within the broader term '*maritime law'*.[12] A particular type of claim peculiar to admiralty laws, in rem claim, gives a claimant the right to arrest and auction the *'ship'* and satisfy its claim against the proceeds if the shipowner fails to provide alternate security in court.[13] Conclusively, a vessel that fails to come within the definition of a '*ship*' under a country's admiralty laws, would destroy the right to claim in rem.

Let's take a look at two definitions from different countries to understand this better.

India: "*(l) "vessel" includes any ship, boat, sailing vessel or other description of vessel used or constructed for use in navigation by water, whether it is propelled or not, and includes a barge, lighter or other floating vessel, a hovercraft, an off-shore industry mobile unit, a vessel that has sunk or is stranded or abandoned and the remains of such a vessel[14].*"

Korea: "*(1) The term "ship" in this Act means the sort of ships that are used or can be used for navigation on or under the water, and their classification shall be as follows:*

1. *Steam ship: A ship propelled by an engine (including any ship that has an engine attached to outside the hull which may be separated therefrom, and any ship using both engines and sails which mainly uses an engine, and a wing-in-ground craft (a ship that exploits "wing-in-ground" effect to fly just above the water surface);*
2. *Sail ship: A ship propelled by sails (including any ship that uses both an engine and sails, with the latter used mainly);*
3. *Barge: A non-self-propelled ship that needs to be towed or pushed by another ship.*

(2) The term "small ship" in this Act means any ship falling under any of the following subparagraphs:

1. *Steam ships and sail ships of less than 20 gross tonnage;*
2. *Barges of less than 100 gross tonnage.[15]*"

From reading the above two definitions, it becomes clear that parties in these two jurisdictions would view different things as '*ships'* from a legal viewpoint. For example, the Indian definition of a '*vessel'* does not include a wing-in-ground craft ("WIG"), whereas the Korean definition expressly includes it under the definition of a '*ship'* and classification of a '*steam ship*'. This entails that where a WIG would be subject to the admiralty law of Korea, this may not be the situation in India. A corollary of this analysis would mean that a claimant trying to initiate an *in rem* action against a WIG in India would

IOP Conf. Series: Materials Science and Engineering **929** (2020) 012005 doi:10.1088/1757-899X/929/1/012005

have to persuade the court that the particular vessel falls within the broader definition of a vessel as given above while in Korea this would not be a hurdle to the claim. Similarly, a hovercraft may be subject to almost identical treatment in a Korean court of law.

In fact, one can also understand the importance of definitions when studied through the lens of WIG in the collision regulations. When WIGs were introduced, they were a unique kind of craft which existed at the intersection of the maritime and aviation industries.[16] Their characteristic 'over-the-water' movement posed a regulatory conundrum. However, acknowledging that the WIG would operate and exist around other waterborne vessels, the IMO adopted it as part of the international maritime regime even with its multimodal nature.[17] Therefore, it lends support to the argument that when a technologically different craft is introduced, it shall benefit the regulatory framework to expressly include it within the definitions.[18]

3. Literature Review

3.1 The law of unmanned merchant shipping – an exploration[19]

Dr. Hooydonk in this article has explored the legal status of commercial, surface-operating unmanned cargo and passenger vessels. He states how maritime law has withstood the test of time and technological advancement. He brings to light the stable nature of maritime law by presenting the reader with a range of developments – from times of sail to nuclear propulsion. At one point, Dr. Hooydonk writes about the distinction between Remotely Operated Vehicles (ROVs) and Autonomous Vehicles (AVs). He emphasizes that a clear distinction should be made between an unmanned vessels, waterborne robots and the like. This clarity lies in whether the waterborne robot can be considered part of a mother ship or its equipment. However, thereafter he uses the term 'unmanned underwater craft' while stating several civil purposes for which these are employed. But this use is unsupported by any reference to a mother ship (or lack thereof), mystifying, whether these should be considered as unmanned vessels or waterborne robots or both.

It is Section 2 of his article, that is of immense consequence to this discussion.[20] He observes that the maritime convention dealing with flag and port state rights and responsibilities, the UN Convention on the Law of the Sea ("UNCLOS"), does not define the term '*ship*' and uses it interchangeably with 'vessel'. He points out that public law maritime conventions apply their own definitions tailor made to serve the Convention's purpose. He thereafter states that major private maritime law conventions; such as those on collision, arrest and limitation of liability; do not provide a definition. Few private maritime law conventions define a '*ship*' in broad terms. It is Dr. Hooydonk's statement, "*National maritime laws give an equally heterogeneous picture*", which resonates with this discussion. He concludes with a fair amount of certainty that having a crew on board is not an essential requirement for the notion of a '*ship'*.

It is imperative to note that where private maritime law conventions are silent and a challenge is made as to the identity of the '*ship*', a court would be left with no choice but to derive it from definitions found in national laws.[21]

3.2 The integration of unmanned ships into the lex maritimea[22]

The other text which has contributed immensely to the discussion argues that the definition conundrum does not pose a major impediment since the presence of onboard seafarers is not an express prerequisite to '*ship'* status under the international regulatory regime. The authors have identified two crucial issues which need answering; whether unmanned crafts can be operated as safely as conventional '*ships'* and can enforcement of claims be at least as effective. It is this second issue which is of relevance to this article. They correctly note that for the unmanned craft to be operated within the territory of that state, it must fall within the national legal framework of that state. Interestingly, the authors have used the word 'craft', without its legal connotations, in referring to unmanned vessels. There is a further distinction, that this article identifies, that some crafts perform a '*predetermined nautical course without any human interaction whatsoever'*, referring to them as 'unmanned autonomous craft'. However, the

IOP Conf. Series: Materials Science and Engineering **929** (2020) 012005 doi:10.1088/1757-899X/929/1/012005

article acknowledges that an unmanned craft may be capable of both, remote as well as autonomous operations.

The authors also concur with Dr. Hooydonk, that UNCLOS has used the terms '*ship*' and 'vessel' interchangeably without realizing that under some jurisdictions these terms are viewed differently[23]. In one of its footnotes, the article underscored the gravity with which the definition conundrum must be seen. This footnote describes how a US-China dispute in respect of US Navy's 'unmanned underwater vehicle ("UUV")' saw the two sides debating whether the UUV was a '*ship*' or an 'unidentified device'[24].

The authors, in citing Gahlen, touch upon the essential aspects of a '*ship*'. These being; floatability, capability of controlled movement on water, capability in the carriage of person or goods beyond its own mass; and sea going capability[25]. It argues that the question of what is a '*ship*' becomes increasingly irrelevant when the unmanned craft '*performs functions more closely akin to conventional carriage operations*'. In its conclusion on the international regulatory regime discussion, the article notes that there is no '*international definition of a ship*' and this question may substantially be answered by national laws. The article provides an insight into US and some European jurisdictions, where with the exception of France, it does not find a major concern in reading an unmanned craft as a '*ship*'. It does, however, indicate that a dependence on national laws would result in a situation where one's state law does not identify another states' unmanned '*ship*'.

It is submitted that the article is a good starting point on the discussion regarding national laws for private claims. In this respect, it is submitted that it matters not whether the human is controlling onshore or the computer is running autonomously. For what is of consequence is that the vessel is unmanned and must be so seen when interacting with manned vessels or entities (like ports) and therefore, giving rise to possible claims *in rem*.

4. Case Review

This article now looks at some English law cases on the subject.

4.1 Steedman v Scofield and Another, 1992 2 LR 163 (QBD)

This was a case involving a collision between a Kawasaki model jet ski and a speedboat. The Plaintiff, owner of the jet ski, had filed a case for damages against the defendants, owners of the speedboat. The defendants contended that by virtue of the late issuance of the writ by the Plaintiff, the case was time barred. To rely on this time bar, the Defendants had to bring the case within the 2 year time limitation provided by the Maritime Conventions Act 1911. However, the application of the Maritime Conventions Act, 1911 depended on whether the jet ski could be called a 'vessel', the definition of which was found in the Merchant Shipping Act, 1984 and is laid out below:

> *"742. "Vessel" includes any ship or boat, or any other description of vessel used in navigation:*
>
> *"Ship" includes every description of vessel used in navigation not propelled by oars…".*

The judge while dealing with this issue pointed out some highly technical details. He contemplated whether the jet ski could be a boat and answering in the negative he held that a boat has its own buoyancy which that jet ski lacked, it could not even be boarded when it was stationary. The judge then considered meaning of the words '*used in navigation*' as it became apparent that this was the dominant idea behind holding a watercraft to be a vessel. The judge opined that 'ordered movement in water' as against simple movement in water and 'transportation of persons or property' were the determining factors. He stated that the purpose behind the jet ski was to enjoy the thrills of waterskiing and so, even if it could be navigated per se, it was not meant for that purpose.

It is submitted that this judgment brings to fore some interesting observations. Firstly, the judge relied heavily on the manufacturer's advertisement i.e. Kawasaki, to come to his findings about what the jet ski was. He specifically refers to the advertisements' nomenclature of calling it a 'personal watercraft'. He refers time and again to the thrill-seeking purpose mentioned in the advertisement. In his declaration that the jet ski is not a boat he refers to the lack of buoyancy but this distinction can no longer be maintained since jet skis now are fully buoyant and capable of being mounted by 1-3 persons when stationary. However, till date Kawasaki advertises its jet skis as 'watercrafts'[26].

4.2 R v Goodwin, 2006 1 LR 432 (CoA)
Interestingly, the second case in our discussion represents the pace of technological developments and the effect it can have on legal interpretations. In this case, which involved a collision between two jet skis, a Yamaha 'Waverunner' jet ski was the defendant's vessel. This jet ski unlike the jet ski in the *Steedman* case, was buoyant enough to seat 3 people when stationary or moving. This was enough for the Recorder, at first instance, to hold that the Waverunner was a '*ship*'. The *Steedman* case had such a bearing upon this judgment that one can read the Recorder's narration of the characteristics of the Waverunner as one contrasting with the characteristics of the *Steedman* jet ski.

On appeal, the Court of Appeal, was asked to answer the issue "was the 'Waverunner' a ship?" One of the novel arguments put forward was that jet skis can be registered as "wet bikes" in the register of British ships and therefore, must be construed as '*ships*'. The court of appeal after looking at the purpose of registration of ships and going through the *travaux preparatoires[27]* held that registration of a jet ski does not make it a '*ship*' for the purposes of the Merchant Shipping Act, 1995. Per the Court of Appeal, registration of a jet ski was closely linked to reasons of financing for purchase.

The court was now supposed to check if the jet ski fell within Section 313 of the 1995 Act which defined '*ship*' as follows:

> "*"ship" includes every description of vessel used in navigation*".

Like the Steedman case, here too the court was required to interpret the words "used in navigation". Acknowledging the diversity in the type of vessels that were in use, the court declined to agree with the Steedman case that "carriage of goods or passengers" was of consequence to this term. The court of appeal relied on two cases; one, where a backhoe dredger, and; the other, a jack up rig; both non-good/passenger carrying vessels, with very different purposes and construction were held to be ships. Effectively, this meant that "*used in navigation*" no longer limited the scope of the law from governing vessels which carried out other specific activities on the water. In fact, in *Von Rocks[28]*, the backhoe dredger could be said to be an anomalous '*ship*' for several reasons; it wasn't self-propelled, it's not normally manned by a crew and it has no rudder or other steering mechanism.

It must be noted that the Court of Appeal, came to the conclusion that the Waverunner was not a ship on the basis that it lacked the other requirement for navigation i.e. ordered movement in water. To reach this conclusion, the court considered the purpose behind the legislation itself and it stated that the purpose was to regulate shipping as a trade or business and not as a pleasure sport.

4.3 Michael v Musgrove [2012] 2 Lloyd's Rep. 37 (QBD)
In this case involving a rigid inflatable boat ("RIB"), the question was whether the Athens Convention[29] for passenger liability would apply. The English court had to construe whether the RIB fell within the definition of a '*ship*' as mentioned in the Athens Convention. This definition is as follows:

> "*'ship' means only a seagoing vessel, excluding an air-cushion vehicle*".

As apparent on the face of it, this definition does not provide a description or even parameters for defining what a *'ship'* is. It indeed lacks any extensional or intensional factors which may lead one to infer what the possible physical characteristics of a *'ship'* could be[30]. However, thanks to the application of this convention in UK by force of law, the court could take assistance of the English statutes as well as cases; some of which we have covered above.

One must make note of the fact that the court has elaborately stated the physical characteristics of the RIB, including what kind of material was used in building it and what paraphernalia form part of the various onboard equipment; the manufacturer, and; the users of RIBs[31]. While the court studied various precedents, we are concerned with what it said on the physical aspects of a vessel considering *Steedman* and *Goodwin[32]*. The claimant's lawyer submitted that the RIB was unseaworthy to be classed as a vessel basis the fact that it had no cabin and thus, not safe for use as a vessel. The court correctly denied accepting this argument as consequential to the seaworthiness of the vessel. Further, in construing "*used in navigation*" it relied on the manufacturers' description of the roles of the RIB, specifically stating that he test of ordered movement on water as opposed to "messing about in boats" had been satisfied.

5. Case Focus: Guardian Offshore AU Pty Ltd v Saab Seaeye Leopard 1702 Remotely Operated Vehicle Lately On Board The Ship 'Offshore Guardian' [2020] FCA 273

This Australian[33] case is the focal point of this paper for several reasons, these are mentioned here in brief:

a. This case involved a remotely operated vehicle ("ROV") which was unmanned[34]
b. The case involved arresting of the ROV for a claim between two private parties.
c. The case adjudicated whether admiralty law would apply to the ROV.

To offer context to our discussion, it is essential that we highlight some key technical aspects of the ROV:

a. It was deployed from a ship, referred to in the judgement as "main vessel".
b. It was tethered to a metal frame known as the launch and recovery system (LARS) which in turn was winched to the main vessel.
c. The ROV is operated by two pilots onboard the main vessel. While one pilot controls the movement of the ROV, the other controls lighting, cameras and other equipment inside the ROV.
d. In words of the court, "*Its physical appearance lacks the attributes that would usually be associated with a structure described as a ship or vessel as a matter of ordinary parlance.*".

Interestingly, the judgment does not reveal what the particulars of the claim between the parties was. However, what is clear is that the plaintiff was owner of the main vessel onboard which the defendant ROV was once used and the claim was against the owner of the ROV. *Interalia,* the court framed the following three issues:

a. Was the arrest warrant issued for the vessel that has been arrested?
b. Is FCDS the owner for the purposes of s 19 of the Admiralty Act?
c. Are the ROVs ships? The reference to ROVs in plural is because the writ named two similar ROVs as first defendant and second defendant. This difference is not an issue in context.

This article is concerned with the third issue i.e. "Are the ROVs ships?". In order to proceed against the ROVs *in rem*, the Australian admiralty law was to be applied. This application depended on whether the definition of a *'ship'* as mentioned in the Admiralty Act of Australia would cover the ROVs. This definition is reproduced herein below:

"***ship*** *means a vessel of any kind used or constructed for use in navigation by water, however it is propelled or moved, and includes:*

(a) a barge, lighter or other floating

vessel; (b)a hovercraft;

(c) an off-shore industry mobile unit; and

(d) a vessel that has sunk or is stranded and the remains of such a vessel;

but does not include:

(e) a seaplane;

(f) an inland waterways vessel; or

(g) a vessel under construction that has not been launched."[35].

The court started its deliberation on this issue by first looking into the historical reasons for incorporating the definition of a '*ship*' into the act[36]. It is submitted that these reasons would also hold true for other admiralty jurisdictions around the world which are common law jurisdiction[37]. The court explored two of these aspects; first, '*ship*' became a convenient means of describing the subject matter of admiralty jurisdiction, and; second, the type of property/*res* against which *in rem* proceedings could operate. It is this second aspect which is of relevance to the discussion. The court further noted that the provision '*should be interpreted liberally and without imposing limitations not found in the express words'[38]*. It would not take one long to realise the commonality between the general words of this section and the words used in the English, Indian and Korean sections mentioned above.

From the words of this section, the court came to the conclusion that a ship must have three attributes. These were: it must be a vessel; it must be used (or constructed for use) in navigation by water, and; it must be moved by means of water and not just be an aid to navigation. This last attribute, the court said, was implicit in nature; found in the words "however it is propelled or moved" in order to strike out fixed structures from the ambit of the definition. Connecting this implicit attribute to the need to include all rigs which are mobile so as to afford the protection of admiralty laws.

The court then looked at the inclusion of the hovercraft and exclusion of the seaplane from the definition of '*ship*'. In reading purpose into the definition, the court made it very clear that what mattered was the extent to which admiralty jurisdiction would be applicable to these structures. It stated three principles of admiralty jurisdiction: salvage, liability for collision, and; *in rem* jurisdiction to arrest. The extent or degree to which these will be applicable to the structure would determine whether it should be included. This is seemingly the only justification why a hovercraft, which operates on water rather than in it, was included, but a seaplane, which would majorly be in-flight, was excluded.

The claimants, Guardian Offshore, presented the court with a Canadian judgment in which an ROV was arrested and held to be a ship under the broad definition therein provided[39]. This decision rests on two considerations: the Canadian definition is very broad and therefore, inclusive[40], and; the purpose of Admiralty law *qua* jurisdiction.

The court thereafter makes some conclusions concerning the definition of a '*ship*'. These conclusions may be summarised in the following manner:

a. Various jurisdictions present a plethora of definitions, none of which are comprehensive in respect to the characteristics of a vessel. However, the definitions do not represent the necessity of carrying cargo or passengers.
b. Navigation '*by*' water is broader than navigation *'in'* water and would not exclude a submersible which moves through water.
c. There are judicial authorities which review the purpose, attributes and capabilities of a

vessel to determine whether the vessel lacks the usual attributes of a '*ship'*.

d. The application of admiralty law to vessels that can be readily removed from a jurisdiction seems to be a significant factor in determining the coverage under a particular definition.

In holding that the ROV is not a *ship*, the court stated that "*It is a submersible piece of equipment, a vehicle for undertaking remotely-controlled activities underwater.*". However, there are several anomalies in the judgment which betray this conclusion. First and foremost, the general words of the statutory definition which the court holds to not cover an ROV seemingly are broader than the English law definitions where the word '*in'* is used. Secondly, the courts' reliance on 'usual attributes/characteristics' seems misplaced since it itself reviews cases which deviate from the *Steedman* requirement for carriage of cargo or passengers. It does not seem justified that, therefore a general connotation which may be ascribed to the words 'usual attributes' should have been a factor. The courts' finding that the ROV cannot sustain its own path without the main vessel if detached from the umbilical, its reliance on LARS, limitations on its propulsion, possible unseaworthiness on high seas, characteristic of being carried as cargo on board the main vessel, small size, looks; are possible insights into what the court considered to be a lack of 'usual attributes' of a '*ship'*. Additionally, the court reiterated that the ROV lacks the capability to face the 'risks and dangers' which give rise to admiralty jurisdiction of the court. Seemingly, the court was referring to the ability of a vessel to leave the territorial waters of a country, on its own.

6. Concluding remarks

The above discussion has made it clear that national laws with varying definitions of a '*ship'* present a treacherous terrain for maritime claimants if they have to prove to a court that an autonomous vessel is a 'ship'. It is arguable that a broad definition such as the one in Canadian law would allow for autonomous vessels to be considered '*ships'[41]*. This conclusion shall have two parts, first, the conclusion will summarise findings from the above literature and case review in order to show that technological changes are not a challenge. Thereafter, the second part will deal with reasons of why an express inclusion of autonomous ships in the national laws is required.

6.1 Part 1 - Findings

It is worth noting that the discussion has not led to any conclusive finding that a crew or manning of a vessel is an essential requirement for a '*ship'*. Which in itself is self-explanatory, for application of Admiralty laws which govern private disputes, the *'ship'* is considered as security/asset for satisfaction of the claim and the presence of crew is inconsequential. The *Guardian Offshore* case reinforces this lack of dependency on manning, where the discussion on 'usual characteristics' of a '*ship'* finds no mention of manning. This is also in tandem with the *Von Rocks* case. The *Guardian Offshore* case states 3 principles of admiralty jurisdiction, all of which, foreseeably, can apply to autonomous ships. As these are not of a technological nature, they do not pose a threshold hurdle for autonomous ships to be governed by the admiralty laws of a country. They can be arrested/auctioned in satisfaction of maritime claims. While this is a positive factor for acceptance of an autonomous ship within the definition of a '*ship'* in the domestic laws of a country, it remains to be seen how international conventions which mandate crewing/manning on board will adopt. Resultingly, where a country's ship registration rules or compliance rules are based on those international conventions, there would be an impediment to the operation of autonomous ships.

One thing is certain, that courts have relied upon the manufacturer's descriptions about the disputed vessels and the industry should maintain its reliance on the term '*ship'*. The growing pace of developments is a clear signal that the definition conundrum will find its day in court. It is submitted that legislatures around the world and especially, maritime countries, should stay updated with these developments and when possible make an express inclusion of autonomous vessels in their legislations.

Weight for this argument is found in the *Steedman* case and manufacturers/operators may be well advised to keep the word '*ship*' in the autonomous ships in development.

There is one common element which finds itself manifesting in the argument on 'usual characteristics of a '*ship*' but also, otherwise; buoyancy. However, the projects in development and those envisioned would not seemingly fall obtuse of this requirement[42]. Although, one would fare better if they keep in mind that buoyancy as a factor did impact the court's analysis of two similar kind of products, jet skis, in the *Steedman* and *Goodwin* cases. This also showcases the way opinions are formed in the legal sphere where reliance is placed on the product developers' representations. Possibly, there may be a *Steedman* to *Goodwin* transition phase for autonomous ship cases.

The *Steedman* requirement of a *ship* being 'used in navigation' has gone through juridical transformation and does not bring up any technical challenges for an autonomous ship till the same can perform functions closely akin to conventional ships. This must include the ability to make ordered movement in water.

6.2 Part 2 - Reasons for express inclusion of 'autonomous ships' in national laws
It is the author's viewpoint that an express inclusion of autonomous ships will benefit the countries so making this decision, for several reasons:

a. It will make administering of the admiralty jurisdiction a streamlined and efficient process. This will be one arena where the law can move at the pace of development. Given the fact that *in rem* actions are time sensitive, threatened by the chance of the *ship* sailing away from a jurisdiction and subject to wrongful arrest claims, it will better suit a jurisdiction's legal system to have prior clarity on the issue. This reason does not come without precedent, as mentioned above, India included hovercrafts in its definition and South Korea included WIG. Hence, there does exist a practice of express inclusion of technologically different ships.
b. Court's will save valuable time which would otherwise be spent debating which interpretation to apply, the public law conventions in maritime are purpose built and the private law conventions are silent. The definitions in national laws of each jurisdiction are somewhat different and even within a jurisdiction, there may be a chance that a specific law defines a '*ship*' differently than another law. There is case law wherein English judges have had to differentiate between definitions in two different statutes[43].
c. Refer to the discussion in *Michael v Musgrove* above, there is proof that even when interpreting international conventions, a domestic court will rely upon the definitions mentioned in the national statutes or case laws.

Acknowledgments
The author would like to recognise the contribution of The Institute of International Shipping and Trade Law, Swansea University, specifically Professor Baris Soyer & Professor Simon Baughen who have been instrumental in creating a pedagogy of maritime law that addresses future issues. This paper acknowledges the work of three authors who have written on the subject: Professor Dr Eric Van Hooydonk, Robert Veal & Michael Tsimplis. Their works have formed the basis of the literature review in this paper. The author would like to thank Ms. Tanvi Kulkarni, LLM, Southampton University for her assistance in research.

References
[1] Marr B 2019 The Incredible Autonomous Ships of the future: Run by Artificial Intelligence Rather than a Crew Forbes.
[2] The Minister of Maritime, United Kingdom very recently stated that "Our goal: to make the UK a major international hub for maritime autonomous technologies, which meet the highest safety levels." See, Parliamentary Under Secretary of State for Keynote January 2020, https://www.maritimeuk.org/media- centre/publications/maritime-autonomous-systems-regulatory-conference-2020-presentations/.

[3] For example, See, Inmarsat: https://www.inmarsat.com/press-release/inmarsat-partners-with-rolls-royce-on- autonomous-ship-project/; Kongsberg: https://www.kongsberg.com/maritime/about-us/news-and-media/our- stories/autonomy-is-here--powered-by-kongsberg/; Marine AI Ltd: https://marineai.co.uk/products/.
[4] For example, See, OrbitMI, https://www.orbitmi.com/about.
[5] For a closer understanding of the discussions at the international level, See, Robert Veal, Shipping and Trade Law, Unmanned ships on the IMO work agenda.
[6] Megumi Shiokari and Susumu Ota 2019 J. Phys.: Conf. Ser. 1357 012005, Considerations on the regulatory issues for realization of Maritime Autonomous Surface Ships.
[7] Jokioinen E et al 2016 Remote and Autonomous Ship – The next steps, https://www.rolls-royce.com/~/media/Files/R/Rolls-Royce/documents/customers/marine/ship-intel/aawa-whitepaper-210616.pdf.
[8] Schuler M 2020 Norwegian Autonomous Ship Project Secures Major Funding, https://gcaptain.com/norwegian- autonomous-ship-project-secures-major-funding/#:~:text=%E2%80%9CThe%20AUTOSHIP%20project%20gives%20Northern,both%20in%20transport%20and%20aquaculture; The Eurpean Union's funding fact sheet for this project may be accessed at: https://cordis.europa.eu/project/id/815012.
[9] A. Akbar et al. 2020, An economic analysis of introducing autonomous ships in a short-sea liner shipping network. Intl. Trans. in Op. Res. 00 1–25, "In this paper, we do not aim at investigating the technical and design aspects of autonomous ships, but rather on analyzing the economic consequences of utilizing autonomous over conventional ships in future maritime logistics systems.".
[10] A. Akbar et al. 2020, An economic analysis of introducing autonomous ships in a short-sea liner shipping network. Intl. Trans. in Op. Res. 00 1–25, "We consider a real case of a Norwegian shipping company that transports containerized goods between Europe and several ports along the Norwegian coast.".
[11] It is worth acknowledging that some of the key players promoting autonomous shipping are from the continent. See, Grieg Connect 2020 Autonomous Vessels Meet Autonomous Ports, https://griegconnect.com/blog/news/autonomous-vessels-meet-autonomous-ports/.
[12] See, for United Kingdom Civil Procedure Rules, Part 61 - Admiralty Claims; for Australia, Admiralty Act 1988; for India, The Admiralty (Jurisdiction and Settlement of Maritime Claims) Act, 2017.
[13] For example, See, The Admiralty (Jurisdiction and Settlement of Maritime Claims) Act, 2017, India, S. 5. "Arrest of vessel in rem.—(1) The High Court may order arrest of any vessel which is within its jurisdiction for the purpose of providing security against a maritime claim which is the subject of an admiralty proceeding..".
[14] Indian - The Admiralty (Jurisdiction and Settlement of Maritime Claims) Act, 2017, Section 2(l).
[15] Korean - Ship Act, Article 1-2.
[16] International Maritime Organisation 2002 Interim Guidelines for Wing-In-Ground (Wig) Craft (IMO Ref. T4/3.01), http://www.imo.org/blast/blastDataHelper.asp?data_id=6713&filename=1054.pdf.
[17] Amendments To The International Regulations For Preventing Collisions At Sea, 1972, 2001, Rule 3 (m), http://www.imo.org/en/knowledgecentre/indexofimoresolutions/assembly/documents/a.910(22).pdf.
[18] Another author who states that MASS (maritime autonomous surface ship) should be defined in the collision regulations. See, Frederick J F 2019 J. Phys.: Conf. Ser. 1357 012011, A Review of the current Collision Regulations to accommodate Multiple Ship Situations and MASS.
[19] Hooydonk E V 2014 JIML 403-423, The law of unmanned merchant shipping – an exploration.
[20] Hooydonk E V 2014 JIML 403-423, "Is the unmanned ship a ship?"
[21] Showcased in the English judgment of 'Michael v Musgrove [2012] 2 Lloyd's Rep. 37 (QBD)',

which is discussed ahead.
[22] Veal R & Tsimplis M, 2017 LMCLQ Issue 2 303, The integration of unmanned ships into the lex maritima.
[23] Indeed, the authors point out that such is the case in English Law.
[24] The article provides these two sources: See, Statement by Pentagon Press Secretary Peter Cook on Return of U.S. Navy UUV 2016 (https://www.defense.gov/Newsroom/Releases/Release/Article/1034224/statement-by-pentagon-press-secretary-peter-cook-on-return-of-us-navy-uuv/), and; Foreign Ministry Spokesperson Hua Chunying's Regular Press Conference on December 19, 2016 (www.fmprc.gov.cn/mfa_eng/xwfw_665399/ s2510_665401/t1425479.shtml).
[25] S Gahlen, (2014) 20 JIML 252, Ships revisited: a comparative study. Veal and Tsimplis caveat this reference by informing that Gahlen's finding are primarily based on European case law.
[26] Kawasaki, https://www.kawasaki.com/Category/WATERCRAFT.
[27] In this case, guidance published by the Registry of Shipping and Seamen under the heading "Registering Pleasure Vessels".
[28] The "Von Rocks", [1998] 2 Lloyd's Rep. 198.
[29] Athens Convention relating to the Carriage of Passengers and their Luggage by Sea (PAL), 1974.
[30] In concurring with Dr. Hooydonk, this author believes that private international law conventions offer few resources to the discussion on definition of a ship.
[31] In the judgment, the court has made an internal heading titled 'Particulars of the vessel and the available evidence about operation of Sea Eagle'.
[32] Since the issue turned on the discussion of the term 'used in navigation'. Refer to discussion in Steedman and Goodwin above.
[33] An ex-colony of England and that inherited the common law legal system. See, Australian National University, Library Guides, The Australian Legal System.
[34] For a detailed understanding of size, specification and pictures of the ROV, one may visit the manufacturer's page at: https://www.saabseaeye.com/uploads/seaeye_leopard.pdf.
[35] Australian - Section 3, The Admiralty Act, 1988.
[36] The Law Reform Commission Report No 33, Civil Admiralty Jurisdiction (1986) (Report).
[37] For discussion on this aspect by the Supreme Court of India. See, M.V. Elisabeth And Ors vs Harwan Investment And Trading, 1993 AIR 1014.
[38] It relied on the judgment of Elbe Shipping SA v The Ship 'Global Peace' [2006] FCA 954; (2006) 154 FCR 439 at [74] (Allsop J).
[39] Cyber Sea Technologies Inc v Underwater Harvester Remotely Operated Vehicle 2002 FCT 794.
[40] Section 2 of the Canadian Federal Courts Act defines a 'ship' as follows: "any vessel or craft designed, used or capable of being used solely or partly for navigation, without regard to method or lack of propulsion, and includes…".
[41] There is support to this argument from voices in Canada. See, The Canadian Maritime Law Association, CMI Questionnaire On Unmanned Cargo Ships, https://comitemaritime.org/wp-content/uploads/2018/05/CMI-IWG- Questionnaire-Unmanned-Ships-CANADA.pdf.
[42] For technical specifications of one such project, See, Kongsberg, Autonomous Ship Project, Key Facts About Yara Birkeland: https://www.kongsberg.com/maritime/support/themes/autonomous-ship-project-key-facts-about- yara-birkeland/.
[43] The Environment Agency v Mr Christopher Gibbs, Mr Glen Parker [2016] EWHC 843 (Admin).

Develop and evaluate of intelligent autonomous-ship framework

Kyeong Deok Moon[1], Chi Yoon Jeong[1], Moo Seop Kim[1], Yun Kyung Park[1], Kwangil Lee[2*]
[1] ETRI, 218 Gajeong-ro, Yuseong-gu, Daejeon, 34129, KOREA
[2] KMOU, 727 Taejong-ro, Yeongdo-Gu, Busan 49112, South Korea

* E-mail:leeki@kmou.ac.kr

Abstract. The Autonomous Ship(AS) is a dream ship which navigates the sea on its own without any captains or crews. Many related studies on AS and the advent of deep learning technologies enable ASs voyage in the near future. They have been focused on proposing and verifying feasibility of their concept. However, there have not been little tried to compare results of ASs to that of captains. In this paper, we propose an Intelligent Autonomous-Ship Framework(IASF) which enables to develop ASs based on deep-learning while supporting fundamental functions necessary for ASs such as collecting navigational information, recognizing navigational status, and controlling route. The IASF-based AS was evaluated through simulation with compared to results of captains.

1. Introduction

An Autonomous Ship(AS) is a dream ship that is a human continuous challenge to develop. An AS safely voyages to its destination without any human intervention. In particular, the AS is gaining importance as a new breakthrough to overcome the crisis of the shipbuilding industry due to stagnation that has persisted over the past several years and to reduce ship accidents due to human errors, which have been rapidly increasing. Because an AS must recognize navigational situation and maneuver throughout its voyage, artificial intelligence technology that recognizes and situations makes decisions like humans will be one of the major factors in deciding whether AS will be the successful or not. With the advent of deep-learning technology, artificial intelligence technology has made breakthroughs, and expectations for the success of ASs with artificial intelligence technology are also increasing. In an AS, as the number of crew members on board decreases, the space for crews can be used as space for cargo. This leads to changes the structure of the ship. The AS will also bring a paradigm shift that has not been experienced in the shipbuilding industry. For example, data that are collected during navigation will have added value, and various services will be deployed which causes new businesses.[1]

Research on AS has been actively began conducted in Europe since around 2010. Since the mid-2010s, semi-ASs have been deployed that are controlled from the shore centers by using navigational conditions determined by ships. In recent years, ASs have been tested in a restricted environment. However, most of these studies have been performed to verify the feasibility of proposed their concepts of ASs. ASs must navigate the sea where not only other ASs but also human maneuvered ships are exist. Thus, it is important to evaluate that ASs act as like maneuvered by captains. Because they have just tried to navigate their ASs in a test area, it is difficult to say whether the performance of ASs is better than that of captains. There wewe not any tries to compare performance of ASs to that of captains. [2]-[7]

In this paper, an Intelligent Autonomous-Ship Framework(IASF) is proposed. The IASF makes it possible for an AS to safely and economically navigate by itself from departure to destination without the involvement of a captain or crew. The IASF provides fundamental functions for performing the process of "collecting navigational environment information→recognizing navigational conditions→determining and controlling routes" performed by ASs.[8] The IASF is composed of two layers, namely, the Functional Layer(FL) and the Abstraction Layer(AL). The FL provides fundamental functionalities for navigation without human intervention. The FL is composed of 3 sub-systems: Navigation Information Aggregator(NIA), Navigational Situation Recognizer(NSR), and Navigational Path Controller(NPC). The NIA collects information on the ocean, weather, and topography that affects the navigation of ships from various sources. The NSR analyzes the status of the own ship by determining the performance indexes of restoration and movement of the own ship, which are affected by external forces, such as the weather and sea. The NSR also recognizes navigational patterns, such as the trajectories of other ships, by fusing information collected from various devices. The NPC determines the risk of collision using the navigational situation obtained from the NSR, and then maneuvers the voyage of the own ship. The IMB which is a component of AL, provides interworking through message exchange among the subsystems of the IASF and between subsystems of the IASF and devices on the own ship. We evaluate performance of AS based on IASF by comparing to that of 6 captains by simulation on 3 defined scenarios.

The remainder of this paper is organized as follows. Related researches on ASs is summarized in Chapter 2. Chapter 3 describes the details of IASF, and Chapter 4 reports experimental results obtained through simulations that were performed to evaluate the navigational performance of the IASF-based AS. Finally, conclusions are presented with directions for future works.

2. Related works

Research on ASs has been conducted in Europe since 2010, and related studies have been increasing. Those projects have proposed concepts of ASs and have tried to verify their feasibility. They have not focused on comparing performance to performance of captains.

Maritime Unmanned Navigation through Intelligence in Networks(MUNIN) developed the AS technology since 2012. The AS, which is autonomously executed on a ship, is a consisted of advanced decision support systems and capabilities for remote and autonomous operation. The monitoring and controlling functionalities are executed in the Shore Control Centre. MUNIN defines the following systems and entities for AS. An Advanced Sensor Module handles the lookout duties continuously fusing data from navigational devices, such as radar, Automatic Identification System(AIS), and cameras. An Autonomous Navigation System follows a pre-defined navigational plan, but with adjusting the route in accordance with legislation, collision risks or weather changes. An Autonomous Engine and Monitoring Control system enables ship engine automation systems while maintaining the optimal efficiency. It also handles additionally installed pump-jet that function as a certain rudder and propulsion back up. The Shore Control Centre, which continuously monitors and controls the AS after it is released from its crew by its skilled nautical officers and engineers. It also assesses its technical, economic and legal feasibility.[2]-[5] The MUNIN project focused on the proof of feasibility of the concept of an AS. As they found it feasible, DNV GL ReVolt and YARA Birkeland perform took one step further with the learning from MUNIN. The AS of YARA Birkeland is the world's first fully electric autonomous container ship, with zero emissions. YARA Birkeland will be operated semi-autonomously near future, but fully autonomously lastly.[6][7]

3. Architecture of Intelligent Autonomous Navigation Framework

IOP Conf. Series: Materials Science and Engineering **929** (2020) 012006 doi:10.1088/1757-899X/929/1/012006

The IASF allows ships to determine their navigating environment and their route in order to safely arrive at their destination without human intervention. The IASF is composed of the FL and the AL, as mentioned earlier. The FL is a layer that provides fundamental and practical algorithms necessary for the vessel to determine its course and voyage on its own, and the AL is a layer that provides interworking among algorithms provided by the FL and between algorithms and navigational devices. This chapter describes the details of structure of the IASF in detail.

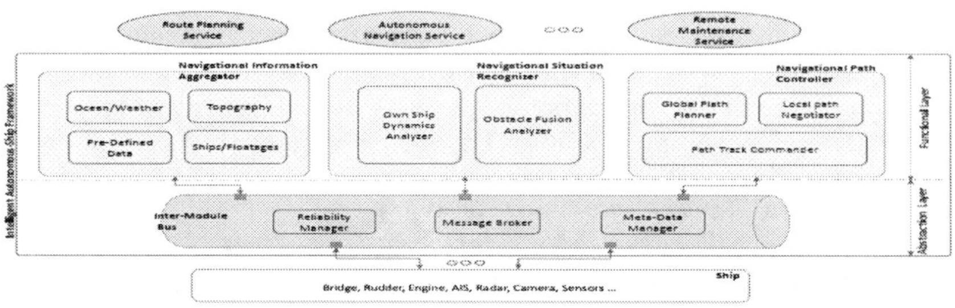

Figure 1. Architecture of Intelligent Autonomous-Ship Service Framework

3.1. Navigational Meta-Data(NMD)

The information collected by the IASF can be classified into two types: static information and dynamic information. Static information does not change during a voyage. It includes the voyage plan and ship information. Dynamic information which continuously changes while the ship is navigating along its route. It can be divided into the following types: information that changes geographically such as sea, weather, and charts; and information that changes over time, such as a navigation route or operation status of an own ship or other ships. The ocean information managed by the IASF is composed of weather information, chart information, and risk information. Weather information includes waves, water temperature, density, and wind tide, which are acquired from weather forecasting centers or sensors of the ship. Chart information includes coastline, water depth, and dangerous areas. Risk information includes other ships and obstacles. Moreover, the route of a ship and its navigational situations are changed over time. So far, ship's routes and situations have been managed based on the waypoints, which are intermediate points or places on a route or line of travel. IASF manages information related to the starting position, and distance, direction, and turning radius for entering the next waypoint, and the average speed of a waypoint. Moreover, the IASF manages navigational information by dividing a route into several waypoints and a waypoint into intermediate maneuver points, called action points in this paper. An action point is an intermediate point or place on a waypoint where a ship is maneuvered to maintain its course to next waypoint. The IASF collects and manages the navigational status of its own ships and information on other ships at regular intervals - 0.5 seconds to 1 second - so that the ASs can predict and handle navigation in relation to other ships that violate COLREGs' navigational regulations.

3.2. Abstraction Layer

The framework should support that researchers develop the ASs based on IASF without considering environments. The IASF can adapt various environments through the AL. The AL consists of an Message Broker(MB), Reliability Manager(RM), and Meta-Data Manager(MM). The AL is designed to comply with IEC 61162-450/460 to increase the possibility of future use.

The MB provides basic transmission and receiving functions to allow seamless interworking among algorithms or between algorithms and the ship. Because various physical media and protocols can be used simultaneously, the MB handles a variety of communication protocols by supporting the parsing, encapsulation, and protocol conversion of messages. In addition, in order to deliver a message to the desired algorithm or device, the MB determines the message passing route by managing basic and connection information on algorithms of the FL and devices. The MB also manages group information for group communication. It is important to provide a reliable and stable environment for ASs. To prevent malicious access to an AS, the RM allows access only to modules that are authenticated based on a certificate. It guarantees the reliability of messages by providing access rights management. Access rights management includes a firewall and a function to determine whether the communication is due to an attack by analyzing the communication pattern. The RM guarantees stable message exchange by performance management, error handling, and log management of the IMB. The MM provides a function to store and manage the NMD. The MM registers and manages information related to the vessel in real time. The MM also retrieves the NMD to manage information related to the navigation changes that occur during the navigation of the vessel. The NMD makes it possible to not only manage the condition occurring during the operation, and to review the navigation of the ship after the navigation is terminated, or analyze the navigational situation.

3.3. Functional Layer

A captain uses information collected from navigational devices such as AIS and radar and visually confirms information to analyze the ship's navigational status and maneuver the ship so that it can safely reach its destination. Like captains, ASs should navigate to the destination by collecting information from various devices and determining everything by themselves without any human assistance. A framework for AS should provide the functions that are necessary to maneuver the ship by analyzing the risk factors and the ships navigational performance using the collected information, and deciding the route so that it can safely navigate to the destination accordingly. In the IASF, the FL provides the fundamental functions for the AS. The FL consists of three sub-systems: The Navigation Information Aggregator(NIA), Navigational Situation Recognize(NSR), and Navigational Path Controller(NPC). New functions can be continuously added for functions for various situations through the AL.

Ships are mainly affected by the temperature, current, wind, wave, and swell of the sea around them. The NIA provides functions for collecting information around ship. The NIA collects weather data from NOAA or sensors on navigation areas. In addition to weather information, the NIA enables the collection of data directly from users. Those data are a voyage plan, ship-related data, and chart-related information. In addition, it collects information from marine devices, such as AIS and radar, or from cameras and sensors so that it is easy to collect information regarding surrounding conditions while the ship is navigating.

The NSR analyzes and provides the navigational status of the own ship and that of other ships by using navigational situation information collected from the NIA. The navigational status is used to determine the route. External forces, such as water density, water temperature, wind, waves, tide, swell, and so forth affect the ship's desired speed and direction, which the ship tries to maintain. The Own Ship Dynamics Analyzer(OSDA) provides a basic function that determines the additional resistance and safety indices, such as restoration and wave tuning performance, which are caused by external forces. To determine the risk of collision with other ships, the Obstacle Fusion Analyzer(OFA) analyzes the navigational trajectories of other ships by integrating data from AIS, radar, and images obtained by camera that acts as a captain's eye. In particular, the OFA provides a deep-learning algorithm[9][10], which analyzes information about other ships in the vicinity from camera images.

The NPC determines the route of an AS during navigation, which avoids the risk of collisions of the own ship. The NPC analyzes the risk of collisions of the own ship based on the navigational status of own ship and other ships from the NSR. It then replans the route and finally maneuvers the engine and rudder of a ship accordingly while keeping the COLREG rules and guaranteeing the safety. The Global Path Planner(GPP) of the NPC generates the entire path from the departure to the destination before departure according to the weather forecast and navigational conditions. At this time, a global path is created using a multi-purpose optimization algorithm to consider the navigation time and fuel consumption at the same time, and the route with the lowest fuel consumption is selected among routes that satisfy the prescribed navigational constraints. The Local Path Planner(LPP) analyzes the risk of collision due to head on situation, crossing situation, and overtaking situations. The LPP also replans a path to perform collision avoidance maneuvers complying with COLREGs according to the analyzed situation. The Path Track Commander(PTC) provides the function of maneuvering the ship's operation so that the ship can follow the path created by the GPP and LPP by itself.[11][12]

4. Experiments and Results

ASs should also evaluate their navigational performance in ports, coasts, and oceans, like other vessels. However, in the case of ports, there are too many factors to consider when evaluating, and there are too many variables, such as applying a separate navigation method for each port. In this paper, we evaluate the navigational performance of the proposed AS-based IASF in coastal and ocean travel by using a ship's track chart. It is possible to understand the ability to avoid collisions between ships and understand navigational status as a whole by analyzing a ship's track chart. To this end, we defined scenarios with head-on situations, crossing situations, and overtaking situations that occur frequently in the real world. Because a simulator can regenerate the same conditions repeatedly, we evaluated the navigational performance of the IASF-based AS by using a simulator and comparing the simulation results with those obtained by six captains who have extensive experience in operation.

4.1. Experiments

Figure 3 shows a testbed based on a simulator to evaluate the performance of an IASF-based AS. The testbed consisted of a PC, a simulator, and an IASF-Simulator Interface Device(ISID). The IASF-based AS ran on a PC, and the simulator simulated the own ship, other ships, and the ocean. The ISID connected the IASF-based AS and the simulator. The navigational status of ships, such as position, speed, and heading of own ship and other ships generated by the simulator, was transmitted from the simulator to the IASF-based AS through the ISID. The IASF-based AS sent an engine and rudder control commands to the simulator through the ISID. Because the simulator generated NMEA messages, the ISID and the simulator exchanged data through the NMEA protocol over serial communication. The sub-modules of the IASF-based AS were connected to the IMB over ethernet which supports 61162-450 protocol. The IASF-based AS was connected to the ISID which, ran on the IMB via ethernet. For interworking between the IASF-based AS and the simulator, the IMB installed on the ISID performed NMEA-to-450 protocol conversion. To perform the simulation, Pre-Defined Data, such as Departure, Arrival, ESTIMATED Arrival Time, Specification of Vessel, Policies for Navigation, and Navigational Information, were input as an XML document. Information about the sea on which the ship navigated, which was necessary for the analysis of the navigational characteristics in advance, received the information generated by the simulator, such as information about the sea.

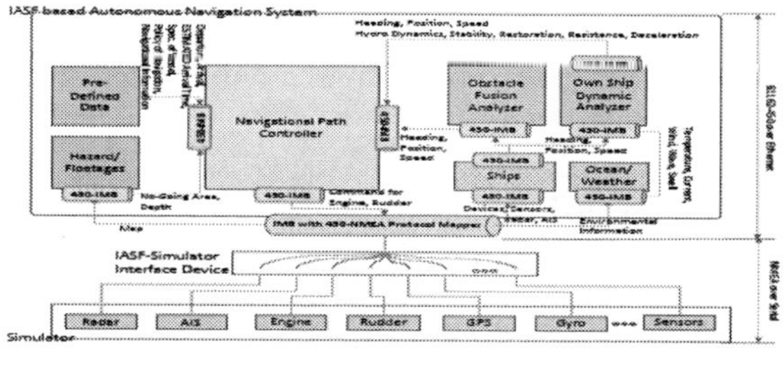

Figure 2. Test-Bed for Experiments

In this paper, the navigation regulations of '72 COLREG's were applied for the performance evaluation. We took into account situations that can frequently occur during voyages to evaluate the navigational performance of the IASF-based AS from the perspective of the ability of the ship to maintain its course and change its way. We have defined 3 scenarios with complex situation which are combinations of head on situations, crossing situations, and overtaking situations. In particular, it has temporal and spatial margins to sufficiently recognize other ships. In addition, we consider the minimum visible distance and situations in which the risk of collision is considerably high although it does not occur often during voyages. We also consider the case of the use of the engine. Scenario 1 is a typical

IOP Conf. Series: Materials Science and Engineering **929** (2020) 012006 doi:10.1088/1757-899X/929/1/012006

head on situation, which occurs frequently and is suitable for evaluation in case of the collision avoidance navigation regulations. Scenario 2 is a typical scenario for the state of overtaking a ship. Scenario 2 is very frequent and is suitable for evaluating compliance with collision avoidance navigation when overtaking. Scenario 3 is used to evaluate the relationship of navigation in a complex relationship in which head-on, crossing, and overtaking all exist. It is a scenario that can evaluate whether or not an IASF-based AS keeps regulations when a complex regulation relationship is established.

We defined simulation environment as follows: The wind blows to the north at a speed of 10 knots; the tide flows to the east at a current velocity of 0.5 knots; and the field of view is set to be relatively calm with a distance of 10 miles. To perform the simulation, we defined three rules. First, the movement of the ship is identified from at least 6 miles where the bow mast, and so forth can be seen. Second, action for collision avoidance is taken within 3 miles which is the minimum distance according to the regulations. Last, the closest distance to the other ships can be more than 1 mile.

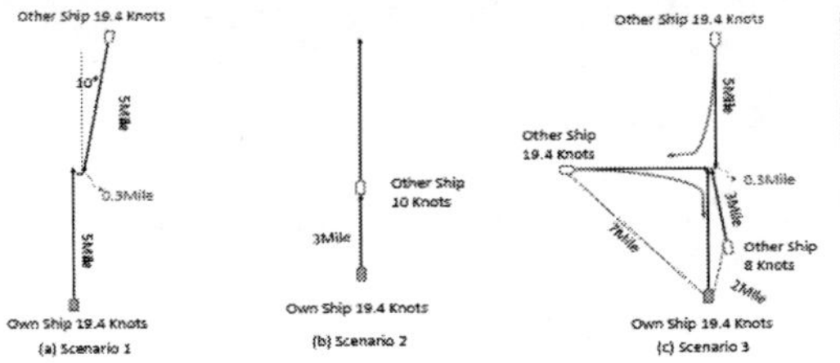

Figure 3. 3 Scenarios for Simulation

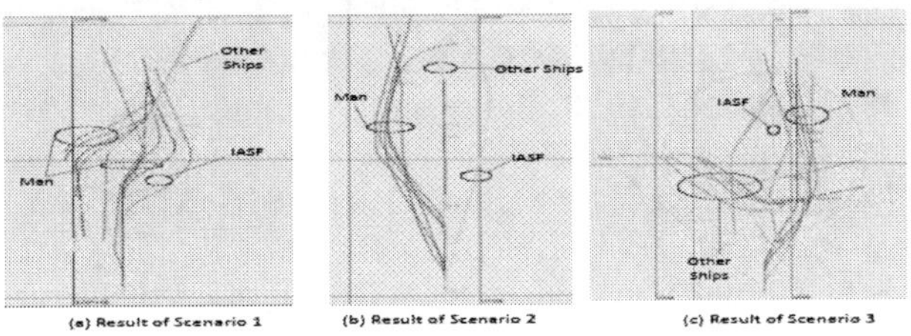

Figure 4. Results of simulation

4.2. Its Results

The navigational performance of the IASF-based AS was evaluated by comparison with the results obtained by experienced captains. For each of 3 scenarios, each of the 6 captains conducted the simulation 6 times. Captains have many experiences on board and have a sailor's license. As a result, the navigational performance of the IASF-based AS showed no major problem in terms of collision avoidance, and the overall navigational performance did not show much difference from the captains'. In the case of collision avoidance, many captains take actions to divert the course of their own ship to avoid the course of the other ship which have a right to keep the course. However, the IASF-based ASs maneuvered toward other ship's course when the proximity distance between ships was sufficient. As a result, proximity distance of the IASF-based AS to other ships was closer than that of captains. In addition, when taking the countermeasures, the usage of the rudder by the IASF-based AS was relatively higher than that of the captains, so the IASF-based AS often took a rapid turn. Because most ships have large displacements, it may be difficult to control the course of a ship in the case of sudden changeover. The captains slowly returned to the planned course after completing the collision avoidance, but the IASF-based AS rapidly returned to the planned course after completing the collision avoidance. Thus, the IASF-based showed a tendency to return rapidly when returning.

5. Conclusions and Future Works

An AS that navigates the sea by itself without human assistance is a dream ship that can only be seen in movies. Many studies have been conducted to realize ASs so far; hence, ASs may be seen navigating in the sea in the near future. As yet, concepts for ASs are developed and applied to ASs to verify the feasibility of AS technology. There has been little research on performance evaluation by comparison to performance of captains. In this paper, we propose the IASF, which collects navigation information necessary to develop ASs, provides a function to determine the navigational status, and determines the route. The IASF provides a message-based architecture that can develop ASs without considering the environments. In addition, we evaluated the navigational performance of the IASF-based AS through simulation. The IASF-based AS avoids collision well and achieves a navigational performance similar to that of 6 captains. The IASF-based AS makes a sudden turn by using a rapid rudder movement to maintain the route in some cases.

For ASs to navigate in ports where traffic is heavy and considerably changed, they must be able to determine the best course in very complex situations. We have a plan to develop a more sophisticated IASF based on a deep reinforcement learning model to enable navigation in port. In addition, we will develop evaluation metrics that can quantitatively evaluate ASs, and conduct research to establish a test environment similar to the real sea by linking multiple simulators.

Acknowledgments
This work was supported by the Technology Innovation Program (20011607, Development of evaluation and validation testbed for autonomous ship performance) funded By the Ministry of Trade, Industry & Energy(MOTIE, Korea)

References
[1] G. Koikas, M. Papoutsidakis, and N. Nikitakos 2019 New Technology Trends in the Design of Autonomous Ships *International Journal of Computer Applications*, Volume 178 – No. 25
[2] Ørnulf Rødseth, Åsmund Tjora and Pål Baltzersen, 2014 D4.5: Architecture specification *MUNIN*, MUNIN-FP7GA-No 314286
[3] Ørnulf Jan Rødseth and Åsmund Tjora 2014 A system architecture for an unmanned ship *COMPIT 2014*
[4] Hans-Christoph Burmeister 2016 AUTONOMOUS NAVIGATION RESULTS FROM THE MUNIN TESTBED *Autonomous Ship Technology Symposium*
[5] Ørnulf Jan Rødset 2016 The unmanned ship according to MUNIN *EMPA 50th General Meeting Unmanned ships: A real challenge for pilotage*
[6] Ziaul Haque Munim 2019 Autonomous ships: a review, innovative applications and future maritime business models *SUPPLY CHAIN FORUM: AN INTERNATIONAL JOURNAL* Volume 20
[7] VOLKER BERTRAM (2013), Towards Unmanned Ship, DnV-GL
[8] Andrzej Felski and Karolina 2020 Zwolak The Ocean-Going Autonomous Ship—Challenges and Threats *Journal of Marine Science and Engineering*
[9] Chiyoon Jeong, Hyun S. Yang, and Kyungduk Moon 2018 Horizon detection in maritime images using scene parsing network *Electronics Letters*
[10] Chiyoon Jeong, Hyun S. Yang, and Kyungduk Moon 2018 Fast horizon detection in maritime images using region-of-interest *International Journal of Distributed Sensor Networks*
[11] Byeong Hyeon Ryoo, and Jin Whan Kim 2018 Ship route optimization considering on-time arrival probability under environmental uncertainty *OCEANS'18 MTS/IEEE Kobe*
[12] Byeong Hyeon Ryoo, and Jin Whan Kim 2018 Ship route optimization under weather uncertainty considering port arrival time constraint *ICMASS 2018*

Human Factors, autonomous ships and constrained coastal navigation

T Porathe[1], K Fjortoft[2] and I L Bratbergsengen[3]

[1]Dep. of Design, NTNU, Norwegian University of Science and Technology, Trondheim, Norway
[2] SINTEF Ocean, Dep. of Energy and Transport, Trondheim, Norway
[3]Kongsberg Seatex, Trondheim, Norway

thomas.porathe@ntnu.no

Abstract. This is a concept paper making propositions for MASS navigation in national coastal waters. The focus is on human factors and the interaction between automation and operators in a remote control centre, as well as seafarers on conventional ships interacting with MASS. Important element is to understand the hand-over process of command, from automation to humans in control, based on the automation on board the vessel as well as the infrastructure surrounding the vessel.

1. Introduction

This is a concept paper making propositions for MASS navigation in national coastal waters. The objective is to provide a non-technical, easy to understand narrative on this matter and how and where humans are envisioned to be involved.

The focus is on human factors and the interaction between the technical automation system and the human operators, not only in a remote-control centre (ROC), but also as seafarers on conventional ships interacting with MASS at sea. An important element is to understand the monitoring-control process as well as hand-over from automation to humans and vice versa.

Autonomous shipping is increasingly coming closer as ongoing projects materialise. Commercial projects such as the Yara Birkeland [1] as well as ongoing research projects have a higher TRL (Technical Readiness Level) level than previous. International MASS voyages are still something for the future awaiting IMO regulations, although some solutions for transoceanic navigation was proposed already by the MUNIN project 2013-2015 [2]. So far this have not led to international regulations. In the meantime, it is important to start concrete work on design of concepts for coastal navigation in national waters. Such work will depend on local features of the nation in question. This paper is looking at the Norwegian context with stretches of open, often exposed coastline as well as confined archipelago navigation, sometimes with heavy commercial traffic as well as leisure crafts and small-scale fishing. The proposed concept will assume unmanned navigation (unmanned bridge) with remote monitoring and (in exceptional situations) remote control of the MASS vessels.

The concept also includes new types of infrastructure presently investigated in the IMAT project (Integrated Maritime Autonomous Transport Systems) funded by the Norwegian Research Council [3]. One element within the IMAT project is to identify the awareness to navigational systems, where the land-based, and space-based infrastructure will be important for building robustness to the MASS operations. This can be awareness to systems for navigation, for communication, and systems for detection of objects that can be critical in a navigating approach of a MASS sailing from A to B.

Using an ergonomic perspective, the paper will start out by looking at the operator and his or her planning and monitoring tasks in the Remote Operation Centre (ROC). We will then continue to look at the interaction between the automation onboard a MASS and ROC, VTS operators and humans onboard other ships and crafts in the vicinity of the MASS. After having spent some time discussing the work situation in the ROC, we will look at the following phases of a typical voyage: 1) maintenance planning 2) navigation planning 3) cast-off 4) transit along open coast and in confined waters, 5) transit in VTS areas, 6) port approach, and finally 7) mooring. What will be the challenges, and what will be possible solutions, is the questions to be answered.

2. The operator in the remote-operations centre (ROC)

MASS is often mistaken for a synonym to *unmanned ships*. A MASS may very well be unmanned but may just as well be manned with maintenance crew or even navigators, even if the autonomous navigation system will do most of the tedious watchkeeping. But whether an unmanned bridge or not, the MASS will almost certainly be monitored from a remote-operations centre (ROC).

2.1. The monitoring processes

The ultimate goal of MASS may very well be a "say goodbye-and-forget, call-me-if-you-have-a-problem" type of vessel. With a mature technology MASS could work invisibly as transport drones just as elevators transport people and goods today without supervision or monitoring. Some would argue that the asset and cargo value is too large to just forget [4], but the point here is not to "forget" the ship, but instead point to the problem of *active* and a *passive* monitoring.

2.1.1. Active monitoring. The MUNIN project envisioned an active monitoring process in an transatlantic voyage, Europe to South America, that was the case conceptualised in that project. We sketched then that one operator was monitoring six different vessels on the open ocean, spending ten minutes every hour looking at key performance indicators at each vessel. These indicators could typically be ship performance (e.g. course, speed, roll, heave, slamming, etc.), environment (e.g. potential traffic conflicts, fishing boats, etc.), weather (wind speed and direction, changes in air pressure, forecast, etc.), engine performance (RPM, vibrations, temperatures in engine, bearings, exhaust, cooling water etc.), energy (batteries, power plants, sun cells, etc.), communication (signal sources and performance, satellite coverage, etc.) to just mention a few [5].

Active monitoring has many advantages. The chance is that many problems can be spotted at a very early stage and even if they appear suddenly the operator will be in the loop with a good understanding of the context and the situation around the vessel.

The backside of active monitoring is that it will be very boring for an operator to monitor perfectly functioning ships on an empty ocean hour after hour, day after day. The risk is that when there eventually is an incident, the operator will be out-of-the loop since long. Much research has shown that humans are not good at monotonous monitoring tasks, e.g. [6]. The risk is that active monitoring soon will drift into passive monitoring.

2.1.2. Passive monitoring. As opposed to active monitoring one might envision a ROC designed for passive monitoring where operators are available on call from the ship when the MASS has detected any problem it needs help to solve. Different levels of such "operator readiness" will inform the operators about the expected response time. With their vessel on the open ocean and no other ships within the 30-mile range, operators might be on call to deal with problems from a home station or mobile devise. Or they might be in an office area close to the control room, ready to step in. This will probably be a much more realistic and desirable work situation in the long run. For a sketch of suggested operator readiness levels see Table 1).

IOP Conf. Series: Materials Science and Engineering **929** (2020) 012007 doi:10.1088/1757-899X/929/1/012007

Table 1. Sketch of possible Operator readiness levels (availability and mental readiness)

Operator readiness levels	Maximum response time	
In control	0	Operator in-the-loop and at control station
High	3 min	Operator should be in-the-loop and close to the control station
Medium	15 min	Operator in the ROC ready to get in-the-loop
Low	1 hr	Operator home on call

The big problem here is the out-of-the-loop performance problem (or syndrome) which leaves operators of automated systems handicapped in their ability to take over manual operations in the event of automation failure [7].

2.1.3. Time-to-get-in-the-loop. A major challenge will be to design a human-machine interface (HMI) that allow minimal time to get in the loop. This will in no case be possible instantly, so there will be a need for a grace period during which automation on the MASS must be able to handle all situations, some way or another. If the operator does not respond within this *response time*, that is, before the *response deadline*, the automation will have to solve the situation on its own [8].

The length of such response time and the automation's measures during the period and after an operator has failed to answer the call will of course depend on the circumstances: For instance, in a busy approach operator might be given a very short response time, requiring her to literally sit in front of the control station ready to take control on a very short call, while on open ocean, with no other ships within the 30-mile range, the control room might be unmanned with the operator on call.

If an operator that fails to answer the MASS request for help before the response deadline, the automation will go into a graceful degradation to a fail-to-safe-mode. The fail-to-safe-model will look different depending on the circumstances: in close vicinity of a lee coast anchoring might be the preferred choice; in a densely trafficked fairway hoovering or heaving to might be the choice and in open waters drifting might be good enough waiting for the operator to get in touch or the communication to be restored.

The design for such automation will be crucial and will probably need to involve machine learning analysing both large numbers of real voyages with manual and automatic ships as well as simulations with all thinkable scenarios. *Black swans*, the unknown unknowns, will be bound to put us up for surprises for many years to come until MASS automatic behaviour eventually will be mature.

An automated HMI should, based of geographical and traffic information be able to, with some accuracy, predict necessary readiness level for the operator.

Another big major challenge will be the design of an *automation transparency* allowing this minimal response time.

2.2. Automation transparency

To facilitate for the operator to quickly get in-the-loop the automation need to be very transparent about its "thinking". Within artificial intelligence (AI) this field is also called *explainable artificial intelligence* (*explainable AI,* or *XAI*). The purpose is that the results of the solution arrived at by an AI system can be understood by humans. It contrasts with the concept of the "black box" in machine learning where even their designers cannot explain why the AI arrived at a specific decision. Cooperation between agents, in this case algorithms and humans, depends on trust. If humans are to accept algorithmic prescriptions, they need to trust them. For that reason, interpretability and explainability are posited as intermediate goals for checking other criteria.[9] Human users should be able to understand the AI's cognition (both in real-time and after the fact), and should be able to determine when to trust the AI and when the AI should be distrusted [10][11].

This is a very complex and difficult area that will need much research in the future. There has been work on decision trees and Bayesian networks, which are more transparent to inspection [12]. A decision tree is a tree-like flow chart with branches representing different courses of action and rules regarding which branch to follow. A trivial example could be how to display the multiple courses of action the in a maneuvering situation when a MASS encounters another ship on a conflicting course. Figure 1 illustrates maneuvering decisions in a close quarters' situation where the operator is shown different maneuvering alternatives that the automation is deciding upon. The certainty calculation is color coded and the present decision made by the system is highlighted in green. By doing nothing the operator accepts the automated decision, but she can also click on any of the less preferred alternatives or take manual control. In this figure the own ship in meeting a stand-on vessel from starboard. The automation signals its decision to turn 45 degrees to starboard and pass behind the other vessel. The operator is offered a selection of other possible actions that the automation does not suggests, but that the operator might click on and thus override the automation. Each option has an index score relating to how "certain" the system is of the option.

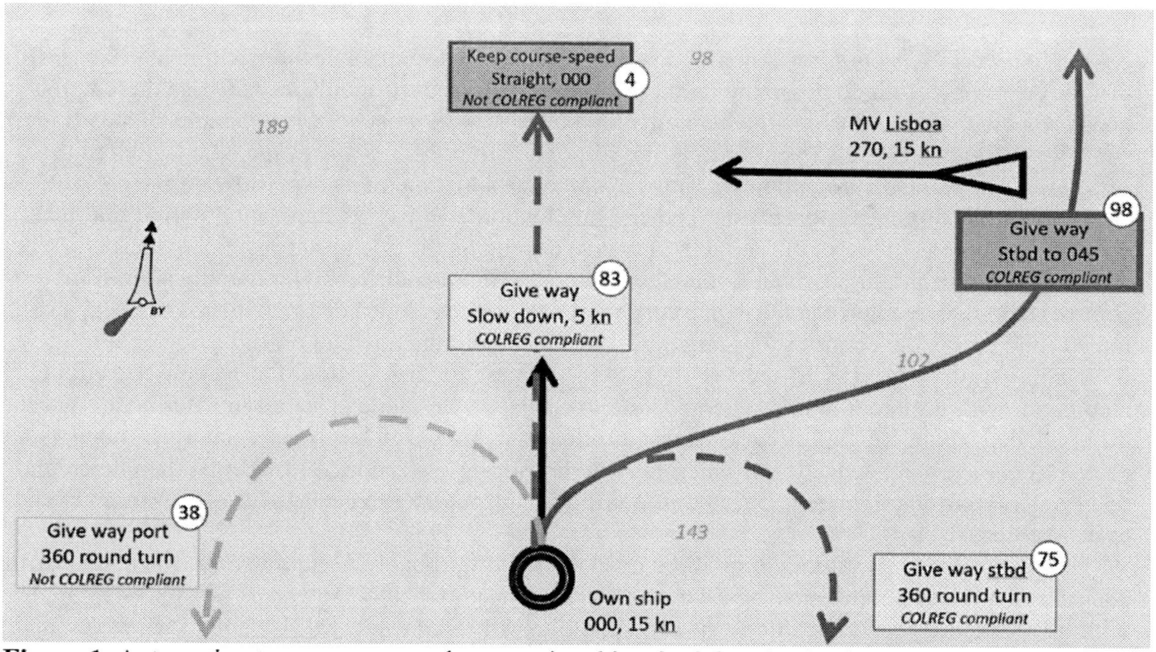

Figure 1. Automation transparency on the operational level might take the form of a decision-tree adapted to the nautical situation. See text for explanation.

More complex, and maybe more realistic situations might borrow methods and techniques from the Explainable AI field. It will be crucial for the decision system of MASSs to behave in a fashion that makes sense for both the operators in the remote-control centre as well as human navigators onboard ships in the vicinity of the MASS. The balance here between information overload and undue simplification for the operator is crucial here and there is an important tradeoff between transparency and completeness of an explanation.

The all-important question is when should the human operator intervene and take over control from the automation? Only when invited, or also in some cases when the automation thinks it is doing OK? What about the timing/response time for handing over control from the automation to humans, should that be a criterion for the different alternatives?

3. Maintenance and readiness

If a MASS is unmanned there will be no-one onboard to fix things that break. Normally the quality of equipment and facilities of modern ships is based on that there is someone onboard that can tighten a leaking valve, change a clogged filter or disconnect and exchange a faulty sensor. If that person is not available a new strategy is needed. This strategy must rest on *better quality, better maintenance plans* and *redundancy*.

In the early days of transoceanic airship aviation, the air mechanic could enter the engine gondola during flight for maintenance and repair. In modern aircrafts the type and quality of engines have changed, and flying air mechanics are gone. Should a problem arise there is an engine redundancy. The same development is needed for MASS: simpler, less vulnerable equipment of better quality and three or four-fold redundancy.

3.1. Preparing the MASS for voyage

A modern manned ship would normally be constantly kept in running order by its engineers. In the best of worlds, wear and tear would be noted and spare parts ordered and replaced before breakage. A MASS cannot rely on failure finding but will need to rely on time-, predictive- or condition-based maintenance. If anything fails a redundant system needs to take over automatically and update the ship's status indicator. Worst case should lead to a fail-to-safe mode. An early proposal for such an *operational status* indicator from the MUNIN project is shown in Figure 2 [13].

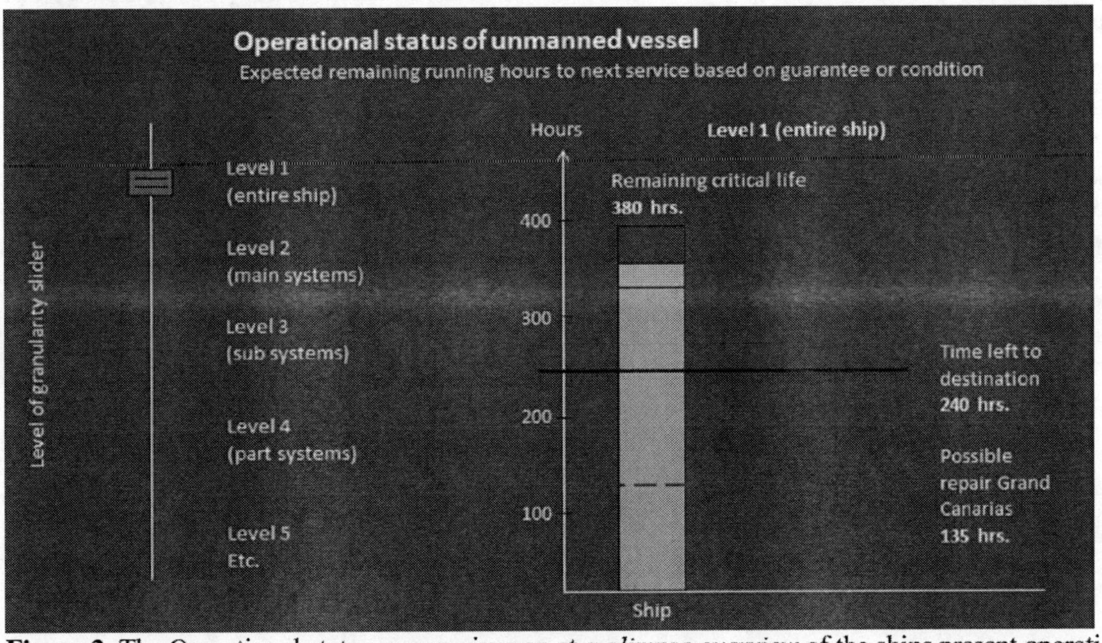

Figure 2. The Operational status screen gives an at *a glimpse overview* of the ships present operational status and the remaining time to destination and to possible repair ports for a voyage from Gravesend in the UK to the Orinoco River in Venezuela. By moving the slider to the left the operator can drill down into all sub-systems for a detailed view of their status [13].

In Figure 2 the top status for a ship preparing for a 240 hours long voyage is shown. This case portrayed a transatlantic voyage from the UK to South America with a possible repair opportunity at the Canary Islands 135 hours away. The indicator will show the operator that the ship is prognosed to have more than 300 hours of critical life with full redundancy. The operator can further drill down into the system to see the life expectancy for any group or single component in the ship. This decision support system should be based on time (certified life expectancy), risk (environmental factors for the voyage) and actual condition from sensors onboard. If anything breaks down during the voyage, the indicator will be updated and show the new situation as a base for decision-making. Such a status indicator will be quite complex to develop, but will give the operator a good, *at-a-glance* indication of the mechanical status of the MASS.

IOP Conf. Series: Materials Science and Engineering **929** (2020) 012007 doi:10.1088/1757-899X/929/1/012007

4. A human factors perspective

Using an ergonomic perspective, we will in the following take a closer look at some selected phases of a MASS voyage. The purpose is to enlighten the human contribution to the autonomous process and potential problems that might follow.

4.1. Navigation planning

Normally the second mate is responsible for the passage planning onboard a merchant vessel. Her job is to prepare a berth-to-berth voyage plan according to IMO regulation [14]. This means that in an electronic chart (ECDIS) draw up the exact route the ship plans to sail from departure berth (anchorage or quay) to finish. The track must also include such things as cross-track distance (a safety margin on each side of the track) and should be checked for under keel clearance. For instance, should every turn during the voyage include a turn radius based on the ship's particulars. This will be a challenge for the navigation planer in a ROC who needs to keep differ ships particulars apart.

In a ROC this task will be done by the operator or a special group in the ROC working with passage planning. In shuttle traffic this task will reuse old tracks. It will be of importance that the tracks use traffic separated routes so that ships do not regularly will meet head on and need to do evasive manoeuvres. In Norway the Coastal Administration is from 2019 launching a system of (mostly) traffic separated "recommended routes" [15].

Once a detailed voyage plan is entered into the navigation system of a MASS the ship will be able to automatically follow the programmed route.

4.2. Cast-off

The cargo is onboard, status of all systems is go and the MASS is ready to sail. A lot of stakeholder has pressed a green "go" button before we are here: a cargo master have signed off that the cargo is loaded and lashed, an agent responsible for that ship is bunkered, repairs has been made and maintenance personal is off the ship, a port master that the paper work are done and all fees are paid, maybe an inspector from the Safety Administration declaring that the ship is seaworthy and the VTS, if applicable, that the vessel has clearance to leave and that the port approach is free. The operator in the ROC should have an easy and unambiguous way of knowing if everything is ready: a red or a green light. If needed he or she should be able to drill down and see all the different entities giving their clearance.

Before ordering the linesmen or the automatic docking system to cast off, maybe the operator also need to do a manual camera check that there are no kayaks or swimmers around the vessel, that no cable has been forgotten and is still attached to land before hitting the cast-off button. In some cases tugboats will be involved, which will be another layer of communication.

During cast-off the MASS will have the full focus from one of the operators in the ROC. The operator readiness level will be high, in an "*In control*" mode.

4.3. Transit along open coast and in confined waters

Once free of its berth the operator will attach the MASS to its voyage plan, time will be synchronised, the estimated time of departure (ETD) will be changed to the real departure time and all passage times along the route will be updated. Potential conflicts with other known traffic along the route will now be flagged up for the operator, maybe in an itinerary like display [13]. As an example, it could be passing ferry crossings with given timetables or places like the Lepsøyrenna north of Ålesund. This is a 100-metres wide and 1.8 km long dredged channel where you do not want to meet another ship. At some point in advance the system needs to flag up and suggest solutions (stand-on or give-way) for potential conflicts with other ships. Of course, there is no point in doing this days in advance. Instead the system will constantly calculate risks based on other ships *intended routes* [16] within a limited time span ahead of the present position, and of course also include weather data in its calculation. While AIS will give the present position, course and speed of other SOLAS vessels along the path, new systems will be necessary for detecting and tracking small leisure crafts and fishing vessels not equipped with AIS. The

IMAT project propose to set up an auxiliary infrastructure in the Trondheimsfjord in Norway with land radars and cameras to be able to track and predict paths of small crafts and to confirm and identify AIS targets. This fused sensor information will be available for the MASS' navigation system and the ROC operator [3].

Once the MASS has left the port area and entered out into open sea the high readiness level (short response time) of the operator in the ROC is relaxed and if no other traffic or dangers are in the vicinity the vessel may be added to a pool of other ships with the same low readiness level monitored by one operator. The ship will follow the traffic separated coastal route recommended by the Coastal Administration [15]. One may even envision special designated routes intended only for MASS, some distance from the routes recommended for manned vessels. All in the interest of keeping conflicting situations to a minimum. And the MASS's navigation system working as much as possible in "path-following mode". Most modern ships can use the autopilot in a mode where it exactly follows the pre-planned track made in the voyage plan. With good GNSS coverage and 5-10 meters precision, a MASS will be able to navigate automatically through a complex archipelago with no uncharted obstacles. But of course, this is not the case, because there are other fellow mariners as we will see.

4.3.1. Path-following. The navigation system of a MASS will have two major ways of working: a dumb path-following mode, and an intelligent, AI governed, automatic or "autonomous" mode (apart from when the operator in the ROC may take manual control). As mentioned above, path-following mode is practiced by many ships already today. The ship uses its autopilot to follow a pre-programmed voyage plan, staying within the set cross-track distance (e.g. 50 metres). If for some reason the autopilot (e.g. strong winds or current) cannot keep the ship within the designated corridor there will be an alarm and the automatic (AI driven) navigation system will take over (or the operator in the ROC). To avoid automation surprises, you would like to keep the ship as much as possible in this path-following mode.

4.3.2. Autonomous navigation. When an anomaly occurs for the MASS transiting along its planned voyage in path-following mode the operator in the ROC will be warned and the ship will switch to automatic-navigation mode, meaning that the automatic system (AI governed) will try and resolve the anomaly. If the anomaly is that the ship cannot keep on its designated path due to wind, waves or current, a new path needs to be designed, or if there is another ship on a conflicting course, a COLREG compliant action needs to be taken. The operator in the ROC may or may not be instantly available depending on the operator readiness level. In any case the automatic-navigation system needs to handle the situation in some way and here is where a lot of research needs to be done in coming years. The important thing from a human factor's perspective is how the automatic decisions are flagged up at the ROC, and some examples of this has already been given above.

4.4. Interaction between MASS and conventional ships

For a foreseeable future MASS will need to coexist with traditional ships with a manned bridge. IMO will need to make a decision on whether a MASS should be identifiable as navigating in "autonomous mode". Such identification could for instance be by displaying a designated flag or light signal. For instance, a *purple* all-around top light has been suggested [17]. An "A" could be added to the AIS symbol. One may argue that a MASS should follow COLREG and behave just like any other ship and therefore do not need to be identifiable as sailing with an unmanned bridge. However, identifying MASS as an automatic ship is not mainly to give her special favours, or that other ships should keep out if her way, instead it can be of interest to broadcast the "limited intelligence" of its automation, and that it always will follow the rules of the road, as opposed to manned ships, where you always need to keep an open mind about intended actions. Ship traffic management (STM) using route intentions will hopefully in the future facilitate conflicting situations such when two MASS meets, the automatic navigation systems on both vessels will negotiate the encounter.

As a MASS transits through the fairways along the coast it is essential that the vessel communicates intentions to other seafarers in the vicinity. This can be done in traditional ways by manoeuvring according to regulations (COLREG), local culture and in exceptional cases by voice transmission. However, manoeuvring in a way that is understandable to other mariners will be of outmost importance. An AI-driven automatic navigation system may well behave in a way that if not understood just because it has the "cognitive power" to look further into the future than the human mind with its limited working memory. In this way it could deal not only with the present conflict situation but also with following situations that a manoeuvre may lead to. This might not always be evident to other human mariners. A trivial example involving just two ships might be a COLREG compliant meeting where the MASS will give-way to a vessel on its starboard side. The automatic navigation system may easily calculate the exact collision position and change to a course just enough to pass behind the stand-on vessel. However, such intention to give way might not be visible enough on the bridge of the stand-on ship. Instead a human watch-keeper would often make an unnecessarily large starboard turn aiming a cable or so behind the stern of the stand-on vessel and then holding this aiming point on the aft while slowly turning back to port until the original course is reached. In this way the give-way vessel might clearly show, both day and night, its intentions of giving way at a cost of some extra distance travelled. This was a trivial example. More complex scenarios with several ships can be investigated in a similar way. However, no matter how many focus groups with navigators, workshops, and automatic randomised simulations that are done to help coding this form of automation transparency we must acknowledge that life will always surprise us with the unexpected. The conclusion being that an automatic navigation system must be programmed to show its intentions in a human-readable way.

4.5. VTS areas
In five areas along the Norwegian coast the Norwegian Coastal Administration has set up Vessel Traffic Service (VTS) stations. The underlying policy is that the Norwegian state grants permission for a ship to traffic its waters and ships entering the area will need to report on arrival and its intended route to the port or through the area. This reporting is today done by VHF voice call, but could, in the future, also be doe automatically by digital messaging. An argument against this is that the traffic situation will be less transparent when other ships no longer hear each other reporting on the designated VHF traffic channel. However, STM and transmission of "route intentions," as mentioned before, might be a solution.

VTS operators spoke to has been very clear that they do not want any extra responsibility for MASS [18]. They want no "emergency stop" button for a MASS transiting its area. Instead they want to be able to call a MASS by voice radio just like any other vessel. Here we have a potential problem: as long as VTS works *analogue* (by manual voice over radio) and do not use *digital* means (as in digital communication and automatic clearance handling and recommendations) we might have a problem that will need much manual handling by operators in a ROC. But eventually this hurdle will be overcome and VTS services will be digital and more automatic.

4.6. Port approach
As the MASS approaches port another set of "green lights" will have to come on from much the same stakeholders as was mentioned in the Cast-off section of this paper. The agent will acknowledge that all is ready, tugs, cranes supplies etc. and the port master that the ordered dockside is available to the right length. If linesmen are needed, they too will acknowledge and someone will need to check that wind, tide and currents are within the envelope the vessel can cope with. When entering a port like Trondheim, for instance, there is a lot of local information that must be considered and that will not be available in nautical charts or publications. Also, the port has a *port control* (much like a VTS) that needs to be informed about the arrival. In Trondheim there is usually a local regatta south of Munkholmen each Thursday during the summer months. It would not be a good idea for a MASS to try and press thought the flotilla of these small sailing boats, often with children at the helm. Instead a route north of

Munkholmen might be preferred on such occasions. This and much more of local data needs to be collected and made available in a machine- and human-readable format for the MASS systems and the operators in the ROC.

The IMAT project was briefly mentioned earlier [3]. The main goal of the Integrated Maritime Autonomous Transport (IMAT) project is to define and describe needed infrastructure and organization to conduct safe and effective autonomous transport operations. The IMAT project is elaborating with a Local Information Centre concept. A place where such local information as regattas, ferry timetables and the tracking of kayaks and small fishing boats is collected and made available for ROC and automatic queries.

4.7. Mooring
The MASS is finally reaching its destination. Linesmen or an automatic docking mechanism is ready. Just before the docking there needs to be a check that there is nothing between the ship and the quayside: a small rowing boat that has moored in the wrong place, a kayaker resting his tired arms, a swimmer or some obstruction debris. A manual camera check might have to be done before the "mooring" button is pressed and the Dynamic Positioning (DP) system pushes the MASS sideways the final meters on to the quay. Lines are fast: check; power is attached: check; propulsion is stopped: check, containers are de-lashed: check. Maybe a gangway is automatically lowered for maintenance personal to go onboard. And many, many more things that will be done manually at the start but eventually also will be automated.

5. Conclusion
This paper attempts to offer an understandable simple and non-technical concept description of the Maritime Autonomous Surface Ship system with a focus on the human element: where and in which capacity will humans interact with the system, in ports, in the remote operation centre, onboard small crafts and on the bridge of other conventional ships. We have done so by describing a typical voyage of a small electrical container shuttle on a coastal voyage on national waters along the Norwegian coast. Much of what has been said are conceptional thinking by the authors and do not necessarily represent official views form maritime authorities in Norway or internationally, but the authors have for many years worked within the autonomous ship domain and possess some degree of domain knowledge.

We hope that this paper may inspire some engineers working with the technical realisation of this complex system to acknowledge the need for human factor's engineering, and that the way forward is facilitating a strong teamwork between automation and humans.

Acknowledgements
The research has partly been conducted within the IMAT (Integrated Maritime Autonomous Transport systems) project, and partly within the LOAS (Land-based Monitoring of Autonomous Ships) project, both funded by the Norwegian Research Council, which is hereby gratefully acknowledged.

References
[1] Kongberg Maritime (2020) Autonomous ship project: key facts about Yara Birkeland https://www.kongsberg.com/maritime/support/themes/autonomous-ship-project-key-facts-about-yara-birkeland/ [Acc. 2020-06-14]

[2] MUNIN project (2015) Maritime Unmanned Navigation through Intelligence in Networks http://www.unmanned-ship.org/munin/ [Acc. 2020-06-14]

[3] IMAT project (2020) Integrated Maritime Autonomous Transport Systems, https://www.sintef.no/projectweb/imat/ [Acc. 2020-06-14]

[4] Wienberg, C. (2018) Maersk CEO on Unmanned Ships: 'Not In My Lifetime', gCaptain, https://gcaptain.com/maersk-ceo-on-unmanned-ships-not-in-my-lifetime/ [Acc. 2020-06-14]

[5] Porathe, T. & Costa, N. (2014) Deliverable 7.4: Organizational lay-out of the Shore Operations Centre. MUNIN project.

[6] Mustapha Mouloua, Peter A. Hancock (2020) Human Performance in Automated and

Autonomous Systems. CRC Press

[7] Endsley & Kiris, (1995) The Out-of-the-Loop Performance Problem and Level of Control in Automation. Human Factors The Journal of the Human Factors and Ergonomics Society 37(2)

[8] Rodseth, O (2020) Constrained autonomy for a better human-automation interface. (in presse).

[9] Dosilovic, Filip; Brcic, Mario; Hlupic, Nikica (2018). "Explainable Artificial Intelligence: A Survey" (PDF). MIPRO 2018 - 41st International Convention Proceedings. MIPRO 2018. Opatija, Croatia. pp. 210–215. doi:10.23919/MIPRO.2018.8400040

[10] DARPA (2020). Explainable Artificial Intelligence (XAI). DARPA. https://www.darpa.mil/program/explainable-artificial-intelligence [Acc. 2020-06-14]

[11] Holzinger, A., Plass, M., Holzinger, K., Crisan, G. C., Pintea, C-M. & Palade, V. (2017). A glass-box interactive machine learning approach for solving NP-hard problems with the human-in-the-loop. https://arxiv.org/abs/1708.01104 [Acc. 2020-06-14]

[12] Bostrom, N., & Yudkowsky, E. (2014). The ethics of artificial intelligence. The Cambridge Handbook of Artificial Intelligence, 316-334.

[13] Porathe, T. & Man, Y. (2014) Deliverable 7.5: HMI layout for the Shore Operations Centre. MUNIN project.

[14] IMO, International Maritime Organization (2000) "Guidance Notes for Voyage Planning". IMO RESOLUTION A.893(21) adopted on 25 November 1999.

[15] NCA, Norwegian Coastal Administration (2019). Routes and Route information from the Norwegian Coastal Administration (NCA). https://routeinfo.kystverket.no/ [Acc. 2020-06-14]

[16] Porathe, T., Brodje, A., Weber, R., Camre, D. and Borup, O. (2015). Supporting Situation Awareness on the bridge: testing route exchange in a practical e-Navigation study. In A. Weintrit, and T. Neumann (Eds.) Information, Communication and Environment: Marine Navigation and Safety of Sea Transportation, CRC Press, 85-92, 2015.

[17] Porathe, T. (2019). Safety of autonomous shipping: COLREGS and interaction between manned and unmanned ships. Proceedings of the 29th European Safety and Reliability Conference 2019.

[18] Porathe, T. (2019) Personal communication, Brevik VTS.

IOP Conf. Series: Materials Science and Engineering **929** (2020) 012008 doi:10.1088/1757-899X/929/1/012008

Meaningful Human Control in Autonomous Shipping: An Overview

J (Hans) van den Broek[1], J R (Jaco) Griffioen[2] and M (Monique) van der Drift[2]

[1] Sustainable Port City Research Centre, Rotterdam University of Applied Sciences. Heijplaatstraat 23, 3089 JB Rotterdam, the Netherlands.
[2] Rotterdam Main Port Institute, Rotterdam University of Applied Sciences, Lloydstraat 300, 3024 EA Rotterdam, the Netherlands.

Corresponding author: j.van.den.broek@hr.nl

Abstract. This paper presents an overview of the aspects related to meaningful human control in autonomous shipping. Autonomy is described as a multi-faceted construct including self-sufficiency and self-directedness, which are related to a range of conditions a ship can deal with. This means that autonomy levels will vary in different operational situations. As a consequence, active operator involvement and workload will fluctuate and will sometimes be unpredictable. In case of supervising multiple ships, it is not feasible to maintain a minimal level of situation awareness of all the ships. In order to facilitate operator situation awareness recovery and other human-automation collaboration principles an Intelligent Operator Support System concept is outlined. Whether the large-scale application of Maritime Autonomous Surface Ships will provide equivalent safety and resilience levels will largely depend on the intelligence of the onboard control systems, the design quality of the human-autonomy collaboration system, and the competence and training level of the SC-operators.

1. Introduction

The International Maritime Organization (IMO) defines a Maritime Autonomous Surface Ship (MASS) as a ship which, in varying degree, can operate independently of human interaction [14]. In order to develop and operate a MASS a radical and complete task and function re-allocation from manned execution to automated execution (mechanized and computerized) is necessary.

For those who regard 'autonomy' as 'perfect automation', the sheer discussion on 'human interaction in varying degree' in itself is regarded as 'void', since 'perfect automation' can do without any human control. Furthermore, in this line of reasoning these systems cease to exist as autonomous systems in case they need 'outside' help.

The above expressed either-or view on autonomy is both wrong and incomplete. It's wrong because the autonomy concept incorporates more than 'perfect automation'. It's incomplete because it neglects the notion that autonomous ships are part of a wider maritime context, consisting of other vessels (autonomous and conventionally manned), vessel traffic service stations, harbour pilots, tugboats, and area specific regulations for example, in which collaboration cannot be avoided. Because of the shared nautical space, a MASS needs the ability to understand (the roles of) the other actors in order to be able to interact and coordinate. Hence, the ability to interact and ask for support based on some kind of world model is an intrinsic quality of autonomous systems and not a disqualifier.

1.1. The structure of the paper
The above shows that different conceptualizations about the meaning of 'autonomy' lead to different ideas in regard to the level of human involvement. Therefore, section two will provide a brief discussion on the concept of autonomy.

Furthermore, this paper takes the view that the performance of an autonomous ship depends on the quality and intelligence of the onboard technical control systems (henceforth the artificial navigator), the complexity of the maritime environment, and the human support that is made possible by means of a human-autonomy collaborative system. For the design of such an overall (socio-technical) system it means that at some level interaction between autonomous systems and human actors need to be taken into consideration. The outline and general specification of such an overall system design is described in section three. Section four discusses the Shore Control Centre (SCC) design principles that, from a human factors point of view, need to be in place in order to enable Shore Control operators (SC-operators) to exercise meaningful human control. Section five will discuss the resilience of a human-autonomy collaborative system as the ability to deal with a range of expected and unexpected maritime conditions. Section six will provide the conclusions of this paper.

2. On autonomy and levels of automation
The concept of 'autonomy' and the concept of 'automation' have been used interchangeably in the literature, e.g. [5][17]. Differences in perception and interpretation of these concepts can create different expectations and principles for the collaborative design between system and humans. Therefore, it is important to have a clear understanding and definition of both concepts and the relation between them.

2.1. Defining autonomy
Despite the fact that automation and autonomy are related in terms of the performance of a ship, they are not equivalent constructs. Automation is physical technology (mechanized or computerized) viable for application in a defined environment. Autonomy is a state of situation awareness [9]. For a ship this implies robustness to the environment, independence in action or function, and self-determination of goals and allocation of its resources. Or to put it another way, autonomy can be a desirable design goal for automated systems [18] and hence for a MASS. In line with [18], this paper considers autonomy as a multi-faceted construct. In particular the following two system facets will be discussed:

- self-sufficiency or 'viability' in a given environment,
- self-directedness or capacity to function and to perform independently from other agents,

These facets are complex. For instance, self-governance is not just the absence of external control, it requires specific cognitive capabilities to learn but also to reason on a strategic level. Also, self-directedness does not mean that self-sufficient ships are independent in action since they will be part of a wider maritime context including among other things legislation, other vessels, pilot assistance, and vessel traffic service stations [3].

2.2. Contextual dependence
Self-sufficiency is related to *the range of conditions the system can deal with* and self-directedness refers to *the range of conditions the system is given authority to conduct autonomously*. Figure 1 depicts four 'autonomy states' based on four possible dimension ratios. Any imbalance should be avoided in system design for different reasons. For instance, when self-sufficiency is higher than self-directedness (bottom right), the system is not used optimally. If it is the other way around (top left), the system is allowed to handle situations that it is not capable of. This kind of overreliance is a dangerous condition and that should be avoided for all critical systems because it could lead to undesired states or even dangerous situations. Effective autonomy is possible when both dimensions are in balance (top right). As self-sufficiency is increasing with more advanced intelligent software, choosing the appropriate level(s) of self-directedness is an important task. Higher levels of self-sufficiency enable higher levels

of self-directedness. With increasingly more intelligent software available, the curve should be followed for optimal human-automation system performance.

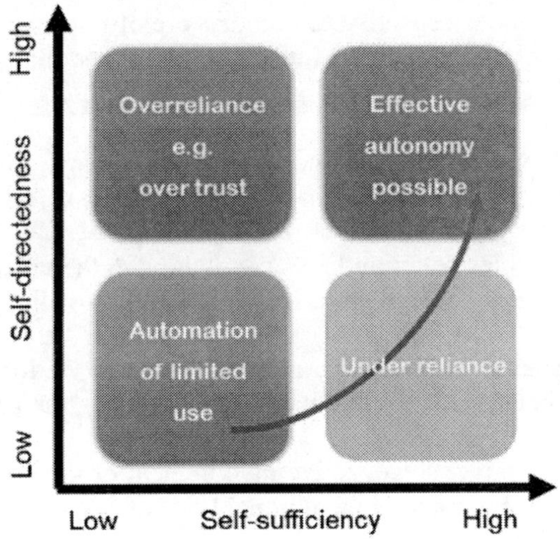

Figure 1. Autonomy consists of two dimensions. The curve shows the trajectory to follow when a gradual approach is chosen (based on [5]).

Because the level of self-directedness and self-sufficiency may vary within different contexts, it means that with a constant (high) level of automation the self- sufficiency of an autonomous ship could be sufficiently high or 'viable' in environment 'A' but sub- optimal or less 'viable' in environment 'B'. Therefore, self-sufficiency and self- directedness cannot be regarded as absolute or given system aspects but instead they depend on the operational complexity the autonomous system is in. Because of the large variation in operational complexity and in environment dynamics [25] a large variation of autonomy states is possible, which in turn require a varying degree of human support and/or support from other autonomous systems. From this it follows that autonomy as a system characteristic cannot be defined in absolute terms, and that the level of autonomy depends on the combination of the level of intelligence of the onboard software and contextual complexity.

3. The challenges regarding the human factor

In dealing with the challenges regarding the human factor, three different stages are distinguished within human-automation collaboration:

1. Supervision stage: The self-direction of the automation is low. The human operator supervises the system(s) 100% of his time and is not involved in other tasks. This strongly relates to Sheridan's model of supervision [27][28].
2. Partial supervision/autonomy stage: The self-directness of the automation is higher. The human operator spends part of his or her time conducting secondary tasks. In case a complex or critical situation emerges that needs attention, the operator switches back to the primary task.
3. Intervener/full autonomy stage. Both self-sufficiency and self-directedness are high. The system is working autonomously 99.99% of the time. The operator is working on other tasks or on other systems. But even fully autonomous systems sometimes fail, and in these exceptional cases the operator does have to intervene or even take over control.

The supervision stage has been well-studied and is already common practice in many work environments. Therefore, this section will only look at the directions for solutions for the latter two less studied and less well-known stages.

The main challenge for partial supervision/autonomy stage, is to keep the operator aware of the status of the critical task and to enable him or her to resume control effectively when required. Both the human operator and the automation develop situation awareness (SA) relevant for the primary task [29]. Automation and human operators will continuously update their situation awareness. Both the union and the symmetric difference of their respective SA models are of importance [2], both for the detection of early signals as well as the need to adjust the human-automation collaboration agreements.

Ways to support human-automation collaboration within partial supervision/autonomy stage are:

1. Support upkeep of operator SA using supervisory displays [30]. This will help the operator decide whether his or her involvement at the primary task is required.
2. Provide SA recovery support after returning to the main task, for example using change detection support [35]. This enables better and faster switching to the primary task.
3. Increase reaction time by detecting early signals and providing on-time alerts for the primary task.
4. Just in time awareness: provide change detection and option awareness support for quick decision making when a critical event has occurred during the primary task.

At the intervener/full autonomy stage, active involvement of the operator is very rare. Therefore, it is not feasible or cost-effective to maintain a minimal level of SA of the primary task. Incidents are just too rare to warrant this effort. Also, regular SA recovery is not required as the operator would not conduct this task anymore under regular circumstances. In this configuration only the last two types of support for the partial supervision stage are relevant. The skill levels of the operator will be much lower than under partial supervision, as he or she hardly controls the system anymore (maybe solely during incident training in simulators). Hence automation support remains essential for safe and effective task completion, even in the intervening mode. This results in partly different types of support:

1. Increase reaction time by detecting early signals and providing on-time alerts for the primary task.
2. Just in time awareness: provide recognition primed decision-making[1], change detection and option awareness support for quick decision making when a critical event has occurred at the primary task.
3. Support operator with lack of skills: prevent the need for fully manual control but deliver lower levels of automation support.

Based on the review of the literature and solution concepts, designing human-automation collaboration at high system autonomy levels should meet the following principles of design:

- All systems, even "fully-autonomous" systems, should be considered from a joint human-automation collaboration viewpoint, because there will be always instances when human action in some form is required. Co-active design [17], focusing on observability, predictability and directability of all actors involved, has been proposed as a sound method to design this collaboration.
- Both human operators and automation need to develop an understanding of the situation, especially at the supervision and partial autonomy stages. This is in line with the distributed SA paradigm that holds the view that SA is not solely 'in the mind' or 'in the world' but is build up 'in interaction'.
- Man, as intervener does not imply that the operator takes over manual control when automation fails. Because, when automation fails, it does not mean that it stops working on all levels. Human

[1] The recognition primed decision-making model is based on the assertion that operators can use their experience to generate a plausible option as the first one they consider.

support can be delivered, for instance, by making decisions in case automation fails to do so. But the execution of actions can be allocated to the automation. Because of the lack of skill of operators in complex future human-automation configurations, this kind of adaptive shift in collaboration agreements and authority provides a more promising approach.

From the above discussed human-automation collaboration principles, it follows that meaningful human control entails more than simply mimicking the bridge layout and conning station in an SCC. Especially SA recovery support and the detection of early signals requires a dedicated operator support tool[2]. At TNO, an Intelligent Operator Support System (IOSS) has been developed (TRL-5) in order to demonstrate the above described design principles in support of dynamic positioning (DP) operators [33].

4. Concepts of Shore Control Centre design

In this section we will discuss how the human-automation collaboration principles described above can be applied within a future Shore Control Centre (SCC). A distinction will be made between a Shore Control Centre, being a kind of back office that provides logistical and technical support, and an SC-operator, being the person responsible for nautical support, and a Shore Control Station (SCS) that functions as the individual physical workstation which presents the necessary information and which enables the operator to exercise control.

The primary task of an SC-operator can be divided into two aspects. One, is to build up SA of the nautical aspects, i.e. the operational picture of the marine area in which the MASS is sailing. The other aspect is that the SC-operator needs to be able to establish whether or not the self-sufficiency is high enough to deal with the situation and in which self-directedness is appropriate and safe. This not only requires an assessment of critical information on the basis of which the artificial navigator bases its decisions, it also requires an assessment whether the ship manoeuvres as it may be expected.

For this reason, the design concept of an SCS must facilitate the presentation of information regarding SA, the decision-making process and remote control for the operator. Also, the SC-operator must be able to communicate with conventional ships that sail in the vicinity of a targeted MASS, but also with other actors within the global maritime system, such as Vessel Traffic Services (VTS), port authorities and perhaps also with pilots and tugboats by using existing communication technologies [4].

4.1. Balancing the SC-operator workload

The general expectation is, that a future SC-operator will have to supervise several MASS-s, of the same or different type. The question concerning the span of control of an individual SC-operator, i.e. the number of individual ships a SC-operator is able to supervise, cannot be expressed in absolute numbers. The span of control highly depends on the level of self-sufficiency and self-directiveness of the ships under supervision and how these levels are distributed over the ships that constitute the case load of the operator. For instance, in case all ships under supervision cross the Atlantic Ocean, their self-sufficiency and self- directiveness levels will be sufficiently high and, as a consequence, the SC-operator workload will be relatively low (and with that, the span of control, i.e. the number of ships under control, could be enlarged). Whereas, if one of the ships under supervision enters a narrow fairway or a busy harbour, the self-sufficiency and self-directiveness levels will (relatively) drop, resulting in higher workload and higher vigilance levels. As a consequence, the span of control of the SC-operator will drop and may even, de facto, drop to even one ship.

The fact that different voyage stages have a particular timeline and can be estimated in advance provides the SC-operator the opportunity to anticipate to these higher workload situations. In [26], the

[2] An important aspect of SC-operator support is the question which and how much information can be transmitted to a SCC in order to adequately support the SA maintenance capability of the SC-operator. However, these technical and bandwidth issues together with latency issues are out of scope of this human factors related paper.

authors discriminate five voyage stages: leaving berth, departure from port, sea passage, and a stage in which exceptional circumstances occur. For each stage the autonomy mode changes for which the workload demands can be planned in advance with exception of the stage in which exceptional circumstances occur.

In addition, communication with third parties (e.g. other ships, vessel traffic service stations etc.) will increase the SC-operator workload significantly. From research on naval frigates it is known that communication, both external and internal, is difficult to combine with a primary task [34][31]. Altogether, it means that workload fluctuations can be predicted to some level but not always. The way forward to mitigate work overload is to apply voyage stage planning as described above, and to establish an adaptive workload balancing approach among several SC-operators to deal with workload fluctuations in a dynamic way [24]. For instance, when the cognitive resources for attention and workload are demanded by a single ship, the other ships that fall within the responsibility of the heavily occupied operator should be transferred to another SC-operator, within the own SCC, or within another SCC that functions as back-up centre. Also, when ships are crossing time zones, it is conceivable, and perhaps even necessary, to convey the supervisory responsibility from one SCC, e.g. in Europe, to another SCC, e.g. in the USA.

4.2. Situation Awareness recovery

The consequence of supervising multiple ships is that the SC-operator must switch between different ships and contexts, which comes with so-called (cognitive) task switching costs [23], being the mental effort and time it takes to reconstruct the situation awareness of the ship state and nautical context to which the attention shifts. However, the key question concerning shift of attention is: how does an operator know or determine which ship under supervision needs attention and which ship doesn't? Is it for instance expected that the operator checks all the critical data from each ship on a regular basis? As argued above, it is not feasible to maintain a minimal level of SA of all the ships under control by cascading constantly from one ship to another. Instead, it requires a mechanism, i.e. a support mechanism, that helps the SC-operator to focus his or her attention.

The following mechanism is based on expected and pre-defined state changes:

- Divide a ship voyage into voyage stages of which levels of self-sufficiency are established or pre-set based on experience.
- This trigger mechanism should be accompanied by a process of SA-recovery, i.e. the build-up of the nautical picture including a risk assessment and the decision to intervene or not.

Additionally, a mechanism is needed to trigger the attention of the SC-operator in case of an unexpected critical situation. The trigger mechanism could be:

- The detection of early signals, i.e. anomalies in ship behaviour.
- The MASS itself triggering the attention of the operator based on a self-awareness capability. For instance, based on feedback that a sensor does not work optimally.
- This trigger class should also be accompanied by a process of SA-recovery, i.e. the build-up of the nautical picture including a risk assessment and to decide on the appropriate intervention.

Because it is envisioned that a single SC-operator has to monitor and has to control several MASS-s, intelligent support of operators should be part of human- automation collaboration design to meet the cognitively demanding task of switching between different ships and contexts. The IOSS [33], introduced above, is developed to support operators in such a way that maintaining a continuous level of SA of autonomous systems is no longer needed because the task of monitoring a ship is handed over to the IOSS. This makes it possible for an SC-operator to pay attention to another ship or to execute a secondary task. When the situation of a ship changes in such a way that the involvement of the SC-operator might be required, the IOSS will trigger the attention of the SC- operator. The critical conditions under which it is necessary to bring the SC-operator back into the loop can be pre-defined by the SC-operator as so-called cooperation rules. Furthermore, in dealing with a possible critical situation, the SA recovery of the SC-operator is supported by offering context specific information instead of a situation in which the SC-operator self must search for information by means of an interface.

4.3. SC-operator competence and skill enhancement

Currently, not much research has been done into the required competences and skills of the SC-operator of the future. To some extent they are regulated by the International Convention on Standards of Training, Certification and Watchkeeping for Seafarers (STCW) [16]. Additionally, the IMO is currently carrying out a scoping exercise assessing IMO-instruments, including reviewing STCW regulations, to see how they may apply to ships with varying degrees of autonomy [15]. The interim guidelines for trials with autonomous ships stipulate that "onboard or remote operators of MASS should be appropriately qualified for operating MASS subject to the trial (2.3.2, p.2)". The guidelines further state: "For the safe, secure and environmentally sound conduct of MASS trials, the human element should be appropriately addressed. The trial should consider the human-system interface as harmonization between human centred design and automation is a key component of MASS (2.4, p.2)". Other organisations, such as DNV-GL [8], Bureau Veritas [6], and the UK Maritime Autonomous Systems Regulatory Working Group [21], published guidelines with equivalent qualifications and requirements as mentioned in STCW. The Institute of Marine Engineering, Science & Technology evaluated future skills requirements by a series of surveys and of roundtable discussions. The final conclusion of the report [10] is that "for the foreseeable future, autonomous systems will remain unable to make critical decisions based on sound judgement. So, the challenge lies not in teaching humans to trust machines, but in providing them the skills and competencies to know when to stop trusting and overrule them (p. 23)". However, the report fails to address the required skills and competencies.

In order to gain understanding of the additional competencies of the remote operator, simulator experiments were conducted on the ship bridge simulator at the nautical school in Rotterdam, using the instructor station as a Shore Control Centre. Twenty nautical students, nearly graduated, performed the task as an SCC operator of three autonomous vessels with same manoeuvring characteristics, sailing in three different coastal areas with navigational obstructions, vessel traffic zones, medium traffic (about four vessels in each area) and good visibility. The experimental design was, that due to a failure of the autonomous systems the operator had to control the autopilot, rudder and thruster of the three autonomous vessels. The scenarios with a runtime of about 30 minutes contained traffic situations where the operator had to perform manoeuvres to avoid collision according to the COLREGS, taking into consideration navigational obstructions. All other vessels in the scenario were visible on radar and AIS information was received.

Despite the relative low complexity of the task (not all watchkeeping tasks were performed, e.g. plotting the position of the vessel in the ECDIS, VHF communication with shore stations and all vessels complied with the COLREGS), from observations and briefings it became clear that the cognitive workload was high during the control of the three vessels. The central problem was to regain situation awareness when switching between the autonomous vessels, especially when there is a close quarter situation and the operator has to focus for a longer time than planned. This is in line with [36], who report the 'Constant reorientation to new tasks (p. 1042)' in the context of remote container handling. Due to the high cognitive workload, information from AIS was most times not cross checked with radar information. These findings support the above described necessity of workload balancing mechanisms for SC-operators.

A MASS will have similar manoeuvrability characteristics and must apply to the COLREGS as any other conventional manned sea going ship. Therefore, it is necessary for the SC-operator to master the STCW-skills and competencies (both knowledge and practice). However, remote navigation and supervisory control possess additional challenges compared to a navigator who is located at the bridge because the behaviour of a ship and the environmental circumstances (wind, waves, currents) can only be assessed in an indirect way by means of a two dimensional display, instead of being immersed and tangibly connected with the ship and the elements. This requires additional competencies:

- The ability to built-up situation awareness of the nautical situation, e.g. vessel traffic, distance, speed and circumstances, e.g. wind, waves, currents from a 2D-interface.
- The ability to translate required manoeuvring changes, e.g. speed and course, in adequate high-level instructions for a MASS.
- The ability to deal with different ship types and conditions that result in a large range of manoeuvring characteristics. Normally, it takes time to learn how particular ships behave under different conditions.
- Thorough knowledge, including the limitations of various autonomous control systems, including Advanced Sensor Modules, Remote Manoeuvring Systems, and Deep-Sea Navigation Systems.
- Resource Management, Teamwork and Leadership in order to improve safety [16]. These so called non-technical skills, i.e. human factors training or soft skill training, are not new in the maritime industry, but were recommended for ship operating companies, and training was therefore carried out on a voluntary basis. In line with the renewed emphasis on non-technical skills for onboard bridge teams, this should also apply for SC-operators. Because not only will SC-operators be responsible for the external communication with third parties (ships, VTS-stations etc.), it is also very likely that a SC-operator will be part of a SCC-team.

In section 4 it is stated, that in the intervener/full autonomy stage the skill levels of the operator will be much lower, in comparison with the partial supervision/autonomy stage, as he or she hardly controls the system anymore. Is this contradictory with the above claim that SC-operators should have and maintain STCW-skills and additional competencies? No, it's not, because it is stated that the nautical skill level will be much lower, but that does not mean that it may be entirely absent. The question, which competencies will be essential in the long term and which less in this context, must be determined by future human factors research. In the far future however, it can be expected that the SC-operator's competences will likely shift from skill level (hands-on experience) to domain knowledge level and general nautical competences. In the short run, SC-operators still need the ability to shift between the intervener/full autonomy stage, partial supervision/autonomy stage and even supervision stage operations when circumstances dictate, operations in which STCW-skills are still relevant.

5. The resilience of a human-autonomy collaboration systems

The sailing of a vessel always takes place in a dynamic environment due to different weather and hydrodynamic conditions, and traffic. In a dynamic environment it is important for a system to have the ability to observe, anticipate, and respond, i.e. to be resilient. In the definition of Eric Hollnagel [12] a system is resilient if it can adjust its operation before, during or after events (changes, disruptions and opportunities), and thereby maintain the required operations under both expected and unexpected circumstances. In terms of the above, resilience is the ability of a system to deal with a range of conditions. In conventional shipping, resilience is a characteristic of a socio-technical system. It means that a ship system involves a combination of people (socio) - such as individuals, teams and organizations - and technical elements - such as automated systems, computers and instruments – that interact and collaborate to support organizational activities. The basic assumption of the resilience approach is, that accidents occur when the resilience (ability) of a system falls short in dealing with certain conditions. Hence, resilience engineering is focused on extending the resilience ability of socio-technical systems.

Shipping is perhaps the most international of all the world's great industries - and one of the most dangerous [15]. Worldwide, almost every week there are accidents with commercial ships [7]. According to the IMO and various insurance companies, seventy-five to eighty per cent of the number of accidents is caused by the human factor [1]. The claim that autonomous shipping will be safer than conventional shipping, which now can be rephrased as the claim that autonomous shipping will be more *resilient* than conventional shipping, is based on the assumption that when human crews are replaced by technology the accident rate will be reduced with eighty per cent. Firstly, this assumption is fuelled by the fact that the label 'human error' is always interpreted differently and has become a collective term for every involvement of 'humans' in situations where 'things are going wrong' [12] Also, to conclude that if eight out of ten accidents are partly due to humans, that therefore sailing without a crew will be safer is not valid. Statistically speaking, only one out of ten thousand events turns into a failure [13]. From the ninety-nine thousand nine hundred ninety-nine non failures we do not know in how many cases crews were resilient enough to turn a near-disaster into a safe situation, that otherwise would have been a disaster. And we certainly do not know whether autonomous ships can produce the same level of resilience. Secondly, it's not correct to state that autonomous shipping takes the human 'out-of-the-loop'. Yes, the bridge crew is taken off the ship, but (partially) replaced by SC-operators as part of a human-autonomy collaboration system [21]. Therefore, the safety claim, often presented as rationale for autonomous sailing, should be rephrased into the question whether the resilience of human- autonomy collaboration systems will be equivalent or better than conventional manned socio- technical ship systems.

According to Hollnagel [11], resilience is based on four essential abilities: monitoring, responding, learning and anticipating. The starting point is that a system can be made more resilient by reinforcing these four essential abilities. These abilities differ hugely between humans and artificial systems, both in terms of internal mechanisms, and in range and richness of their world models. This will be illustrated with an example.

A Dutch start-up, called Captain AI, recently sailed autonomously in the port of Rotterdam. The artificial navigator was trained through virtual simulation to recognize elements in the environment and to react as learned. Because SA starts with perception of the elements in the environment, the actual sensor suite that is implemented will have de facto a selective world view and the autonomous vessel will have a selective and limited SA in combination with the learning algorithm. This has consequences for adaptability. To understand this, the distinction between 'work as imagined' (WAI) and 'work as done' (WAD) is introduced [11]. WAI is the 'outside-in' view of the blunt-end of the organization. WAD is about the resources which are actually needed to execute operations at the sharp end of the organization. The danger of designing systems from a limited WAI-point of view, is that the resources, systems and mechanisms are too rigid to adapt to unexpected (from the perspective of WAI) circumstances. For the Captain AI example, it means that the system is not able to react to a set of circumstances that falls outside the scope of what is perceived and what is learned. Also, what the Captain AI algorithm has learned within a geographical specific area cannot be generalized to other distinct areas.

This domain or task specific intelligence is known as weak AI, which, in contrast to strong AI, cannot be generalized to other domains and tasks. This, in contrast to the general intelligence qualities of humans, i.e. the extent to which a type of cognitive skill (e.g. learning a foreign language) is associated with other cognitive skills (e.g. mathematical skills). Paired with the general ability to deal with fussy (ill-defined) situations, to be creative, and to improvise results in high resilience ability. Of course, artificial intelligent systems also have advantages compared to humans. They do not get tired, do not have vigilance problems and can analyse huge amounts of data necessary for applications as face recognition, diagnostic systems etc. Because this form of human-computer dichotomy is considered counterproductive, Licklider [20] argued for the concept of human-computer symbiosis as early as 1960. With the introduction of the human-autonomy cooperation system, this paper endorses this line of reasoning.

Woods [37], calls the limited flexibility of system 'brittleness', i.e. a sudden collapse or failure when events push the system up to and beyond its boundaries for handling changing disturbances and variations. As the opposite of brittleness, it is important that a system has 'graceful extensibility', i.e. the ability of a system to extend its capacity to adapt when surprise events challenge its boundaries. A way to avoid brittleness as much as possible is to select or create a less complex and well controlled area of operations to accommodate a MASS as much as possible. An example of which is the Yara Birkeland the first ever zero emission, autonomous ship [19]. The goal of this challenging project is not only to build a newly designed ships but also to build dedicated terminals and infrastructure (e.g. charging the ship's batteries) to fully accommodate the Yara Birkeland operation[3]. Also, the ship will be tested in the Brevik area of Southern Norway, which provides an ideal marine area to experiment with MASS operations because the Yara Birkeland will sail on two fixed and pre-defined routes. In addition, the Brevik area has VTS which will be used for traffic monitoring and for nautical support. Also, the Yara Birkeland will be monitored and facilitated by two additional SCC-s, one for the logistic processes and one for technical and maintenance assistance [32].

The expected brittleness and rigidity of artificial systems will lead to reduced resilience in dealing with fussy and less well-defined situations, i.e. in terms of the above, could lead to reduced levels of self-sufficiency in unexpected and difficult situations. This means that a SC-operator, being part of the human-autonomy collaboration system, will mostly be involved in critical and difficult situations, that the artificial navigator cannot handle for whatever reason. The frequency with which exceptional situations will occur are expected to be low, resulting in low experience and training opportunities for SC-operators. This adds to the challenge of timely turning critical situations into safe situations.

Currently, not much is known of the resilience ability of the human-autonomy collaboration systems for safe navigation, particularly in realistically complex situations. Hence, at this stage of MASS development it is not possible to prove that autonomous shipping can establish safety levels equivalent to conventual shipping, let alone prove that autonomous shipping is or will be more resilient than conventual shipping.

6. Conclusions

This paper has provided an overview of the aspects related to meaningful human control in autonomous shipping. Autonomy is described as a multi-faceted construct including self-sufficiency and self-directedness which are related to a range of conditions a ship can deal with. This means that autonomy levels will vary in different operational situations. From this it follows that autonomy as a system characteristic cannot be defined in absolute terms and that the level of autonomy depends on the combination of the level of intelligence of the onboard software and contextual complexity. As a consequence, active operator involvement and workload will fluctuate and will sometimes be unpredictable. In case of supervising multiple ships, it is not feasible to maintain a minimal level of situation awareness of all the ships. In order to facilitate operator situation awareness recovery, task switching support, and workload balancing it takes more than mimicking the bridge layout and conning station. An Intelligent Operator Support System concept is outlined as alternative for implementing the human-automation collaboration principles.

Whether the large-scale application of MASS-s will provide equivalent safety and resilience levels compared to conventional manned ships will largely depend on the intelligence of the onboard control systems, the design quality of the human-autonomy collaboration system, and the competence and training level of the SC-operators.

References

[1] Aliance 2018 Safety and Shipping Review 2018: An annual review of trends and developments in shipping losses and safety. Consulted, February 2020, through https://www.agcs.allianz.com/content/dam/onemarketing/agcs/agcs/reports/AGCS-Safety-Shipping-Review-2018.pdf.

[2] Arciszewski, H F, De Greef, T E and Van Delft, J H 2009 Adaptive automation in a naval combat management system. IEEE Trans on Sys, Man and Cyb, Part A: Systems and Humans, **39** (6), 1188-1199.

[3] Due to the Covid-19 pandemic and the changed global outlook, Yara has decided to pause further development of the vessel and will assess next steps together with its partners [19].

[3] Bastiaansen, H J M, Lazovik, E, Den Breejen, E and Van Den Broek, J 2019 A Business Process Framework and Operations Map for Maritime Autonomous and Unmanned Shipping: MAUSOM. J. Phys.: Conf. Ser. **1357** 012017. https://iopscience.iop.org/article/10.1088/1742-6596/1357/1/012017/pdf.

[4] Bastiaansen, H J M, Lazovik, E, Den Breejen, E and Van Den Broek, J 2018 JIP Autonomous Shipping – WP4 Report: Collection, processing and communication of data. TNO-rapport. TNO 2018 R10444. TNO, the Netherlands.

[5] Bradshaw, J M, Hoffman, R R, Woods, D D and Johnson, M 2013 The Seven Deadly Myths of "Autonomous Systems". IEEE Intell Sys **3**, 54-61.

[6] Bureau Veritas 2019 Guidelines for autonomous shipping guidance. Note NI 641 DT R00 E. Consulted, February 2020, through https://marine-offshore.bureauveritas.com/ni641-guidelines-autonomous-shipping.

[7] Transportation Safety Board Of Canada 2018 Annual Report to Parliament 2018–19, Transportation Safety Board of Canada Cat. No. TU1E-PDF. ISSN 1704-1120. Consulted, February 2020, through http://publications.gc.ca/collections/collection_2019/bst-tsb/TU1-2019-eng.pdf.

[8] DNV-GL 2018 Autonomous and remotely operated ships. Consulted February 2020 through https://rules.dnvgl.com/docs/pdf/DNVGL/CG/2018-09/DNVGL-CG-0264.pdf.

[9] Endsley, M R 1995 Towards a theory of situation awareness in dynamic systems. Human Factors, 37, 32–64.

[10] IMarEST 2020 Autonomous Shipping – Putting thc Human Back in the Headlines. Consulted, February 2020 through https://www.imarest.org/policy-news/institute-news/item/4446-imarest-releases-report-on-the-human- impact-of-autonomous-ships.

[11] Hollnagel, E 2018 Safety-II in practice: developing the resilience potentials. Abingdon: Taylor & Francis. https://doi.org/10.4324/9781315201023.

[12] Hollnagel, E 2016 The NO view of 'human error'. Consulted, June the 19th 2017 through http://erikhollnagel.com/ideas/no-view-of-human-error.html.

[13] Hollnagel, E 2014 Safety-I and Safety-II: The Past and Future of Safety Management. Farnham, UK: Ashgate.

[14] IMO 2020a IMO takes first steps to address autonomous ships. Consulted, February 2020, through http://www.imo.org/en/MediaCentre/PressBriefings/Pages/08-MSC-99-MASS-scoping.aspx

[15] IMO 2019 Interim guidelines for mass trials. MSC.1/Circ.1604. Consulted, January 2020 through https://www.register-iri.com/wp-content/uploads/MSC.1-Circ.1604.pdf.

[16] IMO 2010 International Convention on Standards of Training, Certification and Watchkeeping for Seafarers. Including the 2010 Manilla Seafarer Training Amendments. The International Maritime Organization, London.

[17] Johnson, M, Bradshaw, J M, Feltovich, P J, Jonker, C M, Van Riemsdijk, M B and Sierhuis, M 2014 Coactive design: designing support for interdependence in joint activity. J, of Human-Robot Inter, Vol. **3**, No. 1, 2014, pp. 43-69.

[18] Kaber, D B 2018 A conceptual framework of autonomous and automated agents. Theoretical Issues in Erg S Vol **19**/4.

[19] Kongsberg 2020 Autonomous ship project, key facts about Yara Birkeland. Consulted, August 2020 through: https://www.kongsberg.com/maritime/support/themes/autonomous-ship-Project-Key-Facts-About-Yara-Birkeland/

[20] Licklider, J C R 1960 Man-computer symbiosis. IRE Trans on Human Factors in Elec, 1(maart), pp. 4–11. Consulted, 19 June 2017 through http://worrydream.com/refs/Licklider%20-%20Man-Computer%20Symbiosis.pdf.

[21] Lutzhoft, M, Hynnekleiv A, Earthy J V and Petersen E S 2019 Human-centred maritime autonomy - An ethnography of the future *J. Phys.: Conf. Ser.* **1357** 012032

[22] Maritime UK 2020 Maritime Autonomous Surface Ships - UK Code of Practice. Consulted, February 2020 through: https://www.maritimeuk.org/media-centre/publications/maritime-autonomous-surface-ships-uk-code-practice.

[23] Neerincx, M A 2003 Cognitive task load analysis: allocating tasks and designing support. In Handbook of Cognitive Task Design, Erik Hollnagel (ed.). Published by Lawrence Erlbaum Associates.

[24] Post, W.M, Van Den Broek, J 2005 Situationeel ontwerpen van teams [Situational team design]. TNO-rapport. TNO-DV3 2005 A089. TNO, the Netherlands.

[25] Prison, J, Dahlman, J and Lundh, M 2013 Ship sense—striving for harmony in ship maneuvring. WMU J Marit Affairs **12**, 115–27 https://doi.org/10.1007/s13437- 013-0038-5

[26] Rødseth, Ø J, Nordahl, H 2017 Definition of autonomy levels for merchant ships. Technical Report Norwegian Forum for Autonomous Ships, 2017-08-04

[27] Sheridan, T B 2012 Human supervisory control. In Handbook of Human Factors and Ergonomics, Fourth Edition (ed G. Salvendy), John Wiley & Sons, Inc., Hoboken, NJ, USA. Doi: 10.1002/9781118131350.ch34

[28] Sheridan, T B 2011 Adaptive automation, level of automation, allocation authority, supervisory control, and adaptive control: Distinctions and modes of adaptation. IEEE Trans. on Sys. Man and Cybernetics - Part A Systems and Humans **41**(4):662 - 667.

[29] Stanton N A, Stewart R, Harris D, Houghton R J, Baber C, Mcmaster, R, Salmon P M, Hoyle G, Walker G H, Young M S, Linsell M, Dymott R and Green, D 2006 Distributed situation awareness in dynamic systems: theoretical development and application of an ergonomics methodology. Ergonomics **49** (12-13):1287 – 1311.

[30] St. John, M F 2013 Janus Design Principles and the Acquisition Process for a Supervisory Situation Awareness Display, Human Factors and Ergonomics Society Annual Meeting 57/1, pp.2002- 2006, SAGE Publ.

[31] Strobach, T, Wendt, M, & Janczyk, M 2018. Multitasking: Executive Functioning in Dual-Task and Task Switching Situations. Place of publication not identified: Frontiers Media SA.

[32] Van Den Broek, J 2019 JIP Autonomous Shipping – WP5 Report: Design for Unmanned Operations. TNO-report. TNO 2019 R12184. TNO, the Netherlands.

[33] Van Den Broek, J, Van Diggelen, J, Van Der Kleij, R, Hueting, T F, Van Der Waa, J, Van Schendel, J A and Langefeld, J J 2017 Adaptive Maritime Automation: Final Report. TNO-report. TNO 2017 R11325. TNO, the Netherlands.

[34] Van Den Broek, J, Van Delft, J H, Essens, P J M D, Helsdingen, A S and Meessen, W R M J 2004 Eindrapport toekomstige maritieme commandovoering [Final report future maritime Command & Control]. TNO-rapport. TM-04-A068, TNO, the Netherlands.

[35] Van Der Kleij, R, Hueting, T and Schraagen, J M 2018 Change detection support for supervisory controllers of highly automated systems: Effects on performance, mental workload, and recovery of situation awareness following interruptions. Int J of Ind Ergo **66** 75-84. doi.org/10.1016/j.ergon.2018.02.010.

[36] Wahlström, M, Hakulinen, J, Karvonen H and Lindborg, I 2015 Human factors challenges in unmanned ship operations – insights from other domains. Procedia Manuf, **3**, pp. 1038-45. Doi: 10.1016/j.promfg.2015.07.167, Elsevier B.V.

[37] Woods, D D 2006 Essential Characteristics of Resilience. In: Hollnagel, E., Woods, D.D. and Leveson, N., Eds., Resilience Engineering: Concepts and Precepts, Ashgate Publishing Ltd., Aldershot, 21-34.

IOP Conf. Series: Materials Science and Engineering **929** (2020) 012009 doi:10.1088/1757-899X/929/1/012009

Potential benefits of 5G communication for autonomous ships

Stig Petersen[1], Pål Orten[2] and Bård Myhre[3]

[1]SINTEF Digital, Trondheim, Norway

[2]Kongsberg Maritime, Kongsberg,Norway

[3]SINTEF Digital, Oslo, Norway

Corresponding author: stig.petersen@sintef.no

Abstract. 5G, the next generation of mobile networks, aims to deliver wireless connectivity for new applications beyond traditional connectivity and mobile broadband for the smartphone market. For the maritime industry, 5G has the potential to address a wide range of usage scenarios within ship-to-ship, ship-to-shore, and onboard communication. In this paper, we give an introduction to mobile network technologies and systems in general, as well as an overview of technical capabilities and novelties introduced by the forthcoming 5G specifications relevant for maritime operations. Furthermore, perspectives related to applications and opportunities enabled by 5G networks are discussed, along with the potential benefits for unmanned and autonomous ships.

1. Introduction

In recent years, technological advances in the fields of electronics, automation, and machine learning have been the foundation for the introduction of digitalization, Internet of Things, and autonomy in industrial domains. This also includes the maritime industry, where "smart ships" and "smart harbors", as well as remote-controlled, unmanned, and autonomous ships are being introduced. These new concepts may disrupt the whole maritime value chain, redefine the roles within the industry and provide new business opportunities for shipyards, equipment manufacturers and ship owners.

Advances have also been made in communication technologies, and we have seen new standards and technologies addressing different types of communication, ranging from low power sensor networks to high capacity systems with significantly longer range. Common for previous generation of technologies is that they were targeted at specific applications, and custom made for that application. Examples are 3G and 4G targeting mobile broadband communications, and WirelessHART [1] which is custom designed for industrial process automation applications with low capacity demands and high volume of sensors with more deterministic behavior than for instance the very similar ZigBee standard [2]. This traditional way of thinking was challenged in 2015, when the International Telecommunication Union (ITU) released a vision-document, detailing a framework and overall objectives for the future development of International Mobile Telecommunications (IMT) for 2020 and beyond [3]. In this vision, ITU has for the first time expanded the scope of mobile communication systems to include emerging new use cases beyond traditional connectivity and mobile broadband for the consumer and business market. Example applications include wireless control within industrial manufacturing or production process, remote medical surgery, distribution automation in smart grids, and transportation. Mobile networks addressing these objectives are

IOP Conf. Series: Materials Science and Engineering **929** (2020) 012009 doi:10.1088/1757-899X/929/1/012009

formally classified as IMT-2020 systems but are more commonly referred to as 5G.

With 5G, mobile communication networks will, for the first time, address use cases beyond mobile voice and broadband for the consumer and business market. Examples of target applications include industrial control systems, remote medical surgery, smart grid, and transportation. We believe this technology has a potential of taking a key role also in maritime communications in general, and for autonomous ships in particular since such ships will require both high capacity communications as well as highly reliable and low-latency sensor communications.

The structure of the paper is as follows; section 2 gives a short introduction to mobile communication components and technology in general, the target technical capabilities of 5G communication are presented in section 3, while section 4 summarizes relevant novelties in 5G compared to the previous generations of mobile networks. The potential benefits of 5G for the maritime domain are analyzed in section 5, classified into ship-to-shore communication, ship-to-ship communication, and onboard communication, and conclusions and further work are discussed in section 6.

2. Mobile communication components

Mobile communication networks operate in licensed frequency bands which are regulated by national authorities. The telecom operators must typically purchase licenses for the use of the frequency bands through national auctions. In addition to the exclusive rights for operation in a frequency band, a license also includes an obligation to provide a certain nationwide network coverage. When multiple telecom operators purchase licenses in a country, the frequency bands are divided into sub-bands to avoid interference between operator networks. Countries with more than one telecom operator will thus end up with a redundant mobile communication infrastructure, one for each license owner.

A mobile communication network provides wireless connectivity to end devices called *User Equipment* (UE), which is typically a mobile phone. The wireless coverage is provided by base stations called *Radio Access Networks* (RAN). The area covered by a single RAN can be referred to as a cell, which is the origin of the term cellular networks which is often used interchangeably with mobile networks. The RANs of a telecom operator are connected to the operator *Core Network*, which is responsible for network and data management, user authentication, routing, subscriptions, billing, and more. A (simplified) schematic description of the basic components in a mobile network is shown in Figure 1.

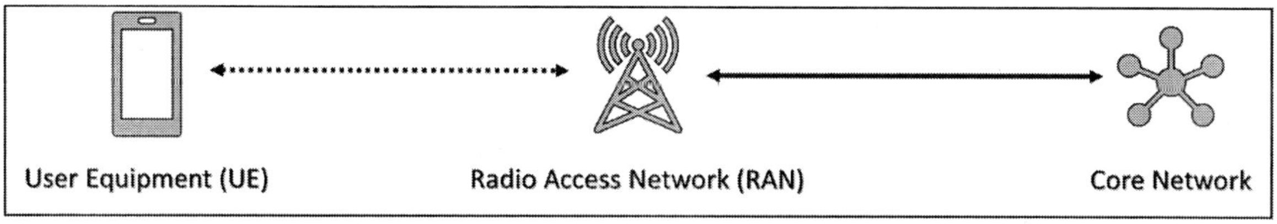

Figure 1. Basic components in a mobile network

To handle network access and user authentication, the telecom operators issue Subscriber Identify Module (SIM) cards to their subscribers. User Equipment (UE) with a valid SIM card can connect to a local RAN and thus make and receive phone calls and send/receive data. From a data flow perspective, all connections, calls and data transfers are routed through the centralized operator Core Network, regardless of the location of the sender and recipient.

3. 5G technical capabilities

As mentioned in section 1, ITU's new vision for IMT-2020 is pushing the telecom industry to address new applications beyond traditional connectivity and mobile broadband for the smartphone market. To easier distinguish between different applications and their requirements, IMT-2020 defines a set of three usage scenarios [4]:

- Enhanced massive broadband (eMBB)
- Massive machine type communication (mMTC)
- Ultra-reliable low latency communication (URLLC)

Due to the diversity of the applications introduced by these new usage scenarios, future IMT-2020 mobile networks must be able to offer a broad variety of capabilities in order to fulfil the various use case requirements. To ensure adequate performance of the future mobile networks, IMT-2020 identifies a set of key capabilities and their target values, which are beyond what current generations of mobile networks can deliver. These key capabilities are listed in Table 1.

Table 1. IMT-2020 key capabilities for future mobile networks [3]

Key capability	Target value
Peak data rate	20 Gbit/s
User experienced data rate	100 Mbit/s
Latency	1 ms
Mobility	500 km/h
Connection density	10^6 devices per km^2
Area traffic capacity	10 $Mbit/s/m^2$
Energy efficiency	100x *(compared to 4G)*
Spectrum efficiency	3x *(compared to 4G)*

It is important to note that it is not possible to achieve all key capabilities simultaneously through a single service or frequency range. New spectrum resources are therefore being assigned for 5G networks, with initial pioneer bands being 700 MHz, 3.6 GHz and 26 GHz bands.

The three usage classes, and their requirements, are presented in more detail in the following sections.

3.1. Enhanced mobile broadband (eMBB)

The main target of eMBB is mobile broadband for the consumer and business market, with the aim of providing high-speed access to multi-media content, services, and data. eMBB can be considered an evolution of 4G, with key capabilities focusing on improved data rates and network capacity.

Selected minimum technical performance requirements for eMBB as defined by IMT-2020 are listed in Table 2.

Table 2. Key requirements for eMBB [4]

Key capability	Target value
Peak data rate	20 Gbit/s downlink 10 Gbit/s uplink
Peak spectral efficiency	30 bit/s/Hz downlink 15 bit/s/Hz uplink
User experienced data rate	100 Mbit/s downlink 50 Mbit/s uplink
Area traffic capacity	10 Mbit/s/m2
Latency	4 ms
Mobility	Up to 500 km/h
Mobility interruption time	0 ms

3.2. Massive machine-type communication (mMTC)

The usage scenario mMTC addresses Internet of Things (IoT), which can be defined as a network of physical objects which communicate and collaborate to generate and share information. IoT consists of so-called 'smart' objects embedded with sensors, electronics, software, and wireless communication. IoT covers a large variety of technologies and applications, both in public, private and industrial domains. Key capabilities for mMTC include support for a very large number of connected devices typically transmitting a relatively low volume of non-critical data. The devices are furthermore required to be low cost and have very long battery life.

The key minimum technical performance requirement for mMTC as defined by IMT-2020 is listed in Table 3.

Table 3. Key requirement for mMTC [4]

Key capability	Target value
Connection density	10^6 devices per km^2

3.3. Ultra-reliable low latency communication (URRLC)

URLLC shall provide fast and reliable communication for use cases within e.g. industry, transport, medical and power generation and distribution. Typical requirements are low latency combined with high availability and reliability.

The key minimum technical performance requirements for URLLC as defined by IMT-2020 are listed in Table 4.

Table 4. Key requirement for mMTC [4]

Key capability	Target value
Latency	1 ms
Mobility interruption time	0 ms
Reliability	99.99%

4. 5G novelties

As discussed in the previous section, IMT-2020 defines three usage scenarios for mobile networks (eMBB, mMTC and URLLC), each targeting applications with different requirements. To achieve the target key capabilities listed in Tables 1,2,3 and 4, 5G will introduce a set of novelties which are beyond the capabilities of previous generations of mobile network technology. The novelties we believe are more relevant for maritime applications are described in the following sections.

4.1. Network slicing

In 4G and earlier generations of mobile networks the available network capacity is in principle shared equally amongst all users connected to a specific RAN. This means that the experienced Quality of Service (QoS) for a specific user will change depending on the activity level of other users in the vicinity. For more critical applications such as factory and process automation, health, transport, and energy, a non-deterministic QoS performance of a wireless network is not acceptable. To address this challenge, a local 5G RAN will have the capability to support multiple service classes with different requirements through a new feature called *network slicing*. A legal entity (e.g. a public or private company, enterprise, or organization) can negotiate QoS agreements with the telecom operator to reserve bandwidth and resources and get a private "slice" of the network in order to get guaranteed communication performance. A legal entity which acquires a network slice is called a *tenant* in 5G terminology. An example of network slicing is illustrated in Figure 2, where an industrial robot can have dedicated QoS despite other user traffic from mobile phones on the same RAN.

IOP Conf. Series: Materials Science and Engineering **929** (2020) 012009 doi:10.1088/1757-899X/929/1/012009

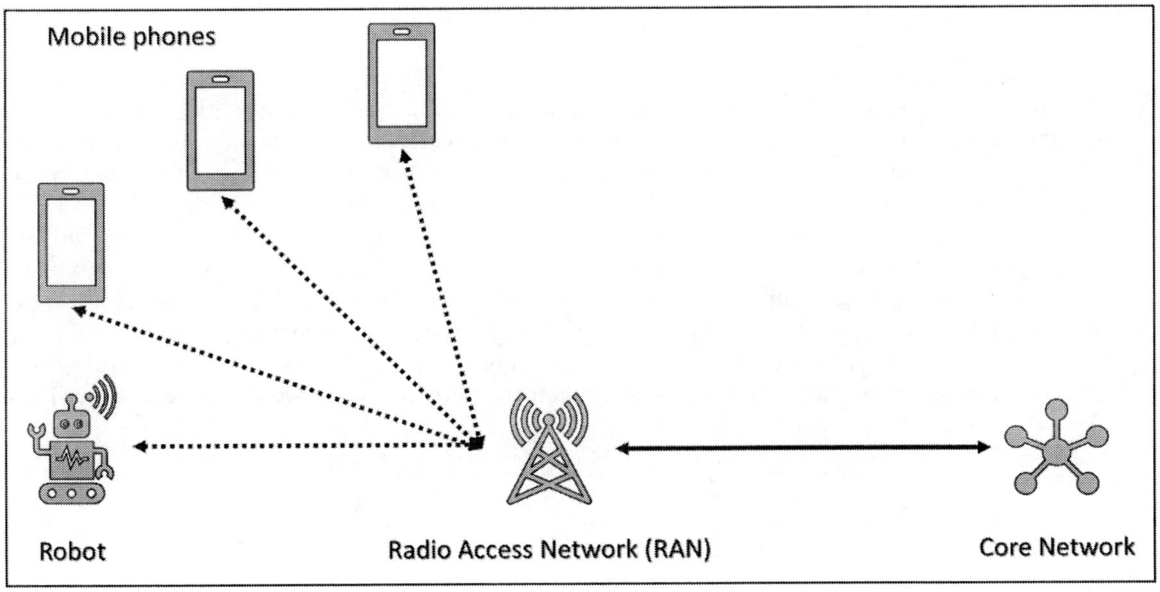

Figure 2. Network slicing

4.2. Edge computing

As described in section 2, data traffic in 4G and earlier generations of mobile networks are routed through the centralized Core Networks of the telecom operators. While this is of no specific relevance in traditional consumer smartphone applications, this model is not always well suited for industrial and maritime applications due to the following drawbacks:

- Sensitive production data is not under complete control of the data owner as it is transmitted from the RAN to the Core Network.
- Routing data through the Core Network introduces unnecessary latency, making the technology potentially unsuitable for low latency applications.

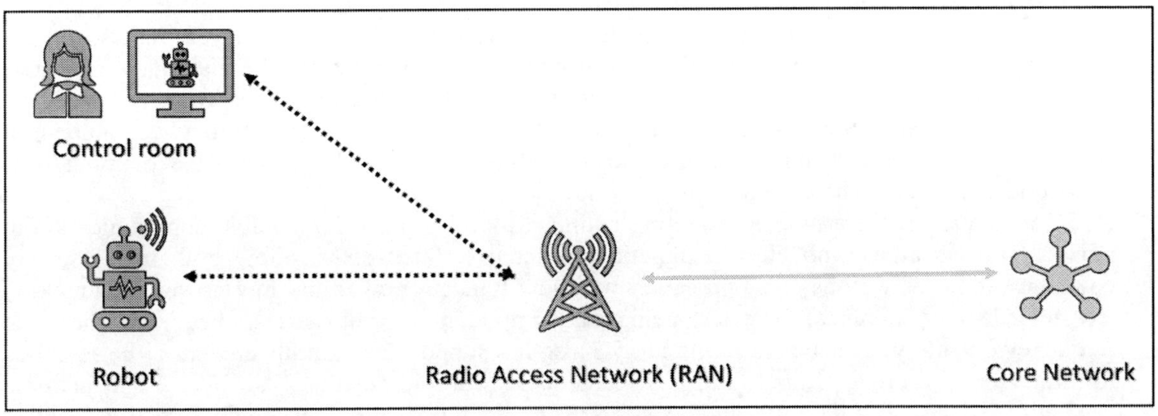

Figure 3. Edge computing

To address these issues, 5G will have the possibility for local network management, thus bypassing the need for all data to go through the Core Network. This feature is called *edge computing*, and it allows for a simplified private Core Network to be installed locally, as illustrated in Figure 7.

Latency will be reduced, and the owner will have more control of valuable operational data (provided the RAN does not also send data to the operator Core Network).

4.3. Private frequencies

The frequencies dedicated to mobile networks are managed by national authorities, and licenses for operation are traditionally auctioned off to telecom operators, under the obligation to build a nationwide infrastructure providing an adequate area and population coverage. To ensure a more open competition and hopefully enable 5G connectivity in broader range of applications and domains, ITU has recommended that 5G should allow for specific frequencies to be reserved for local, geographically limited, frequency allocations. A licence to operate in these frequencies can thus be purchased by actors other than the national telecom operators. Support for local frequency allocations must naturally be approved and managed by national telecommunication authorities.

The combination of edge computing and local frequency allocation will allow industrial actors to purchase a local frequency license and deploy their own private 5G networks, as depicted in Figure 4.

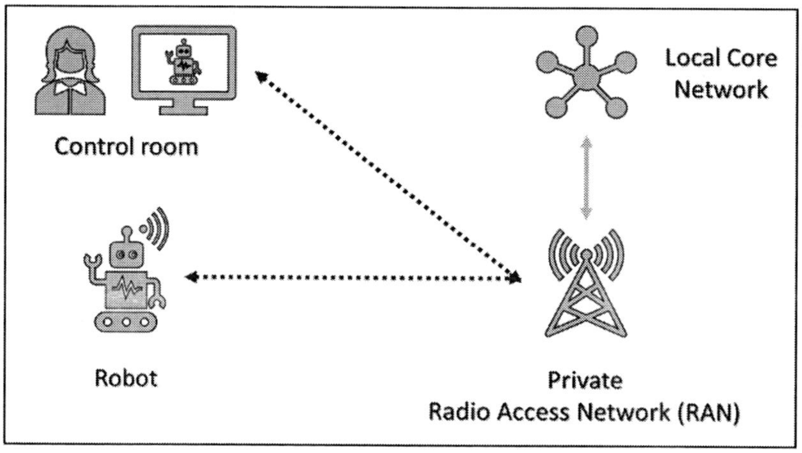

Figure 4. Private network

4.4. Radio communication and signal processing

We have already mentioned network slicing dedicating a set of radio and network resources to a user or an application enabling more deterministic performance and capacity. There are also other new and/or state-of-the art methods introduced in 5G to facilitate both even higher capacity broadband communications, and ultra-reliable low latency, and communications to and from a huge amount of extremely low power consuming sensors. The aim in this paper is not to go in detail explaining these techniques, but we give here a very brief overview.

5G uses state of the art error correcting coding which is crucial for reliable communication in noisy areas and areas with electromagnetic interference. Such codes allow both high capacity communications as well as communication with low transmit power and low power consumption (for long battery lifetimes). The last being very important in industrial applications where the frequency of battery replacement is critical. 5G also has support for multiple antennas, the so-called Multiple Input Multiple Output (MIMO) techniques. Multiple antennas can be used to increase capacity, to increase reliability, and to steer the radio propagation in certain directions, called beamforming. Furthermore, 5G can be applied in many different radio frequency spectrums. Each frequency range has different properties, typically lower frequency having higher range than higher frequencies, whereas at higher frequencies, although shorter range, there is larger available useful bands that give higher possible data rates. Another feature of 5G is that it supports non-orthogonal multiple access. This means that a transmitter is not allocated a certain frequency slot or a certain time slot for transmission. Instead, the transmitted waveform is constructed and coded such that advanced signal processing in the receiver will be able to distinguish the different data despite "colliding in the air". This has some capacity gain, but most of all at gives the opportunity to transmit without prior signaling to allocate a dedicated channel. This saves capacity, and even more importantly, it reduces time until transmission since prior signaling to reserve a channel is not necessary.

5. 5G communication for autonomous ships

The ongoing digitalization of the maritime industry opens new opportunities for products, applications, and services, but the full potential can only be gained if an efficient, flexible, and secure communication network for exchange of various types of data is in place. In the maritime area we see several dimensions that pave the way for new innovations within three different communication scenarios: ship-to-shore communication, ship-to-ship communication, and onboard communication. The next sections address the potential of 5G solutions within each of these three categories.

5.1. Ship-to-shore communication

Traditional bi-directional ship-to-shore communications for operation and leisure is a requirement for most ships regardless of type and mission. Future autonomous operations will put new requirements on this communication, where a secure and reliable connection to a Shore Control Centre (SCC) is needed for supervision and remote operation of unmanned and autonomous ships. In addition to navigation and control information, it is expected that various sensor feeds from the ships, such as live audio/video, radar, and LiDAR, must be available to the SCC operators.

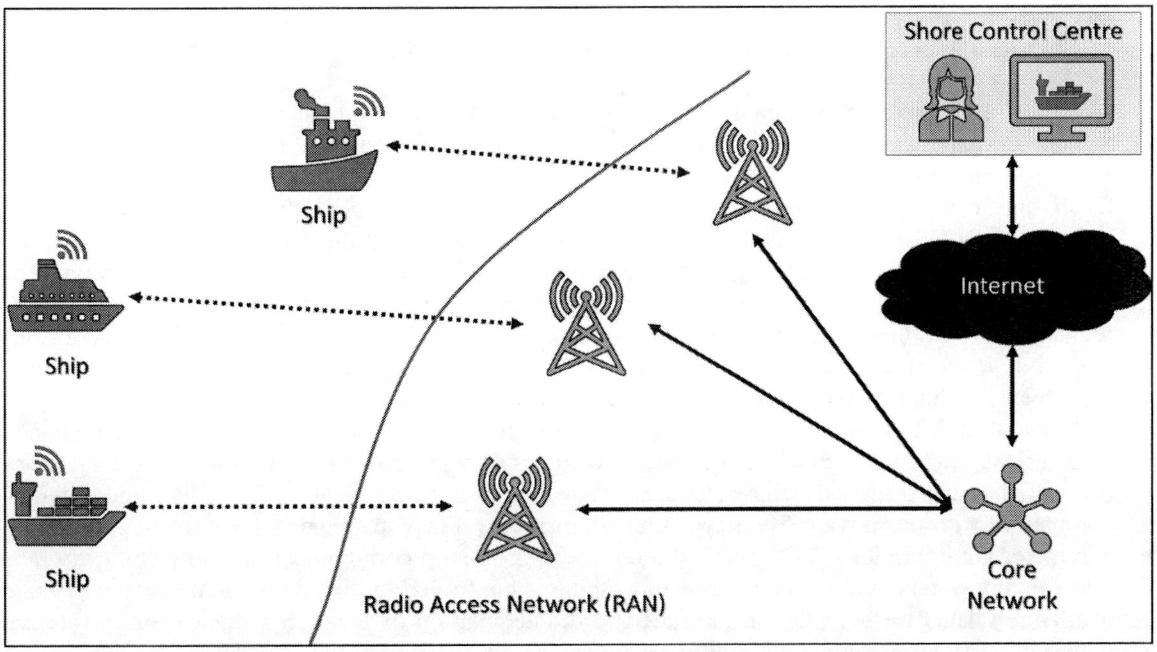

Figure 5. Ship-to-shore communication using 5G infrastructure

The applicability of using 5G for ship-to-shore communication is highly dependent on the type of maritime operation. Mobile communication networks are built on the premises of providing coverage by having many RANs, each with a communication range in the order of a few kilometers. Ship-to-shore communication using 5G is thus only relevant for ships operating in coastal-near waters or inland waterways. This includes quays and harbors with 5G infrastructure as well. For coastal-near waters, it will potentially require dedicated RANs located along the coast, with antennas directed towards the sea. For inland waterways, coverage will be provided by national 5G infrastructure in urban areas, while some remote rural areas could end up without 5G coverage. An example of ship-to- shore communication using 5G is illustrated in Figure 5. Here, several ships are in range of a set of land-based RANs, and the 5G infrastructure provides internet access (and communication to a SCC) through the Core Network of the 5G network operator. Note that is this scenario, each ship is considered as a User Equipment (UE) in 5G terminology. In order to communicate with a RAN, the ship must have a 5G UE modem with an antenna, as well as SIM card from the relevant network operator. For ships operating across international borders, it will be necessary to facilitate international roaming agreements with telecom operators.

From a communication requirement perspective, all communication between ship and shore must have high reliability and low latency. For navigation and control data, the date rate requirement is low, while the sensor feeds, on the other hand, require a high data rate. Ship-to-shore communication is thus expected to be best addressed with 3.6 GHz eMBB (see Table 2 for key eMBB requirements), although communication range may be a challenge. To get increased range, the 700 MHz band could be considered, but this will come at a cost of data rate, as this frequency is primarily dedicated for mMTC and IoT applications with many devices with low data rates.

5.2. Ship-to-ship communication

Communication between ships is relevant for scenarios like simultaneous operations and navigation in narrow channels and waterways and has the potential to increase efficiency and traffic safety when ships are in the vicinity of each other. Ship-to-ship communication can also be used for non-operational data exchange, e.g. one ship can offer internet connection to another, and non-satellite communication to shore can be extended by setting up mesh networks between multiple ships in coastal-near waters.

Ship-to-ship communication with 5G can be solved in two ways:

A. Communication through a land-based 5G infrastructure as shown in Figure 5.
B. Communication through a private 5G network located onboard a ship as shown in Figure 6.

In option A, the communication between two ships are routed through a land-based 5G infrastructure. This has inherent drawbacks as the ships must be within range of a RAN on land, and relevant scenarios are thus limited to operations in coastal-near waters and inland waterways. The upside to this solution is low requirement on communication infrastructure, as each ship only needs to install a 5G UE modem and antenna with a SIM card from the relevant telecom operator. For ships operating across international borders, it will also be necessary to facilitate international roaming agreements with telecom operators.

For option B, a *primary* ship participating in the ship-to-ship communication must be equipped with infrastructure for a private 5G network, comprising a private RAN and a local Core Network connected to the bridge systems. The *secondary* ships only need a 5G UE modem and antenna, compatible with the private 5G network of the primary ship with regards to frequency and radio parameter configurations. With this solution, the ship-to-ship communication can technically take place in any waters, and the communication setup is controlled by the ships themselves. The main challenge related to this solution is related to the acquisition of local frequencies for private 5G networks. Current regulations by national authorities in different counties reserves frequencies on either a national level for the telecom operators, or on a regional/local level for private 5G networks within a *geographically limited area*. So far, the concept of a mobile RAN has not been addressed, and it is still an open issue whether it will be resolved soon. The complexity of this situation is further increased if ships are in international traffic, as regulations for private 5G networks are handled different in different countries.

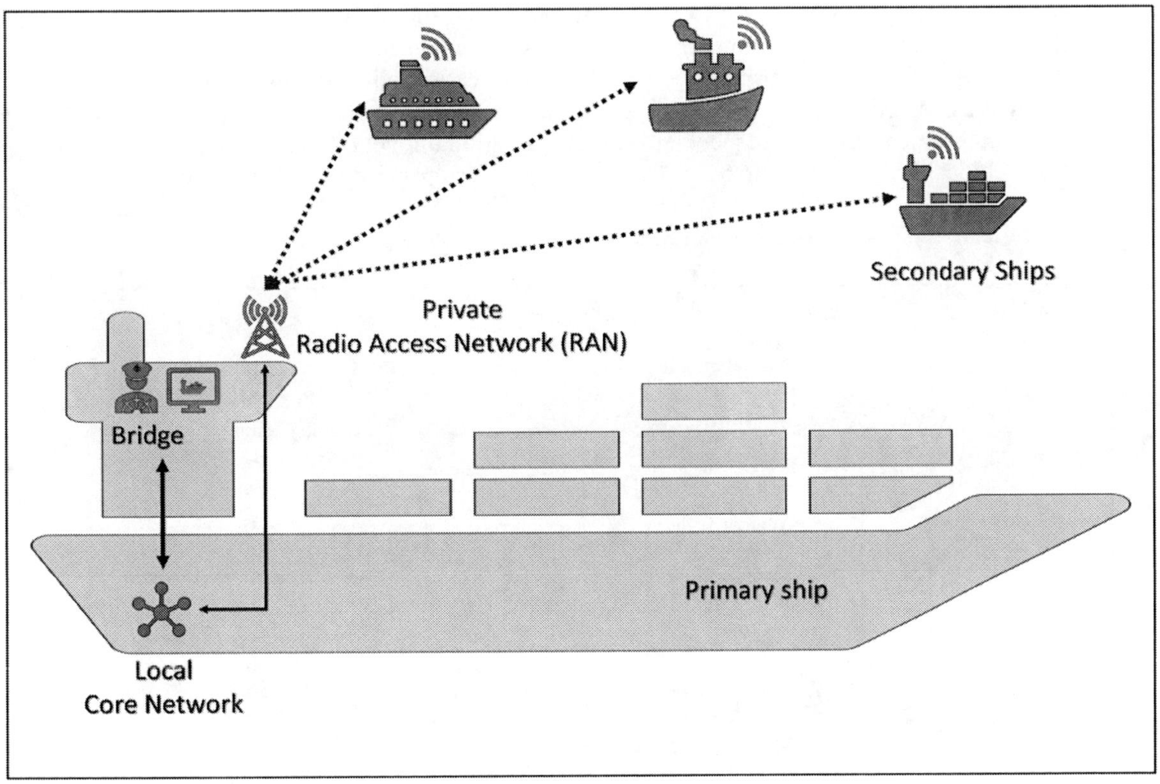

Figure 6. Ship-to-ship communication using a private 5G network

From a requirement perspective, using ship-to-ship communication for simultaneous operations and navigation in narrow channels and waterways will require a communication channel with high reliability and low latency. The operations require ships to exchange data related to position, orientation, heading, speed etc., but the bandwidth requirements can be considered low. From a 5G usage scenario perspective, 3.6 GHz eMBB is the preferred communication configuration. To get increased range, 700 MHz mMTC can be considered, as the data rates should still be sufficient for the envisioned applications.

5.3. Onboard communication

Unmanned and autonomous ships will have increased demands of sensors and instrumentation onboard the ships. In this regard, wireless communication can bring benefits such as simplified engineering and commissioning, reduced cabling, and increased flexibility. The usage areas for wireless communication onboard vessels are many, including:

- Sensor feeds for SCC operators (e.g. cameras, radar, LiDAR)
- Non-critical monitoring (e.g. wind, temperature, humidity, weather)
- Autonomous operation support (e.g. loading crane, mobile robots)
- Instrumentation for automation and control systems (e.g. sensors and actuators in machine room)

For onboard communication, a private 5G infrastructure is the only sensible architectural choice. The local Core Network can be connected to bridge systems, while one or more private RANs provide wireless coverage throughout the vessel. Since metallic structures block electromagnetic waves, multiple RANs may be required below decks, while a single RAN should be sufficient for above deck. An example of a private 5G network infrastructure for various applications onboard a ship is

IOP Conf. Series: Materials Science and Engineering **929** (2020) 012009 doi:10.1088/1757-899X/929/1/012009

illustrated in Figure 7. Here, one RAN provides coverage for equipment on deck, while a separate RAN is installed for wireless communication for sensors and actuators in the machine/engine room.

Using a private 5G network for onboard communication will be subject to similar regulatory challenges as described in the previous section on ship-to-ship communication since regulations by national authorities reserves frequencies on either a national level (for the telecom operators), or on a regional/local level for private 5G networks within a *geographically limited area*. The open issue on how to regulate frequency allocations for a mobile RAN must thus be solved before onboard communication using 5G becomes a reality.

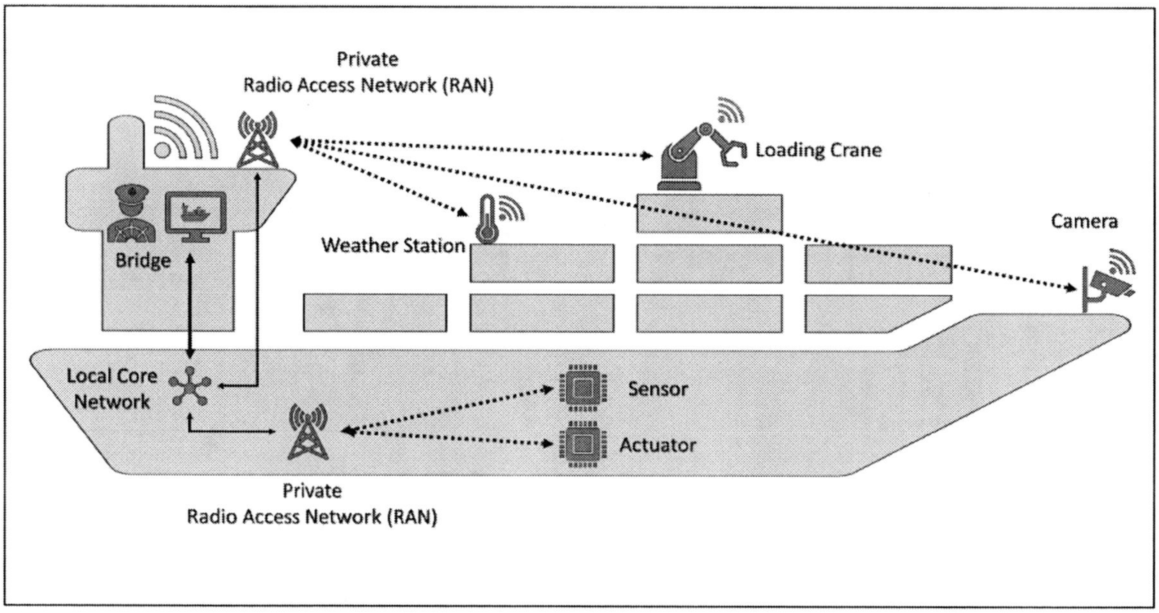

Figure 7. Using a private 5G network for onboard communication

The requirements for onboard communication vary depending on the usage area, and in this setting the flexible nature of 5G networks can be used in its entirety. eMBB with its high data rate can handle video, radar, and LiDAR streams, URLLC with high reliability and low latency is capable of supporting machine room closed-loop control and operation of autonomous cranes and robots, while mMTC can be used for non-critical monitoring where long battery life is desirable.

6. Conclusions

This paper has introduced target technical capabilities and novelties with 5G, the next generation mobile network technologies. Furthermore, the potential applicability of 5G within ship-to-shore, ship- to-ship and onboard communication has been analyzed, with the conclusion that 5G should be capable of addressing a large set of use cases and applications within the maritime domain. In addition, the possibility of having one technological framework for a multitude of applications is on its own a great benefit, especially compared to today's situation where maritime technology and solution providers have to provide equipment and support for close to one hundred different wired and wireless communication protocols.

At the time of writing, the main challenge with 5G is related to frequency regulations for private 5G networks. For onboard communication and many scenarios with ship-to-ship communication using national infrastructure is not feasible, and the only practical solution is to establish a private 5G network on the ship. However, the regulatory challenge related to frequency allocation for a mobile RAN must be addressed by the authorities. This is further complicated for ships in international traffic, which will encounter regulatory issues in many countries.

In addition to exploration of regulatory possibilities, a more thorough analysis and study of concrete business models and their requirements for various maritime 5G applications are a natural next step for further research.

Acknowledgments

The authors acknowledge the support of the H2020 GSA H2H project (Grant 775998), and the Sacomas project funded by the Research Council of Norway (Grant 269510).

References

[1] HART Field Communication Protocol Specification, Revision 7.0, HART Communication Foundation, Austin, Tx, Sept. 2007.
[2] ZigBee Specification, ZigBee Alliance, 2015.
[3] Recommendation ITU-R M.2083-0, "*IMT Vision – Framework and overall objectives of the future development of IMT for 2020 and beyond*", 2015.
[4] Report ITU-R M.2410-0, "*Minimum requirements related to technical performance for IMT-2020 radio interface(s)*", 2017.

LTE massive MIMO (Pre-5G) test with an updated boat user terminal solution for land-to-boat scenarios in Oslo fjord

Kun Yang*, Ning Zhou*, Terje Røste*
***Super Radio AS, NO-0556 Oslo, Norway.**
Email: kun@superradio.no

Abstract. Autonomous shipping is considered to be one of the most important technologies in the maritime industry. Remote control of autonomous shipping is supported by communication solutions featuring long coverage, high throughput and low latency. Since 2018, together with the Norwegian Internet Service Provider (ISP) Telia, we have performed tests with a massive MIMO (Pre-5G) LTE base station, using the 3.7 GHz band, in Horten close to the Oslo fjord. The first Pre-5G tests with user terminal mounted on a 40 feet sailboat performed in 2018, showed great potential for maritime applications in terms of throughputs [1]. The Customer Premise Equipment (CPE) used in 2018 was a LTE Pre-5G terminal from Telia for fixed LAN use, and a modified version of it in 2019. However, since both the base station and the user terminal/CPE are not originally designed and optimized for maritime radio propagation environments, the test results in 2018 unveiled that a CPE designed for land usage experienced limited coverage and instability caused by received unstable signal level. Since 2019, a new user terminal prototype dedicated and optimized for maritime propagation conditions, was designed by Super Radio AS. This new CPE was tested in autumn 2019 under the similar setups as in 2018. The test results of the Reference Signal Receive Power and throughput are analysed and compared the test results in 2018. The comparison proved that the new user terminal solution showed considerable improvement with respect to stability and throughputs in the harsh maritime environments, which means that the new designed solution can be a potential maritime terminal solution for the remote control of autonomous shipping. The prototypes mounted at different places of the sailboat also show different system performances at the same TX-RX distances, which indicates the potential of exploiting the system diversity in the similar propagation environments.

1. Introduction

Reliable and high throughput maritime communications are considered to play an important role in maritime activities involving various types of ships and vessels, offshore installations, unattended buoy platforms, autonomous underwater utilities, offshore and onshore observation sites, etc.
Autonomous ship and the digitalization of numerous maritime activities including oil exploitation, maritime transportation, fish farming, environmental surveillance and other activities need to be supported by communication solutions with low latency, long coverage, high throughput and low cost.

All these performance criteria drive the needs of broadband maritime communication technology. In 2018 the deployment of LTE base stations with Massive MIMO capability began in Norway. For maritime applications, the radio propagation conditions from land-to-sea will change considerably from the open sea to close to the coastline with narrow fjords, straits and islands. For maritime applications, the question is the following: Will conventional 5G massive MIMO systems originally designed for the terrain propagation environments (Urban, Suburban, etc) still demonstrate high performance for maritime applications? To answer this question, a pre-5G LTE Massive MIMO base station (BS) (8T8R)

provided by the MAMIME [2] project partner, Telia Norge AS, was installed, in Horten at a site close to the Oslo fjord. This BS installed by Telia for pilot test purposes, covers the 3.7 GHz band (band 43). The pre-5G, 8T8R BS supports the features beamforming soft split of beams, and beam adjustment like beamwidth adjustments in azimuth, elevation together with tilt adjustments. The BS during the tests had a fixed configuration that was optimized for the test routes of the ship on which the CPEs were mounted. LTE massive MIMO tests with CPEs mounted on a 40 feet sailboat were conducted in Oslo fjord in autumn 2018 and autumn 2019. In 2018 user terminals made for fixed installation on land were installed on the ship, and in 2019, optimized user terminals for the maritime propagation environment were installed. It also needs to be mentioned that the BS at Horten provides a very good coverage for a national authorized autonomous ship test area in the Oslo fjord outside Horten. This will provide the possibility for future cooperation between autonomous ship companies and the Norwegian research community to investigate pre-5G system performance for autonomous ship applications.
The paper is organized as follows: In Section II, a discussion of the land-to-sea propagation conditions for different scenarios is given. In Section III, selected Reference Signal Receive Power (RSRP) measurement results from 2018 and 2019, and corresponding analysis results are presented. In section V, the measurement results and analysis of throughput in Mbit/sec in 2018 and 2019, are presented. Finally, in section VI, conclusions are drawn, and further research work is suggested. In Table 1, a list of abbreviations and definitions used in this paper is given.

Abbreviation	Definition
RSRP	Reference Signal Receive Power
LOS	Line of Sight
LTE	Long Term Evolution
MIMO	Multiple in Multiple out
BS	Base Station
MS	Mobile Station
CPE	Customer Premises Equipment
MAMIME	LTE, WIFI and 5G Massive MIMO Communications in Maritime Propagation Environments
MIMO	Multiple In Multiple Out
8T8R	8 Transmit and 8 Receive antenna elements

Table 1. Abbreviations and definitions

2. Propagation conditions land to sea

2.1. Radio propagation between a BS and a MS on land has been subject to extensive research over many years. Models and RSRP prediction tools for coverage analysis have been developed and extensively used to roll out the different generations of mobile communications systems on land, including urban, sub-urban and rural areas. When it comes to the land-to-sea radio propagation, the research has been limited compared to terrain environments. The MAMIME research project [2] funded by the Research Council of Norway has been dedicated to developing a better understanding of the radio propagation between land and sea with focus on 4G and 5G mobile systems. Further, the gain of using MIMO for terrestrial mobile systems compared to mobile systems for the land-to-sea environments, may be different.
A fixed BS is placed at the coast. Land-to-sea mobile radio communications may be categorised into the following main scenarios:

- Scenario 1: A vessel (i.e. a MS) moving in fjords, narrow straits and in-between islands
- Scenario 2: A vessel in the open sea with LOS to the BS
- Scenario 3: A vessel in the open sea with non-LOS where the terrain (either island or hills on land) obstruct the radio propagation
- Scenario 4: A vessel in the open sea where the round earth will obstruct the propagation

- Scenario 5: A fixed object (i.e. a sensor either on the sea surface or a fixed installation) instead of a moving vessel and with the 4 above mentioned scenarios

In scenario 1, there will be propagation conditions close to rural areas on land, with LOS, non-LOS and reflections, either many or few reflections depending on the topography.

In scenario 2, the propagation will be a LOS component together with reflections and scattering from the sea surface. The movement of the ship will heavily depend on the ship size and sea wave size (i.e. weather conditions). The ship movement and the sea surface condition will cause small scale fading. The LOS component will change caused by ships angular movements (i.e. roll, pitch, yaw) and the antenna pattern properties. The reflections and scattering from the sea will change with the change of the part of the sea surface that reflect the radio waves towards the antenna. Finally, the antenna placement on ship is important.

With scenario 3, the ship will move in and out of conditions with non-LOS and LOS causing loss of communication due to shadowing. Diffraction loss will be significant. In addition there may be reflected components coming from nearby islands or hills along the coastline. A multiple element antenna may be configured to high gain steered towards the LOS component or the non-LOS component undergoing diffraction loss. The latter may be important when the 700 MHz band is used

With scenario 4, the ship is far out on the sea with LOS to the BS. In this case, the loss may be determined by the earth curvature. It is referred to [3] where a propagation model for this scenario is presented.

Scenario 5, is an important case, but is outside the scope of this paper.

As the measurements reported here were made by using a 40 feet sailboat, and with the topography in the Horten – Moss area of the Oslo fjord, Scenarios 2 and 3 are most relevant and will be focused hereafter.

Small scale fading will occur due to the angular and linear movements of the ship, antenna properties and its placement on the ship, the sea surface condition (wave height etc.), and sea surface movement surrounding the ship. Large scale fading will occur when the ship changes the distance between the ship and the BS along its route, or when the ship moves between LOS and non-LOS conditions caused by e.g an island.

As discussed above, with scenario 2, the impact of sea waves on propagation are heavily depending on the size of the ship, the placement of antennas and number of antennas. In our case with a small vessel, the ship movements are significant in sea conditions experienced in the Oslo fjord. Ideally, the antennas should be placed high up in the masts or ships construction, which is not always practical or possible. Further, with scenario 2, the reflections and scattering of radio waves from the sea surface sea, will change considerably with time causing RSRP small scale fading. When the sea waves become large, in case of a small vessel, the sea waves may even shadow the LOS component (See Fig. 1)). The following discussions and results keep this as a focal point.

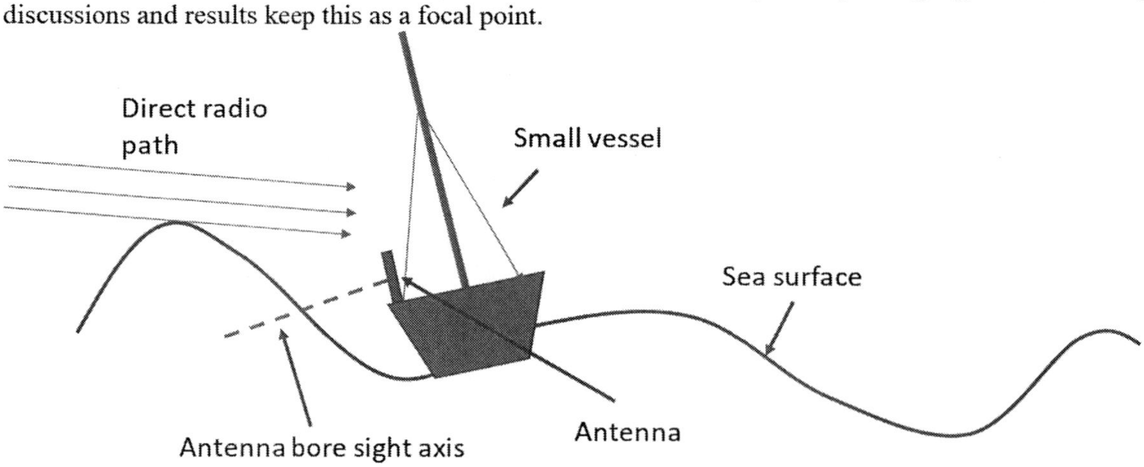

Figure 1. Small vessel on the sea surface. Possible shadowing by sea waves.

3. Result RSRP

The measurement campaign was described in detail in [1], from which the radio coverage prediction, campaign setups and facilities can be found. It needs to be mentioned that the CPE we used for test in 2019 has been updated according to our research of maritime radio propagation environments. The test routes in 2018 and 2019 shown in figures Fig 2 (a) and Fig 2 (b), respectively, are very similar, which makes it meaningful to compare with each other to demonstrate the enhancement of the updated CPE system performance.

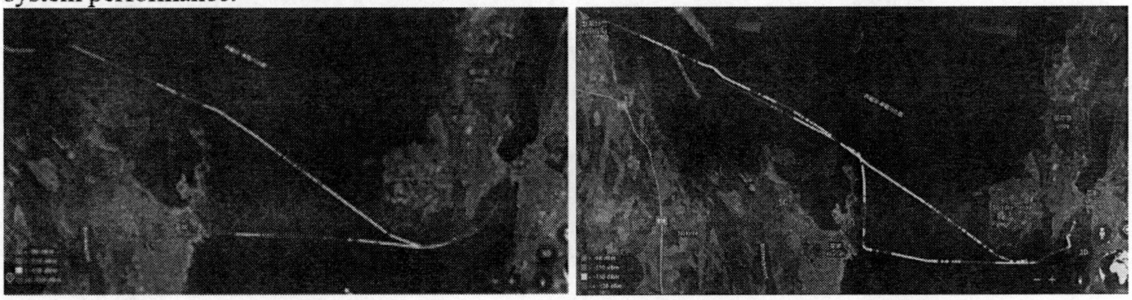

Fig 2. (a) Test route 1 in 2019 **(b) Test route 1 in 2018**

In the 2018 measurement campaign, two antennas were mounted one at the bow and one at the rear, 10 m apart, and both 1 m above the sea level. In the 2019 three 3 antennas were mounted on the ship. The placement of the antennas was the following:

- One at the bow 1 m above the sea level
- One at the rear ca 1 m above the sea level and ca 10 m apart from the bow antenna
- One up in the masts pillar ca 2 m above the sea level

During the test route 1 in 2019, the sea conditions were calm, and boat speed was about 5-6 knots. The wind conditions on that day was 1-7 m/sec. The wave height on that day, typical wave bottom to top around 0.25 meters. The measurement data structure and RSRP analysis method are described in detail in [1]. In this paper, only the RSRP results of outbound trips in 2018 and 2019 are shown and compared in Fig. 3, from which it can be found that the RSRP of the new CPE shows much better stability than the old CPE in the route 1. It needs to be mentioned that large fading of the new CPE RSRP at the TX-RX distance of 10.5 km is due to the blockage of the LOS when the sailboat moved into the bay area of Moss. Therefore, our comparison analysis focuses on the TX-RX distances between 4 km and 9.5 km. To have a better understanding of the stability improvement, we further study the small-Scale fading and the corresponding the standard deviation (STD), which are shown in Fig. 4 (a) and Fig. 4 (b), respectively. The staircase diagram of Fig. 4 (b) is caused by the 150 sec block length of the STD analysis time window. It can be found that the STD in 2019 is around 3-14 dB less than in 2018, which proves the great improvement on stability in our new updated CPE, compared with the old CPE.

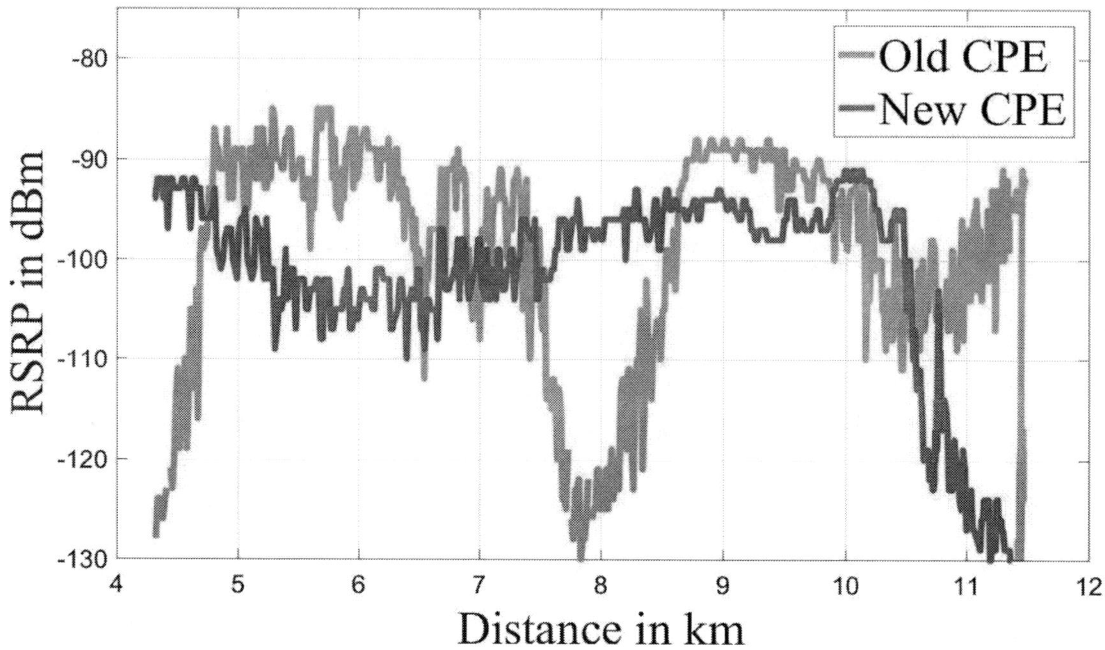

Fig 3. RSRP comparison between measurement data in 2018 and in 2019, route 1

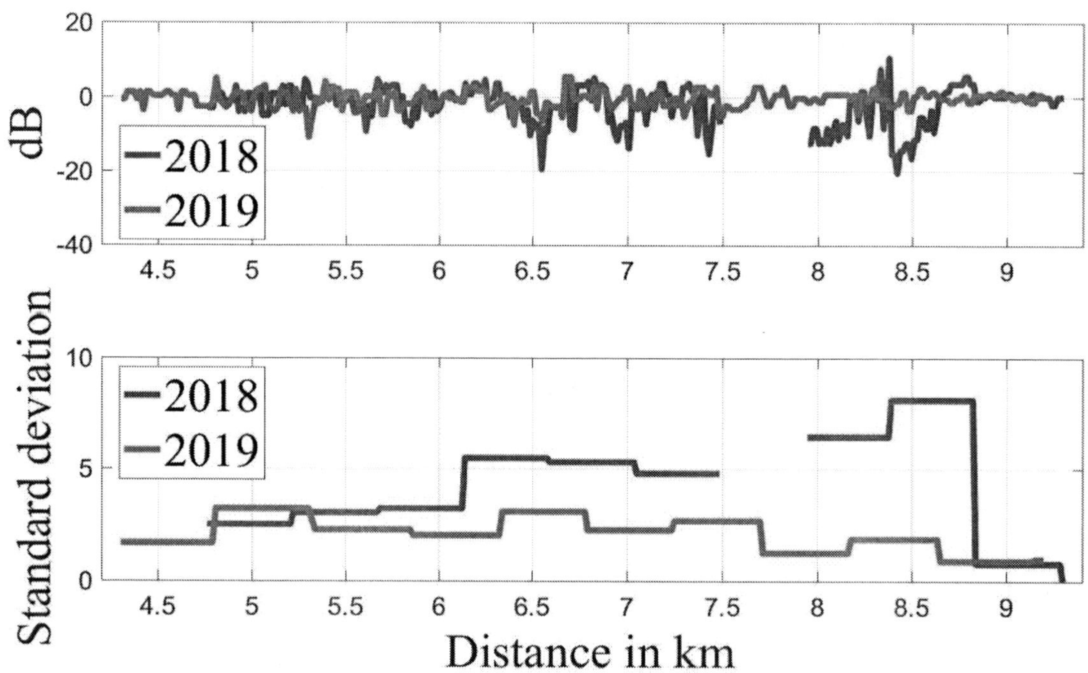

Fig4. (a) Comparison of Small-scale fading. (b) Comparison of standard deviation of the small-scale fading.

4. Results throughput

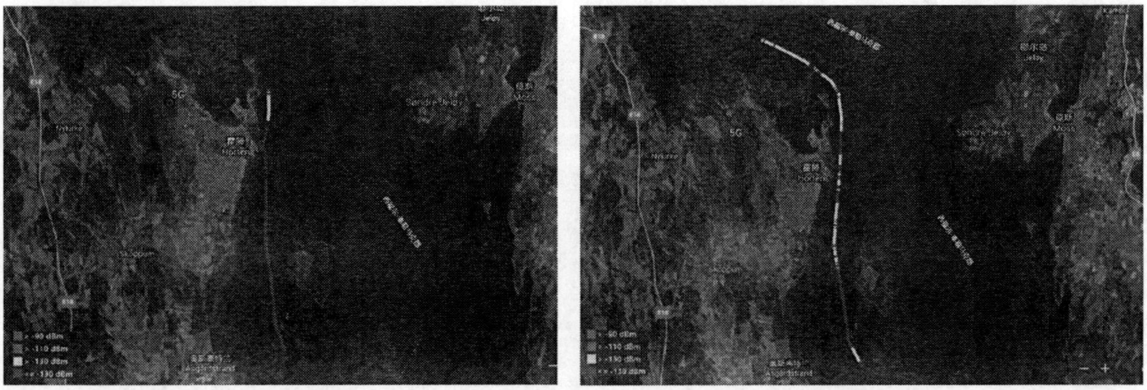

Fig 5. (a) Test route 2 in 2019 (b) Test route 2 in 2018

The system throughput (Mbit/sec) is one of the most important factors for the maritime applications. In our test campaign, we used Iperf which is an open-source software and widely used to measure the transmission throughput. It can measure the throughput between the two end users for both uplink and downlink. The transmission protocol can be set to either TCP or UDP. In this paper, the UDP throughput was recorded in the route 2 shown in Fig. 5, from which it can be found that the test routes 2 in 2018 and 2019 are close to the same so that we can use them for comparison. It also needs to be mentioned that the weather condition in route 2 in 2019 was very harsh with heavy rain and wind. The boat speed was about 6-7 knots. The wind conditions on that day was 6-16 m/sec. The wave height on that day, typical wave bottom to top around 0.75 meters . This is why the RSRP in 2019 is less than in 2018, which is shown in Fig. 6. Further study shown in Fig. 7 also indicates that the STD of the small-scale fading of the RSRP in 2019 is within [1, 7] dB while the STD in 2018 is within [4, 13] dB. Even though the test in 2019 was conducted in harsh weather, the stability of RSRP obtained in the updated terminal is higher the old terminal in 2018. The corresponding UDP throughput results are shown in Fig. 8, from which it can be found that the data rate in the new updated CPE terminal is around 90 Mbps while the data rate in the old CPE terminal is around 30 Mbps. It needs to be addressed that the stability of the throughput has been greatly improved as well. Since the CPE mounted on the top of the sail boat experienced much severe boat movement than the CPE mounted on the rear of the sail boat, the throughput variation in the new CPE on the top is larger than the new CPE on the rear, which is also can be found in the Fig. 8. It also suggested the difference of the CPE placement will caused different system performance, which may be exploited for the system diversity. It can be concluded that the stability is highly related to the throughput, which should be paid great attention in the maritime communication system design.

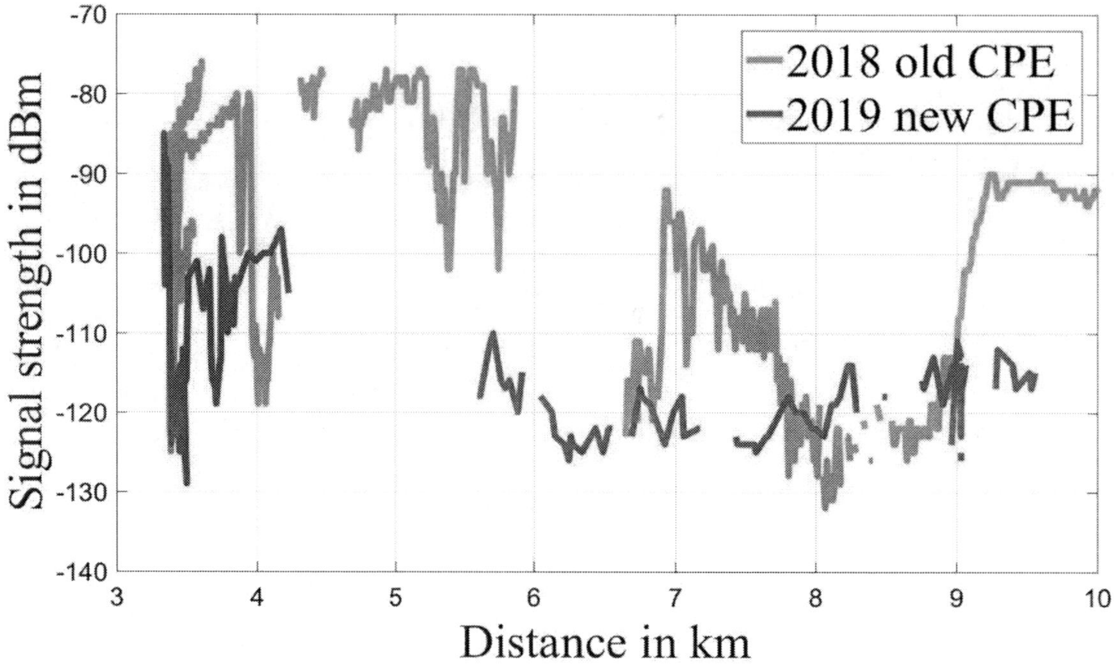

Fig 6. RSRP comparison between the route 2 measurement data in 2018 and in 2019

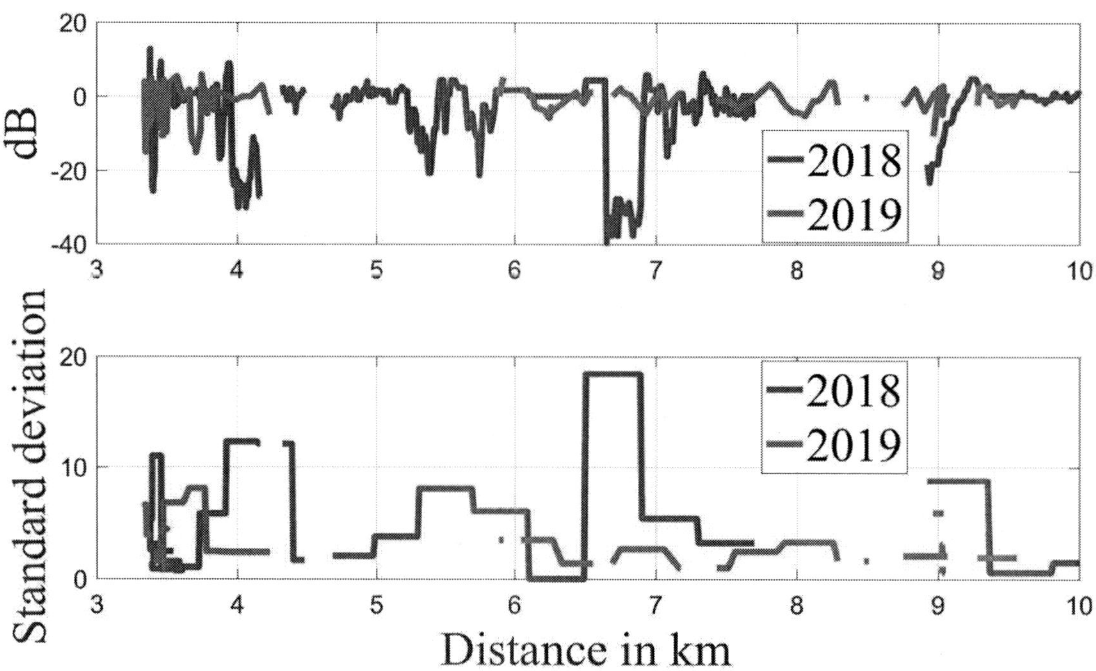

Fig7. (a) Comparison of Small-scale fading. (b) Comparison of standard deviation of the small-scale fading.

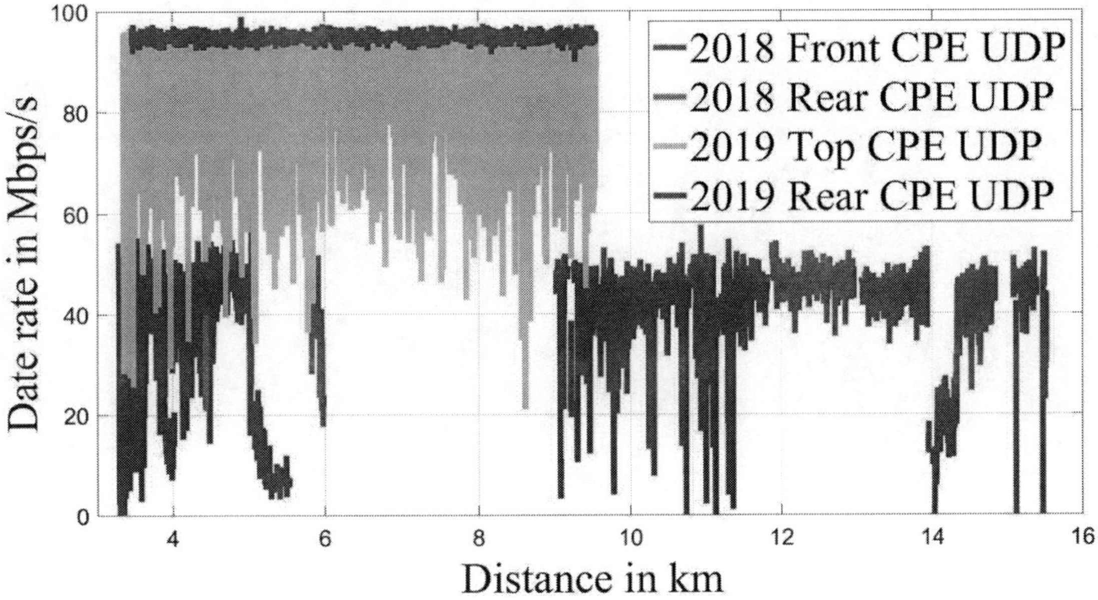

Fig 8. Throughput comparison between the route 2 measurement data in 2018 and in 2019

5. Conclusion

A Pre-5G LTE system measurement campaign with an updated CPE terminal at 3.7 GHz has been carried out in maritime propagation environment in Oslo fjord, Norway. The obtained RSRP is compared with the previous LTE test results obtained by using the old CPE terminal. The comparison of RSRP and the small-scale fading STD prove great enhancement on the stability by using the updated CPE terminal. Similarly, another comparison of the RSRP and STD in the Route 2 shows the improvement on the stability. Besides, the corresponding throughput is greatly increased, which further proves the importance of the stability and the effectiveness of the updated CPE terminal. The system diversity of the different CPEs on the same sailboat is also shown in the variation of the throughput large, which shows the potential of the multi-user diversity under the current solutions.

Acknowledgments

Telia Norge AS is acknowledged for providing their BS type (EasyRET 3.5 GHz 8T8R antenna) to the measurement campaign. Heiki Neskvern and John Eriksen are acknowledged for their support to set up and carry out the measurement campaigns in the Horten – Moss area. This work is supported by the MAMIME project under Grant Agreement No. 256309 by the Norwegian Research Council.

.

References

[1] Kun Yang, Ning Zhou, Terje Røste, Heiki Neskvern and John Eriksen, "LTE massive MIMO (Pre-5G) test for land-to-boat scenarios in Oslo fjord," The 2nd International Conference on Maritime Autonomous Surface Ship (ICMASS 2019), Trondheim, Nov 2019.

[2] http://superradio.no/a/home/NATIONAL/

[3] K. Yang, A. F. Molisch, T. Ekman, T. Røste and M. Berbneau,"A Deterministic Round Earth Loss Model and Small-scale Channel Properties for Open-Sea Radio Propagation," IEEE Trans on vehicle Technology, 2019.

IOP Conf. Series: Materials Science and Engineering **929** (2020) 012011 doi:10.1088/1757-899X/929/1/012011

Emergency ship positioning and distribution system for radio disturbance

Seongmi Mun [1], Kwaingil Lee[2*]

[1] Seanus, 846, Jungang-daero, Busanjin-gu, Busan, Republic of Korea
[2] KMOU, 727, Taejong-ro, Yeongdo-gu, Busan, Republic of Korea

[*] E-mail: leeki@kmou.ac.kr

Abstract. Global Navigation Satellite System (GNSS) has been used as a key position, navigation and timing (PNT) input in applications or services provided by ships. However, GNSS is vulnerable to space-vehicle failures, solar disturbances, unintentional radio interference and deliberate jamming. Despite these weaknesses, GNSS has become a key element not only of maritime navigation, but also of critical national infrastructure, often without any backup being provided. In autonomous ships, the use of various integrated systems increases and the dependence on the location data of GNSS increases. Therefore, in the event of a GNSS failure, it is impossible to guarantee the safe operation of the ship and it has various accident possibilities. In this paper, we propose and describe the system newly that can collect PNTs from various equipment to detect anomalies, and measure and distribute PNTs from digital sextants, which are backup systems, when anomalies occur. In the future, we will verify the performance of the implemented system by a test with the full mission bridge simulator provided by the Kongsberg.

1. Introduction

As the need for autonomous ships is increasing worldwide, various research and development in related fields are underway. Accordingly, various technologies for the safe operation of ships have been developed. Global Navigation Satellite System (GNSS) is used as a key position, navigation and timing (PNT) input in applications or services provided by ships. However, GNSS is vulnerable to space-vehicle failures, solar disturbances, unintentional radio interference and deliberate jamming. Despite these weaknesses, GNSS has played an important role as a key component not only of maritime navigation, but also of critical national infrastructure, often without any backup being provided[1][2].

In autonomous ships, the use of various integrated systems increases and the dependence on the location data of GNSS increases. Therefore, in the event of a GNSS failure, it is impossible to guarantee the safe operation of the ship and it has various accident possibilities[3]. Therefore, although it is compatible with GNSS, it needs a complementary system independent of GNSS, and if an abnormality of GNSS is detected, a system that can provide the PNT and provide the information to systems that need it on board or off board is needed.

we propose and describe the system newly that can collect PNTs from various equipment to detect anomalies, and measure and distribute PNTs from digital sextants, which are backup systems, when anomalies occur.

2. Configuration of system

The proposed Emergency Ship Positioning and Distribution System (EPDS) enables rapid identification and use of ship positions without GNSS in radio disturbance situations. The system consists of Position Comparator (PC), Digital Sextant Unit (DSU) and Position Distributor (PD) as shown in Figure 1.

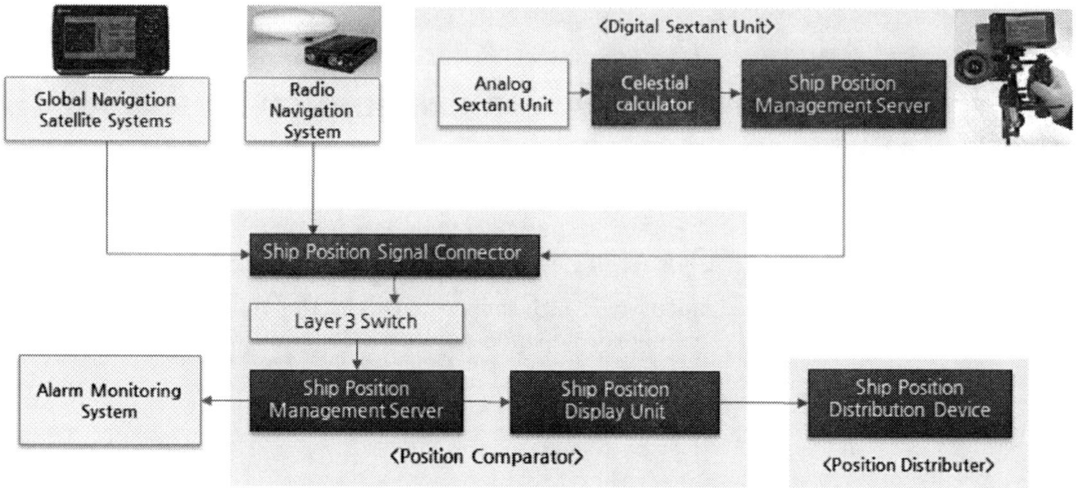

Figure 1. System Configuration

2.1. PC

The PC acquires various ship positions in connection with GNSS and DSU, and displays and compares the acquired ship positions on an ENC. In the case of coastal voyage, ship position is calculated by geo navigation through electronic drawing. It also includes an algorithm that recognizes an abnormal situation such as a distinct location difference or jamming for each identification means and analyzes the cause. If it is judged to be an abnormal situation, an alarm is generated in connection with the General Alarm System (GAS).

2.2. DSU

The DSU uses the sextant to calculate the location information based on celestial navigation. In the traditional way, the officer measures the altitude value of the celestial body with a sextant and then draws it on the chart to calculate the position of the ship.

DSU is equipped with an algorithm that obtains the current position using the measured altitude value and the DR(dead-reckoning) position to quickly calculate the position. In addition, it is equipped with a wired/wireless communication interface so that the position and measurement time can be stored in a database for information about ships.

2.3. PD

It is equipped with an algorithm that determines the priority by considering all the acquired ship positons for each –position fixing means. Also, in case of an emergency, it is possible to automatically transmit position and distribute it manually if necessary. PD is developed to be compatible with each other to interface with the legacy device. It is developed to be connected to the navigation system based on the commercial ship network standard IEC61162-450.

3. Design of EPDS

Table 1 shows the key factors for EPDS design.

Table 1. Key factors in EPDS design

Unit	Key factors
PC	- Module for receiving ship position and data format conversion for each navigation system - Server support for log management such as storage, deletion, and inquiry of position logs - Automatic/manual module for optimal ship determination - GAS interlock - Electronic drawing calculation module - Display of ship position according to the system operation mode
DSU	- Ability to calculate/store/transmit a position more quickly than a legacy sextant - Compute and process celestial-based positon and display it on the monitor - Positon saving function over 2 weeks - Transtmit position to PC
PD	- Provides data compatibility with IEC 61162-450 data for data receiving equipment data format

3.1. PC Design

The PC should be able to interconnect with the existing navigation system of the ship, and calculate and display the position of ship. In addition, the PC has the function of supporting decision-making by identifying and analyzing anomalies caused by jamming and excessive errors by comparing the fixed position from various means.

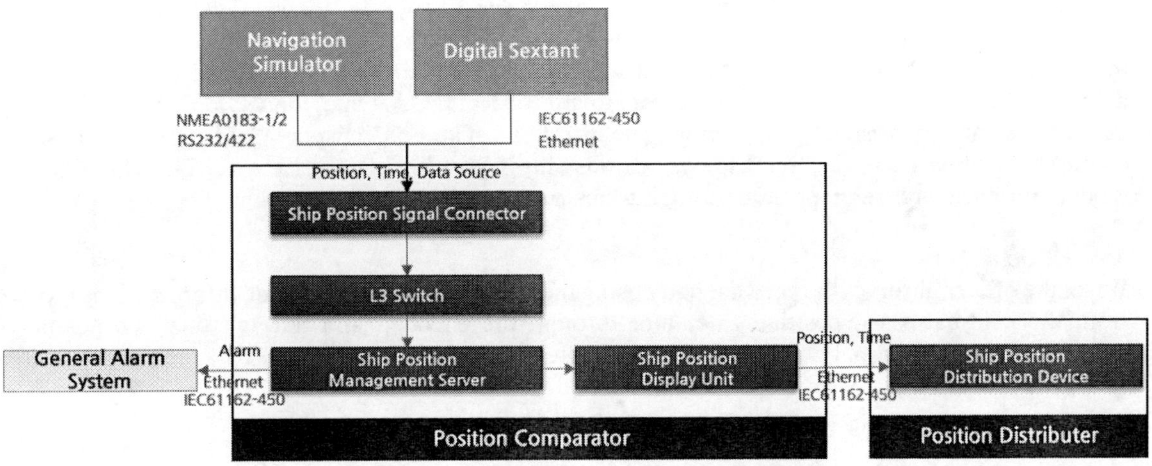

Figure 2. Configuration and data flow of PC

NMEA0183-1/2 and IEC61162-450 interfaces must be provided to communicate with the existing navigation system, PD and GAS. Figure 3 shows the Computer Software Configuration Item (CSCI) of the PC.

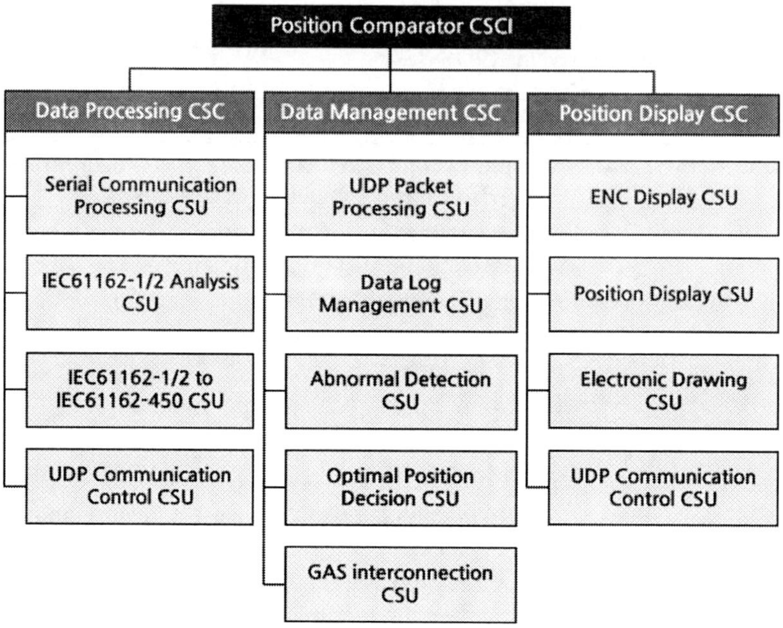

Figure 3. CSCI of PC

Data Processing CSC(Computer Software Component) is a module for receiving and analyzing various signal signals, converting them to IEC61162-450, and controlling UDP communication. Serial Communication Processing CSU(Computer Software Configuration Unit) receives the IEC61161-1/2 format data for each navigation system through serial communication and analyzes it in the IEC61162-1/2 Analysis CSU. The IEC 61162-1/2 to IEC61162-450 CSU converts the analyzed IEC61161-1/2 format data into IEC61161-450 packets. UDP Communication Control CSU performs IEC61161-450 packet transmission management function from Ship Position Signal Connector to L3 switch.

The UDP Packet Processing CSU processes the IEC61162-450 packets received from the L3 switch and performs DB management functions for log management such as storage, deletion, and inquiry of position information logs in the Data Log Management CSU. The Abnormal Detection CSU compares the position of each navigation device to detect anomalies, and the Optimal Position Decision CSU determines the position. When an abnormal situation occurs, GAS Interconnection CSU generates an alarm for an abnormal situation and propagate it to GAS. The ENC Display CSU expresses these data on ENC. Position Display CSU displays various ship's positions, and Electronic Drawing CSU is a module for calculating ship positon through terrestrial navigation.

3.2. DSU Design

When the officer observes the celestial body using the legacy sextant, the altitude information is received from the celestial-based position calculator through the encoder, and through this, the position is calculated through internal processing. The computed positon and observation data are transmitted to a storage server through wireless communication and stored and managed.

Figure 4 shows the signal and system diagram of the DSU.

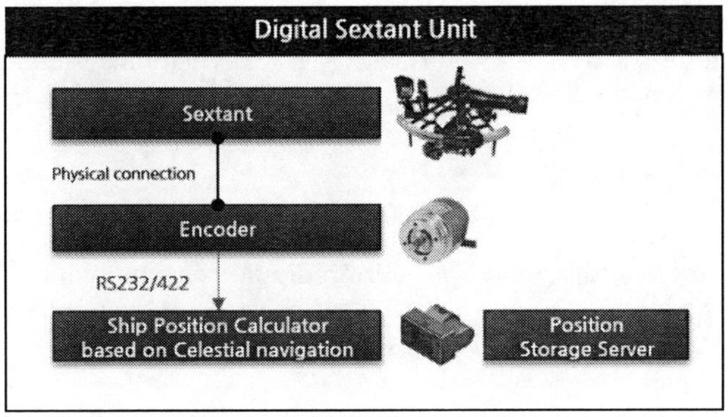

Figure 4. Configuration of DSU

The CSCI of DSU is shown in Figure 5.

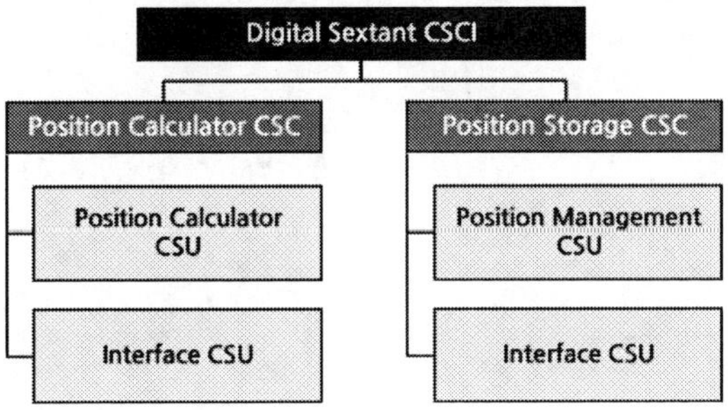

Figure 5. CSCI of DSU

The Position Calculator CSU calculates the position line with the observation altitude, then processes the correction value through the altitude measurement of the celestial body, and calculates and displays the ship position through the observation altitude and the current time. Interface CSU performs interworking function between ship position storage servers. The Position Management CSU performs the function of storing the calculated information on the position information storage server. Interface CSU transmits the measurement time of the celestial body to the position calculator to calculate the position and performs the function of transmitting the position to the PC.

3.3. PD Design

The composition of PD is shown in Figure 2, and the CSCI is shown in Figure 6.

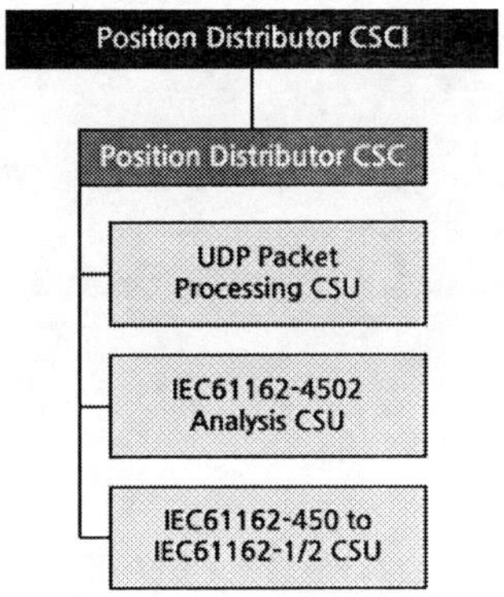

Figure 6. CSCI of PD

IOP Conf. Series: Materials Science and Engineering **929** (2020) 012011 doi:10.1088/1757-899X/929/1/012011

The configuration is similar to PC. UDP Packet Processing CSU can receive IEC61162-450 format data for each navigation system through serial communication, and IEC61162/450 Analysis CSU can analyze received data for IEC61162-450 format for each navigation system. Finally, the IEC61162-450 to IEC61162-1/2 CSU functions to convert the analyzed IEC61162-450 format data into IEC61161-1/2 packets.

4. Conclusion

We propose and describe the system newly that can collect PNTs from various equipment to detect anomalies, and measure and distribute PNTs from digital sextants, which are backup systems, when anomalies occur. In the future, we will verify the performance of the implemented system by a test with the full mission bridge simulator provided by the Kongsberg.

References

[1] Daniel Medina, Christoph Lass, Emilio Pérez Marcos, Ralf Ziebold, Pau Closas and Jesús García 2019 22th International Conference on Information Fusion (FUSION) *On GNSS Jamming Threat from the Maritime Navigation Perspective*

[2] A. Jafarnia-Jahromi, A. Broumandan and J. Nielsen and G. Lachapelle 2012 International Journal of Navigation and Observation *GPS vulnerability to spoofing threats and a review of antispoofing techniques* vol. 2012

[3] International Maritime Organization (NAV 57/6/4) 2011 *DEVELOPMENT OF AN E-NAVIGATION STRATEGY IMPLEMENTATION PLAN* 8 April 2011

Prediction of vessel propulsion power from machine learning models based on synchronized AIS-, ship performance measurements and ECMWF weather data

Q Liang[1], H A Tvete[1] and H W Brinks[1]

[1] Group Technology and Research - Maritime, DNV GL, Veritasveien 1, 1363 Høvik, Norway
Email: Qin.Liang@dnvgl.com

Abstract. In this paper, AIS (Automatic Identification System), ship performance measurement and ECMWF (European Centre for Medium-Range Weather Forecasts) weather data were synchronized as a complete dataset. Detailed processing steps and methods are introduced which can be used as best practice for future related studies. All the data preparation was processed on a spark cluster. The optimization and turning of cluster performance will also be introduced. The synchronized dataset was adapted to train different machine learning models to predict ship propulsion power. The dataset includes 228 container vessels covering a time scale of 50 months. The performance of deep learning models with different architecture was compared and discussed. Compared to previous paper [1] in the same project, this paper is an extended scenario which combines the data adapted in scenario 1 and 2. The analysis of the models' performance in different scenarios was discussed. More features were included in scenario 3 (this study). Hence, the best performing model from scenario 3 has more complex structure compared to scenario 1 and 2. The overall absolute R^2 score for test data is slightly lower than scenario 2. However, the performance for individual ships (relative R^2 score) is much better. This means, models that consider more features (operation, ship characters and environment) are beneficial for individual analysis. For general fleet or wide scope analysis, models that require less data and fewer features are better.

1. Introduction

Shipping is facing the introduction of the global 0.5% sulphur cap this year, one of the important changes in its recent history. Up to 70,000 ships will be affected by this regulation [2]. Shipping is the main logistics channel for the global economy and is responsible for the carriage of around 90% of world trade. With the development of worldwide transaction, the emissions from ships show a significant and growing trend. According to the third greenhouse gas study (GHG) from International Maritime Organization (IMO) [3], shipping emitted 938 Mt CO_2 in 2012 which accounts for around 2.6% of global CO_2 emissions. It is already showing a reduction trend compared to 2007 (1100Mt and 3.5% of global CO_2 emissions) which can be attributed to slow steaming and the increase of vessel size [4]. This means appropriate measures are beneficial for the reduction of GHG emissions. IMO has adopted regulations to address the emission of air pollutants from ships and mandatory energy-efficiency measures to reduce the emissions of greenhouse gases [5]. As a result of the increased international attention to global warming and air pollution, different stakeholders are beginning to take actions to reduce local and global emissions. E.g., the use of more environmentally friendly fuel, weather optimized routing, power optimization with battery and the possibility of using alternative energy to propel ships. According to [6], the road to greener shipping will see significant uptake of gas-driven propulsion. Most of the fuel will be LNG (liquefied natural gas) and LPG (liquefied petroleum gas), but some will be LBG (liquefied biogas). In order to evaluate the efficiency of these ship emission reduction measures accurately, an emission inventory model is vital.

One step before the calculation of emissions is to calculate the energy consumption or ships' propulsion power. Various methods have been proposed to estimate ship energy consumption. These methods can be divided into two categories: top-down and bottom up. In [7], the authors presented the definition and an overview of these two methods. The common denominator is that they rely on the relationship between influencing factors and energy consumption. E.g., a detailed ship propulsion model needs to include hull dimension parameters, hydrodynamic performance and different resistance applied to the ships. Many factors can affect the calculation process, e.g. the selection of functions and coefficient. In contrast to these physics-based models, the primary objective of this paper is to evaluate the performance of different machine learning models' ability to accurately predict ship propulsion power from AIS-, ship performance measurements and weather data.

In the previous paper [1] on this project, the performance of various machine learning models was benchmarked against measurement data and existing physical AIS-based model. In both scenarios in the previous paper [1], generally machine learning models perform better than existing physics-based models. While both scenarios have some limitations. For scenario 1, only 1 container vessel was investigated. Due to the limited access to weather data, only AIS, ship dimension characters and measurement were adapted for scenario 2. For this paper, a comprehensive dataset including AIS, ship dimension characters, weather and measurement was adapted as an additional scenario to explore the performance of machine learning models.

Machine learning is a powerful tool to handle complex problems that using normal methods would be difficult or even impossible. But machine learning should be used properly, it is not a solution to every problem. Data processing and preparation is a crucial step in training a good machine learning model. If data with lots of noise and errors are used for training, the performance of the model is certainly not qualified. In this paper, the data synchronization and processing will be introduced. It includes an illustration of how to combine maritime datasets, best practice for tuning cluster performance and verification of data.

There are many related studies trying to develop data-driven ship performance prediction models. However, most of them adapted a limited amount of data. For example, [8] developed a semi- empirical model on noon reports (manual reports) and sea trial data from 2 oil tankers. [9] analyzed the CO2 emission with AIS data through categorizing of ships. [10] explored the relationship between ship's engine power, fuel consumption and the vessel speed by using 1 engine test data. [7] developed a Gaussian Process model considering both operation conditions and the impacts of weather but lack of ship dimension information. This paper aims to fill the gaps in these studies by applying machine learning models with data covering all aspects of vessel operation.

2. Methods and data sources

2.1. Methods

Machine learning, especially deep learning, has attracted a lot of attention in recent years. Especially its ability to solve complex problems. Both traditional machine learning and deep learning models have been adapted in previous study [1], and deep learning models show better prediction results than the traditional machine learning methods. Hence, in this study will focus on exploring deep learning's ability further.

Artificial Neural Network (ANN), also called deep learning, has been applied widely in recent years, e.g. mathematical prediction, image recognition and speech recognition. There are different types of architecture for ANNs, including Multilayer Perceptron (MLP), Convolutional Neural Network (CNN) and Recurrent Neural Network (RNN). ANN architecture is always improving and global researchers are trying to develop different kinds of ANNs for different applications. The problem being solved in this paper is a regression problem. The model aims to find the relationship between different input features and ship propulsion power. Hence, MLP is capable of solving it properly. The reason why the other complex ANNs were not adapted is already introduced in [1].

2.2. Data sources

In this paper, several main data sources were adapted, including AIS data, IHS Fairplay, ship performance measurement data and ECMWF weather data.

The AIS is a mandatory collision avoidance system required by IMO and the several countries' Maritime Safety Administrations. For internationally operating ships with a gross tonnage (GT) of 300 or more, passenger ships of all sizes, domestic vessels with a GT of 200 or more traveling in coastal waters, and inland ships with a GT of 100 or more, it is mandatory to equip an AIS system.

IHS Fairplay provides the most thorough details on all sea-going, self-propelled merchant ships of 100 GT and above. For example, ship characteristics, ownership, port, installed engine power, design speed, etc. IHS Fairplay can be combined with AIS data to track live ship positions with unrivalled AIS coverage, analyze the risk profiles of ships' operation, especially entering or leaving port.

The ship performance measurement data is obtained from a DNV GL in-house database comprised of 4209 ships that have reported their performance data at different granularities, from noon reports to 15-minute measurement intervals.

ECMWF process data including satellite and available observations from non-satellite sources to generate weather forecasts [11]. The forecasts rely on accurate observations of the current weather. From ECMWF data, the sea environment condition can be obtained, e.g. significant wave height, mean wave direction, neutral wind speed, wind direction, etc.

3. Data Processing

3.1. Data processing environment

With the development of new technology, more and more data has become available, and also the computing-power to process it. Different large-scale data processing engines have been under development and improving continuously. Among them the top-two popular engines are Apache Spark and Apache Hadoop. In recent years, Spark has become more popular than Hadoop because of its processing speed. For example, Spark is 100 times faster than Hadoop in solving a logistic regression problem in [12].

The core principle of big data processing is parallel-processing as shown in figure 1. Compared to traditional single node processing which is always limited by the hardware. Parallel-processing has no bottleneck on individual performance. If the computing power or memory was not enough, additional worker nodes will be added automatically by the master node. Both Spark and Hadoop have their advantages and disadvantages. The reason why Spark becomes more popular than Hadoop is because of its in memory processing principle and cloud providers' flexibility. When spark is processing data, all of the data is loaded into the cluster's memory. Thus, it requires much more memory, however, the cloud providers have filled this gap.

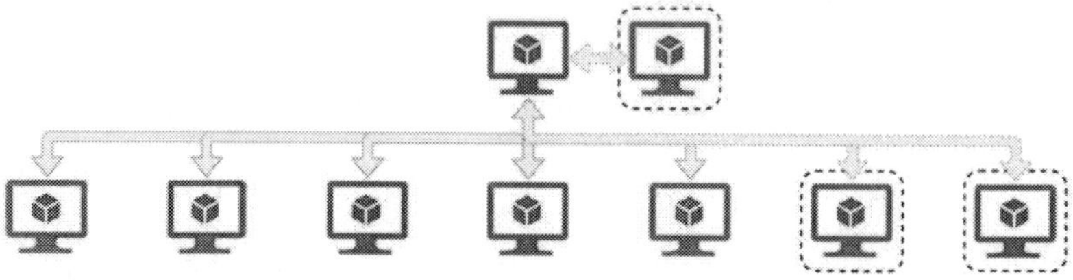

Figure 1. Cluster for parallel computing

There are plenty of computers and associated components ready to be used in a cloud providers' data center. Different hardware can be ordered and accessed by customers with simple steps. It provides both flexibility and scalability. E.g., a huge cluster including 30 worker nodes, 240 cores, 840 GB memory can be set up in 10 minutes.

All the large-scale data processing in this study is running on Azure Databrick from Microsoft. Databricks is a unified data analytics platform also a company founded by the original creators of Apache Spark. It is available both on Microsoft Azure and Amazon Web Services. Google Cloud has its own similar processing platform called Google Cloud Dataflow.

In this study, a cluster with 2 master nodes and 20 worker nodes was deployed. Master node and worker node will adapt the same type of virtual machine: Standard DS4 V2 with 8 cores and 28 GB memory. The Databricks Runtime Version is 6.2 with Spark 2.4.4. Pyspark is the Python API written in python to support Apache Spark.

3.2. Synchronization of AIS, IHS Fairplay and measurement data

As mentioned in chapter 2.2, there are four data sources that need to be synchronized. The AIS data and IHS Fairplay data are the easiest to synchronize. These two data sources are saved together as Delta Lake tables. For big data storage, normally parquet or orc will be selected as the storage format. Delta Lake is an open source storage layer that brings reliability to data lakes based on parquet format. Delta Lake offers a variety of additional features compared to parquet. E.g., ACID (atomicity, consistency, isolation, durability) transactions, metadata handling, and unifies streaming and batch data processing [13] which normal parquet does not provide.

In DNV GL's AIS and IHS Fairplay Delta Lake tables, there are a primary key and a foreign key that define the relationship between these two tables. These two tables can be easily connected with a join function based on these two keys.

DNV GL's performance measurement data is stored on a SQL server. In order to make it accessible with the other data sources, all the available performance measurement data was exported as csv file then loaded into Databricks as Delta Lake table. Because AIS data and measurement data have different sampling times and sampling frequency, it is difficult to synchronize the data without pre- processing. In addition, measurement data is the core dataset in the data processing operation which means no modification will be adapted. Hence, appropriate interpolation was applied to AIS data. Both data sources use the same timestamp format like 'YYYYMMDD hh:mm:ss'. For normal AIS sampling situations, sample rates can vary from several seconds to several minutes depending on the type of transponder and vessel operation status.

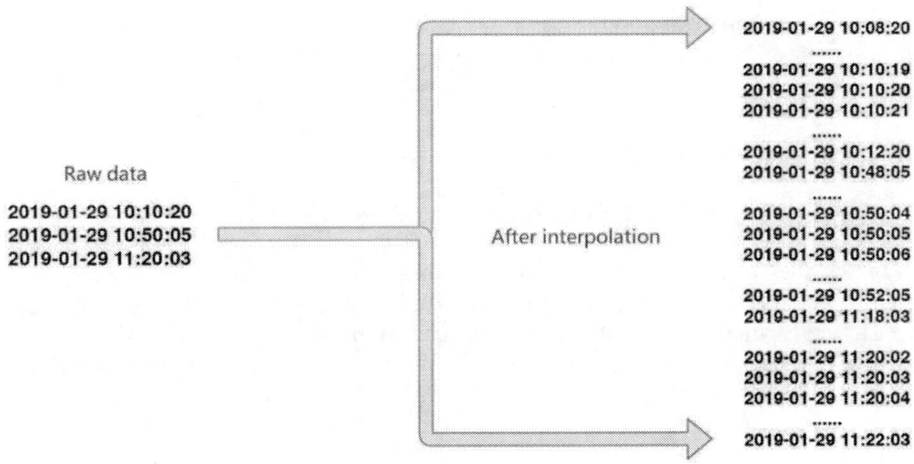

Figure 2. Interpolation method

In order to keep all the available records in measurement data, the AIS data has to be expanded to match with AIS data. Typical AIS data for 1 ship and a time scale of 50 months will have around 150,000 records. If this data was interpolated to cover every second for 50 months, there will be 129,600,000 records. In this study 228 container ships were adapted which means around 30 billion records will be generated. This is a huge data set which makes the processing both difficult and time- consuming. On the other hand, when interpolation is introduced, accuracy is also reduced. Several experiments were adapted to check the interpolation accuracy. Based on the analysis of experiment result, the best practice is to interpolate 120 seconds before and after available sample as shown in Figure 2.

After the pre-processing, the AIS data was left joined with the measurement data. The synchronization quality was checked based on the operation speed profile. If the speed profile is not similar or the speed difference between AIS and measurement data was bigger than 2 knots, the measurement was deleted. The synchronized dataset from this step is called the operation dataset in this paper.

3.3. Synchronization of Operation dataset with ECMWF data

ECMWF weather data have different components covering different aspects of weather. In this study, wave and wind information will be adapted. Before synchronization, pre-processing is applied to the operation dataset. The location information in the operation dataset is longitude and latitude which need to be converted to Cartesian coordinate system. After this step, x, y, z location information in Cartesian coordinate system becomes available.

The ECMWF wave and wind data has a measurement point for every hour or three hours. Interpolation was adapted to give full time scale coverage of every hour. The resolution for weather dataset is 0.36° which equals around 40 km. The next step is to create a mesh table, saving the unique location information of wave and wind data. The wave and wind data is saved as a daily parquet file with a file size slightly more than 1 GB. A year of wave and wind data (360 GB) was adapted to generate the mesh table to make sure it covers all the available zones.

All the wave and wind data will be loaded as pyspark dataframe from the daily parquet files with a size of 1800 GB. A special 'Key' column was created based on the time and location information. The time information adapted here is hours since 1900. An example is explained in Figure 3.

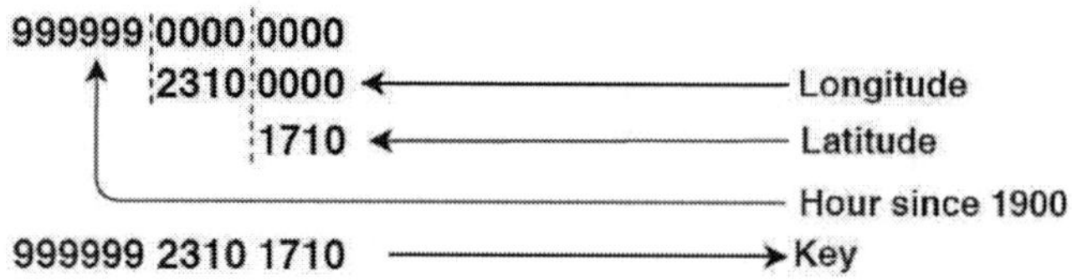

Figure 3. Synchronization key explanation

In order to match the operation dataset with the processed wave and wind data, the same 'Key' scheme should also be adapted. The location information in wave and wind data is fixed like the mesh table produced. However, the location information of operation dataset can be anywhere a ship operates. Hence, if wave and wind data need to match with the operation dataset, the operation dataset should have the closet weather information where wave and wind data is available. The mesh table was adapted to add weather location information to the operation dataset. The detailed steps were illustrated in Figure 4.

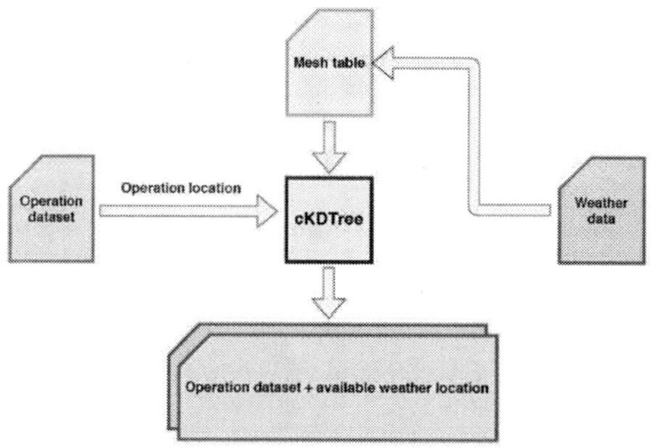

Figure 4. Synchronization dataflow

In addition, the cKDTree function adapted here is based on pandas library and should be applied to pandas dataframe. In order to apply the function with pyspark dataframe, either the pyspark dataframe being converted to pandas dataframe or the cKDTree function being applied as pyspark user defined function like showed in Figure 5.

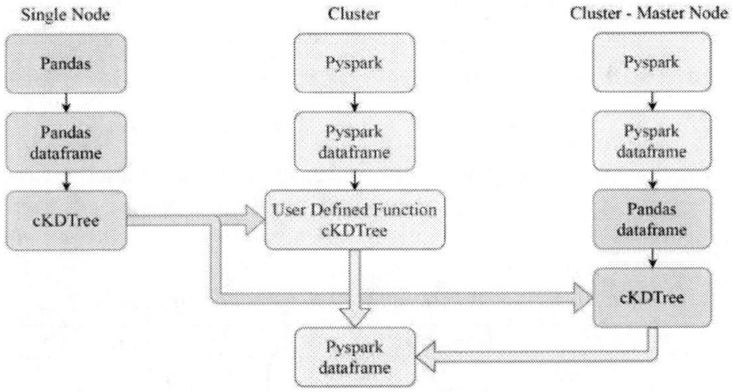

Figure 5. Third-party library application to Pyspark and Pandas

The first method needs to collect all the data from work nodes to the master node which will overload the master node. The master node was adapted to coordinate workload between worker nodes, not for processing the data. If this method was adapted, the virtual machine size for the master node should be increased. Adapting this method can increase the compatibility between pyspark and pandas, but require more time and resources. In the worst scenario, it might not be possible to collect all the data from work nodes to master nodes. Because the number of worker nodes can be unlimited, but the size of master node is limited. The second method works well when the function is not so complex. If another third-party library was adapted or function with complex calculation, some compatibility problems might arise. The available function packages in spark is expanding continuously. The best practice is attempt to adapt spark original functions and avoid using user defined functions.

After this step, the operation dataset contains closet weather location information and a 'Key' can be generated. Until now, both operation dataset and weather data is ready to synchronize. In pyspark there are different join operations as in a normal database, but it also supports other efficient join operation optimized for distributed computing. In this case, weather data has 5.4 million rows per day, in total around 10 billion for 5 years. It is a huge dataset compared to the operation dataset, hence, spark broadcast join was adapted.

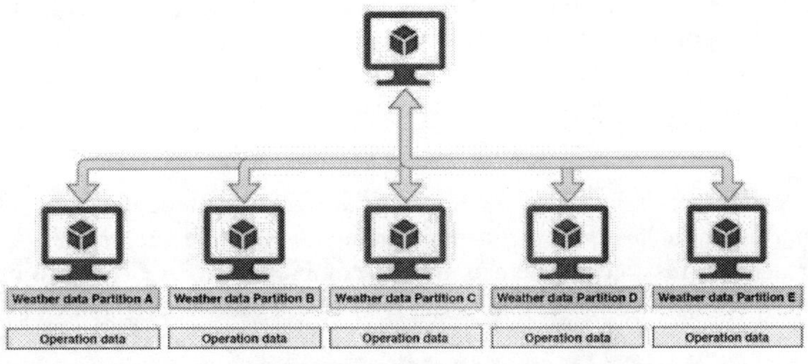

Figure 6. Broadcast data processing

Spark broadcast is ideal for joining a large dataframe with a small dataframe, but cannot be used for two large dataframes. Spark will "broadcast" the small dataframe by sending all the data in the small dataframe to all the worker nodes in the cluster like showed in Figure 6. After the small dataframe is broadcasted, spark can perform the join operation without shuffling the data in the large dataframe. On the other hand, the traditional left join operation could also be implemented but takes more time. Both methods have been tested and got same results in this study.

4. Models performance and result analysis

4.1. Feature selection

For scenario 1 and 2 in the previous paper, limited features were adapted in each scenario due to the limitation of data. After synchronization, a comprehensive dataset with operation, vessel characters and environment becomes available. In total 13 features were adapted in this study as presented in the following Table 1.

Table 1. Features comparison for Scenario 1,2 and 3

Scenario 1		Scenario 2		Scenario 3 (This study)	
Features	Unit	Features	Unit	Features	Unit
Speed over ground	knots	Speed over ground	knots	Speed over ground	knots
Significant wave height	m	Speed through water	knots	Heading	degrees
Wave direction to vessel	degrees	Trim	m	Course over ground	degrees
Wind speed	m/s	Draft fore	m	Moulded breadth	m
Wind direction to vessel	degrees	Draft aft	m	Moulded depth	m
		Design draught	m	Design draught	m
		Moulded depth	m	Length between perpendiculars	m
		Moulded breadth	m	Deadweight	ton
		Length between perpendiculars	m	Significant wave height	m
				Mean wave period	Hz
				Wave direction to vessel	degrees
				Wind speed	m/s
				Wind direction to vessel	degrees

The neural network architecture adapted in this study is MLP (Multi-Layer Perceptron). The correlation between the features and propulsion power was calculated in [1]. It is obvious that the direction related features are not closely related to propulsion power. The reason to include these features is the neural network's ability to extract insights from multi-dimensional or complex relationships.

4.2. Models adapted

As in the previous study, the same number of models with the same architecture were adapted. Compared to the previous 2 scenarios, more features mean more parameters inside the neural network, and longer training times. The present regression problem is not as complex as image recognition, which requires complex layer and deep structure. In general, MLP does not need to be very deep. One to three hidden layers are generally enough to summarize abstract findings from data. The models adapted in this study were shown in Table 2. The model name refers to number of neurons in each layer.

Table 2. Models adapted in the study

Model	Model name	S1 parameters	S2 parameters	S3 Number of parameters (This study)
1	M_1000	7001	12001	22001
2	M_100	701	1201	1301
3	M_50	351	601	651
4	M_30	211	361	391
5	M_20	141	241	261
6	M_10	71	121	131
7	M_5	36	61	66
8	M_500_100	53201	55701	56201
9	M_100_50	5701	6201	6301
10	M_50_30	1861	2111	2161
11	M_10_5	121	171	181
12	M_50_20_10	1541	1791	1841
13	M_10_8_4	189	239	249
14	M_10_8_6_4	235	285	295

4.3. Model performance and analysis

The prediction of propulsion power is a regression problem. Hence the R^2 score, also called coefficient of determination, was selected as the metric to compare the performance of the models. It ranges from 0 to 1, and the higher the R^2 score, the better and more accurate prediction was made. If it is negative, it means the prediction does not follow the trend of the data and fits worse than a horizontal line. The detailed function is shown below. SStot is the total sum of squares. SSrec is the regression sum of squares. fi is the prediction result from the model.

$$SS_{tot} = \sum_i (y_i - \bar{y})^2 \qquad SS_{rec} = \sum_i (y_i - f_i)^2 \tag{1}$$

$$R^2 = 1 - \frac{SS_{rec}}{SS_{tot}} \tag{2}$$

There are 27 container ships in test data, but they are not equally distributed. Some of the major ships represent more than 50% of the test data. The minor ships represent 1% of the test data. Therefore, the absolute R^2 for the entire test data is close to the major ships. As showed in Figure 7, the R^2 score of the whole test data (absolute R^2) for all the models in scenario 2 is higher than scenario 3. But it does not present the real prediction ability that the model demonstrated for individual vessel. In order to present models' ability for individual vessel, relative R^2 is introduced. The absolute R^2 is the R^2 for the whole test dataset. The relative R^2 is the average of R^2 for each test vessel.

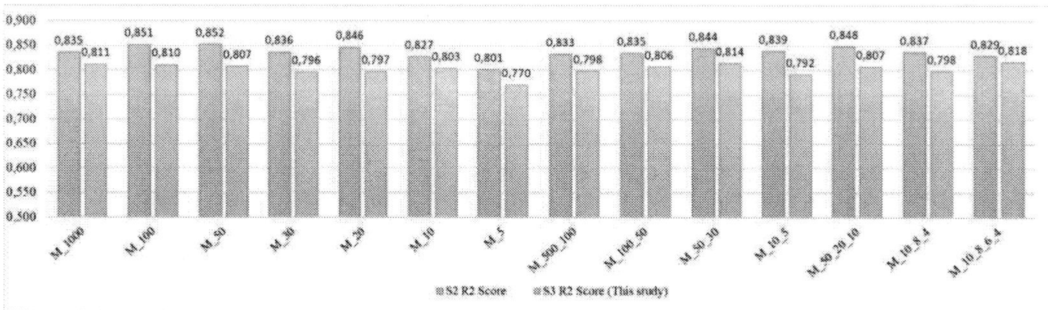

Figure 7. Absolute R^2 for ANN models

The number of neurons in the hidden layer should typically be the average of input and output parameters. The model with more neurons than required by the general principle performed well. Hence, the best practice from previous study to choose an ANN model proved correct. Model M50 was selected to benchmark the performance. In addition, model M10_5 was also selected with its performance and number of parameters. Since more features were adapted, especially environmental features, 2 hidden layers can extract deeper relationship than 1 hidden layer.

Table 3. Absolute and relative R^2 for Scenario 2 and 3

	Scenario 2 M20	Scenario 3 M10_5	Scenario 3 M50
Absolute average R^2	84.6%	79.2%	80.7%
Relative average R^2	40.7%	61.3%	64.7%

In Table 3, the average of absolute and relative R^2 were compared. Even though the absolute R^2 for scenario 2 is higher than scenario 3, the relative R^2 for scenario 3 is much better than scenario 2. The R^2 for all the test vessels is shown in Figure 8.

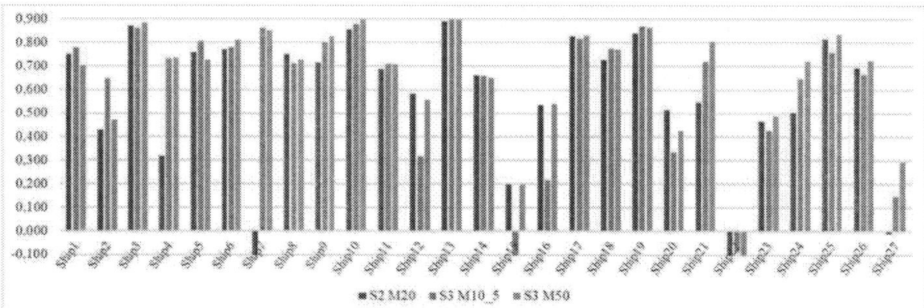

Figure 8. Relative R^2 for ANN models

Model S3_M10_5 performed better on 14 of 27 ships than model S2_M20. Model S3_M50 performed better on 19 of 27 ships. Irrespective of the number of ships or the relative average R^2, both models in scenario 3 performed better than the model in scenario 2. It means that more features adapted in the training data can help the model learn more complex relationships from it. The added resistance from waves and wind is a major part of resistance affecting the ships' performance. To calculate the added resistance requires a lot of effort, e.g. adapting empirical functions and statistics. Many factors and index are introduced during this calculation process. It also means that uncertainty and assumptions are introduced. On the contrary, the data-driven methods only rely on the input features and output propulsion power, which avoid bringing in uncontrollable factors. Hence, the data- driven models are more reliable than the physics models.

5. Discussion

The performance of physics-based models and were compared in previous study. In both scenario 1 and 2, the data-driven models performed better. In this paper, an extended study – scenario 3 was applied. The operational data, ship characteristics, performance measurement data and environmental data were synchronized to create a comprehensive dataset.

The comprehensive dataset was used to train ANN models. More features were adapted in scenario 3 which provides not only the ship operation information but also the environment within which the ship operates. Compared to the best performing model (M20 in scenario 2), models with more layers (M10_5) and more neurons in the hidden layer (M50) performed better in scenario 3. Model M10_5 performed better due to the inclusion of environment features, especially features not directly correlated with output, mean that the regression model needs a higher degree of nonlinear regression. Scenario 3 has more features than scenario 2. Model M50 has more neurons in the hidden layer which can summarize and memorize more learnings from the training data.

Because the test data is not evenly distributed, the absolute R^2 score for the test data leans towards the major ships. Therefore, both absolute and relative R^2 score were calculated and compared. The relative R^2 score could represent a model's ability in handling different ships. The ANN modes in scenario 3 performed much better in relative R^2 score than scenario 2. On the other hand, the ANN models in scenario 2 are also good models. First they performed much better than the physics-based model introduced in [1]. Second, they require less data to predict. The weather data is not always available. In addition, even though the weather data is available, it still requires a lot of effort on data synchronization and data quality check. Hence, the ANN models in both scenario 2 and 3 are valuable. When focusing more on the performance prediction of individual vessels and weather is available, ANN models in scenario 3 could be adapted. Models from scenario 2 could be used when less data is available or when trying to predict the performance of a fleet of ships.

The data-driven models provide a short cut for performance prediction when there is a lack of domain knowledge for feature engineering. It also avoids empirical functions and correction factors being introduced which reduce the reliability of the model. The data-driven models trained with different data can be utilized in different scenarios. The best practices for processing and synchronization of the data are introduced in this paper. With the help of cloud technology and the Spark data processing engine, big data processing becomes much easier and applicable. The data- driven model demonstrates its ability to handle complex prediction problem. It is strongly reliant on the data it has been trained on. If the training data is not representative or the data quality is not good, the data-driven models will not necessarily make good predictions.

In this study, container ships were selected as the pilot group. Other ship types will be explored in the continuation of this work. The data-driven models will also be implemented in DNV GL's production Databricks cluster as a reference to check its ability in a continuous manner. In addition to this, a more detailed physical model adjusted for added resistance is under development in DNV GL [14]. It will provide further comparison for the performance of data-driven models and physics-based models with the same data.

Acknowledgements
This work was funded by the VERDE (short for VERification for DEcarbonisation) project, which is funded by the Research Council of Norway by grant #282293. Special thanks to Vidyunmala Veldore in DNV GL Digital Solutions for help with initial synchronization work and Håvard Nordtveit Austefjord in DNV GL Maritime Advisory for providing the environment data.

References

[1] Liang Q, Tvete H A and Brinks H W 2019 Prediction of vessel propulsion power using machine learning on AIS data, ship performance measurements and weather data J. Phys. Conf.
[2] Ser. 1357
[3] Chryssakis C, Brinks H, Brunelli A C, Fuglseth T P, Lande M, Laugen L, Longva T, Raeissi B and Tvete H A 2017 Low Carbon Shipping Towards 2050 32
[4] IMO 2014 Greenhouse Gas Emissions
[5] Lindstad H and Eskeland G S 2015 Low carbon maritime transport: How speed, size and slenderness amounts to substantial capital energy substitution Transp. Res. Part D Transp.
[6] Environ. 41 244–256
[7] IMO 2019 Low carbon shipping and air pollution control
[8] DNV GL's Group Technology & Research 2019 Energy transition Outlook 2019 Waddenl.
[9] Outst. 269–282
[10] Yuan J and Nian V 2018 Ship energy consumption prediction with Gaussian process metamodel Energy Procedia 152 655–660
[11] Lu R, Turan O, Boulougouris E, Banks C and Incecik A 2015 A semi-empirical ship operational performance prediction model for voyage optimization towards energy efficient shipping Ocean Eng. 110 18–28
[12] Rashidia M and Jaswar 2014 Prediction of CO2 emitted by marine transport in Batam-
[13] Singapore channel using AIS J. Teknol. (Sciences Eng. 69 121–126
[14] Kowalak P, Kasyk L and Borkowski T 2011 Assessment of ship's engine effective power, fuel consumption and emission using the vessel speed J. KONES 18 31–39
[15] ECMWF 2020 ECMWF Official website
[16] Apache 2020 Apache Spark
[17] Lake D 2020 Introduction to Delta Lake
[18] Tvete H A, Guo B, Liang Q and Brinks H 2020 A modelling system for power consumption of marine traffic OMAE2020-18651 (Fort Lauderdale, FL, USA: Conference on Ocean, Offshore and Arctic Engineering) 1–10

Considerations on the common regulatory issues among the IMO instruments for realization of maritime autonomous surface ships

Shiokari, M [1] and Ota, S

Centre for International Cooperation, National Maritime Research Institute, National Institute of Maritime, Port and Aviation Technology, 6-38-1, Shinkawa, Mitaka, Tokyo, 181-0004, Japan

[1] Email: shiokari@m.mpat.go.jp

Abstract. The Maritime Safety Committee (MSC) of the International Maritime Organization (IMO) included an agenda item "Regulatory Scoping Exercise (RSE) for the use of Maritime Autonomous Surface Ships (MASS)" in order to determine how the safe, secure and environmentally sound operation of MASS may be introduced in the regulations. The results of the RSE have been reported to the MSC, and they will be discussed in the 102th session. The purpose of this study is to provide suggestions for making decisions on the necessary future work at the MSC. The authors have reviewed all the results of the RSE under the purview of the MSC and made discussions on the ways of addressing some important common issues. The suitability of the specific ways of addressing these issues may be turned out in the actual works for amendments. On the other hand, regarding the priority of the discussions, the authors consider that it is appropriate to give priority to the discussions on the partial automation and remote operation with seafarers on board over the fully autonomous or remote operations without seafarers on board. Also, it seems appropriate to give priority to the discussions on the navigation systems and tasks.

1. Introduction

1.1. Background

The Maritime Safety Committee (MSC) of the International Maritime Organization (IMO), at its 98th session (MSC 98) held on June 2017, agreed to include a new agenda item "Regulatory Scoping Exercise (RSE) for the use of Maritime Autonomous Surface Ships (MASS)" in order to determine how the safe, secure and environmentally sound operation of MASS may be introduced in the regulations considered by IMO. The MSC, at its 100th session (MSC 100) held on December 2018, finalized the framework for the RSE and agreed that, as a first step, the review of the IMO instruments, i.e. conventions and regulations under the purview of IMO, would be conducted to identify the provisions which prevent MASS operations or which may need amendments or clarifications and, as a second step, the analysis would be conducted to determine the most appropriate way of addressing MASS operations (See chapter 2). It was also agreed that the initial review and initial analysis should be conducted by volunteering Member State(s) in these two steps.

The authors, as representatives of Japan, undertook the initial review and initial analysis of Chapters II-2, VI, VII, XII and XIII of the annex to the International Convention for the Safety of Life at Sea (SOLAS Convention) and associated codes and standards which were made mandatory by these chapters of the annex to the SOLAS Convention.

The results of the review in the first step of the RSE were reported to the MSC Intersessional Working Group on MASS held on September 2019, and the group agreed on the completion of the first step and commencement of the second step. Subsequently, the results of the analysis in the second step of the RSE have been reported to the MSC, and they will be discussed in the next session (MSC 102). Table 1 is the list of documents reporting the results of the first step of the RSE and these documents are not open to the public. Table 2 is the list of documents reporting the results of the second step of the RSE and these documents are accessible through IMO document website (https://webaccounts.imo.org). After discussion on the results of the second step of the RSE, the MSC will decide the future work to realize MASS, e.g. amending regulations, developing new instruments, etc.

1.2. Previous study

The authors indicated the results of the initial review of the first step of the RSE for Chapters II-2, VI and VII of the annex to the SOLAS Convention and the associated codes in detail, in the previous paper for ICMASS 2019 [1].

1.3. Purpose

The purpose of this study is to provide suggestions for making decisions on the necessary future work at the MSC, by considering the results of the RSE for all the IMO instruments, in respect of safety. As mentioned above, the authors undertook a lot of the chapters of the annex to the SOLAS Convention. Also, the authors have reviewed all the results of the first and second steps of the RSE under the purview of the MSC. Through these experiences, we found some important issues which are common among a lot of the IMO instruments. In this paper, we introduce these important common issues and provide suggestions for making decisions on the necessary future work at the MSC, only from technical viewpoints.

2. Outline of the RSE

2.1. Definitions of MASS and degrees of autonomy

In the framework for the RSE for the use of MASS, the definition of MASS is mentioned as the following text: “For the purpose of the regulatory scoping exercise, ‘Maritime Autonomous Surface Ship (MASS)’ is defined as a ship which, to a varying degree, can operate independent of human interaction.” [2].

Also, the degrees of autonomy are organized as shown in Table 3, which does not represent a hierarchic order [2]. It should be noted that MASS could be operating at one or more degrees of autonomy for the duration of a single voyage [2].

2.2. Methodology

The methodology of the RSE consists of the two steps as below.

2.2.1. First step

As a first step, the RSE identifies provisions in the IMO instruments which, as currently drafted: A: apply to MASS and prevent MASS operations; or

B: apply to MASS and do not prevent MASS operations and require no actions; or

C: apply to MASS and do not prevent MASS operations but may need to be amended or clarified, and/or may contain gaps; or

D: have no application to MASS operations [2].

Table 1 Documents reporting the results of the first step of the RSE (Main title of these documents was "Consideration of the result of the first step of the regulatory scoping exercise").

Document ID	Sub-title of document	Submitter
ISWG/MASS 1/2	Results of the first step of the regulatory scoping exercise analysing possible gaps in SOLAS chapter IX and the ISM Code in relation to the safe operation of Maritime Autonomous Surface Ships (MASS)	Norway
ISWG/MASS 1/2/1	Summary of results of the first step of the RSE for SOLAS chapter II-1	France
ISWG/MASS 1/2/2	Summary of results of the first step of the RSE for the International Convention on Maritime Search and Rescue, 1979	France & Spain
ISWG/MASS 1/2/3	Summary of results of the first step of the RSE for SOLAS chapter II-2 and associated codes	Japan
ISWG/MASS 1/2/4	Summary of results of the first step of the RSE for SOLAS chapter VI and associated codes	Japan
ISWG/MASS 1/2/5	Summary of results of the first step of the RSE for SOLAS chapter VII and associated codes	Japan
ISWG/MASS 1/2/6	Findings and common issues identified in the initial review of chapters II-2, VI and VII of the annex to SOLAS 1974 and the associated codes	Japan
ISWG/MASS 1/2/7	Summary of results of the first step of the RSE for SOLAS chapter XII and associated standards	Japan
ISWG/MASS 1/2/8	Summary of results of the first step of the RSE for SOLAS chapter XIII	Japan
ISWG/MASS 1/2/9	Summary of results of the first step of the RSE for CSC 1972	Japan
ISWG/MASS 1/2/10	Summary of the results of the first step of the RSE for STCW-F 1995	Japan
ISWG/MASS 1/2/11	Summary of results of the first step of the RSE for SOLAS chapter III and the LSA Code	Belgium & Netherlands
ISWG/MASS 1/2/12	Summary of results of the first step of the RSE for SOLAS chapter XI-1 and related codes	Finland
ISWG/MASS 1/2/13	Summary of results of the first step of the RSE for SOLAS chapter XI-2 and the ISPS Code	Finland
ISWG/MASS 1/2/14	Summary of results of the first step of the RSE for SOLAS chapter XIV and the related Polar Code	Finland
ISWG/MASS 1/2/15	Summary of results of the first step of the RSE for SOLAS chapter IV	Turkey
ISWG/MASS 1/2/16	Summary of results of the first step of the RSE for SOLAS chapter V	China & Singapore
ISWG/MASS 1/2/17	Summary of results of the first step of the RSE for the International Convention on Tonnage Measurement of Ships, 1969 (TONNAGE 1969)	Liberia
ISWG/MASS 1/2/18	Summary of results of the first step of the RSE for LL 66, PROT 88, IS Code Part A and III Code	India
ISWG/MASS 1/2/19	Summary of results of the first step of the RSE for the International Regulations for Preventing Collisions at Sea 1972 (COLREG)	Marshall Islands
ISWG/MASS 1/2/20	Summary of results of the first step of the RSE for STCW Convention and Code	United States

Table 2 Documents reporting the results of the second step of the RSE (Main title of these documents was "Regulatory scoping exercise for the use of Maritime Autonomous Surface Ships (MASS)").

Document ID	Sub-title of document	Submitter
MSC 102/5	Status report – Progress of the regulatory scoping exercise	Secretariat
MSC 102/5/1	Report of the Intersessional Working Group on Maritime Autonomous Surface Ships	Secretariat
MSC 102/5/3	Summary of results of the second step and conclusion of the RSE for the International Regulations for Preventing Collisions at Sea 1972 (COLREG)	Marshall Islands
MSC 102/5/4	Summary of results of the second step of the RSE for SOLAS Chapter III and the LSA Code	Belgium, China & Netherlands
MSC 102/5/5	Summary of results of the second step of the RSE for LL 1966, LL PROT 1988, IS Code Part A and III Code	India
MSC 102/5/6	Summary of results of the second step of the RSE for SOLAS chapter II-1	France
MSC 102/5/8	Summary of results of the second step of the regulatory scoping exercise for the International Convention on Tonnage Measurement of Ships, 1969 (TONNAGE 1969)	Liberia
MSC 102/5/9	Summary of results of the second step of the RSE for SOLAS chapter V	China
MSC 102/5/10	Summary of results of the second step of the RSE for SOLAS chapter XI-1 and related codes	Finland
MSC 102/5/11	Summary of results of the second step of the RSE for OLAS chapter XI-2 and the ISPS code	Finland
MSC 102/5/12	Summary of results of the second step of the RSE for SOLAS chapter XIV and the Polar Code	Finland
MSC 102/5/13	Summary of results of the second step of the RSE for SAR 1979 Convention	France & Spain
MSC 102/5/14	Development of interim regulatory measures for operation of MASS in the Russian Federation	Russian Federation
MSC 102/5/15	Summary of the results of the second step of the RSE for SOLAS chapter IV	Turkey
MSC 102/5/16	Summary of results of analysis of IMO instruments under the purview of the Maritime Safety Committee	CMI
MSC 102/5/17	Summary of results of the second step of the RSE for STCW Convention and Code	United States
MSC 102/5/19	Summary of results of the second step of the RSE for SOLAS chapter II-2 and associated codes	Japan
MSC 102/5/20	Summary of the results of the second step of the RSE for SOLAS chapter VI and associated codes	Japan
MSC 102/5/21	Summary of the results of the second step of the RSE for SOLAS chapter VII and associated codes	Japan
MSC 102/5/22	Summary of the results of the second step of the RSE for SOLAS chapter XII and associated standards	Japan
MSC 102/5/23	Summary of the results of the second step of the RSE for SOLAS chapter XIII	Japan
MSC 102/5/24	Summary of the results of the second step of the RSE for CSC 1972	Japan
MSC 102/5/25	Summary of results of the second step of the RSE for SOLAS chapter IX and the ISM Code	Norway
MSC 102/5/26	Summary of results of the second step of the RSE for the STCW-F Convention	Japan

Table 3 Degrees of autonomy [2].

Degree of autonomy	Defini tion
One	Ship with automated processes and decision support: Seafarers are on board to operate and control shipboard systems and functions. Some operations may be automated and at times be unsupervised but with seafarers on board ready to take control.
Two	Remotely controlled ship with seafarers on board: The ship is controlled and operated from another location. Seafarers are available on board to take control and to operate the shipboard systems and functions.
Three	Remotely controlled ship without seafarers on board: The ship is controlled and operated from another location. There are no seafarers on board.
Four	Fully autonomous ship: The operating system of the ship is able to make decisions and determine actions by itself.

In the first step, the initial review should be conducted by volunteering Member States [3], and these instruments should be reviewed on a regulation or rule level [2]. In other words, volunteering Member States should choose an option A, B, C or D for each regulation or rule of the instruments that they undertook.

2.2.2. Second step

After the completion of the first step, a second step is conducted to analyse and determine the most appropriate way of addressing MASS operations, taking into account, inter alia, human element, technology and operational factors by:

I: equivalences as provided for by the instruments or developing interpretations;
and/or II: amending existing instruments; and/or
III: developing new instruments; or
IV: none of the above as a result of the analysis [2].

The initial analysis should be conducted, preferably, by the volunteering Member State(s) that conducted the initial review [3]. The initial analysis should be high level and should not be conducted regulation by regulation [3].

3. Review of the results of the RSE

As mentioned in chapter 1, all the results of the first and second steps of the RSE were reported to the MSC (Tables 1 and 2). The authors have reviewed all the results and found the following important issues which are common among a lot of the IMO instruments. The list below shows only the especially important and common issues among a lot of the IMO instruments. The other issues and findings can be found in the documents in Tables 1 and 2.

(1) The definitions of master, crew, etc. should be defined clearly for remotely or autonomously controlled ships.
(2) The requirements for outboard personnel in remote control centers, etc. should be developed for remotely or autonomously controlled ships.
(3) The relation between seafarers on board ships and outboard personnel in remote control centers, etc. should be defined clearly for remotely controlled ships with seafarers on board the ships.

(4) The requirements for remote control centers, etc. should be developed for remotely or autonomously controlled ships.

(5) The requirements for onboard systems and equipment should be reconsidered for remotely or autonomously controlled ships, especially for the systems and equipment which need manual operations in the existing provisions, in relation to firefighting, life-saving, cargo handling, etc.

3.1. Definitions of master, crew, etc.

The definitions of master, crew, etc. are related to almost all the IMO instruments. The definitions of these people should be clarified for degrees of autonomy two, three and four. In section 6.2 of the ISM Code [4], it is required that "each ship is manned with qualified, certificated and medically fit seafarers". For degree of autonomy two, there are personnel both on board and outboard (i.e. remote control centers, etc.). Although onboard personnel are obviously seafarers, it will be needed to clarify whether outboard personnel are seafarers or not. For example, in Part 7 of the IMDG Code [5] *"Provisions concerning transport operations"*, paragraph 7.5.2.13 requires regular inspections of ro-ro cargo spaces during the voyage by an authorized crew member or responsible person for early detection of any hazard. Also, chapter 7.8 refers to the judgment by the master in the event of incident. Regarding these provisions, if the regular inspections can be conducted remotely, it will be needed to clarify whether such personnel who conduct the regular inspections at a remote control center, etc. are crew or not. Also, if the judgement by the responsible person, on behalf of the master, can be made at a remote control center, etc. and communicated to ships in the event of incidents, it will be needed to clarify whether such person who make judgement is the master or not. The same things can be said to the other provisions referring to the actions by seafarers (master, crew, etc.).

Similarly, for degree of autonomy three, since there are no seafarers on board and there are personnel only in remote control centers, etc., it will also be needed to clarify whether such personnel are seafarers or not. On the other hand, for degree of autonomy four, although it is clear that there are no seafarers on board, it is not clear whether there are personnel outboard or not. The authors consider that there are personnel for supervision of a fully autonomous ship in a remote location, but it is not clearly determined at the MSC. Therefore, it will be needed to clarify whether personnel exist outboard for supervision at first, then, clarify whether such personnel are seafarers or not.

Regarding this issue, there are various opinions among the IMO Members: developing interpretations (option I in the second step of the RSE), amending each provision one by one (option II), and developing new instruments (option III). Although the possible means for addressing this issue are various, the aim is identical: to apply the consistent and unified approach over all the IMO instruments. Indeed, it may not be turned out that which means would be best for addressing this issue since it may only be determined during the actual works for amending instruments. However, the authors consider that it may be more appropriate to prescribe clearly in the regulations (amending each provision one by one or developing new instruments) than to make interpretations because these provisions impose obligations on a person.

3.2. Requirements for outboard personnel in remote control centers, etc.

As mentioned in the previous section, it will be needed to clarify whether outboard personnel are seafarers or not for degrees of autonomy two and three, and for degree of autonomy four if there are outboard personnel for supervision. In addition, it will be needed to clarify the required abilities of such outboard personnel, for example as mentioned in the previous section, personnel at a remote control center, etc. who conduct the regular inspections of ro-ro cargo spaces carrying dangerous goods according to paragraph 7.5.2.13 of the IMDG Code, or who make judgement on behalf of the master according to chapter 7.8 of the IMDG Code. Therefore, no matter whether outboard personnel are determined as seafarers or not, the requirements for such personnel should be developed.

The ways of developing the requirements for outboard personnel will depend on the result of the clarification. One way is to consider such personnel as a kind of seafarers, who are not on board but acting as seafarers in practice. In that case, it will be needed to amend the definitions of seafarers for adding such personnel (e.g. "remote operator" as a kind of seafarers) to the existing definitions and to prescribe the roles and responsibilities of such personnel in the respective provisions if they are different from the existing ones for the seafarers on board.

Another way is not to consider such personnel as seafarers, but to develop other definitions for such personnel. In that case, it will be needed to develop new definitions for such personnel separately from the existing definitions, and to prescribe the roles and responsibilities of such personnel in the newly developed definitions (e.g. defining the new terminology of "remote operator" and prescribing the roles and responsibilities of a remote operator in that definition). It needs to be discussed carefully which way is better at the MSC, taking into account merits and demerits for each way of addressing this issue.

3.3. Relation between seafarers on board ships and outboard personnel in remote control centers, etc. As mentioned in the previous section, it will be needed to define the roles and responsibilities of personnel in remote control centers, etc. no matter whether they are defined as a kind of seafarers or not. When defining the roles and responsibilities of such personnel, it will also be needed to prescribe the relation between seafarers on board ships and outboard personnel in remote control centers, etc. for degree of autonomy two. It is quite important to clearly and specifically define the relation between them, i.e. assignment of their roles and responsibilities in normal and emergency situations. If there is an ambiguity in the relation between them, it may possibly lead to significant accidents due to human errors such as inconsistency between their assumptions or prejudices.

However, regulations to be developed for addressing this issue might only refer to the necessity of clear and prescriptive definitions for the relation between them, and require that such definitions should be developed by shipping company, etc. prior to operations of remotely controlled ships with seafarers on board. It also needs to be discussed how this issue should be addressed in regulations at the MSC.

3.4. Requirements for remote control centers, etc.

In addition to the requirements for outboard personnel in remote control centers, etc., it will also be needed to develop the requirements for systems and equipment in such places for degrees of autonomy two and three, and for degree of autonomy four if necessary.

Especially, the requirements for communication between ships and remote control centers, etc. should be developed for ensuring the safe operations. If communication link would be cut off between ships of degree of autonomy three and remote control center, etc. during a voyage, remote control would become unavailable any more. It will also be needed to develop the requirements for emergency responses to ensure safety even if such a situation occurs. For degree of autonomy two, the requirements for emergency responses should also be developed for onboard seafarers to ensure safety without communication with remote control centers, etc. For degree of autonomy four, the requirements will be needed to ensure safety without supervision in remote locations.

Also, the requirements for alerting systems in remote control centers, etc. should be considered so that the personnel in such places can be notified when any abnormal conditions occur in the remotely or autonomously controlled ships. The requirements for addressing abnormal conditions from remote control centers, etc. should also be considered.

3.5. Requirements for onboard systems and equipment for remotely or autonomously controlled ships The requirements for onboard systems and equipment should be reconsidered for remotely or autonomously controlled ships, especially for the systems and equipment which need manual operations in the existing provisions, in relation to firefighting, life-saving, cargo handling, etc.

Among these existing provisions, some provisions clearly require that a human, e.g. a crew or a master etc., shall conduct the operations regarding firefighting, life-saving, cargo handling, etc. In this case, these provisions should be amended since they will definitely prevent the realization of MASS, especially for degrees of autonomy three and four.

On the other hand, other provisions do not clearly require the operations by a human, but they require that some systems and equipment shall be installed, which need manual operations at present. For these provisions, it may be resolved with novel technologies for automatic or remote operations of systems and equipment which are manually operated at present. However, some of these provisions require that some automatic systems shall also be capable of being manually operated, and other provisions require local manual operations on board. In these cases, it is necessary to understand the intentions of the provisions and consider the alternative means for achieving the intended functionalities.

4. Discussions

Since the issues mentioned in chapter 3 are especially important and cross-cutting issues among a lot of the IMO instruments, consistent approach is necessary to realize MASS. The authors suggest that these issues should be addressed considering all the related provisions in all the related IMO instruments under the purview of the MSC.

When considering the amendments of the related provisions, discussions on partial automation and remote control of the existing onboard functions and tasks with seafarers on board (for ships of degrees of autonomy one and two) should be proceeded at first, taking into account the recent technological developments in the world. Partial automation and remote control will be the first step to realize fully autonomous operation and remote operation without seafarers onboard.

In addition, the priority of the scope of discussions should be considered before starting discussions. It seems that the provisions related to navigation are the most significant issues to realize MASS. Also, most of the technologies under developments seem to be related to navigation systems and tasks conducted by seafarers at present. Taking into account these situations, it is appropriate to give priority to the discussions on the partial automation and remote control of the navigation systems and tasks over the other issues.

5. Conclusions

In this paper, the authors introduced the important common issues among a lot of the IMO instruments under the purview of the MSC and made discussions on the ways of addressing these issues. The suitability of the specific ways of addressing these issues should be considered carefully in the actual works for amendments at the MSC. On the other hand, regarding the priority of the discussions, the authors consider that it is appropriate to give priority to the discussions on the partial automation and remote operation with seafarers on board over the fully autonomous or remote operation without seafarers on board. Also, it seems appropriate to give priority to the discussions on the navigation systems and tasks over the other systems and tasks.

References

[1] Shiokari, M and Ota, S, "Considerations on the regulatory issues for realization of Maritime Autonomous Surface Ships", Journal of Physics: Conf. Series 1357 (2019) 012005
[2] IMO document MSC 100/20 "Report of the Maritime Safety Committee on its one hundredth session", Annex 2 "Framework for the regulatory scoping exercise for the use of Maritime Autonomous Surface Ships (MASS)"
[3] IMO document MSC 100/20 "Report of the Maritime Safety Committee on its one hundredth session", Annex 2, Appendix 3 "Plan of work and procedures for the regulatory scoping exercise"
[4] IMO, International Management Code for the Safe Operation of Ships and for Pollution Prevention adopted by resolution A.741(18), as amended
[5] IMO, International Maritime Dangerous Goods Code adopted by resolution MSC.122(75), as

amended (2016 Amendments to the International Maritime Dangerous Goods Code adopted by resolution MSC.406(96))

IOP Conf. Series: Materials Science and Engineering **929** (2020) 012014 doi:10.1088/1757-899X/929/1/012014

Study on the potential gaps and themes identified by IMO Regulatory Scoping Exercise(RSE) for the use of Maritime Autonomous Surface Ships(MASS)

M.C Jo[1], A.R Lee, Y.D Kim and J.S Seo

Maritime Safety Policy Office, Korea Maritime Transportation Safety Authority, 27, Areumseo-gil, Sejong-si, 30100, Republic of Korea

[1] jmc723@komsa.or.kr

Abstract. Currently, IMO discusses to develop international standard regulations for the use of Maritime Autonomous Surface Ships(MASS), and as a preliminary work to develop them, it carries out RSE work to identify measures that might arise when the existing conventions are applied to MASS. The RSE work is done by the Maritime Safety Committee (MSC), the Legal Committee (LEG), and the Facilitation Committee (FAL) for the conventions under the purview of each committee. In this paper, based on the results of the RSE work performed by each committee, a comprehensive analysis of potential gaps and themes that require further discussion and consideration in the next stage of international standard regulations for the use of MASS is presented.

1. Introduction

Along with the development of ICT technology, smart, digital and eco-friendly technologies are also a major trend even in the marine field. Reflecting this, IMO is also making efforts to develop international regulations of a new paradigm by reflecting these major trends in the existing hardware-centered international maritime conventions. Among them, MASS is currently being discussed as the most important issue in the field of IMO maritime safety.

Currently, IMO is discussing to develop unified international regulations and guidelines for the use of MASS, and as a preliminary work to develop these regulations, carrying out the Regulatory Scoping Exercise (RSE) to analyze the potential gaps and themes that could occur when the existing convention is applied to MASS.

The IMO RSE was carried out by the Maritime Safety Committee (MSC) for maritime safety conventions, the Legal Committee (LEG) for liability compensation conventions, and the Facilitation Committee (FAL) for the maritime traffic facilitation convention.

This paper comprehensively analyzes the results of the RSE conducted by IMO to identify the potential gaps and themes that should be preemptively considered before discussing regulations and technology development for the commercialization of MASS, and by classifying them into definition, ship equipment, navigation, and infrastructure sections, the potential gaps and themes are identified and analyzed, and items that need to be considered in the future are presented.

2. IMO Discussion Trend

2.1 Maritime Safety Committee (MSC)

In IMO, MSC 8 was the first meeting that the concept related to ship automation was suggested in 1965. At this Committee, the word 'ship automation' is stated as a word that encompasses both 'complete automation system', 'partial automation system' and 'remote control', which shows that ship automation has been discussed and reviewed from a long time ago.

Since then, with the rapid development of information and communication technology and autonomous navigation technology, the use of unmanned ships in real seas has rapidly increased, and the need for guidelines and regulations to ensure their safety has emerged. Accordingly, in MSC 95(June 2015), starting with the information document from the UK, and in MSC 98(June 2017), the Committee agreed to include in the 2018-2019 biennial agenda of MSC and the provisional agenda for MSC 99, an output on "Regulatory scoping exercise for the use of MASS" with a target completion date of 2020(from MSC 99 to 102).

For RSE, IMO defined Maritime Autonomous Surface ship(MASS) as 'a ship which, to a varying degree, can operate independent of human interaction', and the degrees of autonomy were organized, rather than the autonomous level, following boarding of seafarers and remote control, with the four levels of autonomy as follows. Steps 1-2 are for the seafarers to board, and Steps 3-4 are for the seafarers not to board [1].

(1) Degree one: ***Ship with automated processes and decision support:*** Seafarers are on board to operate and control shipboard systems and functions. Some operations may be automated and at times be unsupervised but with seafarers on board ready to take control

(2) Degree two: ***Remotely controlled ship with seafarers on board:*** The ship is controlled and operated from another location. Seafarers are available on board to take control and to operate the shipboard systems and functions.

(3) Degree three: ***Remotely controlled ship without seafarers on board:*** The ship is controlled and operated from another location. There are no seafarers on board.

(4) Degree four: ***Fully autonomous ship:*** The operating system of the ship is able to make decisions and determine actions by itself

The procedure of RSE is divided into two steps. In the first step, the regulatory scoping exercise identified provisions in IMO instruments which, as currently drafted: 1) apply to MASS and prevent MASS operations, 2) apply to MASS and do not prevent MASS operations and require no actions, 3) apply to the MASS and do not prevent MASS operations but may need to be amended or clarified, and/or may contain gaps, 4) have no application to MASS operations. [2]

a second step was conducted to analyze and determine the most appropriate way of addressing MASS operations, taking into account, inter alia, human element, technology and operation factor by: 1) equivalences as provided for by the instruments or developing interpretations, 2) amending existing instruments, 3) developing new instruments, 4) none of the above as a result of the analysis. [2]

The first step was completed at the intersessional Working Group on MASS in September 2019, and the second step is expected to be completed in MSC 102(November, 2020).

2.2 Legal Committee (LEG) and Facilitation Committee (FAL)

In April 2018, LEG 105 decided to carry out RSE on conventions under under the purview of the Legal Committee, as the necessity of considering liability and compensation conventions other than the technical conventions discussed by the Maritime Safety Committee was recognized. LEG 106(march 2019) approved a framework for the RSE, and LEG 107 to be held in November 2020 will discuss the results of the 1st and 2nd step of RSE.

The 43rd Facilitation Committee (FAL) agreed to use the framework for RSE for the use of MASS approved by MSC 100, and the work will be finalized at FAL 44 to be held in September 2020.

3. Matters requiring clarification of terms and scope of application

3.1 Definition of remote operator, captain, seafarers, responsible person, etc.

In order for MASS to safely operate at sea, a new manpower called "remote operator" is needed to remotely control and manage them on shore. This person is a manpower that does not appear in the existing manned ships, and is a man power necessary to realize the autonomous level of remotely controlling the ship from shore. Therefore, in order to introduce remote operators, it is necessary to consider the insertion of definition, training and education, and qualification requirements of "remote operators" to STCW 1978.

The concept of a remote operator emerges in the degrees of autonomy 2 and 3 level, and a fully autonomous ship that artificial intelligence is in charge of all decision-making will emerge in the degrees of autonomy 4 level. In this case, there will be many changes in the roles and responsibilities previously performed by the master and seafarers, so it is necessary to clearly define the roles and responsibilities of the master and seafarers by each degrees of autonomy level.

For example, the current STCW 1978 defines a master as a "person in command of the ship," but when the autonomous navigation system makes the final decision, the role and responsibilities of the master will need to be re-established.

In addition, since the subject of responsibility according to the liability and compensation convention is determined depending on who is ultimately responsible for the ship, clarification of these terms is important.

3.2 Definition of remote operation/control center

The remote operation center is a concept that appears with the "remote operator," and refers to a space where remote operators work on shore and remotely control and manage ships. The remote operation center will perform a function of remotely controlling a ship on land or monitoring a sailing ship through data such as sensor-based video, audio, and sea conditions sent from the ship. Since the remote operation center is closely related to the wireless communication and navigation safety of the ship, it will be necessary to review the establishment of the definition of the remote operation center, design requirements, necessary facilities, and performance standards for facilities in SOLAS 1974 Chapter 4 (Radio Communication) and Chapter 5 (Sail Safety).

3.3 Consideration on MASS application scope and subject is needed for specific types of ships

The ship type is determined according to the cargo and purpose, and the structure of the ship and the cargo management method during voyage are very different by ship type. Therefore, it will be necessary to consider whether how to include nuclear ships that can lead to large-scale accidents in the event of an accident, passenger ships with a large number of people on board, and chemical tankers that require careful tank cleaning and cargo management in the scope of application of MASS.

4. Considerations related to ship equipment and technology

4.1 Need for sensor-based approach to enable remote operation

In International Regulations for Preventing Collisions at Sea(COLREG) 1972 Rule 5, it is prescribed that 'Every vessel shall at all times maintain a proper look-out by sight and hearing as well as by all available means appropriate in the prevailing circumstances and conditions so as to make a full appraisal of the situation and of the risk of collision.' In addition, SOLAS Ch.5-19 stipulates the carriage requirements for shipborne navigational systems and equipment.

In existing ships, the final decision was made by the navigation officer based on information such as RADAR, GPS, compass, Eco Sounder, and AIS when determining the optimum route setting and any action to avoid collision, but from the degrees of autonomy 2 level, the task which was performed by humans will be done by the autonomous navigation system. Therefore, in addition to the existing navigation equipment, it will be necessary to install additional equipment such as cameras and LiDARs that can play a role equal to the visual and hearing of the navigator, and it will be necessary to consider the performance standards and mounting regulations such as the visibility of such equipment.

4.2 Technical Review Matters for COLREG

COLREG 1972 Rules 20-31 regulate that the ship's lights and shapes are differently hoisted according to the ship's size, maneuvering status, and work conditions so that the conditions of the ship could be identified through them, and in the Rules 32-37, it is regulated, through sound and light signals, to exchange control and warning signals between ships, sound signals within a limited view, and caution or distress signals.

Even if autonomous ships are used, the time when MASS and existing manned ships coexist and sail will inevitably come. Therefore, in such a case, lights and shapes indicating that the vessels are autonomously operated will be needed to distinguish whether they are manned or autonomous vessels.

In addition, in the existing manned ships, the lights and shapes of other ships are identified with the human's vision and hearing, and the sound and luminous signals are recognized, while, in regard to the degrees of autonomy 3 and 4 level without a seafarers, it is necessary to find a way to identify and respond to the lighting and shapes, sound and luminous signals of other ships, by mounting a camera, LiDAR, and acoustic sensor, with the cognitive ability on the same level of sight and hearing as humans.

4.3 Radio communication and distress signals

In the case of existing manned ships, radio communication and distress signals are transmitted and received between seafarers members on board the ship and shore-based person. However, in the case of autonomous level 3 and 4 ship in which seafarers are not on board, it will be necessary to consider a plan for appropriately transmitting distress signals to shore and surrounding ships, when the ships are in distress.

4.4 Lifesaving Equipment and Lifesaving Supplies Arrangement Regulations

SOLAS Chapter 3(Life-saving Appliances and Arrangements) provides requirements related to personal lifesaving equipment, survival craft and rescue boats, personnel assignment of survival craft in case of emergency, emergency drills and exercises. In the case of the degrees of autonomy 3 and 4 level, because the seafarers will not be on board, a large number of provisions on lifesaving equipment and equipment will not apply.

Therefore, it will be necessary to review the provisions for the arrangement of life-saving equipment and life-saving supplies for autonomous level 3 and 4 ship.

4.5 Review of fire fighting requirements and equipment

SOLAS Chapter II-2(Construction- fire protection, fire detection and fire extinction) stipulates ship structures, facilities and training for detection of fire and extinguishing in case of fire. In the case of life-saving equipment, since it is for seafarers members, many provisions do not apply when seafarers members are not on board, but in the case of firefighting requirements, there is always a possibility of a fire, regardless of whether or not seafarers are on board, and since all fire extinguishing works that were performed by the seafarers must be unmanned, in the case of unmanned ships, ship structures and facilities for enhanced fire detection and fire extinguishing will be required.

In particular, different types of fire extinguishing equipment should be used depending on the type of fire (Class A~E), so it is necessary to devise a plan to extinguish the fire within a prompt time in an appropriate manner by identifying the type of fire from an unmanned ship.

4.6 Design of a bridge without seafarers from the design perspective

SOLAS Chapter 5(Safety of Navigation) Regulation 15 prescribes the principles relating to bridge design, design and arrangement of navigational systems and equipment and bridge procedures. For example, 15.1 of the same rule stipulates that the bridge should be designed to 'facilitate the task of the mission team and pilots to fully understand the situation and to operate the ship safely in all operational conditions.'

However, in the case of autonomous level 3 and 4 ships, the seafarers are not on board, so instead of the existing bridge, a bridge design principle suitable for unmanned ships should be established. In addition, in the case of unmanned ships, equipment that can replace the sight and hearing of the navigation officer should be mounted on the vessel, and the visibility distance and mounting position of the equipment that the equipment can play an optimal role should be considered.

5. Considerations related to ship and port operation

5.1 Ship maintenance

Conventionally, many regulations related to maintenance apply to various equipment of the ship, which implicitly assumes that the seafarers are on board. Therefore, it will be necessary to establish maintenance requirements for new types of equipment that are fixed on MASS or necessary for the function of remote control center in the degrees of autonomy 3 and 4 level in which the seafarers are not on board. Moreover, new requirements for preventive maintenance and early failure detection will be needed to prevent accidents from occurring in unmanned ships. Therefore, when considering these changes, the revision of many existing provisions related to manual operation will have to be accompanied at the degrees of autonomy 3 and 4 level.

5.2 Search and rescue

International support obligations for distressed ships should also be applied to MASS, but since seafarers are not on board in the degrees of autonomy 3 and 4 level, it will be impossible to participate in search and rescue operations like existing ships. Accordingly, it will be necessary to clarify the requirements and expectations for autonomous ships to provide assistance and support to people and other ships in the long term.

In addition, when rescue operations are carried out by remotely controlled or fully autonomous vessels, it is necessary to review whose actions or actions can be attributed to the "rescuer". In the case of rescue by remotely controlled vessels, it should also be clarified whether the regulation on the cost of the person "equipment and personnel" includes on-board equipment and personnel[3].

5.3 Cargo operation

Since the current cargo operation procedure (loading, securing, etc.) is based on human supervision, it is necessary to review the applicability of these regulations in the case of unmanned ships, which should be include that to which extent onshore personnel may be required to perform some duties.

In addition, it will be important to review regulations on safety and handling of cargo during voyage when dangerous goods are loaded on autonomous ships.

5.4 Pilot

The situation that ships can encounter mainly in the ocean and the situation that can be encountered in the port are very different. For example, in the case of a port, the amount of traffic is high due to the influence of ships entering and leaving the port, ships waiting for entry, and fishing vessels operating locally, and compared with the ocean, due to the influence of tide, shallow water zone, and the unique environment of the port, there are some high navigational hazards.

In the current situation, since the pilot is to board and operate the ship by oneself, it does not mean that the ship is remotely controlled by the remote control center. Therefore, from the stage of being remotely controlled, it will be necessary to discuss new regulations or other alternatives to allow remote control in relation to pilots.

In addition, in cases where autonomous level 3 and 4 ships that seafarers are not on board are docked in the port, discussion on the use of pilots should also be conducted.

5.5 Document storage and port state control

In the case of autonomous level 3 and 4, it is difficult to manage certificates, record books and other documents required by the existing rules, and therefore, in this regard, a clear regulation on appropriate digital documents, keeping in ships, and management is required. In addition, in the practical aspect of conducting inspections (especially port state control) of unmanned ships, it will be necessary to develop special guidelines and regulations, including solutions related to the detention of ships.

5.6 Safe manning and management

It will be necessary to review whether or not the regulations related to the boarding of seafarers, such as SOLAS Chapter 5/14 Regulations(Ship's Manning) and STCW regulation I/14 (Responsibilities of Companies), will be applied to shore-based persons such as remote operators. For example, whether onshore personnel should also be considered seafarers, or whether onboard personnel competency requirements stipulated in the STCW Conventions and Codes can be applied to them, and, if applicable, how to create and apply new regulations.

5.7 New certificates and procedures related to the information to be provided to the administration

In Standard 2.1, Section 2 of the FAL Convention(Arrival, Stay and Departure of the Ship), documents that the administration may request from ships are stipulated when they are entering and departing. Representatively, there are general declaration, cargo declaration, ship's stores declaration, crew list, passenger list and dangerous goods manifest, but in case of MASS, there is a need to submit additional information to the administration, such as the autonomous level, so it will be necessary to amend the general declaration to be entered or to establish a new form of entry and departure documents.

In addition, since seafarers do not board ships in the degrees of autonomy 3 and 4 level, it will be necessary to review the plan for preparing the seafarers list. [4]

5.8 consideration a new electronic format for various certificates required under the Convention

Ships engaged in international voyages shall have various certificates on board in accordance with international conventions. In addition, FAL 42 approved the guidelines for the use of electronic certificates (FAL.6/Circ.39/Rev.2), encouraging member states to use electronic certificates. In the case of autonomous level 3 and 4, an e-certificate must be used as seafarers do not board the vessel, and new regulations to safely exchange such e-certificates with various stakeholders such as administrations and ports will be required. [4]

5.9 Others

It is necessary to review procedures for responding to emergency situations to apply to unmanned ships where seafarers do not onboard, cyber security and countermeasures to prevent cyber crimes against ships, and to prepare ship security plans. In addition, the establishment of a new regulation on the treatment and treatment of stowaways on board unmanned ships is a matter to be considered.

6. Conclusion

In this paper, based on the results of the RSE conducted by IMO, the potential gaps and themes that should be considered in advance prior to future related regulations and technology development was comprehensively analyzed.

In the discussion of MASS, the review of these potential gaps and themes is considered to be a very important factor, and it is expected that IMO also will conduct detailed reviews of these potential gaps and themes in the future. Therefore, in order to lay the foundation for establishing an organized system that allows MASS to operate safely, it will be important to draw an international consensus through professional and detailed discussion and consideration on the potential gaps and themes identified in this paper.

References

[1] IMO document MSC 100/20/Add.1 "Report of the Maritime Safety Committee on its one hundredth session", Annex 2 "Framework for the regulatory scoping exercise for the use of Maritime Autonomous Surface Ships (MASS)"

[2] IMO document MSC 100/20/Add.1 "Report of the Maritime Safety Committee on its one hundredth session", Annex 2, Appendix 2 "Plan of work and procedures for the regulatory scoping exercise"

[3] IMO document LEG 107/8 " Summary of results of analysis of IMO instruments under the purview of the Legal Committee"

[4] IMO document FAL 44/14 " Report on the results of the regulatory scoping exercise on the FAL Convention"

Technology assessment on the astonomous ships: key findings and recommendations

Ikroh Yoon

Korea Institute of Marine Science and Technology Promotion, Seoul, Republic of Korea

E-mail : iryoon@kimst.re.kr

Abstract. Maritime autonomous surface ship (MASS) has been developed rapidly. MASS could be disruptive technology that will bring about a paradigm shift in related industries as well as the maritime transport system. Due to huge implications and uncertainties, it is required to understand complex socio-economic behaviour induced by MASS for the international society. This paper presents the key findings and some recommendations from technology assessment on MASS which investigated the potential effects from MASS in terms of economic, social, cultural and ethical aspects. The results can be used to design a future driven by MASS in a desirable direction.

1. Background

Recently, advances in technology spanning digitalization, big data and Artificial Intelligence (AI) have reached a level where commercialization of not only autonomous vehicles, but also autonomous ships are imminent. The development of technology controlling ships from shore has been progressing at a rapid pace, especially driven by the industry [1-5]. In 2017, the International Maritime Organization (IMO) decided to adopt as one of its seven Strategic Directions to be pursued for the 2018-2023 timeframe, "Integrate new and advancing technologies in the regulatory framework" [6]. In June 2018, the Maritime Safety Committee (MSC) of IMO decided to embark on the Regulatory Scoping Exercise (RSE) to adopt and operate Maritime Autonomous Surface Ships (MASS) [7].

As a result, the global community has been called to make various efforts in technical, political and social aspects to accommodate MASS. MASS is not merely one of many sides of technological progress. It will be an innovation that disrupts, reconstructs and induces a paradigm shift in the shipping industry and maritime transport system as a whole. In order to guide the future in a desirable direction under such change, the international society should be able to understand the economic and social dynamics encompassing maritime safety, environmental protection, seafarer welfare, training and education, liability, insurance and ethical concerns. In addition, communication and cooperation of multiple stakeholders will be a prerequisite to ensure a safe, effective and efficient maritime transport system with MASS. In this regard, We have carried out technology assessment to not only provide useful guideline for policymakers but also contribute to forming social discourse surrounding MASS.

The remainder of the paper is as follows: Section 2 presents the brief procedure and methodologies of technology assessment; Section 3 describes key findings and disscusions from the technology assessment; some recommendations are suggested in Section 4; and finally results are summarized in Section 5.

2. Technology Assessment : Procedure and Methodology

We conducted a technology assessment on MASS as per Article 15 of the Marine Science and Technology Cultivation Act that took effect in June 2017. Main purpose of this assessment is to explore the economic, social, cultural, ethical and environmental impact of the new technology, thereby maximizing benefits and preparing against negative side effects. The outcome would serve as a guideline for reasonable policy-making. Also, the assessment is expected to facilitate communication among diverse members of the society and contribute to forming social discourse to ease the ripple effect of new technology and promote social acceptance.

The assessment was led by a working group that consists of 27 experts from a number of areas including technology, economics and society. Additional contributions were also made by 11 groups of key stakeholders from wide-ranging fields such as labour unions of seafarers, marine pilot groups, key industries including shipping, shipbuilding and equipment and device businesses, the public sector including Vessel Traffic Centers, and students who will drive the development of maritime transport in the future. The assessment addresses various effects that may arise upon the introduction and operation of MASS including industry stakeholders, changes in risk factor, legal issues, education and training, ethical concerns, liability and insurance.

The working group which was divided into three sub-groups, i.e. technology-, economy- and society-subgroup, assessed cross-disciplinary issues surrounding MASS via a total of twelve meetings for focus group interview (FGI), stakeholder surveys and media analysis. The details of those procedure and methodologies can be found in [8].

A definition on "Autonomous Ship" that is universally agreed upon in the international society is yet to exist. "Autonomous" and "Unmanned" are clearly different ideas. However, the "Autonomous Ship" studied in the technology assessment is defined as "all ships equipped with full-scale or partial autonomous systems operated without human intervention". Thus, it includes "Unmanned Ship" such as ships remotely operated from shore without any seafarer on board.

The technology assessment has been focused on surface cargo ships such as containerships and bulk carriers. However, ships that transport hazardous materials including oil tankers were precluded from the study in that their acceptance levels are predicted to be relatively low. Passenger ships, cruises, fishing vessels and warships were also excluded.

It is worthwhile to mention that the technology assessment, as a multidisciplinary review, has been implemented to forecast diverse impacts in a wide-ranging area including the economy, society, culture, ethics and environment. Basic assumptions must be set for experts from numerous fields and members of the society to explore a single theme together. The following are the key technology of the future and subsequent changes that were presumed for the assessment [8,9]:

- Ships will be operated by remote control from shore or even without remote control from shore (fully autonomous), in which case minimum human intervention is made only in emergency.

- Key information on ships will be monitored from the shore. Frequency of maintenance and repair will be optimized through the use of big data. Even during operation on waters, minimum maintenance and repair can be conducted with the help of technology such as drones.

- Bridge and accommodation will be removed to expand cargo space and improve arrangement of on-board equipment and devices.

- Exterior design of ships will change. Closed structure design could be adopted to prevent hijacking of cargo by pirates. But external access for maintenance and repair, and ease of loading / unloading will also be considered.
- Every equipment and device installed on board will be interconnected and integrated to enable collection, management and analysis of data. They will be equipped with a high level of redundancy and durability, and will be highly modular to avoid failure.

IOP Conf. Series: Materials Science and Engineering **929** (2020) 012015 doi:10.1088/1757-899X/929/1/012015

- Not only navigation, but also docking and maneuvering will be remotely controlled or fully automated. In support, port infrastructure will be transformed.

3. Key Findings

3.1. Ecosystems and Stakeholders

In the long run, MASS will bring about sea change to shipbuilding, equipment and device, and shipping and port industries. Especially, shipping and port industries (see Figure. 1) form a dynamic ecosystem, in which numerous stakeholders participate compared to shipbuilding, equipment and device industries (see Figure. 2). For large parts of this ecosystem, public and private sectors are intertwined through complex regulations. On top of this, a variety of new stakeholders related to MASS are expected to emerge in the ecosystem such as telecommunication service providers and maritime cyber security businesses. The safe, effective and efficient adoption and operation of MASS hinges on communication and collaboration among such stakeholders, especially the shipping and port industries. Detailed procedure and information for preparing Figs. 1 and Fig. 2 can also be found in [8].

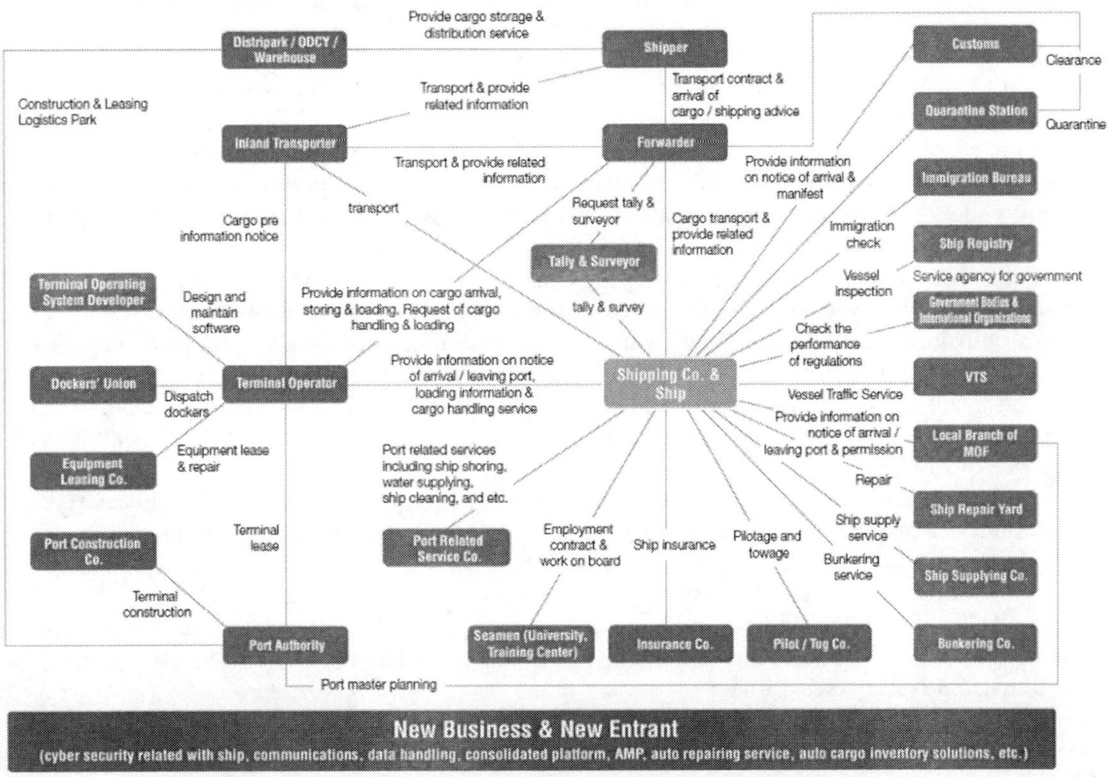

Figure 1. Stakeholders of Shipping and Port Industries of the ROK.

IOP Conf. Series: Materials Science and Engineering **929** (2020) 012015 doi:10.1088/1757-899X/929/1/012015

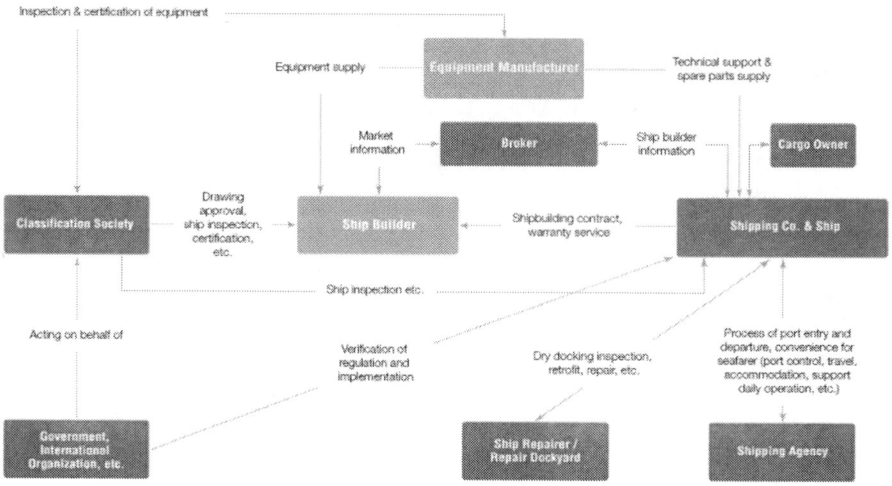

Figure 2. Stakeholders of Shipbuilding, Equipment and Device Industries of the ROK.

3.2. Changes in Risk Factor

The main cause of maritime accidents is human error, which in fact takes account for around 79 % of maritime accidents in the ROK (see Table. 1). However, the introduction of MASS is predicted to change the causes of maritime accidents. MASS reaching autonomy levels 3 (remotely controlled ship without seafarers on board) and 4 (fully autonomous), based on IMO's preliminary degrees of autonomy [10], are widely anticipated to reduce human error since they are operated without seafarers on board [11].

Yet, the absence of seafarers exacerbates the concern that there will be no skilled manpower capable of responding in the event of an accident. According to a study, while risks of navigational accidents such as collision and stranding may decrease, risks of non-navigational accidents including fire, explosion and flooding may rather increase on autonomous ships [12]. To ensure a safe maritime transport system, technical and institutional measures to respond to diverse types of accidents should be derived rather than vaguely anticipating that the occurrence of human errors will decline with the emergence of MASS.

In the future, causes of accidents that have been minimal in number or that have not been recognized as important risk factors could stand out as well. In particular, cyber security is recognized as one of the prerequisites to enable practical operation of MASS. It should be understood that autonomous ships equipped with the same or similar system could be exposed to cyber threats simultaneously. Defects in equipment and devices including autonomous operating systems, information error and distortion, difficulty of recognizing accidents and challenges in cargo management could be all potential threats against maritime safety (see Table. 2).

MASS may drive changes in the patterns of pirate, terrorist and criminal activities. Cases of human loss including hostage situations and kidnapping by pirates and armed robberies may decrease. But attempts at abducting the ship itself targeting valuable cargo may increase due to the absence of seafarers.

There is also the inherent risk that MASS may be abused for crime such as transport of illegal cargo including arms and drugs. Thus, the existing Customs, Immigration and Quarantine (CIQ) procedure centered on human may turn out to be ineffective for MASS. Technical and institutional considerations should take place to strengthen port security by developing new inspection mechanisms or changing the place of inspection if needed. In this regard, the CIQ using electronic means need to be taken into consideration with MASS.

Table. 1. Cause of Maritime Accidents of the ROK for the past 5 years (2013-2017)

Cause		Cases	Ratio
Navigational Faults	Insufficient preparation of departure	17	1.1 %
	Insufficient survey of nautical conditions	0	0.0 %
	Poor management of voyage plan	4	0.3 %
	Negligence of evaluating ship location	39	2.6 %
	Inappropriate maneuvering	55	3.6 %
	Negligence of look-out	744	48.9 %
	Insufficient preparation and response to heavy weather	27	1.8 %
	Inappropriate anchoring and mooring	9	0.6 %
	Violation of navigation rules	128	8.4 %
	Negligence of duty and supervision	12	0.8 %
	Negligence of watch keeping	29	1.9 %
	Other navigation errors	18	1.2 %
	Incompliance with safe working regulations on board	120	7.9 %
	Total	**1,202**	**79.0 %**
Deficiency	Deficiency of hull and machinery	81	5.3 %
	Poor maintenance of machinery	98	6.4 %
	Poor management of inflammable equipment and deficiency of wires	12	0.8 %
	Total	**191**	**12.5 %**
Others	Inappropriate stowage of passenger and cargo	14	0.9 %
	Inappropriate management for ship operation	50	3.3 %
	Inappropriate crew management	1	0.1 %
	Inappropriate provision of AtoN service	8	0.5 %
	Weather and force majeure	36	2.4 %
	Others	20	1.3 %
	Total	**129**	**8.5 %**
Grand Total		**1,522**	**100.0 %**

Source: Korea Statistical Information System

Table. 2. Examples of Potential New Risk Factors against Maritime Safety

No.	Risk Factor	Example
1	Rise of cyber security threats	· Hacker attacks to abduct ship or hijack cargo · Leakage of sensitive information on cargo and customer
2	Failure of equipment or device	· Failure of ship due to failure of key operation systems including propulsion system · Failure of information and communication system required for autonomous operation such as failure of communication
3	Error or distortion of information	· Distortion of information communicated with on-shore control center including information on ship operation
4	Difficulty of recognizing accident	· Failure or delay of on-shore ship operator to recognize the occurrence of accident
5	Challenge of cargo management	· Safety-related problems such as cargo being set on fire without seafarer on board
6	Threat against port security	· Weaponization of autonomous ships

3.3. Legal Issues

The legal status of MASS is yet to be determined, but their concept is expected to change. As human seaworthiness is reduced going forward, ships will no longer remain passive subjects operated by seafarers. Instead, they may gradually transform into active agents that make small and big decisions to achieve safety and efficiency of operation.

A legal controversy surrounding MASS may emerge in regards to the United Nations Convention on the Law of the Sea (UNCLOS), most likely in that it would be difficult to secure effectiveness of enforcement when coastal states exercise jurisdiction (see Figure. 3).

As most international maritime conventions such as the International Convention for the Safety of Life at Sea (SOLAS) and International Regulations for Preventing Collisions at Sea (COLREG) will be applied to MASS as well, global society will also need to review the scope of domestic laws alongside the RSE.

3.4. Jobs, Education and Training

The emergence of MASS has been driving concerns on the decline in the number of seafarers and jobs. It is also anticipated that autonomous ships will enhance the quality of life of seafarers. If ships are controlled from the shore, the difficulty stemming from staying on board for a long period of time and the risks of marine accidents will be alleviated. At the same time, high-caliber workforce equipped with the skill for operating MASS from shore may be able to enjoy higher income and improved welfare.

It is clear that the advent of MASS will significantly change the landscape of jobs related to seafarers. The minimum safe manning level is sure to be decreased and the jobs will be replaced by AI and autonomous systems, just to start with. Therefore, operators capable of maneuvering the ships from shore and backed by relevant certification will be in high demand.

In the face of the decreasing number of seafarers, it would be significant to develop qualification standards for on-shore operators of MASS and to provide relevant training and education. Operators must have basic knowledge on issues such as navigation safety, terrestrial and coastal navigation, planning and conducting passage and determining position, navigation equipment, meteorology and emergency procedures. In the case of an emergency on board, operators must be able to clearly recognize the situation and mobilize basic skills and knowledge as deck officers to respond to danger.

Hence, they should be certified as appropriate under the International Convention on Standards of Training, Certification and Watchkeeping for Seafarers (STCW) Convention. If needed, it might be necessary to add new qualification standards in the STCW Convention or new Knowledge, Understanding and Proficiencies (KUP). These efforts will help appeal the attractiveness of working as seafarers or on-shore operators to the youth. Given the short on-board experience of young seafarers as evidenced in the past, hiring them as on-shore operators of MASS may contribute to guaranteeing continuity of their career as seafarers.

3.5. Ethical Concerns

In the process of developing and adopting autonomous ships, a wide variety of ethical issues are expected to be raised. The controversy surrounding autonomous vehicles in the auto industry, which is a step ahead in applying automation, gives a glimpse of what the future holds for autonomous ships. Yet, ethical concerns can be compounded with autonomous ships given the special circumstances of navigating on sea. Thus, such issues should be dealt with in a serious manner.

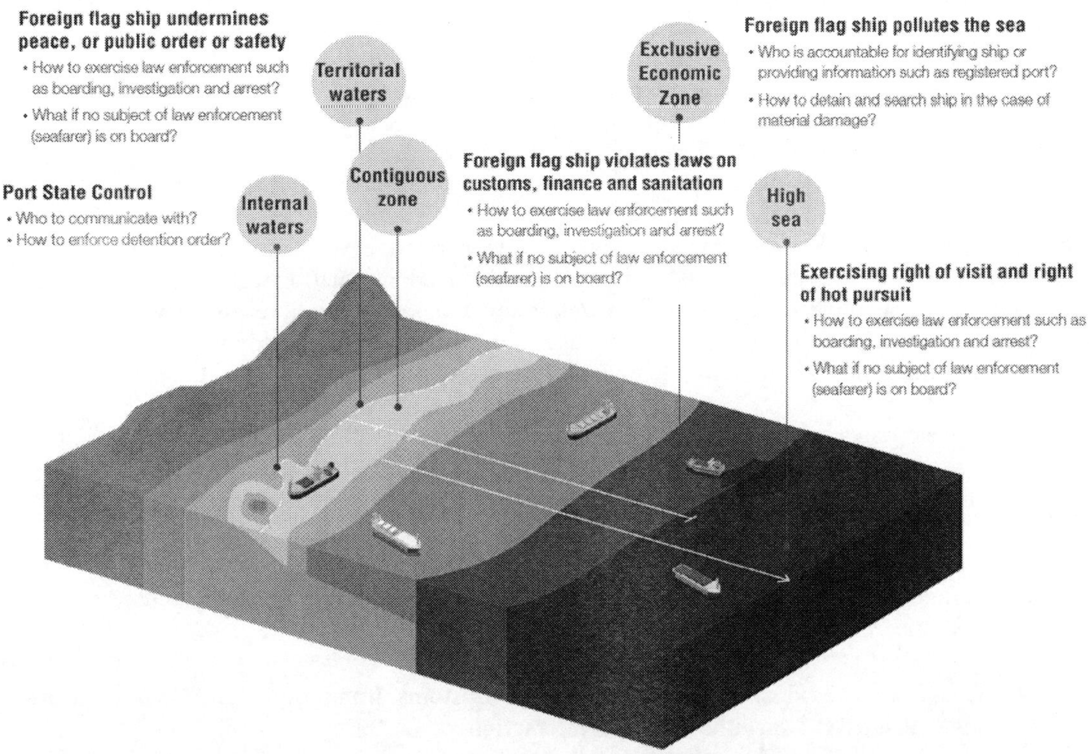

Figure 3. Potential Issues related to UNCLOS Caused by MASS

3.6. Liability and Insurance

Just like other unmanned vehicles, MASS is expected to bring about a shift in the liability structure regarding maritime accidents. It is likely that the liability of manufacturers in the case of MASS will increase in relativity to conventional manned ships. However, pinpointing who to blame can be even more of a challenge due to reasons including modification of ships and failure to update autonomous operation systems in a timely manner.

Especially, setting reasonable criteria and scope on liability between shipowner and manufacturer, and an appropriate security structure for insurance coverage such as who should make compensations first and exercise the right to indemnity afterwards will be tricky issues. In this regard, views of multiple stakeholders should be reflected to consider the public benefit of accommodating MASS while allowing swift remedy for victims in a comprehensive manner. Other issues to be deliberated include whether it is appropriate to maintain the shipowners' limitation of liability system in the case of MASS in light of higher ship price, different liability structure and subsequent burden. It would be worthwhile considering to adopt the mandatory insurance system following certain international maritime conventions including the International Convention on Civil Liability for Bunker Oil Pollution Damage, the Nairobi International Convention on the Removal of Wrecks and the case of autonomous vehicles in several countries.

4. Recommendations for Future Discussion

4.1. Technology and Industry
Once the IMO's Regulatory Scoping Exercise (RSE) is completed, not only establishment and amendment of IMO Conventions related to MASS, but also standardization of core technology should be carried out. There are several key organizations who will possibly address those regulation and standardization issues of MASS. Building a cooperative framework would help raise efficiency as various international organizations such as IMO, International Organization for Standardization (ISO), International Electrotechnical Commission (IEC), International Hydrographic Organization (IHO), International Telecommunication Union (ITU) and other related organizations are expected to work on standardization in a simultaneous manner. In addition, proactive support of international society is required, including the holding of international fora, seminars or workshops on MASS to enable a balanced collection of multi-stakeholder opinion in the process of amending Conventions and developing new regulations as well as new technical standards.

4.2. Maritime Safety
By the time when MASS is introduced and practically operated, new risk factors are expected to emerge such as threats against cyber security and errors in advanced automation systems. In this regard, it is a prerequisite to amend existing Conventions and/or develop new regulations after reviewing RSE. At the same time, international society needs to engage in discussion on the following issues considering the trends of digitization and automation to enhance the safety of MASS:

- Developing safety assessment guidelines by degree of autonomy of MASS.
- Improving cyber security guidelines for various stakeholders including ports as well as ships.
- Reinforcing international cooperation programs for the capacity-building including training and education aimed at ensuring safe operation of MASS, especially focused on developing countries.
- Considering reasonable methods to improve Customs, Immigration and Quarantine procedures given that MASS is without human interaction.

4.3. Legal Issues
As mentioned above, legal controversy emanating from MASS may arise in relation to the United Nation Convention on the Law of the Sea (UNCLOS) as well, most likely in that coastal States will find it difficult to secure effectiveness of legal action in the process of exercising jurisdiction. Hence, potential legal issues in relation to UNCLOS should be widely reviewed.

4.4. Training and Education

In the interest of achieving effectiveness in introducing and operating MASS, it would be imperative to nurture expert personnel capable of operating MASS above all. Especially, in-depth consideration should be given to the qualification of remote control operators on shore who will control MASS on shore. Remote control operators on shore need to be equipped with the basic capability required from seafarers as per the International Convention on Standards of Training, Certification and Watchkeeping for Seafarers (STCW). If needed, the option of adding new qualification criteria or KUP in STCW should be considered. Moreover, international society should look for ways to expand support to developing countries.

4.5. Ethical Concerns
Ethical issues also need to be reviewed. It should be ensured that ethical/moral judgment of humans is not undermined due to overdependence on technology. Even if ships are operated from shore, moral hazard, under which maritime accidents are taken lightly must be checked against with vigilance. Furthermore, a range of ethical concerns need to be considered, including the difficulty of human-machine communication, scope of delegating machines with decision-making authority and privacy protection. , International society should also deliberate ways to come up with ethical norms, which multiple stakeholders such as manufacturers, owners and remote control operators on shore of MASS need to comply with.

5. Conclusion
This paper presents the key findings and recommendations from technology assessment which has been carried out in order to provide useful guideline for policymakers and contribute to forming social discourse surrounding MASS. Key findings are as follows:

- Shipping and port industries form a complex and dynamic ecosystem where numerous stakeholders participate. Communication and collaboration among diverse members of society are highly needed to ensure safe, effective and efficient introduction and operation of MASS.

- Although the current main cause of maritime accidents is human error which takes account for about 79% of entire maritime accidents in ROK, the introduction of MASS is predicted to change the causes of maritime accidents. Six types of potential new risk factors are also proposed.

- The concept of the ship is expected to change. As human seaworthiness is reduced going forward with MASS, ships may gradually transform into active agents that make small and big decisions to achieve safety and efficiency of navigation. Considering the UNCLOS with MASS, legal controversy can emerge due to the difficulty of securing effectiveness of enforcement when coastal states exercise jurisdiction.
- Concerns and optimistic expectations on jobs coexist. Although the number of seafarers' job can decline, quality of jobs can improve and MASS can create new and attractive jobs in the future. Also, since on-shore operators of MASS should have basic knowledge and skills of navigation including the case of emergency situations, it is important to develop proper qualification standards.

- Ethical issues related to MASS are expected to be widely raised and should be dealt in a serious manner. Despite the difficulty of finding universal solutions as with the trolley dilemma, tackling such issues will be significant given that humans will delegate big and small decisions to ships and autonomous systems in the future. Especially, the moral hazard of taking maritime accidents lightly should be prevented as the safety of seafarers will be separated from that of ships by MASS.

- MASS is expected to bring about a shift in the liability structure regarding maritime accidents. The liability of manufacturers would increase compared to conventional manned ships. The public benefit of accommodating MASS and allowing swift remedy for victims should be considered in a balanced way, in the process of setting reasonable criteria and scope on liability between shipowner and manufacturer.

Acknowledgments
This paper is based on original work entitled "Technology Assessment on Maritime Autonomous Surface Ship [8, 15]". The author mentions that the results of technology assessment were also partially presented in [15-17]. The author greatly thanks many experts who participated in the technology assessment project for their helpful and constructive advice on this paper. Details and the full list of experts and contributors can also be found in [8].

References
[1] Levander O. (2016). *Ship Intelligence Era*. Digital Ship Conference, Copenhagen, April 2016
[2] MUNIN (2016). *Marine Unmanned Navigation through Intelligence in Networks*. [Online] Available at: http://www.unmanned-ship.org/munin/
[3] AAWA (2016). *Remote and autonomous ships: the next steps*. Rolls-Royce
[4] Adams, S.D. (2014). *ReVolt – next generation short sea shipping*. [Online] Available at: https://www.dnvgl.com/news/revolt-next-generation-short-sea-shipping-7279
[5] Kongsberg. (2016). *YARA selects Norwegian shipbuilder VARD for zero-emission vessel Yara Birkeland*. [Online] Available at: https://www.km.kongsberg.com/ks/web/nokbg0238.nsf/AllWeb/ C0203EEC2B0E9DC7C12582EB0021D6F5?OpenDocument
[6] IMO. (2017). *Strategic plan for the organization for the six-year period 2018 to 2023* (A 30/Res.1110)
[7] IMO. (2017). *Report of the Maritime Safety Committee on its ninety-eighth session* (MSC 98/23).
[8] Yoon, I. *et al.* (2018). Final Report of Technology Assessment on Autonomous Ships(in Korean), KIMST
[9] Chae, C. *et al.* (2020). A Study on Identification of Development Status of MASS Technologies and Directions of Improvement. *Appl. Sci.* 10(13) pp.4564
[10] IMO. (2018). *Report of the Correspondence Group on MASS* (MSC 100/5).
[11] AGCS. (2017). *Safety Shipping Review 2017*
[12] K. Wrobel, *et al.* (2017). Towards the assessment of potential impact of unmanned vessels on maritime transportation safety. *Reliability Engineering and System Safety*, 165, pp.155-169
[13] Bonnefon, J.F. *et al.* (2016). The Social dilemma of autonomous vehicles. *Science*, Vol. 352, Issue 6293, pp. 1573-1576
[14] Clamann, M. *et al.* (2017). Evaluation of Vehicle-to-Pedestrian Communication Displays 1 for Autonomous Vehicles. *Proceedings of the Transportation Research Board 96th Annual Meeting*, Washington DC, United States
[15] Yoon, I. *et al.* (2018). Technology Assessment : Autonomous Ships, KIMST
[16] IMO. (2018). *Results of technology assessment on Maritime Autonomous Surface Ships (MASS)* (MSC 100/INF.10)
[17] Yoon, I. *et al.* (2018). Key Finding of Technology Assessment on Autonomous Ships, *Proceedings of ICMASS 2018*, November 2018, Busan, Korea (In print)

IOP Conf. Series: Materials Science and Engineering **929** (2020) 012016 doi:10.1088/1757-899X/929/1/012016

It's not all about the COLREGs: a case-based risk study for autonomous coastal ferries

Dag Rutledal, Tore Relling, Tron Resnes

Department of Ocean Operations and Civil Engineering, NTNU Ålesund, Norway

dag.rutledal@ntnu.no

Abstract. Whilst the maritime world is preparing for autonomous vessels or Maritime Autonomous Surface Ships (MASS), as defined by IMO, we suggest an approach to investigate the risks involved with introducing such ships into service. This case study investigates the risks regarding other surface vessels that the MASS can encounter on their voyage between two berths used by coastal ferries in Norway. The ferries in this case, operate within a fjord declared as testbed for autonomous operations. Albeit the technology is available and declared ready, the area of operation is not. This paper seeks to address some of the remaining obstacles with regards to the risks involved.

1. Introduction

The development of Maritime Autonomous Surface Ship (MASS) is a focus area for maritime developers, regulators, and researchers. Despite such focus for several years, no commercial MASS concept is operational. As autonomous vessels will coexist with conventional vessels, it is necessary to have a common set of rules for navigation. As such, all vessels, including the future autonomous vessels, are expected to follow the Convention on the International Regulations for Preventing Collisions at Sea (COLREGs). One major challenge for maritime autonomy is to coexist with conventional vessels and how to ensure safe navigation when no navigators are left on the bridge.

A large part of the COLREGs is prescriptive on how traffic conflict situations should be solved, however, the COLREGs also opens for the navigator making decision on how to resolve a situation based on seamanship. The ordinary practice of seamanship allows the crew to take action to avoid a collision that may not comply with the COLREGs but is necessary in particular circumstances to avoid a collision. While a set of prescriptive rules can be programmed, the human aspect, or seamanship, cannot be included in a pre-programmed solution. Consequently, following the COLREGs is a challenging task for conventional vessel, but even more challenging when autonomous and conventional vessels will have to coexist.

In this study, a geographical area with a large variation of traffic was monitored for a period of 30 days to identify how often conventional vessels deviate from the prescriptive rules of COLREGs to avoid potentially dangerous situations in overtaking, head-on, and crossing situations.

IOP Conf. Series: Materials Science and Engineering **929** (2020) 012016 doi:10.1088/1757-899X/929/1/012016

2. Background

The convention on the International Regulations for Preventing Collisions at Sea (COLREGs), 1972 as amended, regulates the behavior of vessels in marine traffic situations. The COLREGs states that the risk of collision should be carefully assessed at an early state, and if there is a risk of collision normally one vessel is the give-way vessel and the other is the stand-on vessel. However, this is not the case in head-on situations where both vessels are expected to give way. In this it is implied that the officer on the stand-on vessel needs to trust the officer on the give-way vessel and wait for him to conduct an evasive maneuver. The COLREGs do not provide detailed information on when and how an avoiding maneuver should be performed. This leads sometime to uncertainties [1]. According to the COLREGs the stand-on vessel shall maintain her speed and course until she finds herself so close that collision cannot be avoided by the action of the give-way vessel alone. How to determine when this is the case, is up to the navigator on the stand-on vessel. This introduces the human role into the assessment of what and when appropriate action, although sometimes seemingly in contradiction to the COLREGs, should be taken. Porathe, [2] discusses the confusion about which rule is applicable in certain situations. Arguably the COLREGs are a trade-off between safety and efficiency. The rules are open for interpretation with wordings like an evasive maneuver should be carried out in "ample time". What ample time is, is dependent on the state of one's own vessel, the state of other vessels interacting, weather conditions, the condition of the waterways among others. How the navigator perceives these variables will affect the action he chooses.

The Convention on the Regulations for Preventing Collisions at Sea (COLREGs) regulates the behavior of vessels in marine traffic situations through rule 1-19. Depending on the relative bearing between the vessels, different rules describe appropriate action, see Figure 1. In this paper we choose to focus on rules 13 to 17. Rule 13 covers overtaking (vessel D) and states that an overtaking situation is when one vessel is coming up with another vessel from a direction more than 22.5 degrees abaft her beam. This means that the overtaking vessel can only see the other vessel's stern lantern. Rule 14 describes head-on situations (vessel B) and states that this is the case when two power driven vessels are meeting on (or nearly) reciprocal courses. This means that each vessel could see both starboard and port side lanterns and both masthead lights in (or nearly) in line. Rule 15 covers crossing situations and states that this is the case when the crossing are seen to involve a risk of collision, the vessel which has the other on her own starboard side (vessel C) shall keep out of the way and shall avoid crossing ahead of the other vessel. Rule 16, action by give-way vessel, states that a every vessel which is directed to keep out of the way of another vessel shall, as far as possible, take early and substantial action to keep well clear. Rule 17, Action by stand-on vessel states that when one of the vessels is to keep out of the way, the other shall keep her course and speed (vessel A).

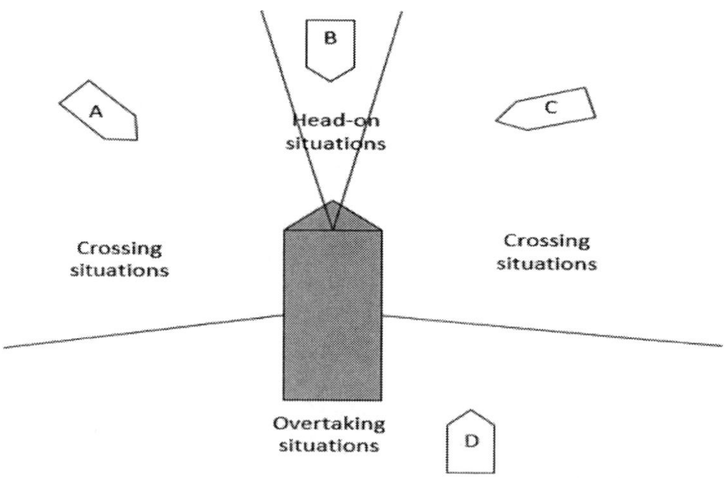

Figure 1: Categorization of situations depending on the relative bearing

The exact time or distance for an avoiding action is not stated. The COLREGs do address this in vague terms in rule 8 "Action to avoid collision", but these discussions lie outside the scope of this paper. This has been discussed on several occasions, among others by Olinderson et. al, [3].

2.1 Maritime autonomy
Autonomous, unmanned vessels are expected to play an important role in future maritime industry. To date, autonomous shipping appears to primarily have been about a technology push rather than considering and providing sociotechnical solutions including re-design of work, capturing knowledge, and addressing human factors in modern shipping, [4]. Several research projects are underway, among others the Autoferry, a small passenger ferry which is expected to be in operation in the near future, [5]. The general idea with this autonomous passenger ferry is to incorporate automatic anti-collision features with detection and identification of other objects in the area, such as leisure boats, kayaks, or drifting debris. The ferry itself is expected to follow the rules of the road stated in the COLREGs.

2.2 Automatic Information System (AIS)
Today it is possible to analyse vessels' actual track with information from the Automatic Information System (AIS), used onboard almost all merchant vessels and many leisure crafts, but not nearly all of them. Messages from the AIS system are sent out with short time periods 2-360 seconds. This interval is depending on the vessel's status (underway, moored, anchored etc.), speed and course change. The accuracy and integrity of positions derived from AIS has been debated earlier, see [6]. Still it is used in many studies as the best way of measuring the track of vessels at a low cost. In this study, in addition to AIS we used a radar tracking system combined with a pan-tilt-zoom (PTZ) camera to also track the vessels without AIS. To our knowledge, this has not been done before.

This projected is limited to look at rules 13 to 17 which dictates appropriate actions by both stand-on and give-way vessels in overtaking, head-on situations and crossing situations. These were the situations most easily recognized with regards to the limitations our equipment represented. The goal was to identify exception to those specific rules. A Maritime Autonomous Surface Ship (MASS) will need to able to recognize such situations warranting deviations from the rules.

2.3 Objective/research question
The aim of this study has been to identify the traffic situations that occurred in a ferry crossing in the north-western part of Norway during a period of 30 days, in the late summer of 2018. The traffic was recorded and during playback, relevant situations were analysed and how they related to the COLREGs were assessed and traffic conflict situations between the ferries and other vessels were identified. In the conflict situations, the study identifies if the evasive actions carried out in were compliance with COLREGs or not. The study is limited to assess the situations against COLREGs rule 13-17, and do not consider if other means, such as the use of radio communication, were used.

The research question is:

How often are the COLREGs apparently breached when making evasive actions to resolve traffic conflict situations in a ferry crossing?

3. Method

Storfjorden, located in the north-western part of Norway see Figure 2, is a designated test area for autonomous vessels. At the point of this ferrycrossing, the fjord is more than three nautical miles wide. Therefore we do not consider this as a narrow channel. In this fjord there are several ferry connections as parts of both national and local roads. At the end of the fjord lies Geiranger, listed on the UNESCO world heritage list. As such, Geiranger is a popular destination for cruise vessels who use the Storfjord enroute to Geiranger.

IOP Conf. Series: Materials Science and Engineering **929** (2020) 012016 doi:10.1088/1757-899X/929/1/012016

Selecting the best location for monitoring purposes was based upon three criteria: Firstly, the crossing should be in the most vessel congested areas of the Storfjorden. Secondly, we needed access to electrical power and thirdly the length of the crossing should be within the range of both the camera and the radar. Based upon these criteria, it was decided to install the monitoring equipment at "Klubben" light located at the "Solavågen-Festøy" ferry crossing, a three nautical miles crossing. Klubben light, in position N 62° 24.5' E 006° 19.28.0' see Figure 2, is a lantern with indirect illumination and has available 220V power as well as a separate storage space in the foundation itself. This storage space was utilized for the monitoring equipment.

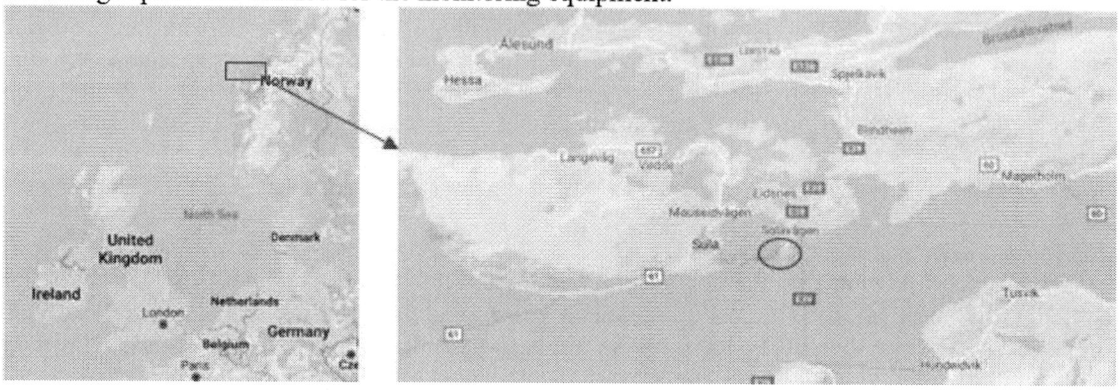

Figure 2: Location of the monitoring equipment

3.1 Monitoring equipment

The monitoring equipment should be able to record and document the traffic in the ferry crossing area. Our aim was to identify the vessels with Automatic Identification System (AIS) installed, as well as the smaller vessels, i.e. kayaks, leisure boats and small fishing vessels that do not have AIS installed on board. In order to do this, a 360° pan-tilt-zoom (PTZ) camera was chosen. In conjunction with a solid state radar with Automatic Radar Plotting Aid (ARPA) and an AIS receiver, this setup would give an almost complete picture of the marine traffic in the area. The complete set-up as installed, and the ferries' normal crossing route is illustrated in Figure 3.

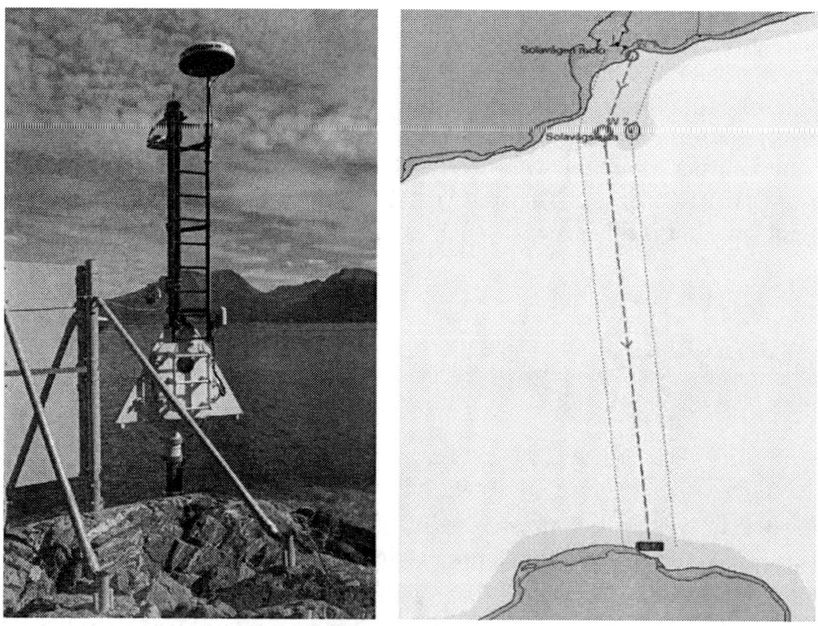

Figure 3: The equipment installed at "Klubben" and the normal crossing route used by the ferry

IOP Conf. Series: Materials Science and Engineering **929** (2020) 012016 doi:10.1088/1757-899X/929/1/012016

The information from these units was integrated, presented and recorded by "Timezero" software. "Timezero" is a coastal monitoring system delivered by Furuno Norway. For further specifications of the equipment used, see **Feil! Fant ikke referansekilden.**.

Table 1: Specifications of the installed equipment on-site

Instrument	Producer	Type
Radar	Furuno	DRS4D-NXT
Camera	Axis	Q6054E PTZ
AIS receiver	Furuno	FA-30
4G LTE Router	D-Link	DWR-921

Albeit the equipment was installed on-site, it was possible to remotely access the installation via TeamViewer software. This access allowed for fine-tuning of radar parameters and adjusting the alarm zones and tracking parameters for the targets. All data was logged on the computer hard drive meaning that when data collection period ended the computer was collected and brought to the NTNU premises for further analysis. A screendump from the Timezero software is shown in Figure 4.

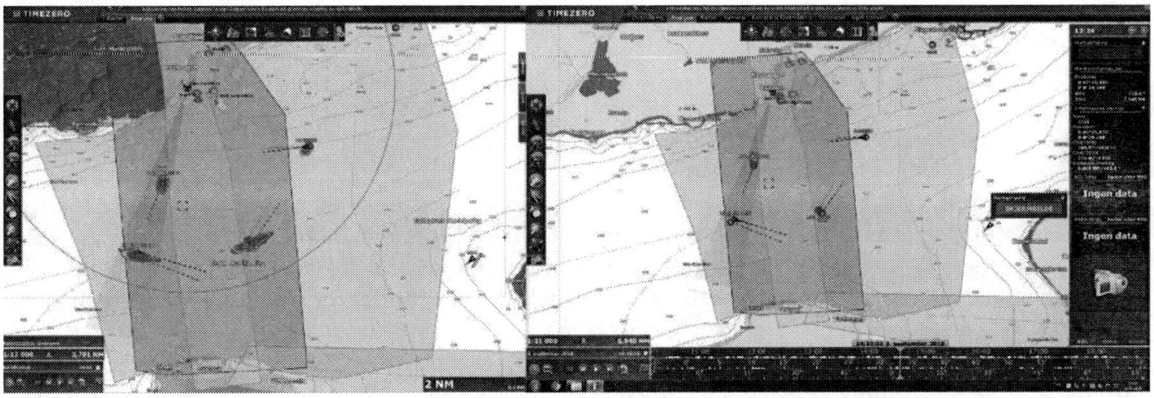

Figure 4: Left, Radar/ARPA. Right, ECDIS with AIS overlay.

3.2 Procedure

To categorize the vessels monitored the area was divided into zones. An area within which any targets entering would necessitate attention from the officer onboard the ferry as well as to tell the camera to focus on this target was defined. This is represented by the pink area in figure 3. The ferry's route was also divided into zones of transit and docking/undocking. In the transit zone the ferry was expected to maintain full speed, whereas in the docking/undocking zones this was obviously not the case.

The sampling period from 21st august to 19th September 2018 was chosen as this period would represent a peak in leisure fishing of crabs as well as it was still within the high season for the cruise vessels transiting the area. The risk of encountering bad weather with reduced visibility, which would reduce the applicability of the camera, was still low.

3.3 Analysis

The actions analysed were done so with no knowledge of what information was available to the navigators, what clues in the environment they searched for and what prospective arrangements they have done with other vessels, i.e. by VHF radio.

The playback of the recordings was reviewed by two licensed Master Mariners. During the laborious review of the playback from the Timezero recordings, targets were identified and categorized. Only targets entering the fairway utilized by the ferries were registered. Targets outside of this would not cause any concerns for the navigators onboard the ferries. The targets within the fairway were all targets which the officer onboard the ferry had to assess whether they would represent a concern for the ferry's transit. These targets were then categorized by what kind of marine vessel they represented and by what source they were identified, either by radar or AIS. Furthermore, where they appeared within the fairway, whether it was in the docking/undocking area or in transit zone, were recorded. Finally, and crucially, the evasive manoeuvres, executed by targets or by the ferry itself, were assessed whether they were conducted in accordance with the COLREGs or not.

3.4 Limitations of the study
Our approach to this project was to identify situations concerning the safe passage of the ferries crossing this fjord. Our installed equipment did not allow us to get the full picture of what information was available to the navigator onboard, neither the ferries nor the targets. Information such as CPA, TCPA and BCR could in hindsight be estimated using the playback function of the Timezero software, this laborious effort could have somewhat complimented our data. The fact that evasive manoeuvres were observed, indicated that the navigators performing these found their threshold limits breached, whether it was CPA, TCPA or BCR. Notwithstanding Part A and B section 1 of the COLREGs, which always should be taken into account, we focus solely on what we could observe with the equipment available to us. This led us to evaluate the manoeuvres solely against rules 13-17 described earlier.

3.5 Reliability and validity
The situations were identified, and the actions analysed by triangulation of sources. The AIS, ARPA and the PTZ camera would all give separate pictures of the situations. I.e. if the AIS indicated a speed reduction as an evasive manoeuvre, this was crosschecked with the speed input from the ARPA. If a course alteration was detected by ARPA, this again was confirmed by AIS and the PTZ camera.

Both analysts had a background as licensed master mariners. Drawing on their experience, situations were rapidly discovered and assessed. Their background as officers on merchant vessels could give them a predisposition to favour the merchant vessels over the leisure vessels though, interpret and criticize their actions differently than say an experienced fisherman onboard a small fishing vessel operating in this area would do.

This study chose a ferry crossing representative of others in Storfjorden. There are at least three similar crossings within a relative short distance. They all have almost similar features, 20-30 min departure intervals, two or three different ferries assigned to the crossing. One could also question the time of year this study was conducted. There were still cruise vessels transiting, albeit the numbers decreasing towards the end of the monitoring period. The leisure boat activity was undoubtedly also on decline as the summer season nearing its end, but the crab fishing season had begun and there were still quite a large number of leisure vessels involved in this.

Even though it was late summer/early fall, the occasional rain shower and wind gusts did reduce the camera's ability to enable us to identify what kind of vessel the ARPA picked up. This was especially the case with vessels appearing near to the end point off the crossing, almost three nautical miles away. At night, targets without AIS could have slipped by without being noticed. As mentioned earlier, one of the analysts has previously worked onboard a ferry operating this crossing. He knows the conditions under which the navigating officers onboard the ferry work. He knew the options available to them with regards to both path selection and speed reserves. This could have affected the study. The other analyst lives nearby and therefore knows this particular area very well. However, we did see this as a representative area and not one where we knew situations would occur more frequently than others.

4. Results and discussion

A total of 1010 targets were identified and categorized as influencing the ferries' transit during the 30-day sampling period. Based upon the analysis criteria and COLREGs (13-17), evasive manoeuvres were performed 85 times (8,42%). The evasive manoeuvres were performed by both the ferry and targets respectively. 925 situations that the target(s) who were crossing the fairway, required no avoiding manoeuvres. Of all the targets tracked, only 41% had AIS installed. Table 2 summarizes the results of the identified evasive manoeuvres.

Table 2: Summary of identified evasive manoeuvres.

Evasive manoeuvres	By ferry	By targets
According to COLREGS (13-17)	49	24
Not according to COLREGS (13-17)	8	3

Further analysis of the manoeuvres not according to COLREG showed that these were either course change, speed change, a combination of the two or delayed departure. To illustrate this, two selected cases will be displayed in detail. These cases are representative of the manoeuvres done by the ferry which, in our opinion was not according to COLREGs.

Case 1: Ferry alters course to starboard and increases speed for a give-way vessel. In this case according to COLREGs the ferry, as the stand-on vessel, should maintain course and speed, and the give-way vessel should alter course and/or speed to resolve the situation. This is not what occurred in this instance.

The ferries have scheduled departure times and in this case the ferry departs at 08:30 LT from Solevågen. It was observed that the ferry increased speed above normal cruising speed and at the same time altered course to starboard in order to increase the CPA. Figure 5 shows a screenshot of the above-mentioned situation.

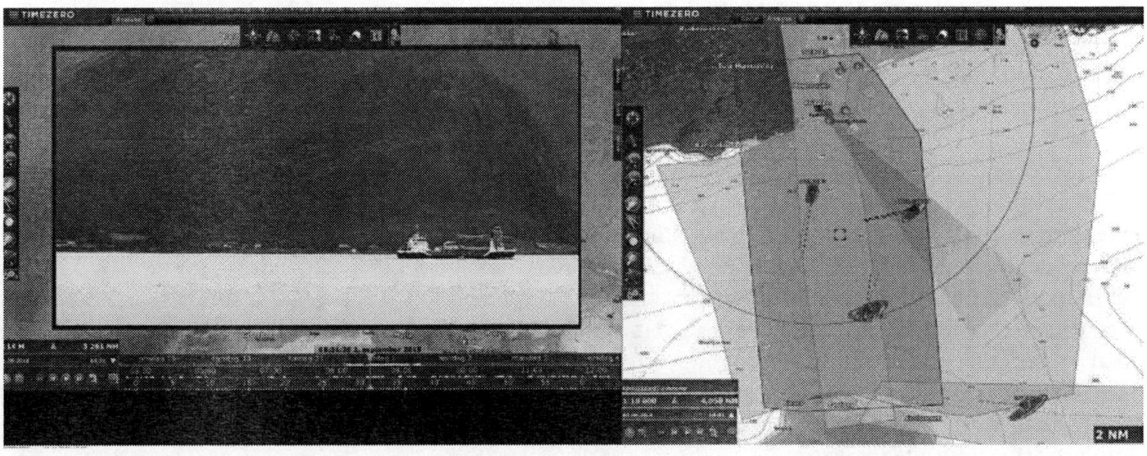

Figure 5 Left: Camera view of give-way vessel. Right: Radar and AIS overlay with give-way vessel and ferries. Note the heading and speed vector of ferry located to the top left in the picture.

Case 2: Ferry alters course to port and at the same time deviates from its normal route for a give-way vessel. Similar to case 1; the ferry, as the stand-on vessel, alters course. However in this case, the ferry also deviates from the normal route to avoid give-way vessel. Normal route for the ferries is west of the "Solevagsfluda" located as a small yellow area top-middle in figure 6. This time the ferry alter course and passes "Solevagsfluda" on the east side to allow passage for the give-way vessel.

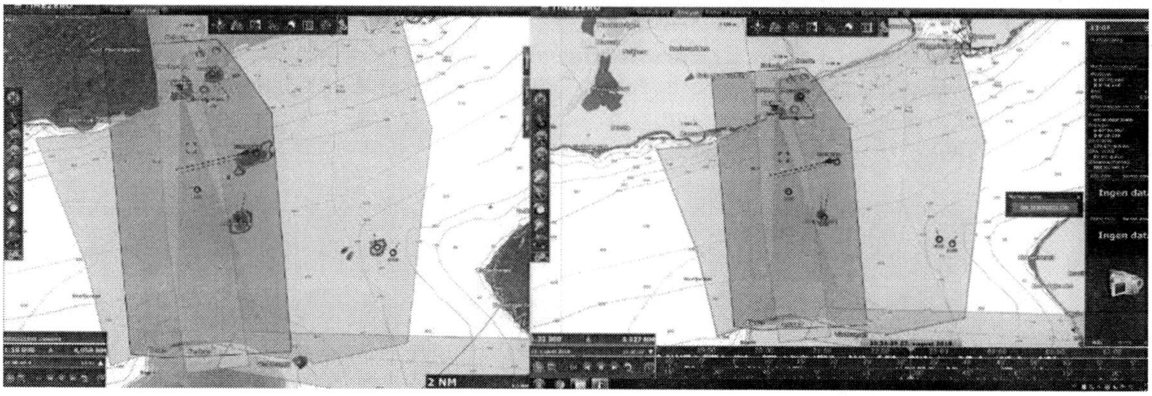

Figure 6 Left: Radar and AIS overlay showing ferry that alters course to port and deviates from its normal route. Right: AIS and ARPA plots of ferry and give-way vessel. Note the ferry located top middle close to shore and give-way vessel in blue.

Case 1 and case 2 are examples of situations that are identified in COLREGs and in both cases the ferry was the stand-on vessel but acted as give-way vessel.

5. Conclusion/further work

The goal of this project was to show that even though the COLREGs are universally accepted as the rules of the road, they are not always strictly followed. Rule 2 in the COLREGs, "Responsibility", also called the seamanship rule, introduces an opening for this. The ordinary practice of seamanship allows the crew to take action to avoid a collision that may not comply with the COLREGs but is necessary in particular circumstances to avoid a collision. Seamanship is not a fixed standard, but a product of factual circumstances. A MASS will have to recognize these circumstances warranting deviation from the COLREGs.

As this study has shown, instances where deviation from the COLREGs appear, are quite frequent. 14% of the evasive manoeuvres done by the ferries were not according to the rules. If MASS are to coexist with conventional vessels, both merchant and leisure, they must demonstrate an ability to recognize and act as expected to these kinds of situations.

It is quite possible, or maybe even likely, that the evasive manoeuvres by the ferry described above were agreed upon, that the action was planned between the give-way vessel and the ferry. This could have been done by VHF. The cruise vessels all have pilots onboard when sailing in this area. It is not abnormal that the pilot calls on the ferry and asks the ferry do give way as it might be easier for the ferry to delay the departure a couple of minutes, or choose a more easterly route when departing Solevågen, than the cruise vessel to reduce its speed or alter course of 300+ meters long ship. Even so, this arrangement needs to be done by someone.

Further studies are needed to verify our findings as we do not know whether they are replicable on other traffic congested areas.

References

[1] Phillip Belcher 2007 *Rule Following Behaviour in Collision Avoidance - a study of navigational practices in the Dover Strait. (PhD)* (Cardiff: Cardiff University)

[2] Porathe T 2017 Is COLREG enough? Interaction between manned and unmanned ships *Marine Navigation - Proceedings of the International Conference on Marine Navigation and Safety of Sea Transportation, TRANSNAV 2017*

[3] Olindersson F, Janson C-E and Dahlman J 2015 Maritime Traffic Situations in Bornholmsgat *TransNav, Int. J. Mar. Navig. Saf. Sea Transp.*

[4] Lutzhoft M, Hynnekleiv A, Earthy J V. and Petersen E S 2019 Human-centred maritime autonomy-An ethnography of the future *Journal of Physics: Conference Series*

[5] NTNU 2017 Autoferry

[6] Felski A and Jaskolski K 2013 The Integrity of Information Received by Means of AIS During Anti-collision Manoeuvring *TransNav, Int. J. Mar. Navig. Saf. Sea Transp.*

IOP Conf. Series: Materials Science and Engineering **929** (2020) 012017 doi:10.1088/1757-899X/929/1/012017

The need for a public key infrastructure for automated and autonomous ships

Ørnulf Jan Rødseth[1,*], Christian Frøystad[2], Per Håkon Meland[2,3], Karin Bernsmed[2], Dag Atle Nesheim[1]

[1]SINTEF Ocean, Trondheim, Norway
[2]SINTEF Digital, Trondheim, Norway
[3]Norwegian University of Science and Technology, Trondheim, Norway

[*]E-mail: OrnulfJan.Rodseth@sintef.no

Abstract. Shipping undergoes rapid digitization, covering safety and security reporting, mandatory ship documentation, electronic port clearance as well as commercial and operational information exchanges. Increasing automation of information processing, including the specific needs for autonomous ships, requires increased "digital trust" to allow humans to remove themselves from the information processing loops. This includes better safeguards against cyber threats such as counterfeiting contents or the originator of critical messages. This paper describes thirteen use cases for maritime services and analyse how a Public Key Infrastructure (PKI) system can provide security barriers to mitigate relevant cyber threats and possible consequences of unwanted events. Such a PKI needs to be designed with the special maritime business constrains in mind; the most important being the international nature of shipping, the lack of connectivity for ships that are far from shore, the network constraints associated with existing communication technologies and regulatory considerations.

1. Introduction

Today, maritime shipping undergoes rapid digitization. This applies to safety and security reporting, mandatory ship documentation, including ship certificates, electronic port clearance as well as commercial and operational information exchanges. There are at least two important reasons for introducing digital information exchanges in the conventional maritime sector. One is to reduce the administrative workload on the involved parties, most importantly the seafarers. This can be done by using computers to automate the processes related to the information transmission, reception, and processing. The second reason is to improve the quality of information used to plan and execute maritime and port operations. Electronic transmissions avoid misunderstandings and simplify the exchange of more complex information. However, even when we assume that electronic communication is error free, there are three issues with this type of data exchange that need to be addressed: One is the possibility of malicious cyber-attacks that may be motivated by commercial gain or attempts to damage life, health, property or the environment or are just random attempts to break into interesting technical systems. The second issue is to establish enough trust in the automated processes so that the need for manual double checks is reduced to a minimum. If not, administrative workload may increase rather than being reduced. A third issue emerges when ships become more automated and eventually without people onboard. This removes the human from the information evaluation and decision-making process, and it becomes critically important to ensure that the digital information is correct and that it can be trusted.

Problems occur if the *safety and security mechanisms* that are inherent in the document-based systems are not replicated and improved in the electronic information exchanges. These mechanisms are:

- *Confidentiality* (sealed and closed envelope): the contents of the document cannot be read by others than the intended receiver(s).
- *Integrity* (broken seal, changes to the printed paper): document tampering will be detected.
- *Authenticity* (signatures, stamps, seals): the identity of the originator of the document can be proven.
- *Availability* (archives, delivery): The document can be easily transmitted, found, and retrieved within a reasonable amount of time.

In addition, an additional mechanism can be derived from the three above, which is *non-repudiation* (registered mail, courier): providing proof that the document was delivered to the recipient, which generally requires an authenticated acknowledgement from the receiver that the specific document has been received.

There is a diverse set of communication interactions in shipping, including ship-to-ship, ship-to- port, ship-to-Remote Control Centre (RCC), ship-to-Vessel Traffic Services (VTS), ship-to-Application Service Provider (ASP), ship-to-Medical Aid Provider (MAP), and ship-to-Search and Rescue (SAR) as well as ship to Maritime Rescue Coordination Centre (MRCC), see Figure 1. The communication directly between the ships and between ships certain land services are likely to be VDES (orange), while other communication between ship and shore might be SatCom (blue). Satellites can be in Low or Geostationary Earth Orbit (LEO or GEO) and communicate via a Satellite Application Service (SAS), to shore entities over land lines (green). GEO-solutions are normally Very Small Aperture Terminal systems (VSAT, i.e. directional dish antenna), but can also be low- directional services such as some of Inmarsat's services. The most common LEO-service today is Iridium, but there are also satellites available that can receive and send VHF digital data. Much of this communication is today done with via Very High Frequency (VHF) voice radio or Satellite Communication (SatCom) telephone links.

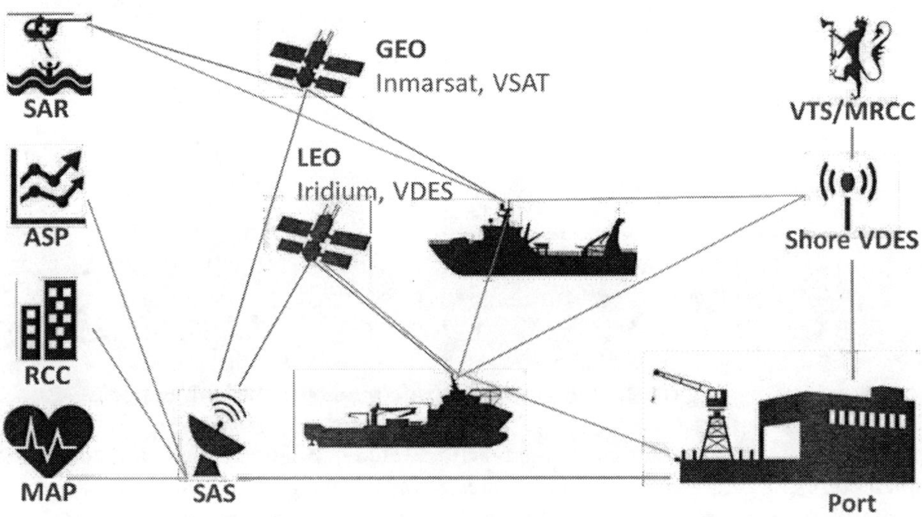

Figure 1. A high-level overview of some communication channels between maritime entities.

VHF Data Exchange System (VDES) [1] is a new two-way radio communication system, which is currently being standardized and validated for various maritime services. By transitioning from analogue voice to digital messages over VDES and making more use of Internet based exchanges over SatCom, the stress on the current communication links will be reduced and new services can be introduced. The use of the different communication links will depend on the ship's location, information to be transmitted, and the stage of the ship's journey. The digital data exchanges need to implement a digital trust model, and the use of a Public Key Infrastructure (PKI) is a common way of realising this. However, with every PKI there are several design and configuration decisions that must be made based on the nature of the operating environment. There are distinct characteristics for this domain that must be taken into consideration for a PKI to be economically, technically, and politically feasible. This paper will elaborate on the business constraints underlying a PKI for the maritime industry and discuss PKI related solutions and initiatives used in a maritime setting. We will use a selection of business cases for maritime digital communication based on the following three groups of ship communication needs:

- Safety related communication over VDES, to other ships or to shore.
- Internet message exchanges associated with the Maritime Single Window [2].
- Commercial and operational services, mainly over SatCom and Internet.

In section 5.1 we present thirteen use cases related to these and continue with an analysis of possible unwanted events caused by cyber threats and possible consequences. We then suggest how a PKI can be used to create barriers for unwanted events related to the use cases.

2. PKI basics

One of the most common technologies to provide secure and trusted digital data exchanges is the use of Public Key Infrastructure (PKI). A functioning PKI solution needs to be able to create, store and distribute cryptographic keys amongst a wide variety of users (including vessels, authorities' shore stations, and organizations) that will need to communicate securely in order to exchange critical information. The PKI can be used for authentication and to establish cryptographic protection of ship-to-shore, shore-to-ship, and ship-to-ship communication, independent of what communication link is being used. The solution can also be used to generate and validate digital signatures of, for example, electronic ship certificates and logbooks.

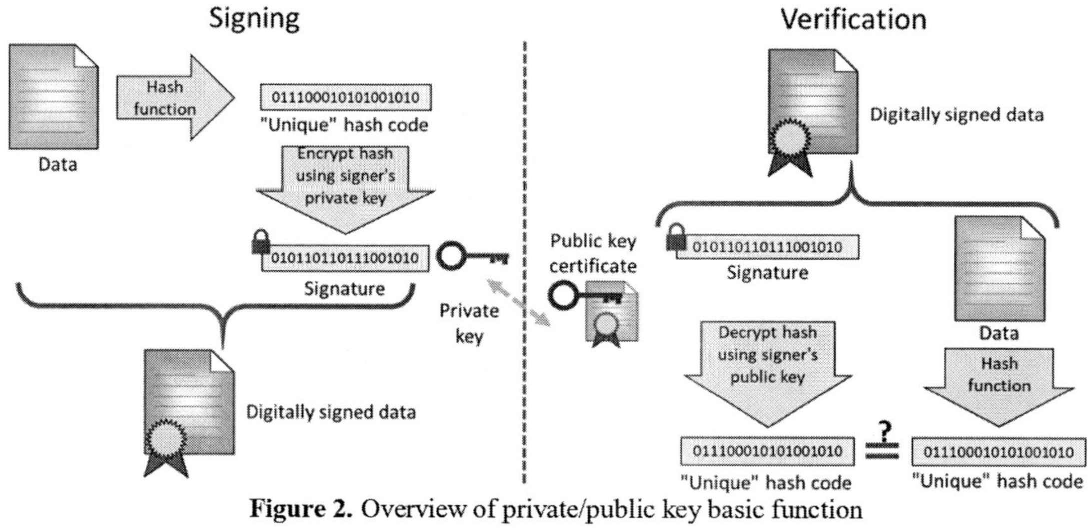

Figure 2. Overview of private/public key basic function

The public-key digital signature mechanism is illustrated in Figure 2, where a securely issued private key and a corresponding public key certificate are the main enabling assets. Once the sending party has digitally signed a message (or document) using his or her private key, the contents of the message can be verified by anyone having a copy of the sending party's public key, which usually is made available in a digital public certificate. The system requires a trusted authority to issue the public certificate of the signing party. The public certificates enable anyone to verify that documents sent by a private key holder are indeed issued by that party and that the content has not been tampered with.

When both parties are part of the PKI, it is easy to send acknowledgements on reception that can be verified by the original sender and used as proof that the original message was indeed sent and received. The sender's signature on the original message serves as proof of the actual content of the message. This implements non-repudiation.

Encryption (confidentiality) of the message contents can be implemented in different ways, also by using the public/private key mechanism. However, normally one will use other and more efficient mechanisms where the PKI is used to negotiate and exchange symmetric encryption keys. These are further useful for securing vessel services with real-time requirements, such as voice communication, system monitoring and remote piloting. Note that it is also possible to encrypt a message with the public key, so that only the owner of the private key can open it.

3. Maritime business constraints

There are many PKI solutions already in operation all over the world, but the maritime sector has some important business constraints that limits using off-the-shelf solutions. This section will go through the most important ofthese.

3.1. The international nature of shipping

Ships regularly call on ports in other countries than where they are flagged. Generally, the ship owner, manager or charterer are also located in other nations than where the ship calls. As there is no common legislation covering all the communication parties, it may therefore be difficult to accept messages or reports from ships to shore as legally or commercially binding under national law. It is of course possible for ships to get nationally accepted public certificates, but this may not be legal in some jurisdictions and will also create a management problem for ships that call in ports belonging to different nations.

A related problem occurs when ships meet at sea (international waters) and need to exchange safety related information, or when a ship passes through a ship reporting or VTS area and needs to exchange information with shore stations.

This constraint means that one internationally managed PKI would be useful to simplify management of ship and shore entity certificates.

3.2. Ships are not always connected to the Internet

Not all ships are continuously connected to the Internet as coverage may be very limited far from shore. This condition makes it difficult to use security solutions that rely on constant network access for their functionality. Existing PKI systems we know from the vast World Wide Web, various blockchain implementations and secondary channel solutions (e.g. SMS codes to mobile phones) would typically fall in this category.

However, within the shipping sector there is a limited number of parties that need to be involved, i.e. between 55 000 and 120 000 ships, dependent on how they are counted, and approximately 110 000 registered ports. This makes it possible to load all trusted root certificates for a PKI, and probably all public key certificates as well, onto the ship when it is in port with a high capacity Internet connection. This cache can be used to verify the electronic signatures received during the part of the voyage when it does not have access to the Internet.

3.3. Limited bandwidth and message sizes

Even when ships have a network connection during the voyage, they usually have a relatively low bandwidth and high communication latency. The cost of communication can also be a limiting factor. Also, some of the new e-navigation services is expected to operate over the VDES data links for some of the message exchanges. This has further restrictions in terms of message size and *Quality of Service* in general (message loss, latency, jitter). These issues have consequences for the cryptographic strength and how the PKI is implemented in practice. One cannot allow large signatures that may exceed the allowed message sizes or that use too much bandwidth.

3.4. Regulatory considerations

It is important to consider the diversity of the applicable jurisdictions at sea when designing the PKI solution. This calls for a solution to be developed in a way that is acceptable for the International Maritime Organization (IMO) and all its member states.

Ships in international trade will have to relate to various international legal frameworks, mainly IMO instruments and the United Nations Convention on the Law of the Sea (UNCLOS) [3]. The latter is of limited relevance in this context except that it will regulate the jurisdiction of relevant regulations and laws that apply to the ship. In practical terms this will be the flag state for most operations on board, port/coast state law when ports are called on and IMO instruments for various regulations applicable for ships on international voyages or for innocent passage through other states' territorial waters.

Flag state law will vary but will generally reflect IMO requirements to safety and security on board the ship. This includes requirements for authentication, e.g. signatures and/or seals on ship certificates, proper signatures on logbook entries, etc. IMO has also published guidelines for use of electronic versions of ship certificates [4]. IMO instruments include provisions for mandatory ship reporting,
e.g. related to ship reporting areas and similar. Today, these requirements do not include any provisions for authentication of sender. However, national legislation, e.g. in Norway [5], can require or recommend that electronic reporting is used which in some cases also may include some form of authentication.

When calling at a specific port, the ship will also be required to follow national legislation related to mandatory reporting before or during the port call. This may or may not include provisions for electronic reporting and possibly requirements for authentication. In Norway, as an example, ships should use the Norwegian *SafeSeaNet single window* where authentication is implicit through a user code and a password. One should also keep in mind that some reports to the port and port services may result in various fees being payable. Errors or omissions in these reports can have direct economic consequences. Finally, one may need to consider export restrictions on certain types of advanced technology, which may make it impossible to fit corresponding technology to certain ships.

It would be very helpful if an international convention, e.g. an amendment to the FAL convention [2], would ensure the international acceptance of the electronic signatures enabled by an internationally managed PKI.

4. Related Work

There are some PKI systems already in operation in the maritime sector. This section presents a brief overview of some of the already established solutions and of the most relevant ongoing initiatives.

4.1. Established solutions

The Long Range Identification and Tracking (LRIT) system [6] collects position reports from ships worldwide and make them available to coastal states that have a legal interest in ships approaching or passing their shores. A PKI is operated by IMO to secure the communication between the distributed LRIT data centres. This is based on public cryptography technology and is used over ordinary Internet connections. The LRIT PKI could in principle be the basis for a wider scoped international maritime PKI, but one may have to change the basic cryptography mechanisms to reduce signature sizes.

SafeSeaNet [7] is a system similar to LRIT, but operated by European Maritime Safety Agency

(EMSA) and covering much more detailed information about ship movements and port calls. The same comments apply to SafeSeaNet as to LRIT, it could be used as basis for an international maritime PKI.

The International Hydrographic Office (IHO) also operates a type of PKI [8] that is used to encrypt and verify the authenticity and integrity of electronic charts. This includes cyber security protection, restricting access to only those elements that a customer has been licenced for; and authentication, i.e. to provide assurance that the data has come from approved sources.

4.2. New initiatives

The Maritime Connectivity Platform (MCP) [9] is an initiative to provide a communication system for the maritime industry, including an identity registry, service registry and a messaging service. Included in the framework is also a PKI, which is intended to be for authentication of, for example, vessels. This is a system that could lend itself to be an international agreed PKI. However, the business model for operating the MCP is still not defined.

ISO/TC 8 *Ships and marine technology* has proposed that a PKI should be used by the issuing party to digitally sign ship certificates [10]. The electronic signatures can then be verified by an inspector by means of computer, tablet or smart phone. In their report [11], ISO proposes the use of X.509 certificates and elliptic curve cryptography for generating and validating the signatures. ISO also envisions the use of a central public key repository, operated by e.g. IMO.

There are various initiatives to provide an authentication service for VDES and the establishment of an international PKI operated, e.g. by IMO has already been suggested [12]. The problem here is limited message length and high overhead resulting from even relatively compact elliptic curve signatures. As an example, the longest Application Specific Messages (ASM) messages in VDES use three message slots and are about 1200 bits long. Thus, a 512-bit long signature would use almost half of the message length. However, VDES also supports longer messages over higher capacity VDE (VHF Data Exchange – several 100 kilobits per seconds) where signature sizes are less problematic.

5. Analysis of unwanted events and PKI benefits

In order to determine the benefit of a PKI in the maritime sector, a set of thirteen use cases undergoing digitalisation were analysed as part of a Norwegian research project on cyber security for merchant shipping (CySiMS) [13]. These use cases have been introduced by Frøystad et al.[14], and are summarized in Section 5.1. In Section 5.2, we show how security experts participating in this project assigned general cyber security threats to unwanted events for maritime communication. With the aid of maritime domain practitioners, these unwanted events were subsequently linked to consequences for the individual use cases as shown in Section 5.3. This whole approach follows the thinking of bow-tie modelling, wherealternativecauses andconsequencesof unwantedevents motives the useof proactiveand reactivebarriers. More details about howthis can be represented and visualised are given by Bernsmed et al. [15]. The final result of the analysis is presented in Section 5.4, showing how the security functionality provided by a PKI benefits the use cases.

The CySiMS project focused on conventional shipping, but it should be clear from the use cases that this problem is even more acute for automated and autonomous ships. Here, there is no operator onboard to do "sanity checks" on messages and the computers rely on being able to trust both content and sender of digital messages. In particular, use-cases 3, 5, 8, 9 and 12 are relevant for autonomous ships.

5.1. Use Cases for the maritime sector

1) **Ship certificates:** Ships are required to carry original versions of a number of safety certificates on board (e.g., International Tonnage Certificate, Safety Management Certificate, and International Anti-fouling System Certificate). These documents must be provided for

inspection by port state control (PSC) and others, such as during vetting by charterers.

2) **Single Window:** Ships entering a foreign port must declare important information on the ship, cargo, and persons on board, normally in good time before they enter.
3) **Safety information:** Different public services can provide Maritime Safety Information (MSI) to vessels in a specified area. This is typically gale warnings, warnings on ships in distress, ongoing search and rescue operations, etc. This is a receive only operation where the ship navigator is responsible for keeping track of these messages and, if necessary, react to them to aid in search operations or to avoid various dangers to safe navigation.
4) **Mandatory Ship Reporting Systems:** Ships entering and leaving ship reporting areas or VTS controlled areas are normally required to report this to the VTS or other reporting authorities.
5) **Nautical Information:** Ships are required to keep critical electronic databases up to date. This includes electronic charts, lists of navigation signals, etc.
6) **Operational exchange:** Ships communicate with owner, manager, charterer, or agents for operational purposes. This includes voyage orders, periodic reports from the ship or performance reports, e.g. in conjunction with charter contracts.
7) **Logbook:** The deck logbook is one example of a logbook kept on board, which can be inspected by the Port State Control (PSC) and that may be used as evidence in case of accidents. An electronic logbook needs to be signed at the time when the log entry was made. It must be impossible to tamper with a recording and recording should as much as possible be automatic.
8) **Traffic organisation advice:** Some services provide advice to the ship to make passage safer or more efficient. The master has to decide if the advice is used and will in principle look at this as any other type of general navigational information. This type of information is commonly sent from Vessel Traffic Services (VTS).
9) **Traffic organisation instructions:** This case is similar to the one above, but here the instructions are stronger than advice. This can be used in port areas and issued from a port VTS. The master is still responsible for safe passage and may refuse to follow instructions if they are deemed unsafe, but not following the orders may have operational or economic penalties (e.g., increased port fee or increased waiting time for pilot or berth).
10) **Telemedicine:** Ships have access to land-based advice during medical emergencies, but this is usually restricted to voice communication. In the future it is foreseen that this may be in part digitalized where e.g., pictures or EKG-data can be transmitted directly to specialists.
11) **Search and rescue:** During search and rescue operations, nearby ships are often used to assist. This can be to searching for persons overboard or to directly assist a ship in distress. The maritime rescue coordination centre (MRCC) or on scene commander can issue detailed instructions, e.g. search patterns, to the ships.
12) **Remote control:** It is possible to, e.g. remotely control a tug from the bridge of the ship being assisted. This will give the pilot and captain better information about both ships responses and improve coordination as well as reduce chances of misunderstanding in voice communication. This function is still on research stage.
13) **VDE Bulletin Board:** When a vessel enters the coverage area of a Terrestrial VDES base station it will receive the Terrestrial Bulletin Board (TBB) message. The TBB information includes important information on the use of VDE in the area. The TBB does not change often and should be transmitted in regular intervals.

5.2. Threats leading to unwanted events

In Table 1 we have enlisted threats towards the communication links between ship-ship and ship-shore. Note that threats related to physical access to ship systems or shore assets, as discussed by e.g. Rødseth et al. [16] and Jones et al. [17], are out of scope in this context.

Table 1: Communication related threats and unwanted events

General threat	Unwanted event
T1: Jamming of terrestrial link	E1: Loss of one or more messages
T2: Jamming of satellite link	E1: Loss of one or more messages
T3: Short DoS attack towards shore-based system	E2: Limited or no communication capacity for a short period of time (1-2 hours)
T4: Long DoS attack towards shore-based system	E3: Limited or no communication capacity for a long period of time (1-2 days, or more)
T5: Wiretapping of terrestrial link	E4: Confidential data overheard by an unauthorized actor
T6: Wiretapping of satellite link	E4: Confidential data overheard by an unauthorized actor
T7: Repudiation of transmitted message	E5: Data is received, but the sender denies having sent the data
T8: Repudiation of received message	E6: Data is sent, but the receiver denies having received the data
T9: Broadcasting of false messages on an open channel	E7: False data received by one or more actors listening to the broadcast channel
T10: Transmission of an unauthenticated message to a single actor	E8: False data received by the ship or the shore
T11: Retransmission of a previously transmitted message	E8: False data received by the ship or the shore

5.3. Consequences of unwanted events

As shown in Table 2, a single unwanted event can have consequences for numerous use cases.

Table 2: Consequences of unwanted events

Unwanted event	Use case consequences
E1: Loss of one or more messages	UC1-2, UC4, UC6-7: Possible delays, ship not seaworthy, ship arrest UC3, UC5, UC8-9, UC12: Possible collisions, grounding, damages to facilities UC10-11: Damages to health or loss of life UC13: VDES not available
E2: Limited or no communication capacity for a short period of time (1-2 hours)	Same as above
E3: Limited or no communication capacity for a long period of time (1-2 days, or more)	Same as above
E4: Confidential data overheard by an unauthorized actor	UC2: Leakage of business sensitive information, data privacy breach, loss of reputation UC6: Leakage of business sensitive information UC10: Leakage of sensitive health information
E5: Data is received, but the sender denies having sent the data	UC2, UC4: Possible delays, breach of contracts, lawsuits
E6: Data is sent, but the receiver denies having received the data	Same as above
E7: False data received by one or more actors listening to the broadcast channel	UC13: Spoofing of VDES services, possible collisions, blocking of harbours

Unwanted event	Use case consequences
E8: False data received by the ship or the shore	UC1-2, UC6: Fraud, smuggling of illegal goods, delays UC3, UC5, UC8-9, UC12-13: Possible collisions, grounding, damages to facilities UC4, UC7: Fines UC10-11: False alerts, damages to health or loss of life

A user survey was conducted to assess the criticality of the different consequences. In the survey, which was conducted on the premise that the ship had ordinary manning, the commercial consequences were generally considered the most severe, i.e. the possibility for detention in port, fines or the consequences of leaked sensitive information. This is probably because with a conventionally manned bridge, it is less likely that fraudulent message or jamming will cause major events as the crew directly or indirectly will verify the saneness of the received information.

However, with more automation onboard, and in particular with fully unmanned ships, the risk for major events may be significantly higher. The automation systems will be relied on more and the probability that false information can lead to unwanted events increases. This should also be seen in conjunction with increasing probability for "cyber terrorism", where enemy parties may want to disrupt international trade by, e.g. blocking an entrance to one of the world's large ports.

The exact nature and scale of the consequence will depend on properties of the ship (e.g. type of ship, size, and cargo) and the context it is operating in (e.g. at shore, in a busy area, geopolitical surroundings, weather conditions).

5.4. Security barriers provided by a PKI

In Table 3, the use cases are mapped to the security functionality that a PKI can offer. The mapping is derived from an analysis of the use case characteristics and their associated cyber security risks.

A special note regarding use case 1 and 7 should be made, since these are related to documents where the PKI is used to sign and verify documents stored on-board the ship and not during the actual communication. Signatures can ensure the integrity and the authenticity of the ship, crew member role and person signing the logbook entries or issuing party of a certificate. These documents are then self- protected, which enables them to be transferred over insecure as well as secure communication channels. However, with events 1-3 the content is lost or unavailable, which means that content protection has no real effect. Instead, a physical inspection of the documents might be needed, causing potential delays and extra costs.

The rest of the use-cases will require secure communication, as indicated by tick-marks in the columns AUT, INT, CON. This is short for authentication, integrity checks and confidentially respectively. E-DOC Sign is indicated where an electronic document kept on board needs a signature.

Use cases 3, 8-9, 11 and 13 share common benefits from message authenticity, availability, and integrity. This is because they typically involve information sent from shore to the ship, and the ship needs to be sure that this information can be trusted. A PKI can ensure the integrity of the message and the authenticity of the sender. Use case 5 is a special case that would also benefit from confidentiality since nautical information is typically bought from a nautical service provider and not classified as *open* or *free* data. Use cases 2, 4, 6 and 10 share similar benefits from a PKI allowing parties on ship or shore to authenticate themselves, to sign the data to be sent and to verify the identity of the recipient. Additionally, a PKI can ensure confidentiality when sending restricted information from ship to shore by either encrypting the data or communication channel. Use case 12 is our only ship-to-ship use case with requirements for a tamper proof communication link. This use case represents a wider range of services where ships need to trust each other, even when they encounter each other at open sea with only local radio communication available. For both this use case and use case 10, it might be more relevant to use the PKI to establish a longer-lasting secure communication session rather than securing individual messages.

Table 3. Mapping of high-level use cases to relevant security functionality offered by a PKI. Actors marked * must be authenticated and arrows indicate the direction of the information flow.

Use Case	**Identification and authentication**	**Secure communication**			**E-DOC Sign**	**Media**	**Unicast / Multicast**
		AUT	**INT**	**CON**			
UC1 Ship certificates	Flag state auth.* → Ship → PSC					offline	N/A
UC2 Single Window	Ship* ↔ Port state authorities*					SatCom	U
UC3 Safety information	MSI provider* → Ship					VDES/ SatCom	M
UC4 Reporting	Ship* ↔ VTS*					VDES	U
UC5 Nautical information	Nautical Service* → Ship					SatCom/ VDES	U
UC6 Op. exchange	Ship operator* ↔ Ship*					SatCom	U
UC7 Log-book	Crew*→ PSC, Auth.					offline	N/A
UC8 TO Advice	Ship* ↔ VTS*					VDES	U
UC9 TO instructions	Ship* ↔ VTS*					VDES	U
UC10 Telemedicine	Ship* ↔ Medical Aid Provider*					SatCom	U
UC11 Search and rescue	Ship* ↔ MRCC*, SAR*					VDES	M
UC12 Remote control	Ship* ↔ Remote Ship*					VDES	U
UC13 VDE TBB	Bulletin Board* → Ship					VDES	M

In addition to required security functionality, Table 3 also outlines which communication channels the use cases will utilize. We can summarize that the PKI solution must be able to support authentication of a wide variety of communicating entities, which can be generalized as being either *Ships*, *Services*, *Organisations*, or *Individuals*. Ships and Services will need to communicate both over VDES and more general communication channels. Organisations and Individuals will primarily use their keys for offline digital signatures of electronic documents.

6. Conclusion

An internationally accepted and widely deployed PKI system will solve many of the security and trust challenges associated with the sharing of digital information in the maritime sector. In this paper we have outlined a number of relevant use cases and showed how a PKI system can provide security barriers to mitigate threats and consequences.

We have also pointed out the most relevant use cases for autonomous ships, which are related to automated processing and decision making based on message information, i.e. use cases 3, 5, 8, 9 and 12. In these cases the importance of source authentication and message integrity is very high.

A PKI needs to be designed with the specific maritime business constrains in mind; the most important being the international nature of shipping, the lack of connectivity for ships that are far from shore, the network constraints associated with existing communication technologies and regulatory considerations.

Acknowledgments
The work presented in this text has been partially funded by the Norwegian Research Council project "CySiMS Service Evolution" under contract number 295969. It has also received funding from the European Union's Horizon 2020 research and innovation programme under grant agreement No 815012 (AUTOSHIP).

References

[1] Lázaro, F., Raulefs, R., Wang, W., Clazzer, F. and Plass, S., 2019. VHF Data Exchange System (VDES): an enabling technology for maritime communications. *CEAS Space Journal*, *11*(1), pp.55-63.
[2] *Convention on Facilitation of International Maritime Traffic* (FAL Convention), Adoption: 9 April 1965; Entry into force: 5 March 1967, as amended.
[3] UNCLOS *United Nations Convention on the Law of the Sea*, Retrieved June 2020 from http://www.un.org/depts/los/convention_agreements/texts/unclos/unclos_e.pdf
[4] IMO, 2016, FAL.5/Circ.39/Rev.2, 20 April 2016, *Guidelines for the Use of Electronic Certificates*.
[5] Department of transport, Norway, *Forskrift om fartøys meldeplikter etter havne- og farvannsloven*, 2015, retrieved June 2020 fr om https://lovdata.no/dokument/SF/forskrift/2015-12-21-1790 (In Norwegian).
[6] IMO, 2020, MSC.1/Circ.1294/Rev.6, 8 April 2020 *Long-Range Identification and Tracking System, Technical Documentation (Part I and II)*.
[7] European Maritime Safety Agency (EMSA), *SafeSeaNet main page*, Retrieved June 2020 from http://www.emsa.europa.eu/ssn-main.html
[8] IHO, 2015, *IHO Data Protection Scheme* Edition 1.2.0, January 2015, IHO Publication S-63, International Hydrographic Bureau, Monaco.
[9] MCP consortium, 2020, *Maritime Connectivity Platform (MCP)*, Retrieved June 2020 from https://maritimeconnectivity.net/
[10] IMO 2017, FAL.2/Circ.131, 19 July 2017, *Requirements for Access to, or Electronic Versions of, Certificates and Documents, Including Record Books Required to be Carried on Ships*.
[11] ISO 2015, FAL 40/6/2, 31 December 2015, *Future Proof and Cost-Effective Standardization of Electronic Ship Certificates*, Input document to IMO FAL Committee.
[12] IALA, 2019, Guideline G1139, *The Technical Specification of VDES*, Edition 3.0, June 2019.
[13] The *Cyber Security in Merchant Shipping* (CySiMS) project, retrieved June 2020 from: http://cysims.no/
[14] Frøystad C., Bernsmed K., and Meland P. H., 2017, Protecting future maritime communication, *Proceedings of the 12th International Conference on Availability, Reliability and Security* (ACM) p 97.
[15] Bernsmed K., Frøystad C., Meland P. H., Nesheim D. A. and Rødseth Ø. J., 2017, Visualizing cyber security risks with bow-tie diagrams, *International Workshop on Graphical Models for Security* (Springer) pp 38–56.
[16] Rødseth Ø. J. and Lee K., 2015, Secure communication for e-navigation and remote control of unmanned ships, *Proc. of the 14th Conference on Computer and IT Applications in the Maritime Industries-COMPIT* vol 15.
[17] Jones K. D., Tam K., and Papadaki M., 2016, Threats and impacts in maritime cyber security, *IET Engineering & Technology Reference*, 2016.

IOP Conf. Series: Materials Science and Engineering **929** (2020) 012018 doi:10.1088/1757-899X/929/1/012018

Impact of cyber risk on the safety of the MilliAmpere2 Autonomous Passenger Ship

Ahmed Amro[1], Georgios Kavallieratos[1], Konstantinos Louzis[2] and Christoph A. Thieme[3]

[1]Department of Information Security and Communication Technology (IIK), University of Science and Technology (NTNU), Gjøvik, Norway
[2]Laboratory for Maritime Transport, School of Naval Architecture and Marine Engineering, National Technical University of Athens, Greece
[3]Department of Marine Technology (IMT), NTNU, Trondheim, Norway
E-mail: ahmed.amro@ntnu.no, georgios.kavallieratos@ntnu.no, klouzis@mail.ntua.gr, christoph.thieme@ntnu.no

Abstract. The digitalization of the maritime sector is continuously growing, leading to increased automation, such as, the development of autonomous vessels. The Autonomous Passenger Ship (APS) is a characteristic instantiation of this development, aiming to transport people on urban waterways. Although emerging technologies deployed in such APS aim to facilitate the functions and operations of the navigation and communication systems, various safety and security risks are inherent to the communication infrastructure due to their interconnectivity. The aim of this work is to study the safety and cyber security of the communication system of an APS, namely the MilliAmpere2 APS. The six step model (SSM) is utilized to facilitate the joint analysis. The application of the SSM enables, among others, the capturing of relationships between cyber attacks and component failures, the assessment of safety and cyber security countermeasures, as well as, the synergies between them. It has been found that most countermeasures in both categories are reinforcing or are conditionally dependent on each other, while few antagonize each another. These findings will allow for improved design and implementation of integrated safety and security management solutions.

1. Introduction

The emergence of the contemporary and interconnected Cyber Physical Systems (CPSs) in the maritime domain and particularly in the autonomous vessels infrastructure, such as, the Autonomous Passenger Ship (APS), is rapid and continuous. To this end, the safety and security analysis of such systems is needed to ensure the vessel's normal and safe operations. Safety and security are interrelated concepts that may face both commonalities and differences in the analysis process since the former is concerned with accidental events while the latter mainly consider malicious actions taken by adversaries [1]. Particularly, security is concerned with the risks originating from the environment impacting the system and typically addresses malicious risks. Whereas safety deals with risks arising from the system that may affect the environment and addresses purely accidental risks. Safety analysis aims to reduce the risks related to systems, humans, and the environment [2] to an acceptable level.

Security analysis aims to minimize the risk related to confidentiality, integrity, and availability of the operational and functional requirements and therefore the data, information, and services of the system [2]. An extended security analysis may also consider the properties of possession or control, authenticity, utility, and non-repudiation. Hazards can be defined as conditions or states that may cause harm [3]. The risk associated with the hazards is a measure of uncertainty with respect to outcomes and may be described through risk sources, events and their consequences [4]. From a security point of view, vulnerabilities are system or software flaws that could threaten the system. Vulnerabilities could be also considered as system weaknesses [3].

The aim of this work is to identify weaknesses related to safety and security of a communication architecture proposed for safe and secure navigation for an APS [5] in order to remove them or reduce the risk associated with them. We apply the Six Step Model (SSM) to analyse security and safety risks and study the implications that security poses to safety. Particularly, leveraging the multidimensional matrices provided in the SSM, the functions, structure, failures, safety countermeasures, cyber attacks, and security countermeasures are identified for the communication, navigation and control systems of the APS. Although various approaches exist in the literature for security and safety co-analysis, the SSM has been chosen as the most appropriate for the case of the APS, due to its holistic approach to assess interdependencies. The complexity of the communication systems and the novel technology used in the communication infrastructure can be appropriately studied by the graphical models of the SSM. Further, the SSM facilitates the collaboration of both safety and security experts towards a more comprehensive safety and security analysis. The SSM and its application to the MilliAmper2 is described in detail in Section 4. The methods being employed in this article and previous works have been carried out as initial steps of a risk management process that is part of the Autoferry project [6]. The risk management process is aligned with the guidelines proposed by the International Maritime Organization (IMO) regarding the inclusion of cyber risk management within the Safety Management Systems (SMS) [7].

2. Related Work

Many safety and security methods that have been developed, do not directly consider each other. However, the combination of security and safety analysis is expected to result in identifying synergies regarding interactions, events and conflicting countermeasures. Various works in the literature examined the interrelation between safety and cyber security [2, 8]. Particularly, Lisova et al. [8] conducted a systematic literature review for safety and security co-analysis and thirty- three approaches have been identified. Further, Kavallieratos et al. in [2] conducted a survey in co-engineering approaches for safety and cyber security in cyber physical systems. Various systematic approaches have been proposed in the literature to analyse safety and security. The System-Theoretic Process Analysis (STPA), is developed to facilitate the safety analysis of complex systems considering the control structure of the targeted systems. The STPA is extended to accommodate security considerations, called STPA-Sec [9]. Further, SafeSec Tropos [10] is a co-engineering methodology for safety and security requirements elicitation in CPSs based on STPA and the Secure Tropos methods from safety and security domain respectively.

Safety related cyber attacks for autonomous inland ships are identified and assessed by Bolbot et al. [11]. In particular, by leveraging from a Cyber preliminary hazard analysis (PHA) method and existing systems vulnerabilities, potential cyber attacks that may compromise the vessel's safety along with a set of general countermeasures are examined. Further, Ba˘ckalov [12] studied the safety of autonomous inland vessels. Namely, the key characteristics of the autonomous inland vessels are analyzed considering the corresponding legislation and standards related to the safety of sea-going ships and inland vessels. Kavallieratos et al. [10] analyzed the safety and security of a cyber-enabled ship that could be either autonomous or remotely controlled. By the application of the SafeSec Tropos method, they identified the necessary security and safety requirements for such vessels.

3. Background

This section summarizes the background on the MilliAmpere2 passenger ferry, an instantiation of an APS which is under development as part of the Autoferry project [6].

3.1 System description and context of operation

The MilliAmpere2 is designed to carry up to 12 passengers over the harbor channel in Trondheim, Norway. The APS is characterized by a high degree of autonomy where the navigational and operational requirements are fulfilled by the APS. A supervisor in a land-based control centre (during the first year located on site) is responsible for actions needed in case of emergencies. Autonomous functions include navigation, docking, passenger registration, charging. Therefore, the communication of navigational and status data to the land-based control centre is vital [6].

3.2 Communication architecture

The communication architecture of the APS enables the communication with the environment through a heterogeneous group of different technologies. There are six main communication gateways in the APS. Particularly, two IP based gateways aim to establish ship-to-shore communication links with the RCC by leveraging several implementation solutions such as mobile communication (4G/LTE/5G) and Wireless Local Area Networks (WLAN) technologies. The third gateway is intended for ship-to-ship communication to enable the vessels in the area to communicate with the APS. Automatic Identification System (AIS) is proposed for the implementation of this module. The fourth gateway is intended to carry emergency communications for the control and navigation of the APS in case of lost communication with the Remote Control Center (RCC) while the fifth and sixth gateways are utilized to receive signals for real time kinematic (RTK) and global navigation satellite system (GNSS).

The internal network architecture of the APS is designed to include redundant communication paths, segregated sub-network, and secure communication. A centralized monitoring and controlling group of servers called the Autonomous Ship Controller (ASC) is interfaced with two network traffic Core and Distribution tiers (C/D), each consisting of main and backup switches with IP routing capabilities. The former tier (C/D A) connects the external gateways with the servers in the ASC, while the latter (C/D B) connects the secondary (i.e., backup) servers in the ASC with the internally segregated sub-networks. Moreover, a centralized component named the connectivity manager is responsible for performing network management functions by configuring and monitoring the network devices in addition to additional functions related to security. The detailed communication architecture along with the corresponding functions are described by Amro et al. in [5].

3.3 Navigation and Machinery Systems

The navigation system is comprised of components able to collect environmental data, establish situational awareness based on the sensing data, and determine safe navigation routes. The navigation system components are arrays of sensors of different types (EO cameras, video cameras, Lidars, and Radars), in addition to RTK GNSS. All these components send their data through the ship internal network to the Autonomous Navigation System (ANS) that hosts the logic to perform autonomous navigation functions, as well, as support remote navigation by the RCC. Moreover, a machinery system implements maneuvers according to the determined route from the ANS. The machinery system consists of a Dynamic Positioning (DP) system, and thrusters, interfaced through Input/Output (I/O) cards. The machinery system is controlled by an Autonomous Engine Monitoring and Control (AEMC) system which host the logic to monitor engine data and determine the appropriate control parameters. Both the ANS and AEMC are hosted in the ASC servers zone.

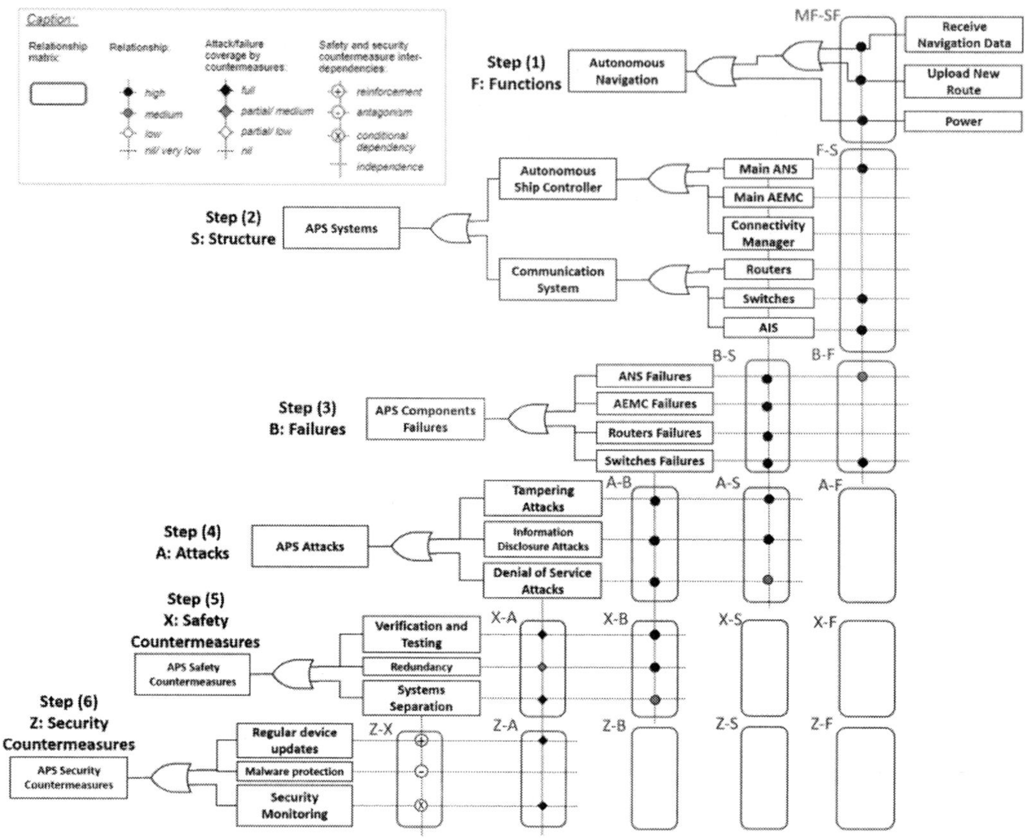

Figure 1. Overview of the applied Six Step Model steps (Adapted from [15])

3.4 Preliminary Hazard Analysis
A PHA for the MilliAmpere2 is presented in [13]. A PHA is a structured hazard identification method, which is guided through keywords that represent possible hazards and their sources [3]. The PHA for the APS was conducted in two sessions with the participation of several experts from several relevant domains. This analysis builds the foundation for identification of safety related issues in this article.

4. The six-step model
This Section summarizes the adopted six-step model (SSM) for the safety and security analysis of the communication, navigation and control systems of the APS. The SSM was proposed by Sabaliauskaite et al. [14] to analyze, both, safety and security aspects of CPS. Furthermore, the applicability of the SSM to and security issues of an autonomous vehicle is examined in [15]. An overview of the SSM that is followed in this work, is depicted in Figure 1. The six steps of the SSM model were performed disregarding some matrices, since their analysis was not considered relevant for the scope of the target analysis.

4.1 Step 1: Functions
The first step of the SSM describes the system's functions. Main and supporting functions are differentiated and their relationships are established. The main functions are cross referenced with the supporting, secondary functions of the system in the MF-SF matrix (Main-functions, Supporting Functions). The matrix provides the basis for further system analysis. Four types of relationships are distinguished, high, medium, low, very low/ none. The relationships description is adopted from [16]. High relationships characterize high dependency of the main function on the supporting function for

proper operation. Medium means that the main functions might be dependent on the supporting functions to operate properly in some operational modes while low relationship means that the main functions are rarely dependent on the supporting functions. Finally, nil/ very low relationships mean that no dependencies on the supporting functions are identified.

4.2 Step 2: Structure
The second step identifies the relationships between the APS components and the APS functions in the S-F matrix (structure: functions). The APS components are identified by decomposing the systems to the appropriate level of analysis. This includes main and supporting systems. This analysis is necessary to determine the relationships between function failures and physical component failures in step 3 (Section 4.3). The rating scheme includes high, medium, low, and nil/ very low levels. A high relationship indicates that the component is highly important for the realization of the function under analysis. A medium relationship indicates that the component might be needed to realize the function in certain operational modes. A low relationship indicates that the component might be needed to realize the function in very specific and rare cases. Nil/ very low is assigned to pairs that have no relationship with each other.

The components are prioritized for threat modeling in Step 4 (Section 4.4) according to their highest effect on the main system functions, considering the relationships studied in Step 2. The scores of the components under analysis are calculated for all the APS's components, taking into account the assigned relationships of each component with the specified system functions following Equation 1. The number of high relationships with systems functions is denoted as "h". Further, the number of medium and low relationships are denoted as "m" and "l" respectively. The components that gathered scores above the average are considered for analysis.

$$Score = 5h + 3m + l \tag{1}$$

4.3 Step 3: Failures
The aim of the third step is twofold; firstly the system failures of main components are identified and secondly the failures' impact on the system's functions are determined. In this step, the B-S matrix (failure: structure) is created. In this matrix, component failures were identified and assessed from the available PHA report [13]. The failures were generalized and the results from the report were used as input to rate the dependencies. The failures are assessed in relation to the system's functions in order to assess the severity of failures on the system's function execution. The information is recorded in the B-F matrix (failures: functions). The rating scheme used for the B-S and B-F matrices included high, medium, low, and nil/very low levels. These indicate the strength of the impact of a failure to either the operation of a component or the implementation of a system function. For instance, a high relationship between a failure and a component means that the latter will most probably not be able to operate, whereas the same level between a failure and a function means that the latter will be potentially severely impaired.

By leveraging the PHA [13] the Function Failure Impact Factor (FFIF) is determined. The FFIF represents the expected impact on the execution of a function that is weighted by potential consequences of the failures. The loss of each function was associated with potential consequence categories that included the following in ascending order of severity: loss of operational/ performance data, loss of remote monitoring and control, and loss of control/ drifting/ grounding/ collision/ injuries/ fatalities. Having calculated the failures' relationships with the system functions (Relationshipi,j (Failurei, Functionj)), the overall impact score for each component failure (i) is calculated using equation 2 where N represents the total number of system functions. Failures with Impact score above the average have been forwarded for analysis in Step 4 (Section 4.4).

$$Impact_i = \sum_{j=1}^{N} Relationship_{i,j} * FFIF_j \tag{2}$$

4.4 Step 4: Attacks

The assessment of cyber attacks is performed in Step 4 by utilizing the STRIDE method. STRIDE stands for Spoofing, Tampering, Repudiation, Information Disclosure, Denial of Service and Elevation of Privilege [17]. The method enables the analysis of complex systems and environments similar to the APS [18, 19].

Potential security threats along with the corresponding attack scenarios are identified and analyzed for the APS's components considering STRIDE. The security analysis takes into account external and internal attackers. The former are able to conduct the attack remotely while the latter perform the attack by infecting system components. Further, malicious passengers are considered as potential adversaries that may attack the APS. The following relationship matrices are generated; matrix A-B (attacks: failures), and matrix A-S (attacks: structure).

The A-S matrix is analyzed through the application of STRIDE. The attacks considered in matrix A-B are a subset of the attacks that are analyzed in the A-S matrix. Using the analysis performed in step 2 (Section 4.2) only attacks against components with most effect on the system's functions have been considered. Therefore, the attacks that are characterized by the highest impact on the main system functions are identified. Further, the A-B matrix includes attack scenarios that violate the security objectives of authenticity, integrity, non-repudiation, confidentiality, availability, and authorization. These objectives aim to ensure the security of the components in the maritime environment [10] and therefore of the APS's infrastructure. The most critical security objectives related to the functions of the APS are integrity, confidentiality, and availability. These ensure the security and reliability of the communication systems.

The A-B matrix reflects the relationships between the prioritized attacks and prioritized failures. The relationships are categorized as high, medium, low, or nil/ very low. A high relationship means that the attack is expected to directly lead to the failure with high possibility. A medium relationship means that the attack triggers the failure with moderate possibility. A low relationship means that the attack may lead to the failure with low probability, while a nil/ very low relationship is considered if no connection between between the attack/ failure pair can be identified.

4.5 Step 5: Safety Countermeasures

Safety countermeasures are identified considering the failures and the attacks in the fifth step. The reasoning for identifying the potential safety countermeasures is based on a high- level consideration for the system design and development process, and strategies to mitigate potential risks. The safety countermeasures include measures that need to be designed into the system (e.g., integrity checks, error handling), measures during commissioning (i.e., testing and verification), and operational measures (e.g., maintenance policies for hardware and software components, minimum risk condition). The matrices assessed in this step are matrix X-B (safety countermeasures: failures), and matrix X-A (Safety countermeasures: Attacks). The X-A matrix enables the identification of synergies between safety and security issues. For X-B and X-A the assessment considered four distinct degrees; full, partial medium, partial low, and nil. A full degree removes the failure or attack and its associated consequences to a large degree or completely. A safety measure assessed as partial medium eliminates the consequences or reduce the consequences to a large degree. A partial low assessment implies conversely a minor reduction in the frequency of occurrence or a minor reduction in the expected consequences. Nil degree describes that no improvement from this measure is expected, or that it has been already implemented.

4.6 Step 6: Security Countermeasures

In this step, the relationship matrices being analyzed are matrix Z-X (Security countermeasures: safety countermeasures) and matrix Z-A (security countermeasures: attacks). The remaining matrices, Z-B (security countermeasures: failures), Z-S (security countermeasures: structure), and Z-F (security countermeasures: functions) were not analyzed. The identification of security countermeasures is needed for the analysis in this step. The considered countermeasures are based on previously established cyber security requirements for the communication architecture [20].

The Z-A matrix described the effectiveness of a countermeasure in mitigating cyber attacks. To this end, three relationship categories; 1) the countermeasure leads to fully mitigating the attack (f), 2) partial mitigation(p), 3) nil (no mitigation or not relevant). By applying equation 3 a score is calculated for each countermeasure to indicate the effectiveness against the attacks based on the analysed relationship.

$$Score = 5f + p \tag{3}$$

The Z-X matrix captures the dependencies between safety and security countermeasures, which is crucial to the study of the synergies between these different sets of countermeasures. In particular, the effects that the security countermeasures may have on the safety countermeasures are represented through four types of relationship as defined in [14]. These are: 1) Reinforcement,
2) Antagonism, 3) Conditional dependency, and 4) Independent.

5. Results

5.1 Step 1: Functional analysis

The proposed communication architecture enables the MilliAmpere2 to perform several functions related to navigation, control, communication, and safety. Figure 2 shows an overview of the functions supported by the communication architecture and reflects their relations with the APS components previously discussed in Section 3 as well as among themselves. These relations highly influenced the analysis in Step 1 and 2 in the SSM model. The main navigation functions provide the situational awareness of the APS for the determination of safe routes. The "Engine Monitoring and Control" functions describe the monitoring and control of the APS's thrusters. Furthermore, autonomous functions are performed by the APS. Remote functions are carried out by the RCC, and emergency functions are executed by the Emergency Control Team (ECT). Further functions are needed to initiate emergency signals by passengers referred to as "Passenger Safety" functions. They are needed to indicate the occurrence of safety-critical events (e.g., passenger falling overboard). Therefore, these functions will only be found in APS or manned autonomous ships, and not in unmanned autonomous ships, since they will not be required. For the purpose of this paper, only emergency functions with respect to passenger communication with the RCC and the emergency services are considered, due to the focus on the communication system. The main communication functions are categorized considering their individual role. The "Ship-to-shore communication" function provides the required connectivity between the ship and the RCC. The "Internal communication" functions provide the needed connectivity between the different components onboard the ship. The "Emergency communication" functions provide the needed connectivity with the ECT. These functions depend on several supporting functions such as power, security, and network system management (NSM).

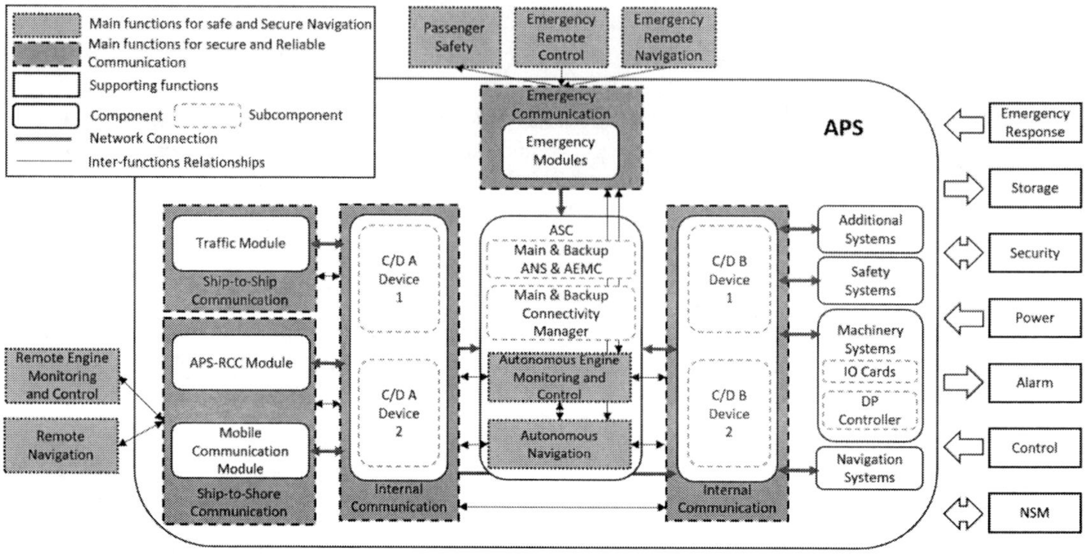

Figure 2. Overview of the relationships between the communication architecture functions and structure (Adapted from [5])

5.2 Step 2: Component assessment

The aforementioned scoring scheme in Step 4.2 was used to identify the components with the highest influence on the system. The components and the number of identified high, medium and low relationships with system functions are shown in Table 1. It can be observed that the components related to communication related have a relatively higher effect over system functions which is logical since they are responsible for the information exchange needed for most functions.

Table 1. Relationships assessment of the components with the most effect on the main system functions

Component	Relationships		
	High	Medium	Low
C/D A Device 1 and 2	30	1	10
Mobile Communication Module	23	1	13
APS-RCC Module	23	1	13
Connectivity Manager	24	1	0
DP controller	9	19	1
C/D B Device 1 and 2	16	3	10
Backup Connectivity Manager	18	1	0
IO cards	10	14	0
Main ANS	15	4	1
Backup ANS	11	4	1
Emergency Module 1	11	1	10

5.3 Step 3: Assessment of failures

As an outcome of step 3, the failures having highest impact according to equation 2 were identified. The failures associated with the connectivity manager are found to have the highest impact score, with C/D part A devices next, and the DP system after them. We argue that the results are plausible since the failures with highest scores would indeed affect the entire APS. For instance, failures in the connectivity manager and network devices would lead to disruption in the information flow within the APS which would lead to total loss of certain functionalities. Additionally, the AEMC, DP, together with the I/O cards and the thrusters highly affect the APS, in case of a component failure Some may even lead to a blackout affecting the whole APS.

5.4 Step 4: Assessment of attacks

Through the application of the STRIDE methodology several attack scenarios were identified as well as their relevance to the system components. Those attacks were analyzed against the failures discussed in Section 5.3. The results reflect which attacks could cause additional failures, as well as, which failures increase the vulnerability of the system to cyber attacks. It was found that attacks against both, the main and backup, connectivity manager components could cause failures with highest impact, followed by the C/D switches. Additionally, it was observed that denial of service attacks cause higher impact failures than others, followed by tampering, while information disclosure attacks have much lower effect on failures. Moreover, susceptibility to cyber attacks is mostly enhanced by failures in the connectivity manager, ANS, and AEMC components.

5.5 Step 5: Assessment of safety measures

Eleven safety countermeasures were identified. The countermeasures are summarized and described in Table 2. Safety countermeasures were identified from the PHA [13] and common risk mitigation strategies, such as laid out in [3]. For the purpose and scope of the article, the safety countermeasures are generic in nature and the specific implementation for the components needs to be defined and described in the further design process.

Table 2. Identified safety countermeasures

ID	Mitigation measure	Description
CSaf1	Choice of communication proto- col	Selecting protocols and bus systems that are robust and suitable for the purpose of communication between the components.
CSaf2	Verification and testing	The component should be tested and its function and behavior verified during different phases of the development process.
CSaf3	Monitoring and trouble shoot- ing through shore operator	The shore operator monitors the system behavior and engages in problem trouble shooting if a problem with the ferry occurs. The ferry design needs to accommodate these trouble shooting abilities.
CSaf4	Component redundancy	A second similar component is introduced in the system design to take over functionalities in case of the failure occurring.
CSaf5	Separate hardware components	The component has its own dedicated computing hardware to run on.
CSaf6	Go to a safe state	A safe state is defined for a failure and will thereby mitigate the consequences of this failure. The ferry needs to be designed such that the safe state can be reached in the failure condition.
CSaf7	Self and status tests of the component	The component must be able to test for correct operation and functioning.
CSaf8	Choice of computing hardware	Sufficient powerful computing hardware needs to be chosen to fulfil the components purpose even under high load conditions.
CSaf9	Cross validation of data inputs for sensor data	The components using sensor data is crosschecked with other data for plausibility. Implausible and invalid data should be rejected.
CSaf10	Hardware maintenance and cleaning policy	A maintenance plan defining preventive and corrective maintenance, including cleaning for the hardware components.
CSaf11	Software maintenance policy	A maintenance policy for the software describing the policy for bug fixing and updating software, and associated tasks.

The assessment of the effectiveness of each safety countermeasure is based on its assumed degree of elimination or mitigation of a failure. However, due to the generic nature of these, their impact cannot be definitively assessed since the implementation of the measure for each component was not specified in detail. For the same reason, it cannot be assumed that all failures are removed from the system. The safety countermeasures address all the failures, as well as, most of the attack scenarios. Only hardware maintenance is not addressing any of the cyber attacks. Most failures and attacks can be addressed through testing and verification efforts (CSaf2) and Self and status tests (CSaf7). Monitoring through an operator (CSaf3) can assist to some degree in identifying safety and security related issues. However, adequate procedures need to be established in order to troubleshoot efficiently and react appropriately. Cross validation of data (CSaf9) and a hardware maintenance policy (CSaf10) is mainly relevant for the sensors and the actuators.

5.6 Step 6: Assessment of security measures

The analysis performed in the Z-A matrix is utilized to assess the countermeasures coverage of the identified attacks. By using equation 3, as depicted in Table 3 various countermeasures are found to mitigate either fully or partially several attacks, such as the application of secure network protocols, and the preparation of incident response plans. The analysis facilitates the prioritization of the countermeasures implementation. It can be observed that the implementation on secure network protocols such as Transport Layer Security (TLS) and Virtual Private Network (VPN) would be most effective for attack mitigation, followed by well planned incident response procedures, and security monitoring. Moreover, it was observed that cyber security training for operators is not of high priority which is logical due to the reduced human involvement in the direct operation of the APS. For the operation of the RCC, the cyber security training may still be of importance.

Table 3. Outcome of the assessment of cyber security countermeasures

ID		At cks a mitigati on		
	Countermeasures	Fully	Partially	None
CSec1	Secure network protocols	14	2	2
CSec2	Incident response plans	7	10	1
CSec3	Security monitoring for detecting malicious and abnormal incidents	8	8	2
CSec4	User access management system	7	7	4
CSec5	Regular device updates	3	9	6
CSec6	Detailed map of IT and network equipment and software	3	8	7
CSec7	Cyber security management framework	2	9	7
CSec8	Regular software security analysis (penetration testing)	1	8	9
CSec9	Backup facilities	4	2	12
CSec10	Periodic inventory of user accounts and their associated privileges	2	5	11
CSec11	Malware protection	0	8	10
CSec12	Cyber security training	1	6	11

In this work the impact of security risk on safety is examined and analyzed. To this end, Table 4 depicts the relationship between the safety and security countermeasures. By leveraging the information depicted in Table 4 most of the security countermeasures are independent (60 relationships) while 40 relationships were assessed as enhancing the existing safety countermeasures. For instance, the cyber security management framework facilitates and strengthens the corresponding safety countermeasures. Additionally, twenty seven countermeasure relationships are characterized as conditionally dependent. Only five relationships between safety and security countermeasures are characterized as antagonism. Namely, the need to separate system components (CSaf5) and specify certain choices of hardware (CSaf8) for safety countermeasures may complicate the implementation of suitable security monitoring solutions (CSec3).

6. Discussion

The SSM analysis identifies the relationships between components and functions and provide a holistic view for the system under analysis. Therefore, the interplay of safety and security is examined in detail, considering the different viewpoints that are provided by the corresponding matrices. Overall the SSM provides an appropriate analysis of a system under development, in the initial steps of the design process where the functional and operational requirements are not defined in detail. Our analysis shows that the SSM provide results that may help to prioritize identified safety and security issues for more detailed analysis.

The analysis of an abstract system architecture is facilitated by leveraging the SSM. The method extracts rigorous and valid results in high organizational and operational levels. However, the SSM scales badly with increasing system complexity, due to the state-space growth in the SSM matrices. The SSM would benefit from additional guidance on how to represent and model the system and the dependencies among its components in a standardized way, similarly to how the STPA defines the safety control structure as a way to model the system. This would help in better determining, for example, common cause failures and assessing the impact of failures on the system structure and functionality. Additionally, further guidance on ranking the relationships described in each SSM matrix and steps for the failure and attack prioritization are needed. This may reduce the effect of subjective expert assessments that may not be justified in a transparent and reproducible manner.

Table 4. Relationships between safety and security countermeasures

Safety	Security											
	CSec7	CSec6	CSec10	CSec4	CSec5	CSec11	CSec1	CSec12	CSec8	CSec3	CSec2	CSec9
CSaf1	●	◑	○	◑	○	○	●	◍	○	●	○	◑
CSaf2	●	●	◑	○	○	○	◑	○	●	○	○	○
CSaf3	●	●	◑	◑	●	●	●	◍	●	●	●	●
CSaf4	●	○	○	○	○	○	○	○	○	○	◑	●
CSaf5	●	◑	○	◑	◍	○	○	●	◑	◍	○	○
CSaf6	●	◑	○	◑	○	○	◑	○	○	●	●	○
CSaf7	●	○	○	◑	◑	◍	○	◑	◑	◑	○	○
CSaf8	●	○	○	○	◑	○	◑	○	○	◍	◍	●
CSaf9	●	●	○	○	◑	●	○	○	●	●	●	◑
CSaf10	◑	○	○	○	○	○	○	○	○	○	◑	○
CSaf11	●	○	●	●	●	○	○	●	●	●	◑	◑

Legend: ● Reinforcement, ◍ Antagonism, ◑ Conditional dependency ○ Independent

Common cause failures, emerging system behavior, and multiple system failures are hard to include in the assessment. This may reduce the ability to identify interdependencies of failures. The identification of safety countermeasures is performed on a high level and detailed risk countermeasures could be developed early in the design phase.Applying the SSM process later, in the detailed design phases, the identified design changes and risk mitigation measures may come with a high cost. Guidelines on how safety and security measures may be identified with the SSM are also desirable.

7. Conclusion

In this work, a joint safety and security analysis of the MilliAmpere2 Autonomous Passenger Ship (APS) has been conducted. The Six-Step-Model (SSM) was applied to capture the different analysis viewpoints, namely, APS functions, structure, failures, cyber attacks, and safety and cyber security countermeasures. The main goal of the analysis was to infer the effect of cyber threats on safety, as well as, the interrelations between safety and security countermeasures, for design and implementation of integrated safety and security countermeasures.

Several conclusions can be drawn from the application of the SSM. It was found that the connectivity manager has most effect on the system functions. It could cause most failures, and is among the most susceptible to cyber threats. Secure network protocols, incident response plans and security monitoring were identified as the most important security countermeasures to be implemented. Moreover, safety and cyber security countermeasures have been found to be mostly compatible. Some measures are contradictory, which is very helpful to know during the design and implementation of both. Further work is needed to establish a security architecture for the APS that considers interrelations with safety. The outcome of this paper is expected to influence the design and implementation of security countermeasures to be adopted in the target security architecture as well as the undergoing design and implementation of the connectivity manager to mitigate it's discovered threats.

Acknowledgements
Thieme acknowledges the support by the Norwegian Research Council through the UNLOCK project, project number 274441.

References

[1] L. Pi`etre-Cambac´ed`es and C. Chaudet. The sema referential framework: Avoiding ambiguities in the terms "security" and "safety". Int. Journal of Critical Infrastructure Protection, 3(2):55–66, 2010.
[2] G. Kavallieratos, S. Katsikas, and V. Gkioulos. Cybersecurity and safety co-engineering of cyberphysical systems—a comprehensive survey. Future Internet, 12(4):65, 2020.
[3] M. Rausand. Risk Assessment: Theory, Methods, and Applications. John Wiley & Sons, Hoboken, New Jersey, USA, 1st ed. edition, 2013.
[4] T. Aven and O. Renn. On risk defined as an event where the outcome is uncertain. Journal of Risk Research, 12(1):1–11, 2009.
[5] A. Amro, V. Gkioulos, and S. Katsikas. Communications architecture for autonomous passenger ship. Submitted for review to Journal of Risk and Reliability (JRR).
[6] NTNU Autoferry. Autoferry - Autonomous all-electric passenger ferries for urban water transport, 2018.
[7] The Maritime Safety Committee. International maritime organization (imo) (2017) guidelines on maritime cyber risk management. http://bit.ly/IMORiskManagement.
[8] E. Lisova, I. Sˇljivo, and A. Cˇauˇsevi´c. Safety and security co-analyses: A systematic literature review. IEEE
[9] Systems Journal, 13(3):2189–2200, 2018.
[10] W. Young and N. Leveson. Systems thinking for safety and security. In Proceedings of the 29th Annual Computer Security Applications Conference, pages 1–8, 2013.
[11] G. Kavallieratos, S. Katsikas, and V. Gkioulos. Safesec tropos: Joint security and safety requirements elicitation. Computer Standards & Interfaces, 70:103429, 2020.
[12] V. Bolbot, G. Theotokatos, E. Boulougouris, and D. Vassalos. Safety related cyber-attacks identification and assessment for autonomous inland ships. In Int. Seminar on Safety and Security of Autonomous Vessels (ISSAV), 2019.
[13] I. Baˇckalov. Safety of autonomous inland vessels: an analysis of regulatory barriers in the present technical standards in europe. Safety science, 128:104763, 2020.
[14] C. A. Thieme, C. Guo, I. B. Utne, and S. Haugen. Preliminary hazard analysis of a small harbor passenger ferry-results, challenges and further work, 2019.
[15] G. Sabaliauskaite, S. Adepu, and A. Mathur. A six-step model for safety and security analysis of cyber- physical systems. In Int. Conference on Critical Information Infrastructures Security, pages 189–200. Springer, 2016.
[16] G. Sabaliauskaite and Jin Cui. Integrating autonomous vehicle safety and security. In Proceedings of the 2nd Int. Conference on Cyber-Technologies and Cyber-Systems (CYBER 2017), Barcelona, Spain, pages 12–16, 2017.
[17] G. Sabaliauskaite and J. Cui. Integrating Autonomous Vehicle Safety and Security. CYBER 2017 : The Second Int. Conference on Cyber-Technologies and Cyber-Systems, (level 0):75–81, 2017.
[18] A. Shostack. Threat Modeling: Designing for Security, volume Wiley Publishing. 2014.
[19] G. Kavallieratos, S. Katsikas, and V. Gkioulos. Cyber-attacks against the autonomous ship. In Computer Security, pages 20–36. Springer, 2018.
[20] G. Kavallieratos, V. Gkioulos, and S. K Katsikas. Threat analysis in dynamic environments: The case of the smart home. In 15th Int. Conference on Distributed Computing in Sensor Systems (DCOSS), pages 234–240. IEEE, 2019.
[21] A. Amro, V. Gkioulos, and S. Katsikas. Connect and protect: Requirements for maritime autonomous surface ship in urban passenger transportation. In Computer Security, pages 69–85. Springer, 2019.

IOP Conf. Series: Materials Science and Engineering **929** (2020) 012019 doi:10.1088/1757-899X/929/1/012019

Risk of electronics failure on digitalization of shipping industry

Dong Hyun Kim[1]

[1]KOMOS, Suite 407, Somerset Palace, 7, Yulgok-ro 2 gil, Jongno-gu, Seoul, Korea (03143)

Abstract. While digitalization drives the development of autonomous ships, the risks rising out of digitalization were reviewed. Adoption of digital technology has many benefits, but the technology itself has its own faults, which will have users suffering from the learning curve. The digital devices onboard are not permanently reliable, and they can fail due to harsh utilization conditions, fatigue, or latent defects. As the shipping industry continues to move down the road towards digitalization, an alert need to be delivered ahead in order to minimize the potential future risks.

1. Introduction

Digitalization is becoming a key word within the shipping industry as various subsystems onboard are now being equipped with digital devices that enable them to measure speed, fuel efficiency, cargo temperature, etc. The onboard devices, such as communication systems, propulsion and machinery management, power control systems, and cargo management systems, are connected to the bridge system and onshore operation center located far away from the ship. This measurement data is then received by the central server and processed in order to produce data needed to operate the ship. The remarkable progress of digital technology is also pushing autonomous ships into development for commercialization in countries such as Norway, Japan, and Korea. The autonomous ship will reduce the number of seafarers on board and consequently human errors. Figure 1 shows various onboard measurement systems and the world's first zero emission autonomous container ship Yara Bilkeland, developed by Yara and Kongsberg, and she will gradually move from manned operation to fully autonomous operation once launched [1, 2].

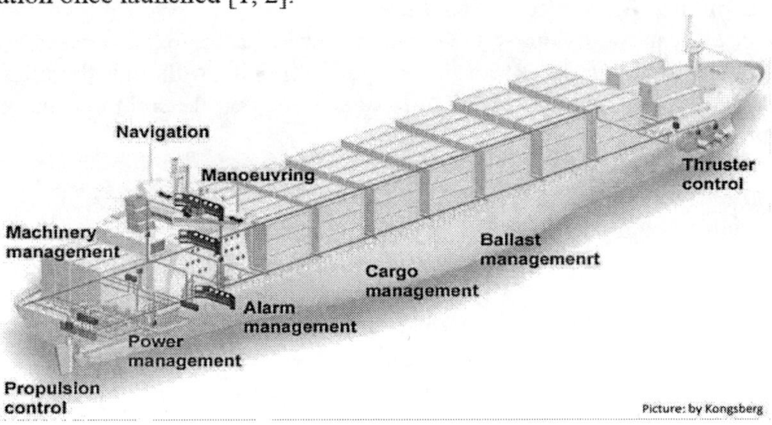

Figure 1. Various onboard management systems (upper) and Autonomous ship, Yara Bilkeland (bottom)

Though the immediate benefits from utilizing digital technology onboard are well recognized, there are a few points to review from a risk management perspective. Cyber threat for example is regarded as a major risk, because cyber attacks can corrupt a huge amount of data that is to be processed for operation, and can also paralyze fleet management systems and automated operation systems. Also, the recovery of data from a cyber attack itself will burden a considerable cost on the shipowner [3].

While there is no doubt about cyber threat being a major uprising risk following the digitalization of ship operations, particularly for those who handle massive amounts of operation data, it is worthwhile to note other potential risks that can rise from digitalization such as the lack of reliability of the electronics onboard.

It may be easy to assume digital products as being reliable or to even think their malfunctions as not serious. For example, troubles in consumer products such as mobile phones or home devices can be resolved by simple replacement. However, if electronic devices onboard malfunction during sail, it could lead to significant casualties, particularly when it is an engine controller that does not work properly while berthing in a harbor. In an autonomous ship, the digitalized automation system will play the role of crew, so the reliability of the electronics onboard is a necessity as crew play a crucial role in the safety of the ship.

A question could arise whether electronics are permanently reliable or whether they are vulnerable to wearing out and getting replaced like mechanical parts in an engine. This paper is intended to introduce the risk of electronics onboard as digital technologies are getting adopted to ships in a broad and swift manner. Various failure modes of electronics and past casualties of transportation modes caused by electronics will be discussed. Cyber threat will not be covered, as the shipping industry is well aware of those risks.

2. Incidents due to Failure of Electronics

It is unknown how many incidents at sea related to electronics failure have occurred in the past. It may be because of insignificant casualties triggered by electronics failures or the difficulty in finding the root cause of the incident in terms of electronic failure.

Starting with a brief of introduction of electronic hardware structure, two cases are shown in order to demonstrate how electronics failures can become catastrophic. Although the aircraft incident case in section 2.3 may be thought as irrelevant to ships at sea, the core lesson in terms of the significance of electronic failure can be seen.

2.1 Electronic Packaging

Electronics are built through an electronic packaging technology which interconnects chips and other electronic components into a circuit board. The effectiveness and reliability of electronics are determined by how they are packaged. Electronic packaging is composed of 1) electronic components such as a semiconductor, 2) Printed Circuit Board (PCB), and 3) solder joints that join electronic components onto the PCB through the soldering process. In the PCB, the solder provides electrical,

IOP Conf. Series: Materials Science and Engineering **929** (2020) 012019 doi:10.1088/1757-899X/929/1/012019

thermal, and mechanical connections. Solder joint reliability is one of the more critical issues in the development of the electronic packaging technologies. As the semiconductor gets smaller and denser, the electronic board assembly gets smaller in size as well. Also, electronic packages are exposed to various environments such as temperature, humidity, dust, shock, and vibration, which can all contribute to the failure of electronics [4].

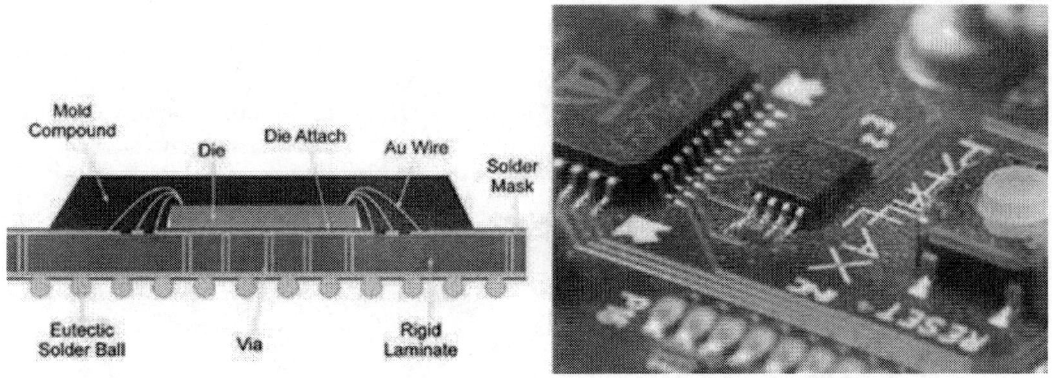

Figure 2. Semiconductor plastic package (left) and electronic components assembled on PCB (right) through solder by surface mount technology

2.2 River Aframax Fire

On 6 Sept 2016, the Tanker River in ballast collided with 2 mooring dolphins in the Houston Ship Channel. As a result, her fuel tank was punctured during the collision, leaking 88,000 gallons MGO. The conning pilot ordered all stops to avoid the collision, but the engine did not respond. Tugboats tried to slow the ship, but in vain. The spilled oil ignited and burned for about 45 minutes, shown in Figure 3. There were no human casualties, but there was damage to the tanker and other vessels [5].

The NTSB (National Transportation Safety Board) report showed that the heat generated from friction during the collision with the dolphins ignited the fuel and caught fire. The report concluded the electronic unit within the main engine governor actuator encountered a momentary abnormality. Therefore, the incident was caused by abnormal function of the electronic module [6].

Figure 3. Tanker River on fire

2.3 AirAsia crash

On 28 December 2014, Indonesian airline, AirAsia Flight QZ8501 (Airbus A320), departed the Indonesian city of Surabaya for Singapore and its communication with air traffic control was cut off just 42 minutes after it took off. The aircraft was later found crashed into the Java sea, and all 162 passengers and crew onboard were killed in the crash [7].

Indonesian NTSC (National Transportation Safety Committee) report showed cracked soldering on the electronic module called the Rudder Travel Limiter Unit (RTLU), shown in Figure 4. The cracked solder resulted in the intermittent failure of RTLU and four warning signals were given throughout the flight. When the pilot powered off the circuit breaker to reset the system to resolve the issue, autopilot mode turned off and the plane crashed due to inability to control the aircraft. The maintenance record showed the RTLU malfunctioned 23 times over the previous 12 months and investigators have concluded the plane crash was initiated by cracked soldering within the electronic module [8]. Therefore, solder joint cracks can lead to intermittent failures or open circuits within electronics.

Figure 4. Crashed AirAsia Flight QZ8501 (upper), Electronic module (bottom left), and Cracked solder joint in the module (bottom right) [8]

3. Failure type of Electronics

There are many different types of failures in electronic packaging due to defects from silicon, electronic components, and PCBs, but only a few are listed in order to provide example. In depth technical details are beyond the topic of paper.

Failure by Use conditions/ Manufacturing/ Material

Electronic packages are exposed to various environments such as temperature, humidity, dust, shock, and vibration. Temperature is the major cause of failure because the differences in thermal expansion between the chip and board induce thermal stress in the solder joints under normal operating conditions. Electronic packages can fail under dynamic loads such as shock or drop impacts and solder joints between chips and PCBs can fail in a brittle manner. The failure of these solder joints under thermal and dynamic loads can lead to an open circuit in electronics [9].

Electronic packages can also fail by poor manufacturing. Some latent defects were shown after using the machine for some time. For example, a severe warpage of electronic board assembly, cold solder joints that did not melt completely, and more can lead to failures during the manufacturing process [10].

Lastly, electronic packages can fail by material. In July 1, 2006, by the ROHS (Restriction Of the use of certain Hazardous Substances in electrical and electronic equipment) directive, lead (Pb) in printed circuit boards were banned. The transition into Pb-free electronic packages impacted the industry. In 2006, a huge batch of Swatch watches were recalled and the Microsoft Xbox360, a video game console device, had a high field failure rate of up to 20% [11,12].

Failure by No Fault Found

No Fault Found (NFF) is a common failure mode that presents symptoms like intermittent failures. Failures do occur, but it works again if the device is rebooted. This instance is called NFF. The two casualty cases introduced in section 2 of this paper can be categorized as NFF coupled with other types of failure. NFF is found broadly in applications such as automotive, avionics, computers, telecom, mobile phones, etc. NFF can be caused by various factors such as sneak circuits, printed circuit boards, connector issues, component-PCB interconnect failures, component failures, manufacturing issues, etc [13, 14].

Failure by Supply Chain Management

Diverse consumer tastes are pushing products into shorter life, lower costs, and faster time-to-market. Supply chain management has become a primary interest to many companies as it involves outsourcing parts in a global supply channel [15].

Effective supply chain management can be achieved not only by a purchasing team to outsource parts, but by a strong lean organization of cross-functional teams where each team will deliver different tasks such as: designing products, qualifying parts, testing products, finding the root cause of the failures from prototypes and the field, manufacturing products, reflecting market changes, auditing suppliers, outsourcing lower costs, handling logistics to meet leadtime, etc. The end goal of supply chain management is to place functional and reliable products to markets, essentially meeting customer needs [16].

Electronics build is challenging because electronics use numerous outsourced parts and services such as contract manufacturers. Supply chain management, which includes qualification processes to screen out poor quality parts and sufficient development time through prototyping and redesigning loops, will be required to ensure a functional and reliable product. It is clear that poor supply chain management, due to pressure for quick time-to-market, will lead to a loss in the reliability of products.

4. Conclusion

Digital technology currently enables operators located far away from ships to monitor various measurement data for efficient automated ship operation using digital devices onboard, and the remarkable progress of this technology has made it to the extent of the introduction of autonomous ships.

The digital devices onboard are built on electronic packaging technology. Electronics are not permanently reliable, and they can fail due to harsh utilization conditions, fatigue, or latent defects caused by poor design or manufacturing. Even more so, the nature of an electronic board being manufactured through a complicated supply chain integration of many outsourced parts makes it difficult to achieve a reliable electronics without effective supply chain management. Finding the root-cause failure of electronics is not always a straightforward task. Smaller dimension factors of electronics and pressure in time-to-market also risk the reliability of electronics. As ships move towards depending more on digitalization and automation, conventional mechanical systems will be integrated even more with electrical devices. Therefore, failure of electronics onboard ships will result in a whole mechanical-electrical system malfunction.

The certain transition into digital technology will launch ships equipped with automated and autonomous equipment, which will then change the way conventional shipping business is done. The effect will impact the legal system, the culture, and the risk. As a whole, the point this paper is intended to deliver is although utilizing the technology may result in benefits such as efficient operation and lesser costs, the shipping industry may suffer from the steep learning curve. This paper also gives the shipping industry an alert of the risks of moving further down the path of digitalization. Along with this alert, the short notes are suggested.

- For the shipping industry to consider prequalification inspection for electronic devices onboard and to maintain use of them according to product life.
- To put efforts in finding the root-cause of electronic failures, so that the industry can collect data to avoid future errors.
- Marine insurers may start to consider product liability for electronic products onboard, which will eventually replace crew in the era of digitalized and autonomous ships [17].

References

[1] Kongsberg, https://www.kongsberg.com.
[2] Yara, https://www.yara.com.
[3] BIMCO, "The Guidelines on Cyber Security Onboard Ships", 2018
[4] Dong Hyun Kim, PhD thesis on Reliability Study of SnPb and SnAg Solder Joints in PBGA Packages, University of Texas at Austin, December, 2007
[5] Professional Mariner, 26 April, 2018
[6] National Transportation Safety Board Marine Accident Brief, 22 February, 2018
[7] The Straits Times, 2 Dec, 2015
[8] Republic of Indonesia, National Transportation Safety Committee report, 2015
[9] Dong Hyun Kim, Tae-Kyu Lee, Sang Ha Kim, Han G. Park, and Kuo-Chuan Liu, "Study on Dynamic Shock Performance of SAC305 Solder Joint after Different Aging Conditions", SMTA, 2008
[10] MANNCORP, Repair Solution for Xbox 360 "Red Ring of Death" Hardware Failure, 10 July, 2007
[11] The Guardian, Within a whisker of failure, 3 Apr, 2008
[12] The Guardian, The truth about the Xbox, 12 Nov, 2008
[13] Bhanu Sood, Michael Osterman, and Michael Pecht, "Tin whisker analysis of Toyota's electronic throttle controls". Circuit World. 37 (3): 4–9, 2011
[14] Haiyu Qi, Sanka Ganesan, and Michael Pecht, "No-fault-found and intermittent failures in electronic products, Microelectronics Reliability", pp663-674, 2008
[15] Managing the Supply Chain, David Simchi-Levi, Philip Kaminsky and Edith Simchi-Levi, McGrawhill, 2003
[16] Lean Thinking, James P. Womack and Daniel T. Jones, Free Press, 1996
[17] John Cusano and Michael Costonis, "Driverless Cars Will Change Auto Insurance. Here's How Insurers Can Adapt", Harvard Business Review, Dec 2017

The 3rd International Conference on Maritime Autonomous Surface Ship (ICMASS 2020) IOP Publishing
IOP Conf. Series: Materials Science and Engineering **929** (2020) 012020 doi:10.1088/1757-899X/929/1/012020

Integrating accountability in the systems design of autonomous and remote-controlled operations

Bård Myhre[1], Ørnulf Jan Rødseth[2] and Stig Petersen[3]

[1] SINTEF Digital, Oslo, Norway
[2] SINTEF Ocean, Trondheim, Norway
[3] SINTEF Digital, Trondheim, Norway

Corresponding author: bard.myhre@sintef.no

Abstract. In a previous paper we have suggested that the transferal of human accountability from an on-site human actor (such as the captain) to a remote human actor (such as the creator of the autonomous control system) could be regarded as the defining characteristic of autonomous systems. In this paper we take this approach one step further, by suggesting a methodology for how accountability can be used as a basis for systems design of autonomous and remote-controlled operations. Furthermore, the suggested methodology is applied on a hypothetical case of a vessel supporting both autonomous and remote-controlled operation.

1. Introduction

The concept of autonomy is by many considered to be the next paradigm shift within transportation. Brought forth by technological advances in sensor technology, machine learning, and computing power, the vision of self-driving cars, vessels, and airborne drones has seemingly only been held back by the final technological break-through. However, over the last couple of years, some concern has been raised about the challenges that remain for the autonomous revolution to happen. In the automobile industry, executives of both Ford and Waymo have stated that fully self-driving cars may be decades away [1]. Within the maritime domain, IMO is performing a regulatory scoping exercise for the use of Maritime Autonomous Surface Ships (MASS), with targeted completion by 2020. The reported results from IMO [2] so far seem optimistic, but several complex challenges are to be resolved before autonomous vessels will be a commercial reality.

From a technical perspective, much effort is put into making self-driving cars and vessels detect obstacles, react properly to various situations, and enabling operation under a broad range of conditions. However, although autonomous devices technically speaking operate by themselves, the responsibility for their actions will still be held by humans. This expectedly means that either a driver, an operator, a technician, or an organization will be held accountable for any accidents caused by an autonomous car or vessel. This aspect is rarely touched upon by the technical community, although the question of responsibility is described as a "fundamental issue to be addressed" by IMO [2].

In a previous paper [3] we have suggested that the transferal of *human accountability* from an *on-site human actor*, such as the captain, to a *remote human actor*, such as the creator of the autonomous control system, could be regarded as the defining characteristic of autonomous systems. Even though questions of responsibility and liability have received increased attention over the last years [4], most research seems to focus on how fast humans can take over in emergency situation, ethical aspects and transparency of machine learning decisions, and how to properly test autonomous systems. Our notion however implies that the concept of autonomy may benefit from also being associated with legal aspects rather than solely technical ones. Also, it may suggest that legal aspects such as certification should be the starting point of any design process involving autonomous systems, instead of being the final hurdle after the technology development is completed.

As classification schemes for autonomous ships are an area in constant development [5], we do not think that an accountability-centered approach towards autonomy will necessarily be the holy grail within the field of autonomy. However, we do hope that our perspectives may contribute to expanding the understanding on how autonomy could be regarded, and on how the design of autonomous systems may be approached. On this background, this paper aims to take the ideas from [3] one step further, by suggesting a methodology for how accountability can be used as a basis for systems design of autonomous and remote-controlled operations. Furthermore, the suggested methodology will be applied on a fictitious case of a vessel supporting both autonomous and remote-controlled operation.

To establish a common ground, section 2 of this paper will first give a brief summary of how accountability could be regarded as a starting-point for autonomy, as introduced in [3]. Section 3 will then suggest some definitions for central concepts related to non-conventional navigation and control. In section 4 we introduce the above-mentioned methodology for integrating accountability in the systems design of autonomous and remote-controlled operations, while section 5 walks us through an imaginary example case illustrating the practical use of the methodology. Section 6 will summarize our considerations and findings, before we conclude the paper in section 7 while also pointing towards some unresolved challenges and future work.

Accountability as it is presented in this paper can thus be seen as a variant of transfer of control as discussed e.g. in the context of automated cars [6]. However, as the paper will show, the novelty here is to use this as a high-level design principle rather than as a consequence of design decisions or as a fallback solution.

2. Accountability as a starting point for autonomous and remote-controlled operations

As mentioned in section 1, our previous paper [3] suggested a definition for *autonomy* that have been modified as follows:

An automated system is considered *autonomous* within a specified operational envelope if it can *lawfully accept accountability* for an operation, thereby assuming the accountability that was previously held by either a *human operator* or another *autonomous system*.

Note: As an automated system itself cannot be held *legally* accountable, the factual accountability for the actions of an autonomous system will necessarily reside with, e.g. the creators, installers or operators of the system, dependent on how the liability regime develops [7]. Furthermore, this definition implies that a human operator most likely can be considered to have exercised due care if she or he leaves control to the autonomous system, as long as the autonomous system has accepted accountability. The human operator will however be required to remain in control if the autonomous system declines to take control, and she or he must retake control if the autonomous system later alerts the operator of a possible future exit from the automatic control part of the operational envelope [5].

From the perspective of a system designer, the suggested definition of *autonomy* may have implications on how to describe, design and implement an autonomous system.

First, the capability of *lawfully accepting accountability* could be handled through a type of certification. Such a certification could be limited, for instance to certain operations, operational conditions, time of day, etc. It also means that some actors could be certified for parts of an overall operation only, and not for the entire operation.

Second, accountability can be transferred but not shared. This interpretation is in line with the concept of "Responsibility charting" introduced in [8], where the accountable person is the individual who is ultimately answerable for an activity or decision. This means that there at any time must be one (and only one) actor holding the accountability for the ship.

Third, as accountability can be transferred between both human operators and other autonomous systems, this means that several actors can be certified to hold the accountability for an operation, although only one at the time.

Fourth, an autonomous system does not have to be physically located on the vessel it controls. Depending on its certification, autonomous systems could also be located remotely.

Fifth, the definition above implies that an accountable system is "on its own" in case of problems, when it is inside the automatic control part of the operational envelope [5], and that it cannot rely on other systems or actors for help if the situation comes out of control.

Sixth, if the accountable actor is not able to fulfil its obligations of controlling the vessel due to communication loss or other failure, the system should have a fallback to a Minimum Risk Condition.

3. Defining central concepts in non-conventional navigation and control

Although terms such as *autonomous* and *remote-controlled* may be used colloquially in everyday talk, defining concepts as stringent as possible is important to avoid ambiguities and misunderstandings. For the remainder of this paper, we will adhere to the following definitions:

- **Navigation and control (N&C) system.** A technical on-board system that enables navigation and control of a vessel, but without the functionality to do this automatically.
- **Classic Automated Ship Controller.** A control system that is technically able to provide input to a vessel's N&C system, but that cannot lawfully accept accountability for an operation. A traditional autopilot would fall into this category, as the captain would still be accountable for the vessel's operation.
- **Autonomous Ship Controller (ASC).** A control system that is technically able to provide input to a vessel's N&C system, and that can lawfully accept accountability for an operation, thereby assuming the accountability that was previously held by either a human operator (e.g. the captain) or another autonomous control system.
- **Remote-controlled vessel.** A vessel whose accountable control is held by someone not physically present on the vessel. A vessel can be remote-controlled both by remote-located human operators and remote-located ASCs.
- **Remote Control Centre (RCC).** An off-vessel location for remote-controlling vessels.
- **Remote Control Centre Operator (RCCO).** A human that is certified to remote-control a vessel under specific conditions and limitations, working physically or logically through an RCC.
- **Crewed vessel.** A vessel with a human operator on board that can lawfully accept accountability for the vessel.
- **Uncrewed vessel**. A vessel without a human on board that can lawfully accept accountability for the vessel.
- **Minimum Risk Condition (MRC).** A minimum risk condition (MRC) is a state that the vessel should enter in situations that are outside those in which it can operate normally, but is still expected to deal with in one way or another [9].
- **Minimum Risk Condition Function (MRC Function, or MRCF).** An MRC Function is an Automated (but not Autonomous) Ship Controller whose task is to navigate the vessel into an MRC if no accountable operator (either human or ASC) is in control of the vessel.

According to these definitions, there exist a wide range of variants of how a ship can be controlled. One could for instance imagine a **crewed vessel being controlled by an ASC,** enabling the on-board captain to be off duty and without any formal accountability for the vessel. Furthermore, in such a case the ASC could either be located on the ship or at the RCC, where the latter in practise would constitute an *Autonomous Remote Controller (ASC)*.

4. A 3+3 step methodology for high-level design of autonomous and remote-controlled systems

To successfully develop autonomous and remote-controlled systems, we have suggested the idea that accountability should be considered a fundamental capability. Integrating accountability in technical design is not necessarily intuitive, and we have therefore attempted to specify a two-phased design process. Each phase consist of three distinct activities, and the entire methodology could therefore be regarded as a 3+3 step procedure, as illustrated in Figure 1.

The first phase is the *concept phase*, whose objective is to establish the overall requirements for the system with regards to accountability. This phase consists of the following three stages: *i) actors and capabilities*, *ii) accountability states*, and *iii) control states*. The stages are logically to be developed in sequence, but iterations should be expected as the insight into the case evolves.

The second phase is the *high-level design phase*, which establishes the top-level design for fulfilling the requirements from the first phase. This phase consists of three more stages: *i) system architecture*, *ii) messaging scheme*, and *iii) module requirements*. Like for phase 1, the stages form a logical sequential order, but iterations should be expected also here.

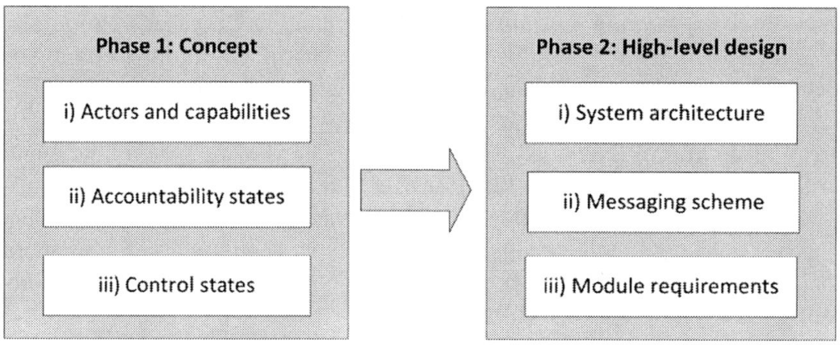

Figure 1. A 3+3 step methodology for high-level design of autonomous and remote-controlled systems

This principle can be embedded in the first design steps, either in a V-type or a more agile development process, e.g. using a spiral model with incremental refinement of the design and implementation.

5. A fictitious case: Autonomous and remote-controlled vessel with contingency function

To illustrate how the methodology can be applied in practise, we will describe and develop a fictitious case along the phases and activities from section 4. This will demonstrate where various design choices have been made during the overall process, and at some points we will also highlight some design choices that we deliberately *have not taken*. Note however that the presented case is not a real one, and that the objective of this section is to exemplify the suggested process. It should therefore be stressed that this paper is more in search of *relevant questions* than *correct answers*, as the actual answers for a real case would inevitably need to be developed with the real-word stakeholders.

5.1. Phase 1: The concept phase

The concept to be developed in this paper could be described as "an uncrewed vessel, with an RCCO that can assume accountability for the entire voyage, an on-board ASC that can accept accountability during sea passage (i.e. outside port), and an MRC Function to assume control from the RCCO if communication with the vessel is lost".

5.1.1. Actors and capabilities

The actors and their fundamental capabilities, along with their interactions, are illustrated in Figure 2:

- One **vessel**, with
  - One **N&C unit**, that enables physical control of the vessel; receiving input from either the **ASC**, the **MRC Function** or wirelessly from the **RCCO.**
  - One **ASC**, that may accept accountability for the vessel *during sea passage only*[1].
  - One **MRC Function**, that shall assume control (without accepting accountability) if input from RCCO is lost.
- One **RCC**, with
  - One **RCCO**, that may accept accountability for the vessel during *the entire voyage*[2].

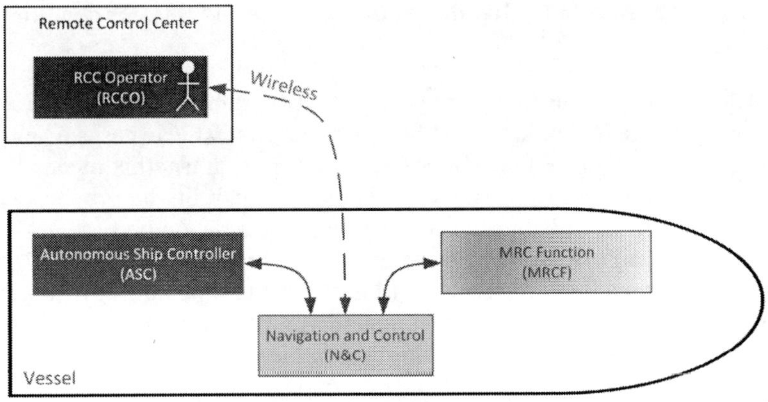

Figure 2. Actors and capabilities

Note that we at this stage only consider overall functions and capabilities, and that the actual system architecture will be left for phase 2. For the remainder of phase 1, we will therefore exclusively consider the entities that have been listed above.

5.1.2. Accountability states

Both the RCCO and the ASC may accept accountability for the vessel, with the ASC having certification for sea passage only. The interaction between the two units is however still to be decided: Will one of the entities have higher rank than the other, meaning that it may reclaim accountability at any time? And if one entity has higher rank than the other; shall the entity with lowest rank be able to offer accountability to the entity with higher rank, or should it simply wait for the entity with highest rank to reclaim accountability?

For the sake of the example case, we have made the following design choices:

- The RCCO shall have higher rank than the ASC, meaning that the RCCO at any time can reclaim accountability from the ASC[3].
- The ASC shall keep accountability until the RCCO has reclaimed accountability, and the ASC shall thus not formally offer accountability to the RCCO.

A visualisation of the accountability states and transfer conditions are provided in Figure 3.

[1] Note that the operational limits of the ASC should be formally defined in a real case.
[2] Note that the operational limits of the RCCO should be formally defined in a real case.
[3] Note that if the RCCO wants the ASC to reassume accountability after a reclaim operation, this requires a new acceptance from the ASC.

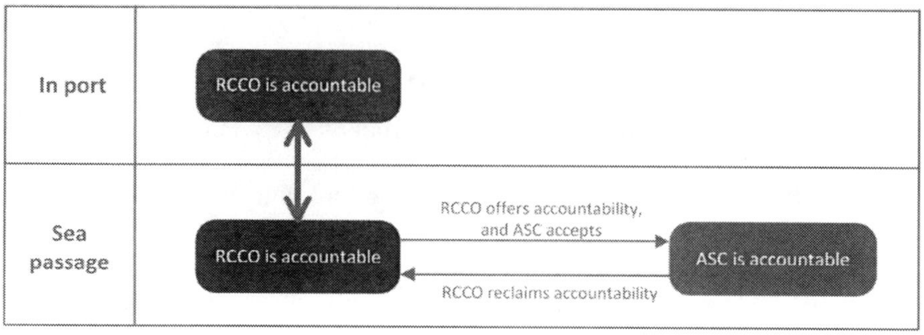

Figure 3. Accountability diagram with states and transfer conditions

Note that the MRC Function is not included in the accountability diagram in Figure 3, as it cannot accept accountability for the vessel. In our imagined case, the MRC Function is solely a backup if the RCCO loses control. Hence, the RCCO should still be considered accountable in cases where the MRC Function is enabled. Also observe that since the ASC cannot enter port areas, this means that the RCCO must actively monitor the situation in order to timely reclaim accountability before the vessel reaches the end of sea passage. If not, the ASC will simply keep the vessel outside the port area. Another design option would be to let the ASC formally request a transfer of accountability back to the RCCO when reaching the end of its operational envelope. In terms of illustrating the methodology, this functionality would however add unnecessary complexity to the accountability diagram in Figure 3 as well as the subsequent design phases without contributing correspondingly in value.

5.1.3. Control states

Having defined accountability as a fundamental capability in an autonomous and remote-controlled operation, it is vital that there is no doubt as to which entity *provides input* to the N&C of the vessel, and which entity holds the *accountability*. As the RCCO, the ASC and the MRC Function may all provide input to the N&C, there must exist a predefined logic of how to select the input for the N&C.

In our case, we define the input requirements as follows:

- Initially, the RCCO holds accountability and thus provides input to N&C.
- If the RCCO provides input to N&C and communication between RCCO and the vessel fails, then the MRC Function shall provide input to N&C.
- If the MRC Function provides input to N&C and communication between the RCCO and the vessel is restored, then the RCCO shall provide input to N&C.
- While in sea passage, the RCCO may offer ASC to assume accountability.
- If, and only if, the RCCO has offered ASC to assume accountability, ASC may accept (or implicitly refuse, by not accepting) the offer for accountability.
- If ASC accepts accountability, ASC shall provide input to N&C.
- If ASC provides input to N&C, the RCCO may at any time reclaim accountability. If the RCCO reclaims accountability, the RCCO shall provide input to N&C.
- The RCCO may at any time withdraw an offer to ASC of assuming accountability. This equals to reclaiming accountability from ASC.

It should be noted that the requirements above do not describe what to do if the ASC would fail. Such a situation could for instance be handled by using the MRC Function as fallback (with the ASC keeping accountability), but one could also imagine that an on-board ASC is required to provide its own, built-in MRC Function. Such considerations would be vital in a real case but are left out of our example to maintain a manageable scope.

The defined requirements can be translated into a more detailed *control state diagram* for the vessel, as presented in Figure 4. Here, the dark blue boxes indicate that the RCCO provides input to the N&C, the medium blue state indicate that the ASC provides input to the N&C, while the light blue states indicate that the MRC Function provides input to the N&C. Note that the RCCO is still considered accountable when the MRC Function provides input to the N&C.

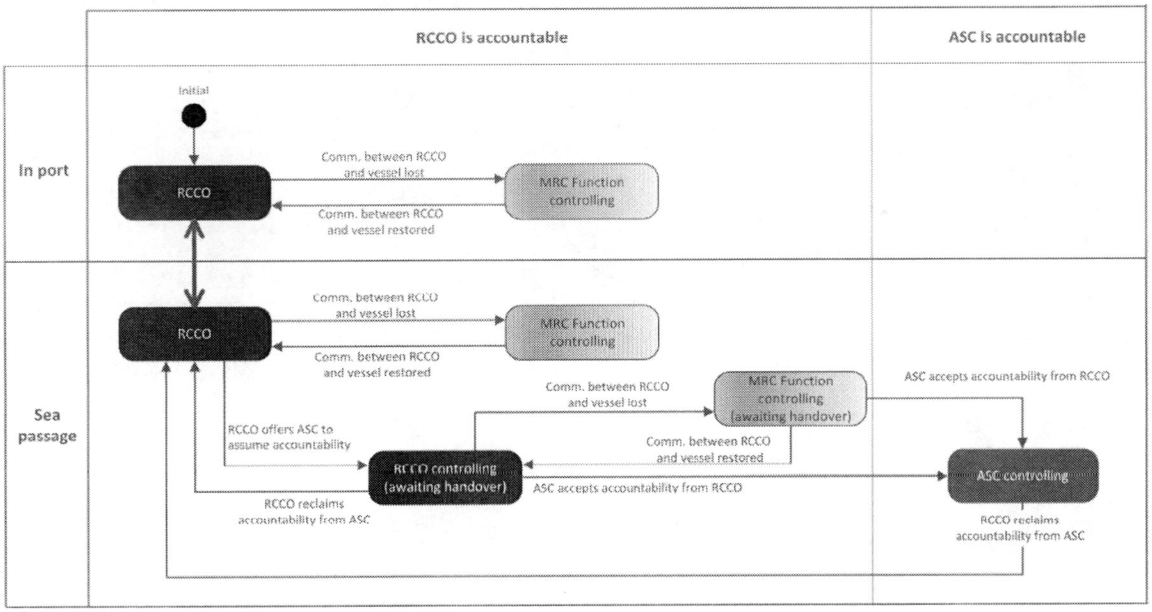

Figure 4. Control state diagram, with states and transfer conditions

In practise, these requirements mean that losing communication between the RCCO and the vessel does not affect a situation when the ASC is accountable.

Note that although these requirements may seem obvious at first glance, there are several other requirements that *could have* been stated. For instance, one could imagine a requirement that enabled the ASC to consider claiming accountability if communication between the RCCO and vessel is lost, instead of simply enabling the MRC Function. This would allow for a smoother operation in the cases where communication is lost during sea passage but is for simplicity left out of our case.

5.2. Phase 2: High-level design

The objective of the high-level design phase is to establish a technical system architecture as well as requirements for module interaction and behaviour. This is achieved by systematically going through the three steps of this phase.

5.2.1. System architecture

There exist numerous possible system architectures for realising the requirements from Phase 1, and it is not within the scope of this paper to perform a thorough evaluation of these. Some architectures still deserve specific attention. Figure 5 shows an architecture where the RCCO communicates directly with the MRC Function, and where the ASC and the MRC Function have separate inputs to the N&C. This scheme is visually intuitive but requires adding logic related to the N&C input. Figure 6 shows another solution, where the MRC Function is included in the ASC, and where the ASC holds the communication with the RCCO. This scheme may be regarded as a natural consequence of the idea that the RCCO and the ASC cooperate in controlling the ship. However, in this case the ASC would serve both as ASC, MRC Function and switching board for the N&C. Figure 7 declutters Figure 6 a bit, by separating the MRC Function and ASC, but still resembles a sequential "switching board" of input signals. This finally leaves us with Figure 8, which is inspired by both the Central Processing Unit (CPU) of modern computers, as well as the philosophy of "separation of concerns" brought forth by e.g. Dijkstra [10]. Here, we introduce the concept of an N&C Input Manager (NCIM), handling the communication between all the various entities as well as doing the selection of input for the N&C. This centralised architecture ensures that all distinct functions from Phase 1 are kept within separate functional modules, and also enables us to keep the responsibility for selecting the correct input for N&C within one single entity, i.e. the NCIM. From a practical perspective, it should be noted that such a distributed architecture does not hinder combining two or more modules into one single product[4].

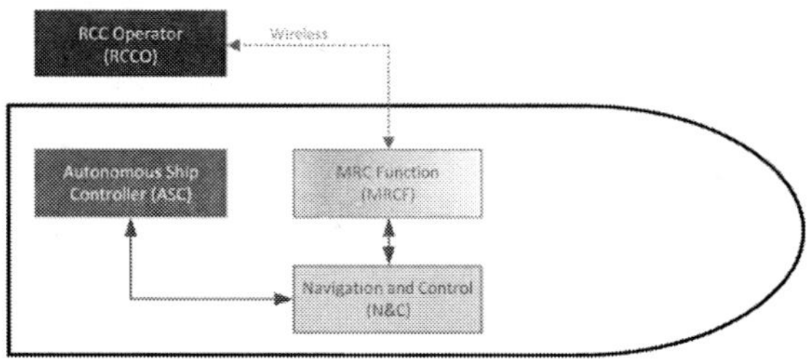

Figure 5. Parallel input management

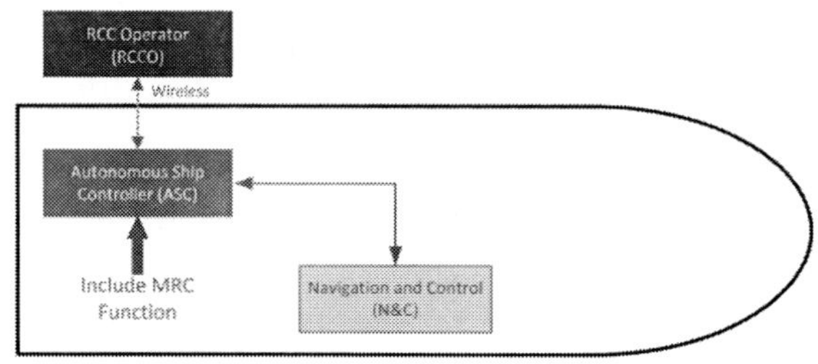

Figure 6. Sequential input management, with ASC providing MRC Function

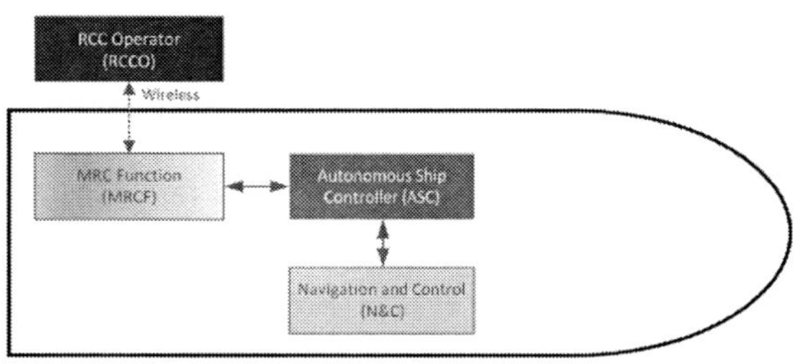

Figure 7. Sequential input management, with separate MRC Function

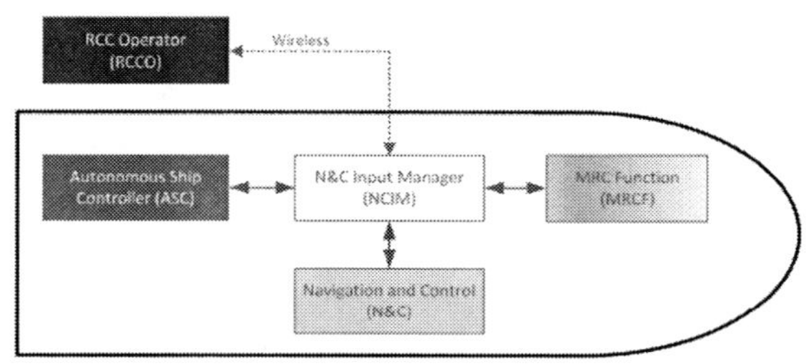

Figure 8. Centralised input management

Within the scope of this paper, we make the following decision for system architecture:

- The requirements stated in Phase 1 shall be realised through a centralised system architecture, where a NCIM will serve as a hub for all communication between the RCCO, the ASC, the MRC Function and the N&C. This architecture is illustrated in Figure 8.

5.2.2. Messaging scheme

One of the benefits of the centralised architecture illustrated in Figure 8 is that the control state diagram from Figure 4 can be directly converted into a state diagram for the NCIM. Each transition must however be investigated with care, to ensure that the relevant triggers are implemented and that corresponding messages are created for the relevant modules. Figure 9 shows one possible implementation of the NCIM state diagram, where each transition is assigned with a triggering event and corresponding action(s). The *triggering events* are denoted in two ways; either with hyphens ("), indicating a decision made within the current module, or angle brackets (< >), indicating a message from another module. The *corresponding actions* are denoted immediately after the slash (/), indicating message recipients (following the @ sign) and a message description.

[4] Interestingly, the architecture in Figure 6 could be regarded as one possible implementation of the architecture in Figure 8, with the former presenting the ASC, NCIM and MRC Function as one combined unit.

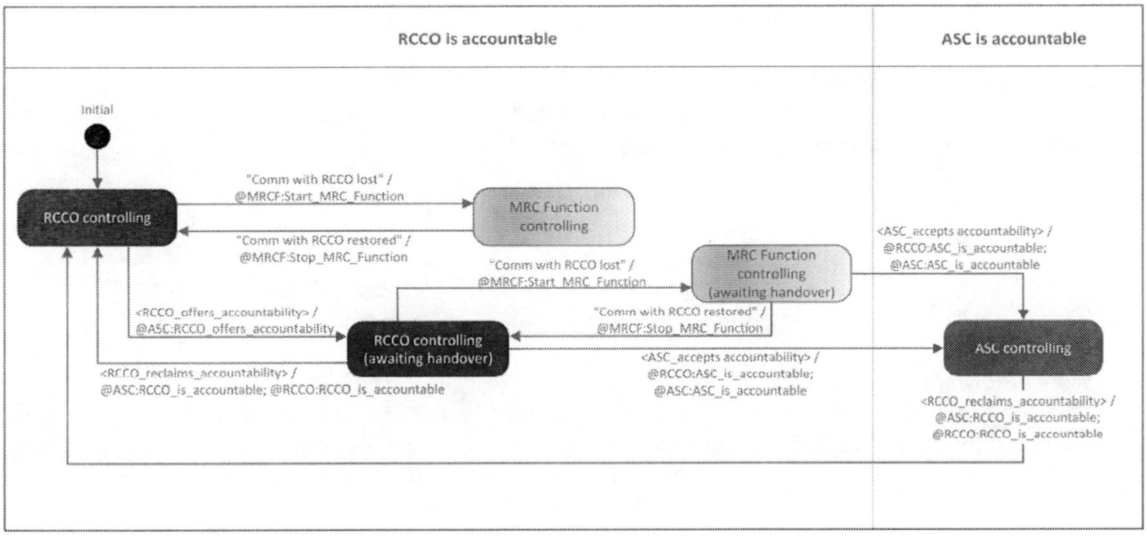

Figure 9. State diagram for the N&C Input Manager (NCIM)

To ensure correct operation, the following decisions must be made by the NCIM:

- "Communication with RCCO lost"
- "Communication with RCCO restored"

The following messages are required from the RCCO:

- <RCCO_offers_accountability>
- <RCCO_reclaims_accountability>

The following message is required from the ASC:

- <ASC_accepts_accountability>

Finally, the following messages must be communicated from the NCIM to the other modules:

- @MRCF:Start_MRC_Function
- @MRCF:Stop_MRC_Function
- @ASC:RCCO_offers_accountability
- @ASC:RCCO_is_accountable
- @RCCO:RCCO_is_accountable

It should be noted that the only two stand-alone decisions to be made by the NCIM are related to the status of the communication with the RCCO. This means that the communication between the NCIM and the RCCO requires a mechanism for timely detection of communication loss, e.g. in the form of a heartbeat. As detecting this communication loss would in most cases be considered safety-critical, one might want to investigate established standards for safe communication within other domains, such as EN 50159 [11] for railway signalling.

5.2.3. Module requirements

The final stage of the 3+3 methodology is to establish requirements for the individual modules. The requirements for the NCIM are already established as a part of the messaging scheme, but the other modules remain. However, their internal logic can for the most part be derived directly from the NCIM state diagram and message requirements as presented in section 5.2.2.

State diagram for the MRC Function
The state diagram for the MRC Function is illustrated in Figure 10. This module has only two operational states, and it is in practise controlled directly by the NCIM.

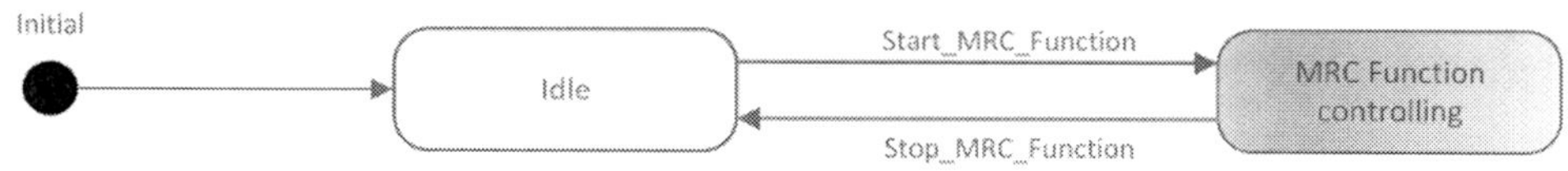

Figure 10. State diagram for the MRC Function

State diagram for the ASC
The state diagram for the ASC is illustrated in Figure 11. This module receives requests for accountability from the RCCO and needs to evaluate whether these requests should be accepted. Note that the ASC will have to provide control information to the NCIM as soon as accountability is accepted, although it will have to wait for the confirmation from the NCIM to know that it has actually received accountability. This is to ensure that an accountable entity will always provide information to the N&C.

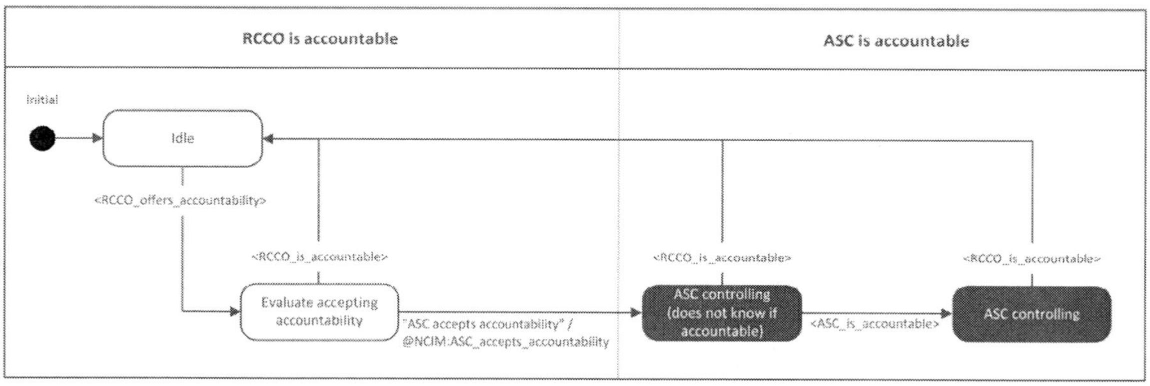

Figure 11. State diagram for the Autonomous Ship Controller (ASC)

To ensure correct operation, the following decision must be made by the ASC:

- "ASC accepts accountability"

Also, the following message must be communicated from the ASC to the NCIM:

- @NCIM: ASC_accepts_accountability

State diagram for the RCCO
The state diagram for the RCCO is illustrated in Figure 12. As the RCCO is a human being, keeping track of the states should ideally be handled digitally by a management system.

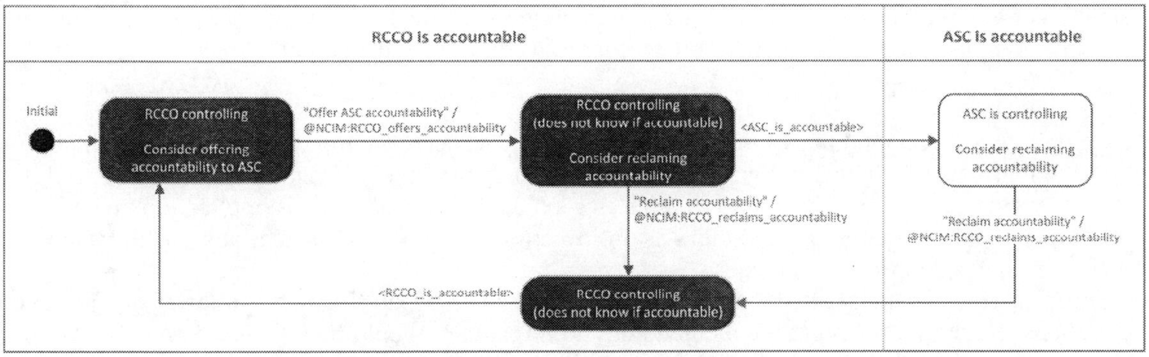

Figure 12. State diagram for the RCCO

To ensure correct operation, the following decisions must be made by the RCCO:

- "Offer ASC accountability "
- "Reclaim accountability"

Also, the following messages must be communicated from the RCCO to the NCIM:

- @NCIM:RCCO_offers_accountability
- @NCIM:RCCO_reclaims_accountability

One might further expect that the RCCO would like to know whether communication with the vessel is working correctly. This can be handled by a separate service (and hence a separate state diagram), as illustrated in Figure 13. Note that *knowing* the communication status will not influence the RCCO's ability to control the vessel.

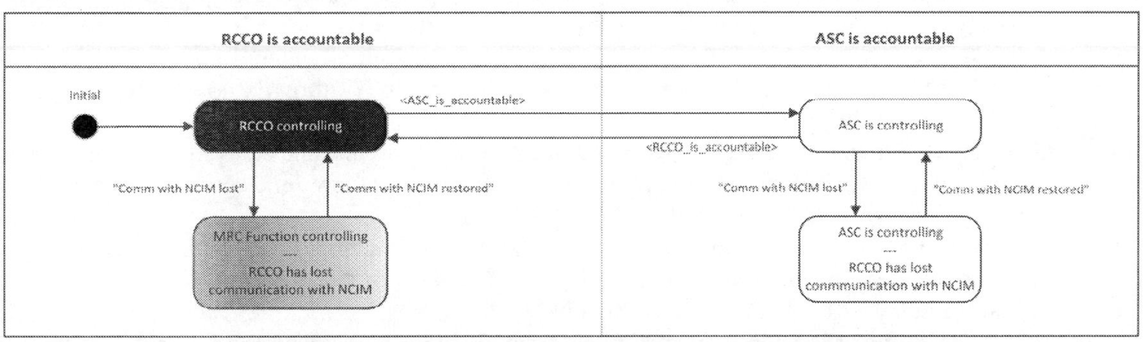

Figure 13. State diagram for the RCCO's communication status

To ensure correct operation, the following decisions must be made by the RCCO:

- "Communication with NCIM lost"
- "Communication with NCIM lost"

6. Considerations on the presented methodology

As mentioned in section 1, much research can be found on specific functions of autonomous systems, and on technologies that are assumed to be relevant for the success of autonomy. This paper has a different approach, and by defining *accountability* as a starting point for technological development, we have aimed to illustrate how it can be possible to integrate various *accountable entities* (or systems) into what could be regarded as an *accountable system-of-systems*. This also means that the presented methodology should not be considered a substitute for other design methodologies, but rather seen as an attempt to combine various aspects such as remote-control, autonomy and contingency operations into one common framework, with accountability as a common denominator.

7. Conclusions

By employing the methodology introduced in this paper, we have attempted to integrate *accountability* into the high-level technical design of a hypothetical autonomous and remote-controlled vessel. One must however expect new challenges to arise as soon as the methodology is applied to real cases, and this paper should therefore be seen more as an investigation into new ways of approaching autonomy than an actual and definite solution.

From a research perspective, the topics presented in this paper can be followed up in several directions. Naturally, applying the methodology to a real case would be a logical first step. Also, our choice of system architecture deserves further evaluation, both from an implementation perspective and from the viewpoint of systems design and analysis. Finally, to ensure compatibility with existing practices, one should also investigate how the proposed methodology could be integrated with existing legal and technical regulatory frameworks.

Acknowledgments

The work presented in this text has received funding from the European Union's Horizon 2020 research and innovation programme's GSA H2H project under grant agreement No. 775998, the AUTOSHIP project under grant agreement No. 815012 and the AEGIS project under grant agreement No. 859992 as well as from the Research Council of Norway's Sacomas project under grant agreement No. 269510.

References

[1] Wired Magazine (2019) Ford taps the brakes on the arrival of self-driving cars https://www.wired.com/story/ford-taps-brakes-arrival-self-driving-cars (Retrieved 2020-04-26)

[2] Tunfors H (2020) IMO regulatory scoping exercise on MASS *5th UK Maritime Autonomous Systems Regulatory Working Group Conference* https://www.maritimeuk.org/documents/544/Henrik_Tunfors_suw9wJK.pdf (Retrieved 2020-04-26)

[3] Myhre B, Hellandsvik A and Petersen S 2019 A responsibility-centered approach to defining levels of automation *J. Phys.: Conf. Ser.* **1357** 012027 http://dx.doi.org/10.1088/1742-6596/1357/1/012027

[4] Alawadhi M, Almazrouie J, Kamil M and Khalil K A 2020 Review and analysis of the importance of autonomous vehicles liability: a systematic literature review *Int J Syst Assur Eng Manag* https://doi.org/10.1007/s13198-020-00978-9

[5] Rødseth Ø J (2018) Defining ship autonomy by characteristic factors *Proc. of the 1st International Conference on Maritime Autonomous Surface Ships (ICMASS 2018)*

[6] Molnar, L. J., Ryan, L. H., Pradhan, A. K., Eby, D. W., Louis, R. M. S., & Zakrajsek, J. S. (2018). Understanding trust and acceptance of automated vehicles: An exploratory simulator study of transfer of control between automated and manual driving. *Transportation research part F: traffic psychology and behaviour*, *58*, 319-328.

[7] Kuro, J. (2018). *Maritime Product Liability The Case of Unmanned Vessels,* Master's thesis from University in Oslo.

[8] Smith M L, Erwin J and Diaferio S (2005) Role & responsibility charting (RACI) *Project Management Forum (PMForum)* **Vol. 5**

[9] DNV GL (2018) Class guideline DNVGL-CG-0264: Autonomous and remotely operated ships

[10] Dijkstra E W (1982) Selected Writings on Computing: A Personal Perspective (New York: Springer-Verlag) pp 60-66

[11] CENELEC (2010) EN 50159:2010 Railway applications – Communication, signaling and processing systems. Safety-related communication in transmission systems.

IOP Conf. Series: Materials Science and Engineering **929** (2020) 012021 doi:10.1088/1757-899X/929/1/012021

Process map for collision avoidance based on information exchange for autonomous navigation of vessel

Ho Namgung[1], Jung Sik Jeong[*2], Kwangil Lee[3], and Sun-Young Kim[4]

[1]Graduate school of Maritime Transportation System, Mokpo National Maritime University, Mokpo, Republic of Korea
[*2]Department of Maritime Transportation Science, Mokpo National Maritime University, Mokpo, Republic of Korea
[3] Department of Control and Automation Engineering, Korea Maritime and Ocean University, Busan, Republic of Korea
[4] Maritime Safety and Environmental Research Division, Korea Research Institute of Ships & Ocean Engineering, Daejeon, Republic of Korea

Email: jsjeong@mmu.ac.kr

Abstract. A reliable collision avoidance method is essential when operating an autonomous vessel. For this to become a reality, a number of researches have been conducted in order that the autonomous vessel could take an action for collision avoidance by herself. However, it may lead to dangerous situation without knowing the navigation intention of target-ships. In order to solve the issue, we developed the process map for collision avoidance based on information exchange via the communication relay system as follows: (i) the navigation situation analysis of the autonomous vessel was conducted based on manned ship voyage definition; (ii) the sequence diagram for information exchange was developed based on the autonomous vessel's navigation context and the communication relay system, and (iii) the process map for collision avoidance based on information exchange was proposed based on the developed sequence diagram and the existing collision avoidance method. It would be expected that the autonomous vessel could avoid collision more safely than taking an action by herself.

1. Introduction

Preventing collision with approaching target-ship is one of the most critical ability to ensure safety and effectiveness during an autonomous vessel's operation. For this to become a reality, trustworthy and effective algorithms for collision avoidance by herself have been developed [1-6]. However, according to "A Guide to the Collision Avoidance Rules" [7], it stated that information exchange between vessels can be useful in circumstances of collision avoidance because of analysing navigation intentions prior to taking an action.

Recently, the communication relay system based on available communication interfaces at sea has been developed to enable information exchange between the autonomous vessel and approaching target-ship [8]. Therefore, in the paper, we propose the process map for collision avoidance based on information exchange between the autonomous vessel and target-ship. This paper has essentially three parts: In the first part, the autonomous vessel's navigation situation context is defined. In the second part, sequence diagram for information exchange is made via the defined navigation situation context and the communication relay system. In the third part, the process map for collision avoidance based on information exchange is proposed via the sequence diagram and collision avoidance method.

The remainder of the paper is organized as follows: Section 2 describes the communications relay system and collision avoidance method. Section 3 presents the process map. Finally, a summary of the work and the conclusion are drawn in Section 4.

2. Related Work

2.1. Communication Relay System

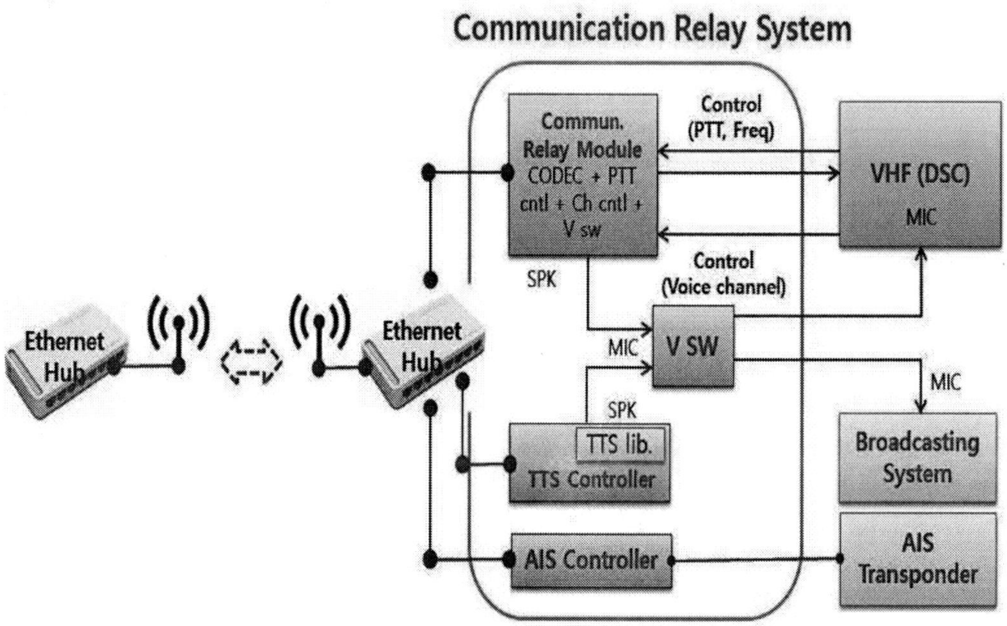

Figure 1. Configuration of the communication relay system [8].

The communication relay system [8] consisted of three type of controller: Very High Frequency (VHF) communication controller can transmit the voice message received from the Vessel Traffic System (VTS) or manned ship to the Maritime Control Station (MCS) via the Voice over Internet Protocol (VoIP), and the MCS respond the VTS or manned ship by sending the voice message generated from the MCS over the VoIP. The TTS controller can receive the voice message composition from the MCS, requesting the TTS server by sending the message converted into the Speech Synthesis Markup Language (SSML) format. Then the TTS server synthesizes the voice as the WAV file, and transmits it to the WAV player to play it. The output is passed through the voice switch to the VHF transmitter or the broadcast amplifier. The AIS controller is responsible for transmitting AIS-ASM messages received from the autonomous vessel to the MCS via VHF Data Exchange System (VDES).

2.2. Collision Avoidance Method

All vessels must take an action for collision avoidance in compliance with the International Regulations for Preventing Collision at the Sea (COLREGs) [9] as presented in Figure 2. First, every vessel shall at all times maintain a proper look-out by using all available means in order to assess of the risk of collision. Second, encounter relation (i.e., head-on situation, crossing situation, and overtaking) is determined if the risk of collision exists. Third, a give-way vessel or a stand-on vessel is determined according to encounter relation. Fourth, any action to avoid collision shall be taken.

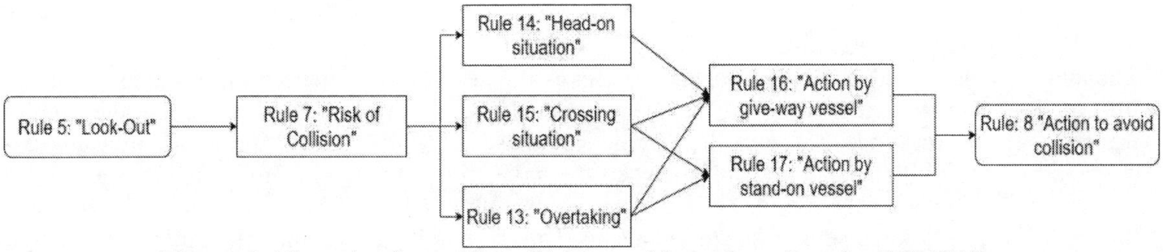

Figure 2. The rules for a ship collision avoidance based on the COLREGs.

In other words, collision avoidance starts assessment of the risk of collision. In particular, it stated that the radar plotting is suitable equipment in order to obtain early warning or the risk of collision from observation of detected objects according to "Rule 7: Risk of Collision" [9].

Therefore, a number of researches have been conducted by using the radar plotting [1-6]. At this time, the risk of collision was assessed by Collision Risk Index (*CRI*), which is calculated by Distance to Closest Point of Approach (D_{CPA}) and Time to Closest Point of Approach (T_{CPA}). In particular, access to the *CRI* calculated by fuzzy logic was not only assess *CRI* at the same time when approaching multiple-ship, but also a point of time for a give-way vessel and a stand-on vessel were defined according to the encounter relation. Figure 3 shows the inference process of Fuzzy Inference System (FIS) [4,5].

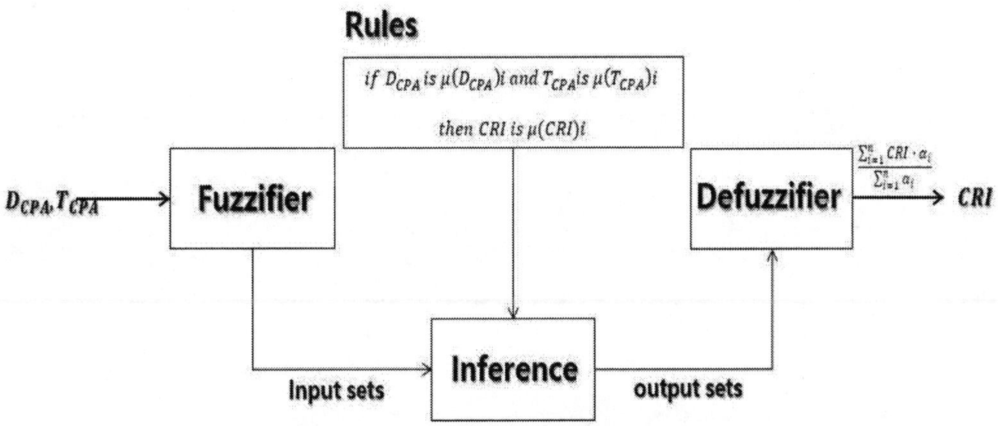

Figure 3. The inference process of FIS.

Range of the *CRI* is from 0.00 to 1.00 [4,5], and a point of time for collision avoidance is as follows: (i) a give-way vessel must take an action for collision avoidance when the *CRI* exceed more than 0.33, and (ii) a stand-on vessel must take an action for collision avoidance when the *CRI* exceed more than 0.66.
where a give-way vessel and a stand-on vessel must take an action according to type of encounter relation [9].

- In case of the head-on situation, both vessels should alter course to the starboard for collision avoidance.
- In case of the crossing situation, the vessel having the other vessel on its starboard is a give-way and should alter its course so it passes behind the other vessel. The other vessel should keep a steady course and speed as a stand-on vessel. But if a give-way vessel does not take appropriate action, a stand-on shall take an action as will best aid to avoid collision.
- In case of the overtaking, the vessel being overtaken should keep steady speed and course. The overtaking vessel could pass the other vessel on both sides.

3. Process Map for Collision Avoidance based on Information Exchange

3.1. Navigation Situation Analysis

The autonomous vessel's navigation phase was set as the departure-navigation-arrival phase. Then, the information exchange message [10] was constructed by identifying and extracting the required information that may occur during the navigation phase. Figure 4 shows the autonomous vessel's navigation situation context. First, we analysed the navigation situations of manned ships and set the navigation situations for the purpose of carrying out the autonomous vessel's navigation. In the case of manned ships, a pilot, who carry the ship to a safe channel, will board the ship and guide the ship to the pilot station. However, in the case of the autonomous vessel, maneuvering by the pilot was unnecessary because a person does not board. Therefore, the autonomous vessel added an access phase to enter the navigation route. Based on this, the autonomous vessel's navigation route was established to construct various situations that can occur in each navigation route, and a standard information exchange message for information exchange was applied.

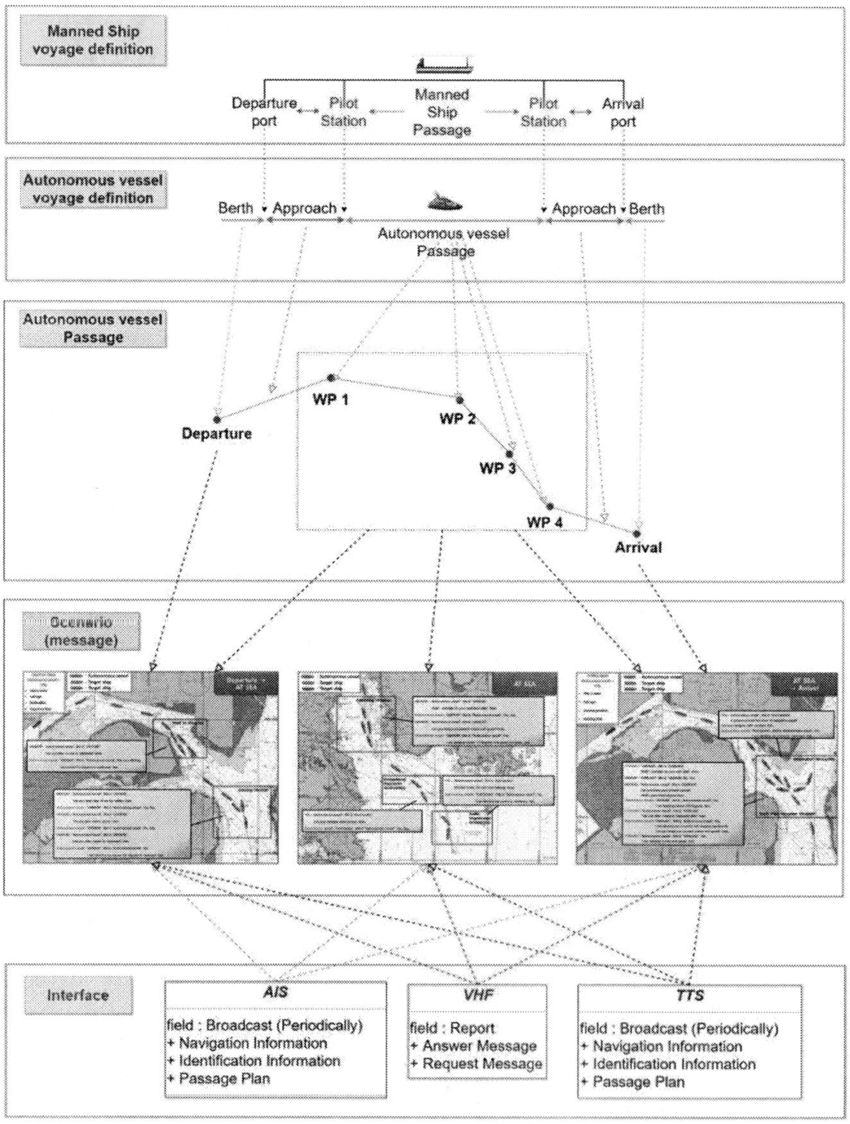

Figure 4. Autonomous vessel's navigation situation context.

3.2. Sequence diagram

Based on the autonomous vessel's navigation context, sequence diagram for information exchange between the autonomous vessel and VTS and/or target ship was configured via the communication relay system and standard information exchange messages as presented in Figure 5. In case of departure and arrival, the MCS sends created report messages to the autonomous vessel by using VHF telephone, AIS, TTS via dedicated communication channel [8] between the MCS and the autonomous vessel. Then, the autonomous vessel transfers created message to VTS and/or target ship by using VHF telephone, AIS, TTS via existing communication channel [8] between the autonomous vessel and VTS. In terms of navigation, target ship sends created navigational message to the autonomous vessel by using VHF telephone and AIS via exiting communication channel. Then the autonomous vessel transfers received message to the MCS by using VHF telephone and AIS via dedicated communication channel. After analysing navigation intention from target ship, the MCS respond to target ship via the autonomous vessel. Like the preceding, the MCS is able to send navigational messages to target ship via the autonomous vessel in case of navigation.

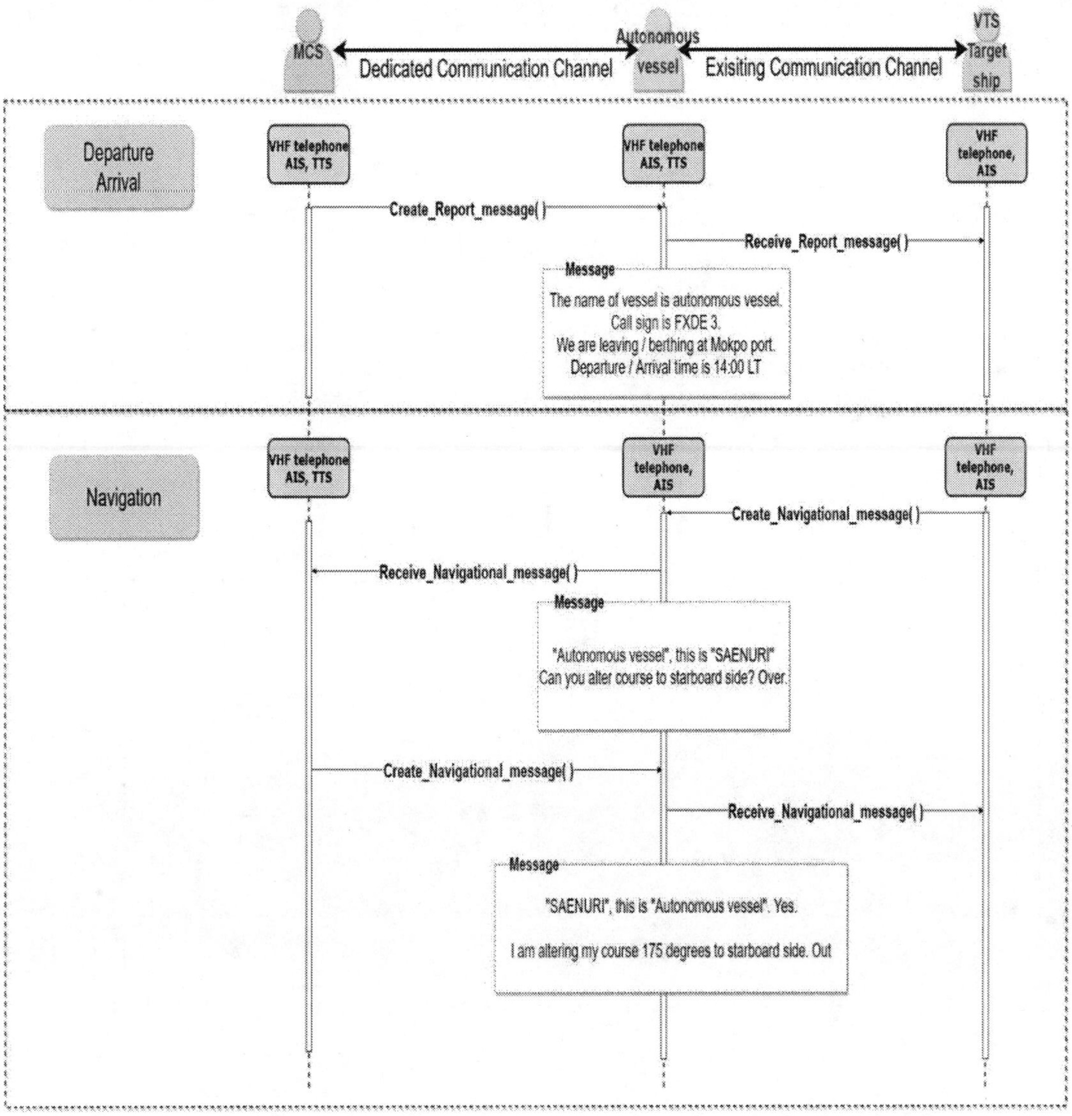

Figure 5. Sequence diagram for information exchange based on navigation situation context.

3.3. *Process Map*

Process map was established as shown in Figure 6.First, the autonomous vessel infer the *CRI* via the FIS. Second, in case of the *CRI* more than 0.33, encounter relation (i.e., head-on situation, crossing situation, and overtaking) is analysed according to the COLREGs. Third, in case the autonomous ship is a give-way vessel, the MCS creates navigation intention message. But if the autonomous ship is a stand-on vessel, the MCS creates navigation intention message according to whether the *CRI* more than 0.66 or not. Fourth, the MCS inputs the created message as a voice data or a text data, and then transfers the data to the communication relay system. Fifth, the communication relay system sends the received data to target ship via VHF radio and AIS controller.

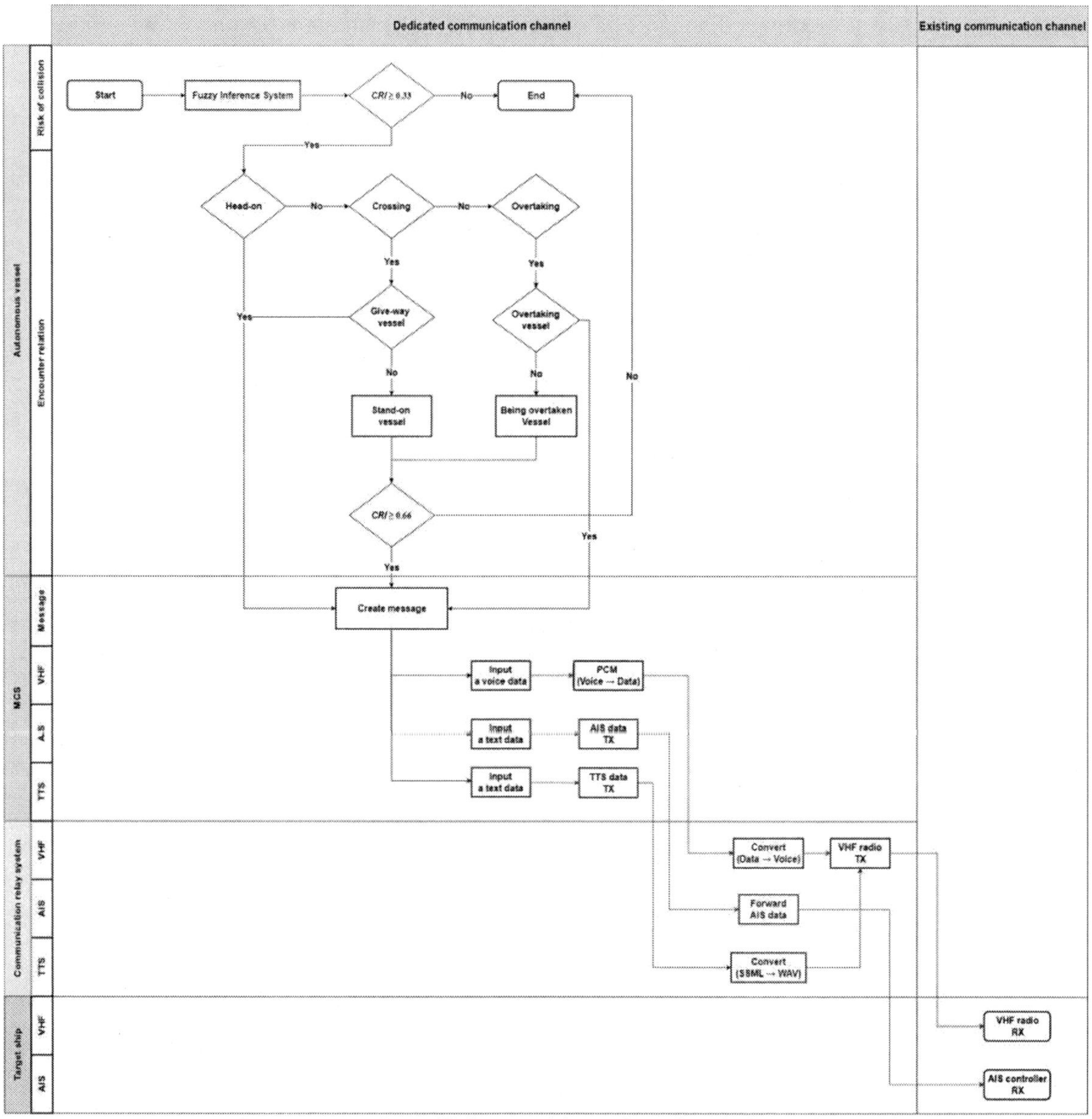

Figure 6. Process map for collision avoidance based on information exchange.

4. Conclusion

To avoid collision between the autonomous vessel and target ship via information exchange, the process map was proposed. In order to develop it, the navigation situation analysis for the autonomous vessel and the sequence diagram for information exchange via the communication relay system were made. Then, by synthesizing the preceding, the process map for collision avoidance based on information exchange was established systematically. It would be expected that the autonomous vessel is able to take and action for a reliable collision avoidance because of analysing navigation intentions of target ship in advance. In the near future, we are looking forward to extending our work on simulation and experiment at sea.

Acknowledgments

This research was supported by Basic Science Research Program through the National Research Foundation of Korea(NRF) funded by the Ministry of Education(2020R1I1A1A01060533).

References

[1] Kearon J., Computer Program for Collision Avoidance and Track Keeping. Proceedings of the Conference on Mathematics Aspects of Marine Traffic, London, UK, 1977
[2] Lee, Han-Jin.; Rhee, Key-Pyo. Development of Collision Avoidance System by using Expert System and Search Algorithm. *ISP.* **2001**, 48, 197–212
[3] Jin-Hyeong Ahn.; Key-Pyo Rhee.; Young-Jun You. A Study on the Collision Avoidance of a Ship using Neaural Networks and Fuzzy Logic. *Appl. Ocean Res.* 2012, 37, 162-173
[4] Ho Namgung.; Jung Sik Jeong.; Joo-sung Kim. Inference System of Collirion Risk using Ship Near-Collision Data. *J. Korean Inst. Intell. Syst.*, **2019**, 29, 395-402
[5] Ho Namgung.; Jung Sik Jeong.; Joo-sung Kim.; Kwang-Il Kim. Inference Model of Collision Risk Index based onArtificial Neural Network using Ship Near-Collision Data. *J. Phys. Conf. Ser.* **2019**, 1357, 012044
[6] Sung Wook Ohn.: Ho Namgung. Interval Type-2 Fuzzy Inference System based on Closest Point of Approach for Collision Avoidance between Ships. *Appl. Sci.***2020**, 10,3919
[7] Cockcroft, A. N.; Lameijer, J. N. F. A Guide to the Collision Avoidance Rules. Butterworth Heinemann, Oxford, UK, 2011
[8] Ho Namgung.; Jung Sik Jeong.; Jo-chen Choi.; Geun-ung Kim.; Sun-Young Kim. An Experimental Result on Information Exchange using USV communication Relay System. *J. Phys. Conf. Ser.* **2019**, 1357, 012043
[9] IMO. *Convention on the International Regulations for Preventing Collisions at Sea, [with amendments 336 adopted from December 2009]*; International Maritime Organization, London, UK, 1972
[10] Ho Namgung.; Jung Ski Jeoung. Design of Message Standard Codes for Safe Navigation of Unmanned Surface Vehicle. Proceedings of the Conference on the 18th International Symposium on Advanced Intelligent Systems (ISIS2017), Daegu, South Korea, 2002

A hybrid kinematic controller for resilient obstacle avoidance of autonomous ships

Mathias Marley[1,*], **Roger Skjetne**[1,†], **Morten Breivik**[2,‡] **and Caroline Fleischer**[3,§]

[1] Norwegian University of Science and Technology (NTNU), Department of Marine Technology, NO-7052 Trondheim, Norway
[2] Norwegian University of Science and Technology (NTNU), Department of Engineering Cybernetics, NO-7491 Trondheim, Norway
[3] Kongsberg Seatex AS, NO-7462 Trondheim, Norway

E-mail: *mathias.marley@ntnu.no, †roger.skjetne@ntnu.no, ‡morten.breivik@ntnu.no, §caroline.fleischer@km.kongsberg.com.

Abstract. Resilience is an important feature of autonomous systems. To be resilient, a control system must be stable, robust, and safe. This paper explores the use of hybrid feedback controllers to ensure robustness towards uncertainties and disturbances in motion control systems for autonomous ships. Motivated by recent developments in control barrier functions (CBFs) for safe maneuvering of autonomous ships, a CBF-based hybrid kinematic controller for obstacle avoidance is proposed. The controller uses course angle as control input, making it suitable for ships with a limited speed envelope. The performance of the controller is illustrated by simulations, using an underactuated ship as a case study.

1. Introduction

Autonomous ships are inherently complex systems [1], consisting of multiple subsystems ranging from high-level mission management to low-level plant control. The control system architecture for autonomous ships may roughly be divided into three layers; mission management layer, executive guidance layer, and reactive feedback control layer, as illustrated in Figure 1. Resilience is an important objective when designing the autonomous system as a whole and the various subsystems [2], [3]. For path following of autonomous ships, the motion controller can be considered resilient if it is stable (the ship converges to and stays on the desired path), robust (with respect to uncertainties and disturbances) and safe (the ship deviates from the nominal maneuvering plan if safety demands it). In this paper we focus on safety and robustness in the reactive feedback control layer, using obstacle avoidance of an underactuated ship as a case study.

For many control problems, robustness towards noise and disturbances can only be achieved by employing some switching logic [4]. When designing control systems for autonomous ships, such problems are frequently encountered. Consider for instance the task of steering a vessel along a path while avoiding an obstacle. If the obstacle is at rest and placed exactly on the path, then passing the obstacle port or starboard is equally valid. While such a scenario cannot occur in real-life (the two options will never be exactly equal), small measurement noise in the vessel or obstacle position measurement can make the commanded action switch back and

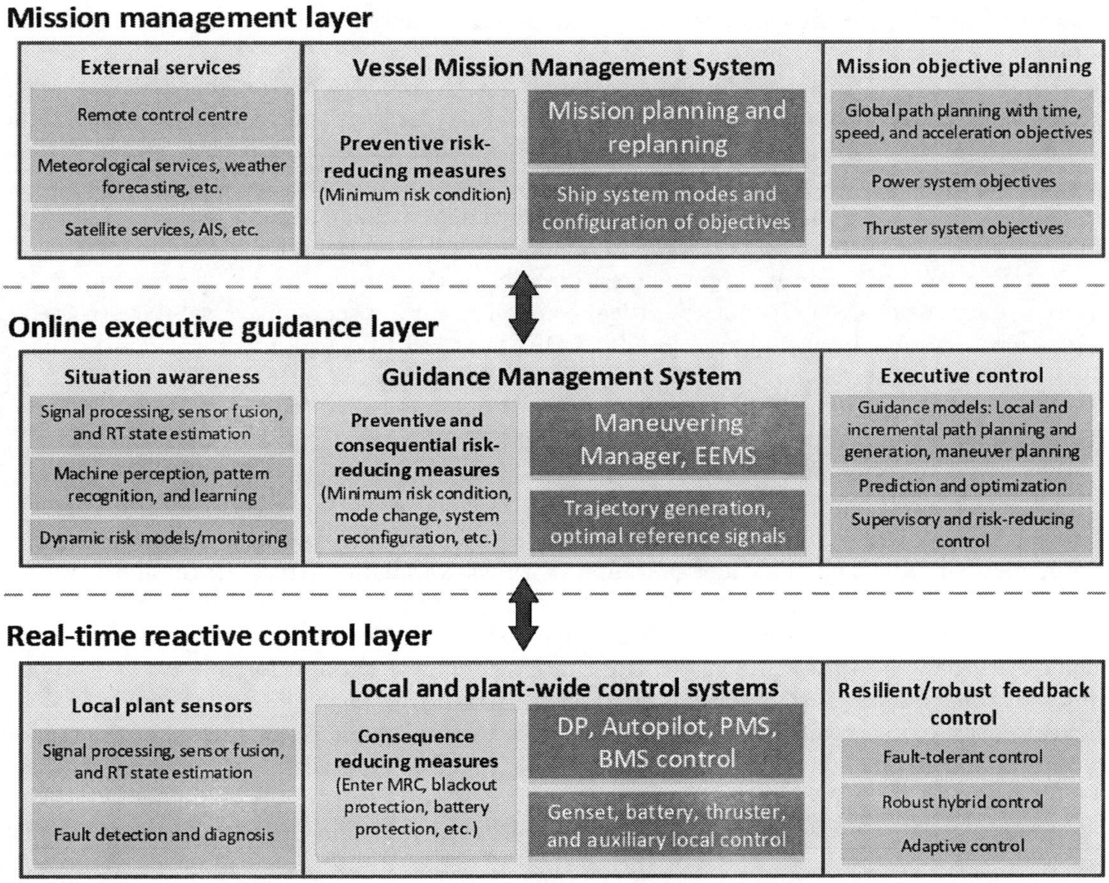

Figure 1: Example of control system architecture for autonomous ships. Abbreviations: AIS; Automatic Identification System. EEMS; Energy and Emission Management System. DP; Dynamic Positioning. PMS; Power Management System. BMS; Battery Management System. RT; Real-Time. MRC; Minimum Risk Condition. Adapted from [1].

forth indecisively between a port and starboard turn, with the result that the vessel is unable to alter the course and instead steers towards collision. In such a scenario, an experienced human captain would decide on an obstacle-free path, stick to that path and clearly show his intention, unless there is a significant change to the circumstances. Similar behavior can be built into autonomous control systems by use of intelligent switching logic. The framework of hybrid dynamical systems [5], enables systematic design of such switching logic, while providing the mathematical tools needed to formally prove stability and robustness – ensuring the overall objective of our research project; resilience in the autonomous control functions.

Significant research on obstacle avoidance has been performed as part of the Autosea project; see [6] and the references therein, where the authors propose to distinguish between global and local collision avoidance methods. Global methods construct a nominal path which is free of obstacles known prior to the mission. Local methods detect and avoid obstacles encountered during the mission, deviating from the nominal path if necessary. It is further proposed that local methods are separated into proactive and reactive methods. In reactive methods, the "evasive control inputs depend directly on the state vector through a functional relationship" [6]. While no clear definition of proactive methods is given, the context suggests that proactive methods cannot have a continuous feedback from the state vector to the evasive control inputs. This implies that proactive methods involve having a predictive component that is used to make a plan for an evasive maneuver.

An advantage of proactive methods is the ability to come up with a consistent and smooth path for an evasive maneuver. However, problems may arise if the vessel is not able to follow the evasive path, or if the circumstances change such that the evasive path is no longer feasible and safe. The former may happen if the environmental disturbances are larger than anticipated, while the latter can occur if the encountered vessel changes its course in an unpredictable manner. In such situations, it may be necessary to abandon the initial plan for evasive maneuver.

Reactive methods may be implemented directly in the motion control system using control barrier functions (CBFs) [7]. Safety-critical controllers for autonomous vessels constructed from CBFs have recently been proposed [8], [9]. CBFs are used to design a control system that only allows control inputs that are considered safe in some sense. A CBF can be constructed such that the resulting safety-critical controller only intervenes when the evasive path planned by some higher-level proactive method is considered unsafe.

The main purpose of this paper is to explore the use of CBFs for underactuated ships with limited speed envelope. While CBF-based continuous feedback controllers can guarantee collision free operation of fully actuated vessels, the possibility of deadlock situations in perfectly symmetric conditions preclude any guarantees for task completion [10], [11]. Similarly, for vessels required to maintain a positive forward speed, safety cannot be guaranteed by continuous control. Safety can be achieved by discontinuous control. However, due to the discontinuity, arbitrarily small disturbances in the state vector can cause switching back and forth between significantly different control inputs, referred to as chattering. To ensure safety *and* robustness, a CBF-based hybrid kinematic controller is proposed. The vessel course angle is considered a virtual control input, similar to the constant avoidance angle algorithm proposed in [12]. Unlike the constant avoidance angle algorithm, a CBF-based formulation enables the use of higher derivatives of the vessel position as control input in the safety-critical controller. This is the topic of current research by the authors.

The remainder of this paper is organized as follows: In Section 2, background on hybrid control systems is provided. In section 3 a kinematic controller for obstacle avoidance is presented. Safety is ensured by using a CBF, while robustness is achieved by a hybrid feedback controller. Merging CBFs with hybrid controllers in this way is the main contribution and novelty of this paper. In Section 4, the proposed kinematic controller is used for safe maneuvering of an underactuated rudder-controlled ship. Finally, Section 5 concludes the paper.

2. Hybrid control systems

It is necessary to briefly review the concept of hybrid dynamical systems, as defined in [5]. Hybrid dynamical systems is a framework for analyzing system states that evolve both in continuous time (referred to as flow) and in discrete time (referred to as jumps). In the context of marine control systems, the physical states such as vessel position and velocity evolve by flow only, while logic variables in the control system evolve by jumps only. In [13], the hybrid systems framework is used to improve the performance of a vessel in dynamic positioning (DP) operation. The DP controller can select between several candidate controllers, depending on the state vector and an estimate of the sea state. This results in supervisory switching logic implemented as a hybrid control system.

2.1. Hybrid feedback controllers

Hybrid feedback control laws can be used for robust global stabilization of systems that cannot be globally stabilized by continuous control. Control of orientation is one such example [14]. When global stabilization by continuous control is not possible, any globally stabilizing feedback controller must have at least one point of discontinuity. Suppose the control objective for a given system can be achieved by two different control laws, where the points of discontinuity are disjoint in the state space. Such a system may be represented by the differential equation

$$\dot{x} = f(x) + g(x)u(x, q), \tag{1}$$

where $x \in \mathbb{R}^n$ is the physical state vector and $u(x, q)$ is the hybrid feedback control law. Here, $q \in Q := \{-1, 1\}$ is a logic variable, which may take on the values $q = 1$ or $q = -1$. The switching between candidate controllers may be determined by

$$q^+ = \begin{cases} q, & P(x,q) < P_{lim}, \\ -q, & P(x,q) \geq P_{lim}, \end{cases} \tag{2}$$

where q^+ assigns the update of q after an instantaneous change (a jump). $P : \mathbb{R}^n \times Q \to \mathbb{R}$ is some indicator function which is continuous in the state vector x. q is toggled when $P(x, q)$ exceeds the threshold P_{lim}. If the indicator function consists of overlapping control Lyapunov functions (CLFs) which satisfy certain properties, rigorous robustness and stability proofs are readily available; see [15] and [16].

The challenge is often to design the indicator function such that q is toggled whenever necessary, while avoiding chattering. Since $P(x, q)$ depends on the state vector x, and not explicitly on time, the result when designed properly is a hysteresis switching mechanism.

2.2. Robust heading control

We will first illustrate the key ideas of this paper by considering heading control of a surface vessel. The yaw angular error is usually mapped to some predefined interval, typically $-\pi$ to π radians (see e.g. [17] and [18]). At the 180 degree error point, the commanded action switches between a clockwise or counterclockwise turn. At this point, arbitrarily small noise can cause chattering, which will delay convergence to the desired yaw angle[1]. To avoid a discontinuous mapping of the yaw angle to some predefined interval, the vessel heading can be represented on the unit circle, as illustrated in Figure 2. This is in many ways a more natural representation of orientation. The unit orientation vector corresponding to the yaw angle ψ, is given by

$$z^{\psi} := \begin{bmatrix} \cos(\psi) \\ \sin(\psi) \end{bmatrix}. \tag{3}$$

For control purposes, and neglecting the superscript, we use the orientation error vector given by

$$\tilde{z} := \begin{bmatrix} \cos(\tilde{\psi}) \\ \sin(\tilde{\psi}) \end{bmatrix}. \tag{4}$$

where $\tilde{\psi} := \psi - \psi_d$ is the yaw angular error. Controlling $\tilde{\psi} \to 0$ is equivalent to controlling $\tilde{z} \to [1\ 0]^\top$. For more details see [19].

[1] In fact, the 180 degree error point will have the characteristics of an extra (unstable) equilibrium point, implying that global stabilization of the 0 degree error point is impossible.

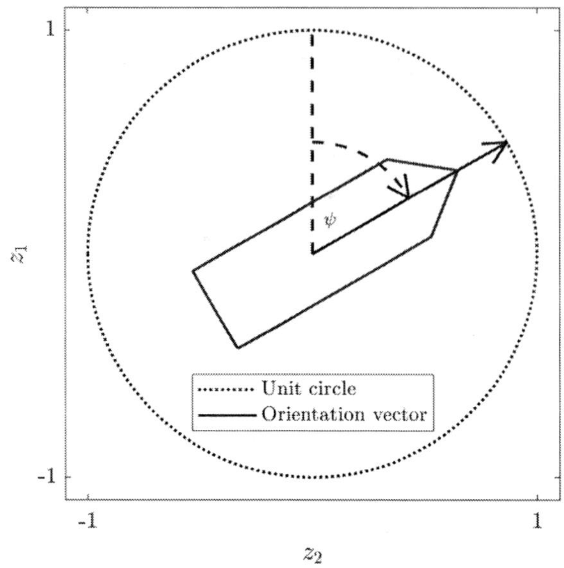

Figure 2: Vessel orientation represented on the unit circle, using the North-East-Down reference frame [20].

To introduce hysteresis, we can follow the procedure in [21] where the switching logic is constructed from two overlapping Lyapunov functions on the unit circle. The 180 degree error point is shifted in the clockwise or counterclockwise direction depending on the value of q. When $\tilde{z}$ enters some neighborhood of the 180 degree error point in the shifted coordinate system, then q is toggled. The resulting hysteresis mechanism is illustrated in Figure 3.

A similar logic may be obtained by mapping the yaw angular error to the interval $(-\pi+\alpha, \pi+\alpha]$ if $q = 1$ and $(-\pi-\alpha, \pi-\alpha]$ if $q = -1$, where $\alpha \in \mathbb{R}$ is a constant selected depending on the vessel dynamic properties and measurement quality. This mapping procedure appear simpler than using overlapping Lyapunov functions. However, since the vessel yaw rate cannot instantaneously change, a dynamic switching condition is desired, where the yaw rate and possibly higher derivatives are taken into account. Immediately the task of ensuring sufficient hysteresis width becomes more challenging. By augmenting the Lyapunov functions with additional terms based on the yaw rate and higher derivatives, a dynamic switching condition may be constructed.

When a switch occurs, there will be a discontinuity at some level in the control system. The control input should be sufficiently smooth for good performance, e.g. for a rudder controlled ship one may require that the commanded rudder angle is continuous. Sufficiently smooth actuator usage while maintaining strong formal guarantess for robustness and stability can be achieved by backstepping [15], [16]. In [22], a hybrid feedback controller for robust trajectory tracking of underactuated vehicles is proposed.

3. A kinematic controller for obstacle avoidance

We return to the example given in the introduction; steering a vessel along a path while avoiding a static obstacle. The control task is formulated as a kinematic problem. The vessel is assumed to have a constant speed, while safety is ensured by restricting the commanded orientation to a safe set, making it suitable for underactuated vessels. A CBF is used to determine the set of safe control inputs. When the nominal control input is unsafe, the CBF is said to be active and an evasive maneuver is performed. The evasive maneuver is determined by a hybrid feedback controller, giving the autonomous steering function the desired feature of resilience.

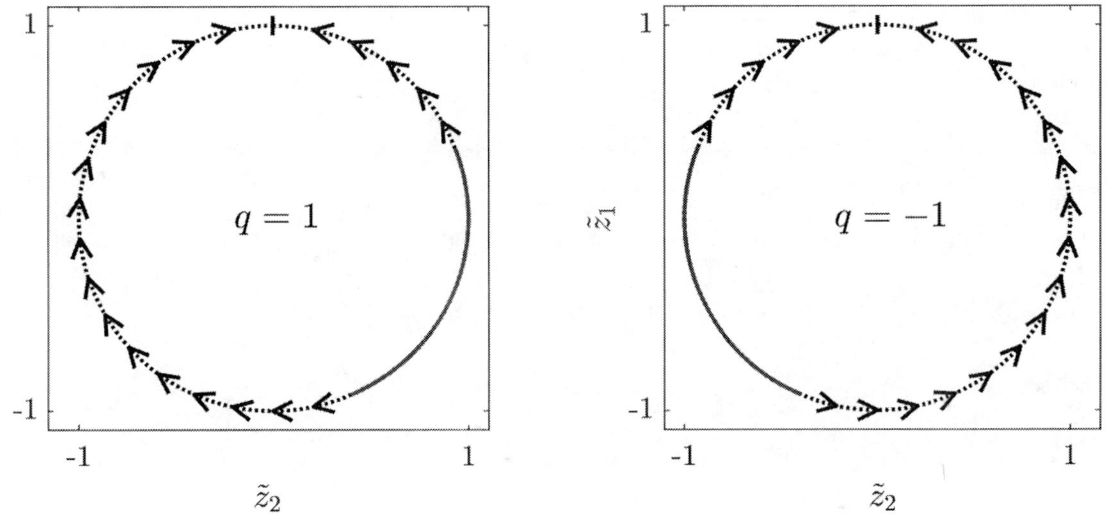

Figure 3: Illustration of the vector field for a static hysteresis mechanism for heading control which depends only on the orientation error vector $\tilde{z}$. The arrows indicate the commanded turning direction. If $\tilde{z}$ enters the sector marked by the solid red line, q is toggled and the state is governed by the other vector field. The illustration shows a hysteresis width about the 180 degree error point of 45 degrees.

3.1. Kinematic equation and problem formulation

Let $p := [x\ y]^\top$ be the vessel position in the horizontal plane, ψ the yaw angle, and $\nu := [u\ v]^\top$ the body-fixed velocity vector. The kinematic equation for motion in the horizontal plane may be written as $\dot{p} = R(\psi)\nu$, where

$$R(\psi) := \begin{bmatrix} \cos(\psi) & -\sin(\psi) \\ \sin(\psi) & \cos(\psi) \end{bmatrix} \tag{5}$$

is the rotation matrix. An alternative representation of the vessel kinematics, used in e.g. [23] and [18], is

$$\dot{p} = U z^\chi, \tag{6}$$

where $U = \sqrt{u^2 + v^2}$ is the total vessel speed and

$$z^\chi := \begin{bmatrix} \cos(\chi) \\ \sin(\chi) \end{bmatrix} \tag{7}$$

is the course orientation vector. Furthermore, $\chi := \arctan(\frac{\dot{y}}{\dot{x}}) = \psi + \beta$ is the course angle, where $\beta := \arctan(\frac{v}{u})$ is the crab angle (sometimes referred to as the sideslip or drift angle [20]).

We will consider a circular obstacle located at $p_o = [x_o\ y_o]^\top$ with radius r_o. The control task is to design a feedback controller for the desired course angle z_d^χ such that the vessel does not collide with the obstacle, deviating from the nominal path if necessary. By controlling the course angle directly, rather than the yaw angle, sideslip is automatically compensated. This is desired when performing an evasive maneuver by turning, since the crab angle during the turn will be in the direction of the obstacle (assuming low current speed and other environmental disturbances) [23].

IOP Conf. Series: Materials Science and Engineering **929** (2020) 012022 doi:10.1088/1757-899X/929/1/012022

3.2. Control system design

A CBF-based control law for obstacle avoidance in the horizontal plane is proposed, arguably in its simplest form. See [7] for an introduction to CBFs. Accordingly, we define the barrier function

$$B(p) := ||p - p_o|| - r_o, \tag{8}$$

which is negative when the vessel is in the interior of the obstacle region and positive when the vessel is outside the obstacle. Safety is ensured by enforcing

$$\dot{B} \geq -\frac{1}{T_b} B, \tag{9}$$

where

$$\dot{B} = U \frac{(p - p_o)^\top}{||p - p_o||} z^X \tag{10}$$

is the time derivative of B, and $T_b > 0$ is a time-constant chosen dependent on the vessel dynamic properties. The constraint (9) allows B to decrease when the vessel is far away from the obstacle, while restricting $\dot{B}$ to be non-negative when the vessel is at the boundary of the obstacle. Combining (8), (9) and (10), the set of safe orientations is obtained as

$$z^X_{CBF}(p, U) := \left\{ z^X \in \mathcal{S}^1 : U \frac{(p - p_o)^\top}{||p - p_o||} z^X \geq -\frac{||p - p_o|| - r_o}{T_b} \right\}, \tag{11}$$

where $\mathcal{S}^1 := \{z \in \mathbb{R}^2 : z^\top z = 1\}$ is the unit circle [21]. When the distance between the vessel and the obstacle is less than $T_b U$ meters, the set of safe orientations is restricted according to

$$z^X = -R(\alpha) \frac{p - p_o}{||p - p_o||}, \quad \cos \alpha \leq \frac{||p - p_o|| - r_o}{T_b U}, \tag{12}$$

where α is the relative angle between z^X and the unit vector pointing from the obstacle towards the vessel. As the vessel approaches the obstacle, the allowable orientations become more restricted; see Figure 4 for an illustration. Define

$$\alpha_0 := \arccos \left(\frac{||p - p_o|| - r_o}{T_b U} \right), \tag{13}$$

where the range of $\arccos(\,\cdot\,)$ is defined as $[0, \pi]$, following the normal convention. The safety-critical controller is chosen as

$$z^X_d = \begin{cases} z^X_{pf}, & z^X_{pf} \in z^X_{CBF}(p, U) \\ -R(q\alpha_0) \frac{p - p_o}{||p - p_o||}, & z^X_{pf} \notin z^X_{CBF}(p, U) \end{cases}, \tag{14}$$

where z^X_{pf} is the feedback controller for path following, while $q \in Q := \{-1, 1\}$ is a logic variable determining if the vessel shall pass the obstacle in the clockwise or counterclockwise direction. To assign the value of q, we first define the indicator function

$$P(p, z^X) := \frac{(p - p_o)^\top}{||p - p_o||} R\left(\frac{\pi}{2}\right) z^X, \tag{15}$$

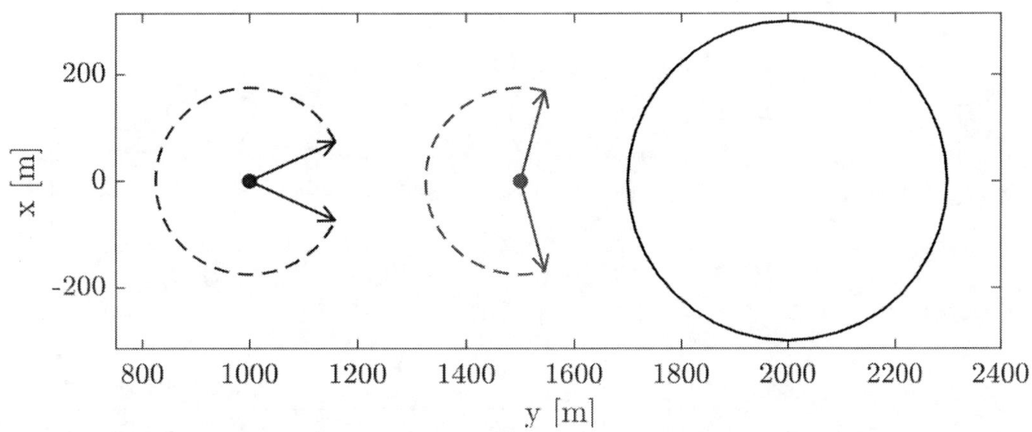

Figure 4: Illustration of the set of safe orientations constructed from the CBF. The obstacle is shown as a black circle, while the dashed circle sectors show the set z^{χ}_{CBF} for two distances (blue: 700m; red: 200m) from the boundary of the obstacle. The arrows correspond to $\pm\alpha_0$. Vessel speed $U = 7.7$m/s and time constant $T_b = 100$s.

which is negative if z^{χ} is pointing in the clockwise direction relative to the obstacle, and positive if z^{χ} is pointing in the counterclockwise direction relative to the obstacle. Choosing $q = \text{sign}(P)$ will ensure safety while minimally interfering with the path-following control objective. However, the resulting controller will be discontinuous at $P = 0$, and consequently not robust towards disturbances. Instead, we let q be given by

$$q^{+} = \begin{cases} q, & P(p, z^{\chi})q \geq -\sigma \\ -q, & P(p, z^{\chi})q < -\sigma \end{cases}, \tag{16}$$

where $0 < \sigma < 1$ is a desired hysteresis width. The combination of (14) and (16) results in a hybrid feedback controller. If σ is set to zero, this reduces to $q = \text{sign}(P)$, where sign(0) can take on both the values -1 and $+1$.

4. Case study

A case study is performed with zero and non-zero hysteresis width, to illustrate the importance of robustness. The numerical model of a Mariner class vessel provided in the Marine Systems Simulator ([24],[25]) is used in the case study. This is a 3 degree-of-freedom maneuvering model [20] of a rudder-actuated ship with length of 160m. The control input to the model is the commanded rudder angle. The model returns the vessel yaw rate and perturbation about the nominal vessel surge and sway velocity $\nu_0 = [U_0 \; 0]^{\top}$, where $U_0 = 7.7$m/s is the nominal speed. No environmental disturbances are included in the numerical model.

4.1. Simulation setup

The desired path is chosen as $x = 0$, traveling in the positive y-direction. For path following, we use the line-of-sight (LOS) algorithm [20]. Following the convention in this paper, this can be formulated as

$$z^{\chi}_{pf} = z_{LOS} := \frac{1}{\sqrt{\Delta^2_{LOS} + x^2}} \begin{bmatrix} -x \\ \Delta_{LOS} \end{bmatrix}, \tag{17}$$

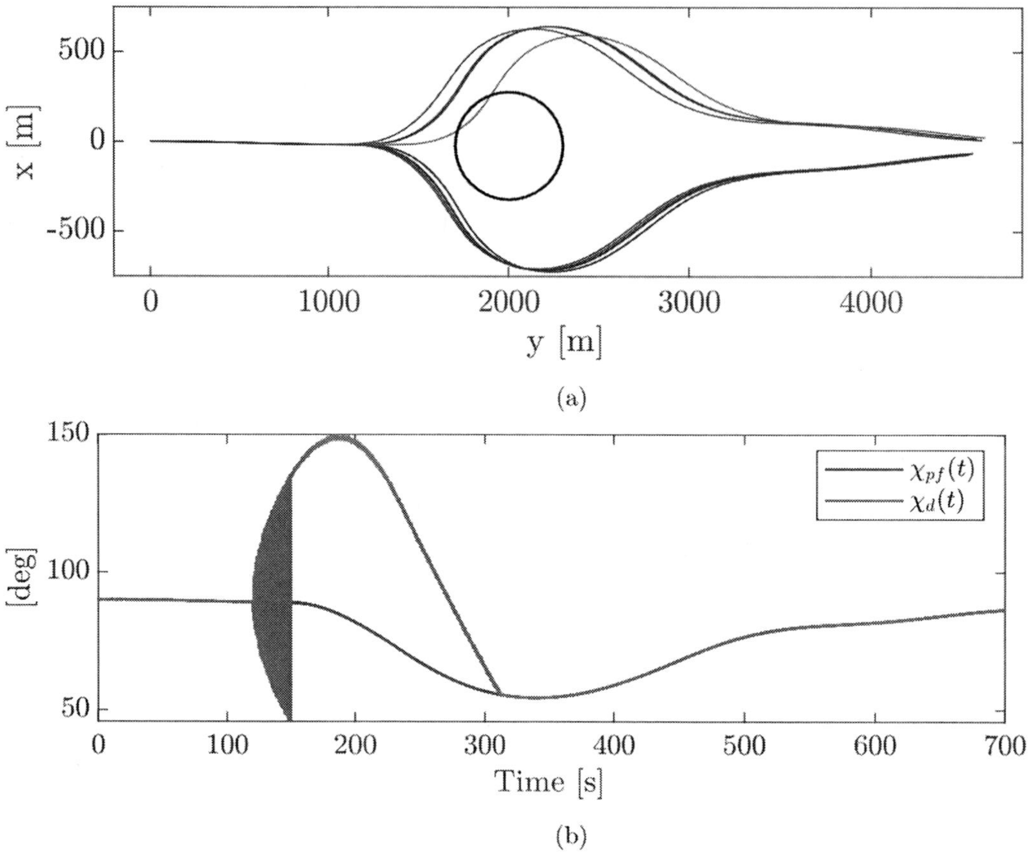

Figure 5: (a) Trajectories for 10 different runs with $\sigma = 0$. The obstacle is shown as the black circle. The ship is traveling from left to right along the positive y-axis. The trajectory which fails to avoid the obstacle is shown in red. (b) Commanded course angle for the trajectory shown in red. There is significant chattering from approximately 120 to 150s, before the ship settles on a direction to turn for the evasive maneuver. The course angle for path following is shown in blue.

with look-ahead distance set to $\Delta_{LOS} = 1000$m.

The desired course angle is given by (14), together with the switching logic (16). The evasive maneuver is initiated when $z^{\chi}_{pf} \notin z^{\chi}_{CBF}$, that is, when the desired course for path-following is considered unsafe.

An appropriate value of T_b may be determined from the maximum forward movement d_{max} of the ship during a hard turn. Choosing $T_b > d_{max}/U_0$ ensures that the set of safe orientations is restricted before the vessel is a distance d_{max} from the obstacle. d_{max} was found experimentally to be approximately 640m, giving a lower bound of $T_b = 640/7.7 \approx 83$s. In the simulations $T_b = 100$s is used, providing extra safety margin.

The control law for the rudder angle is chosen as

$$\delta_d(\tilde{\chi}, r) = -K_p(\tilde{\chi} + T_d r), \tag{18}$$

saturated by $|\delta_d| \leq 30$ degrees. Here, $\tilde{\chi} := \text{atan2}(\tilde{z}^{\chi}_2, \tilde{z}^{\chi}_1)$ is the course angular error mapped to the $[-\pi, \pi)$ interval. We note the obvious discontinuity at $\tilde{\chi} = \pm\pi$, discussed in Section 2.2.

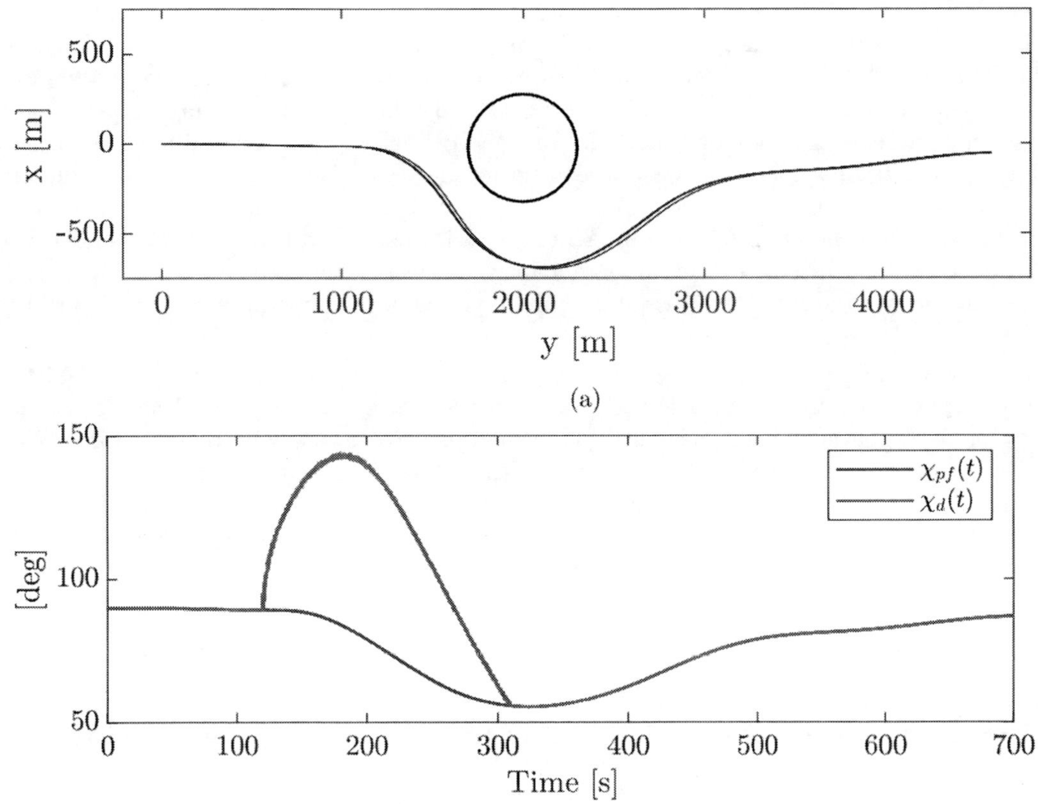

(b)

Figure 6: (a) Trajectories for 10 different runs with $\sigma = 0.05$. The obstacle is shown as the black circle. The ship is traveling from left to right along the positive y-axis. (b) Commanded course angle for one of the trajectories. The course angle for path following is shown in blue. The safety-critical controller is active from approximately 120 to 300s.

Presenting the details of a robust heading controller is outside the scope of this paper, and for the performed simulations this robustness issue did not pose any problems. The controller gains were set to $K_p = 1$ and $T_d = 10$, while $r = \dot{\psi}$ is the yaw rate[2].

Two different trials are conducted:

1) Hysteresis width $\sigma = 0$.
2) Hysteresis width $\sigma = 0.05$ and initial $q = 1$.

With $\sigma = 0$ the evasive control law has a discontinuity when z^{χ} is pointing directly towards the obstacle. For both trials p, U, z^{χ}, and p_o are corrupted by white noise. Since this results in a stochastic simulation, 10 runs are performed for each of the trials.

The obstacle is placed at $p_o = [-20\ 2000]^{\top}$, with radius $r_o = 300$m. The x-position of the obstacle is tuned to trigger chattering when $\sigma = 0$. For all simulations the vessel has initial position at the origin, and initial orientation along the desired path. All simulations are performed for a duration of 700s.

[2] Note that $r = \dot{\psi}$ may be replaced by $\dot{\chi}$, where $\dot{\chi}$ may be estimated from the surge and sway acceleration. Under certain assumptions on the crab angle dynamics, the control law (18) is stable using both r and $\dot{\chi}$ in the argument.

4.2. Simulation results

The trajectories from trial 1 are shown in Figure 5a. Notice the significant spread in the trajectories. For most of the runs, the ship safely passes the obstacle on either the port side or the starboard side. Out of the 10 runs, 1 fails to avoid the obstacle entirely. The commanded course angle for the run that fails to avoid the obstacle is shown in Figure 5b. Chattering is observed for a significant amount of time before the ship finally settles on a direction to turn. The chattering results in a delayed decision, leading to the vessel entering the unsafe region.

The trajectories from trial 2 are shown in Figure 6a, with corresponding commanded course angle for a selected run shown in Figure 6b. All trajectories pass the obstacle at a safe distance. Furthermore, the spread in trajectories is significantly reduced compared to trial 1 (in fact the various trajectories are barely distinguishable), which is a sign of robustness and predictability.

5. Concluding remarks

The main result of this paper is a CBF-based hybrid kinematic controller suitable for obstacle avoidance of underactuated ships. The controller addresses two of the key ingredients for resilient operation of autonomous ships; safety and robustness. The proposed controller uses a single barrier function in combination with a hybrid feedback controller. In future work, multiple overlapping barrier functions will be used, following a similar idea as the hybrid Lyapunov functions discussed in Section 2. This is a more elegant and flexible approach which enables considering higher derivatives of the vessel position as control input, such as rudder angle for rudder-controlled ships.

Acknowledgments

This work was supported by the Research Council of Norway through the Centres of Excellence funding scheme, NTNU AMOS project number 223254.

References

[1] Reddy N P, Zadeh M K, Thieme C A, Skjetne R, Sørensen A J, Aanondsen S A, Breivik M, and Eide E 2019 Zero-Emission Autonomous Ferries for Urban Water Transport: Cheaper, Cleaner Alternative to Bridges and Manned Vessels. *IEEE Electrification Magazine* **7**.4 pp. 32–45. ISSN: 23255889

[2] Baruah S, Lee P, Sarathy P, and Wolf M 2020 Achieving Resiliency and Behavior Assurance in Autonomous Navigation : An Industry Perspective. *Proceedings of the IEEE* **108**.7 pp. 1196–1207

[3] Vachtsevanos G, Lee B, Oh S, and Balchanos M 2018 Resilient Design and Operation of Cyber Physical Systems with Emphasis on Unmanned Autonomous Systems. *Journal of Intelligent & Robotic Systems* **91** pp. 59–83

[4] Sanfelice R G, Messina M J, Tuna S E, and Teel A R 2006 Robust hybrid controllers for continuous-time systems with applications to obstacle avoidance and regulation to disconnected set of points. *Proceedings of the American Control Conference.* Minneapolis, MN, USA pp. 3352–3357

[5] Goebel R, Sanfelice R G, and Teel A R 2012 *Hybrid Dynamical Systems: Modeling, Stability, and Robustness.* Princeton University Press

[6] Brekke E F, Wilthil E F, Eriksen B O H, Kufoalor D K, Helgesen K, Hagen I B, Breivik M, and Johansen T A 2019 The Autosea project: Developing closed-loop target tracking and collision avoidance systems. *J. Phys.: Conf. Ser.* **1357 01202**. ISSN: 17426596

[7] Ames A D, Coogan S, Egerstedt M, Notomista G, Sreenath K, and Tabuada P 2019 Control barrier functions: Theory and applications. *European Control Conference.* Napoli, Italy: EUCA pp. 3420–3431. arXiv: 1903.11199

[8] Basso E A, Thyri E H, Pettersen K Y, Breivik M, and Skjetne R 2020 Safety-critical control of autonomous surface vehicles in the presence of ocean currents. *IEEE Conference on Control Technology and Applications.* Montreal, Canada

[9] Thyri E H, Basso E A, Breivik M, Pettersen K Y, Skjetne R, and Lekkas A M 2020 Reactive collision avoidance for ASVs based on control barrier functions. *IEEE Conference on Control Technology and Applications.* Montreal, Canada

[10] Borrmann U, Wang L, Ames A D, and Egerstedt M 2015 Control Barrier Certificates for Safe Swarm Behavior. *IFAC-PapersOnLine* **48**.27 pp. 68–73. ISSN: 24058963

[11] Panagou D, Stipanovic D M, and Voulgaris P G 2016 Distributed Coordination Control for Multi-Robot Networks Using Lyapunov-Like Barrier Functions. *IEEE Transactions on Automatic Control* **61**.3 pp. 617–632. ISSN: 00189286

[12] Wiig M S, Pettersen K Y, and Krogstad T R 2017 A Reactive Collision Avoidance Algorithm for Vehicles with Underactuated Dynamics. *Proceedings of the IEEE Conference on Decision and Control.* Melbourne, Australia pp. 1452–1459

[13] Brodtkorb A H, Værnø S A, Teel A R, Sørensen A J, and Skjetne R 2018 Hybrid controller concept for dynamic positioning of marine vessels with experimental results. *Automatica* **93** pp. 489–497. ISSN: 00051098

[14] Bhat S P and Bernstein D S 2000 A topological obstruction to global asymptotic stabilization of rotational motion and the unwinding phenomenon. *Systems & Control Letters* **39**.1 pp. 63–70. ISSN: 07431619

[15] Mayhew C G, Sanfelice R G, and Teel A R 2011 Synergistic Lyapunov functions and backstepping hybrid feedbacks. *Proceedings of the American Control Conference.* San Francisco, CA, USA pp. 3203–3208

[16] — 2011 Further results on synergistic Lyapunov functions and hybrid feedback design through backstepping. *Proceedings of the IEEE Conference on Decision and Control and European Control Conference.* Orlando, FL, USA pp. 7428–7433

[17] Wiig M S 2019 "Collision Avoidance and Path Following for Underactuated Marine Vehicles." PhD thesis. Norwegian University of Science and Technology

[18] Hung N T, Rego F, Crasta N, and Pascoal A M 2018 Input-Constrained Path Following for Autonomous Marine Vehicles with a Global Region of Attraction. *IFAC-PapersOnLine* **51**.29 pp. 348–353. ISSN: 24058963

[19] Marley M, Skjetne R, and Teel A R 2020 Heading control of surface ships using unit circle representation. *To appear in the Proceedings of the IEEE Conference on Decision and Control.* Jeju Island, Republic of Korea

[20] Fossen T I 2011 *Handbook of marine craft hydrodynamics and motion control.* John Wiley & Sons

[21] Mayhew C G and Teel A R 2010 Hybrid control of planar rotations. *Proceedings of the American Control Conference.* Baltimore, MD, USA: IEEE pp. 154–159

[22] Casau P, Sanfelice R G, Cunha R, Cabecinhas D, and Silvestre C 2015 Robust global trajectory tracking for a class of underactuated vehicles. *Automatica* **58** pp. 90–98. ISSN: 00051098

[23] Wiig M S, Pettersen K Y, and Krogstad T R 2020 Collision Avoidance for Underactuated Marine Vehicles Using the Constant Avoidance Angle Algorithm. *IEEE Transactions on Control Systems Technology* **28**.3 pp. 951–966. ISSN: 15580865

[24] Fossen T I and Perez T 2004 *Marine Systems Simulator (Mss).* `https://github.com/cybergalactic/MSS`. Accessed: 2020-06-05

[25] Perez T, Smogeli Ø N, Fossen T I, and Sørensen A J 2006 An overview of the marine systems simulator (MSS): A Simulink® toolbox for marine control systems. *Modeling, Identification and Control* **27**.4 pp. 259–275. ISSN: 03327353

IOP Conf. Series: Materials Science and Engineering **929** (2020) 012023 doi:10.1088/1757-899X/929/1/012023

Lipschitz Constrained Neural Networks for Robust Object Detection at Sea

Jonathan Becktor, Frederik Schöller, Evangelos Boukas, Mogens Blanke, Lazaros Nalpantidis

Department of Electrical Engineering, DTU - Technical University of Denmark, Denmark

E-mail: jbibe@elektro.dtu.dk

Abstract. Autonomous ships rely on sensory data to perceive objects of interest in their environment. Deep Learning based object detection in the image domain commonly used to solve this issue. The robustness of such approaches in non-ideal conditions is, however, still to be proven. In this work state of the art methods are applied on the RetinaNet architecture attempting to create a more robust object detection network given noisy input data. The GroupSort activation function and Spectral Normalization is used and the results are compared to the standard RetinaNet network. Our findings show that these modifications perform better and ensure robustness under moderate noise levels, than the standard RetinaNet network.

1. Introduction

Traversing the ocean can present many challenges to an autonomous vessel. A diverse set of weather conditions, equipment malfunctioning, continuous movement all need to be taken into account. The task of classifying boats and buoys at sea, as depicted in figure 1, might work reliably under normal conditions [13] harnessing the descriptive power of neural networks [12]. However, uncontrolled changes of the input—as they are to be expected in an autonomous marine system–can result in radically different outcomes. Some of the inherent weaknesses of neural networks come from their internal workings. A neural network is a data driven, directed non cyclical graph where each level in the graph is a layer. Each graph node has a weight and bias followed by a nonlinear activation function, such as a rectified linear unit (ReLU), hyperbolic tangent, or sigmoid function. Weights and biases are updated with the gradient of

Figure 1. Example image (cropped for better visibility) of annotated data used in this paper.

a distance function, given the network output and the correct target. Thus, given a neural network, very slight changes to the input can have compounding effects on each layer and result in very different output. Exposing the pipeline to noisy input, such as rain, snow, adversarial noise, and varying lighting conditions, can all have detrimental effects on the outcome of neural networks [14] and computer vision systems in general [11].

The *goal* of this paper is to explore architectures that add robustness to object detection against noisy input in marine environments. Lipschitz constrained neural networks [4] have been shown to alleviate a lot of these issues [1]. As a result, in this work, we apply Lipschitz constrained architectures to confirm their applicability and robustness in the autonomous marine domain.

2. Related Work

This work focuses on the use of deep neural networks to achieve robust object detection at sea. Previous approaches has been along the lines of [5] where detection of small surface vessels was achieved by using Scale Invariant Feature Transform (SIFT) in concert with bag-of-features. In [13], a comparison of using RGB, long wave infrared (LWIR), and near infrared (NIR) data within the Resnet-50 RetinaNet [9] architecture is provided, followed by a thorough evaluation of the networks. The RetinaNet architecture, as introduced in [9], is a one stage detector. This means that an input only needs to be inferred once due to a novel loss function, thus increasing the speed. The need for robustness of neural networks is addressed in [14], where the authors constrain the Lipschitz constant of a network to be lower than 1 by constraining each layer's weight matrices to be orthonormal. Furthermore, [1] explains how training neural networks under a strict Lipschitz constraint is useful for provable adversarial robustness, generalization bounds, and interpretable gradients. The paper, introduces a Lipschitz constrained activation function and the Björck convolution, a convolution that ensures layer-wise orthonormality and thereby being 1-Lipschitz. Spectral Norm introduced in [10] introduces another tool to maintain the Lipschitz continuity. The paper describes a layer-wise regularization step that normalizes the largest singular value of each weight matrix to be less than one. In the works of [4], a method of constraining the Lipschitz constant is explored, the authors apply an orothonormality constraint on all weights of the neural network and can thereby regularize the total network to be 1-Lipschitz.

3. Background

In this section, the possible architecture modifications will be presented. This analysis will constitute the base for defining the proposed approach in the following section.

Lipschitz continuity. Given two real values x and y a function is Lipschitz continuous if and only if the norm of the functions subtracted is less or equal norm of the inputs multiplied by some λ see figure 2.

$$|f(x) - f(y)| \leq \lambda |x - y| \tag{1}$$

λ is the Lipschitz constant, which is also what is constrained in a Lipschitz constrained network. For a network to be Lipschitz continuous it suffices to compose 1-Lipschitz layers and activations.

GroupSort. GroupSort, as introduced in [1], is a Lipschitz continuous activation function, it sorts the pre-activated weights into some selected grouping. For example a GroupSort with a grouping size of 2, will split the weights into groups of 2, sort the weights in the groups, as can be seen done in figure 3. This operation is a nonlinear operation which is differentiable as the Jacobian is a permutation matrix which also preserve every p-norm [1]. Selecting a group of 2

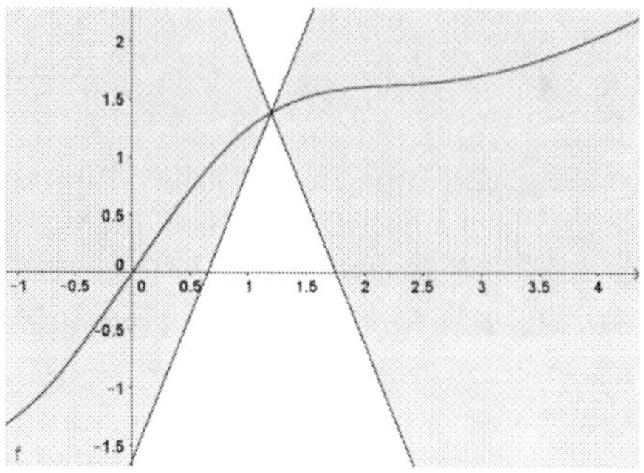

Figure 2. Lipschitz continuous function. Figure from [1]

is equivalent to the Orthogonal Permutation Linear Unit [3] also called MaxMin [1]. Whereas only using 1 group, e.g. sorting the entire slice, the operation is called FullSort.

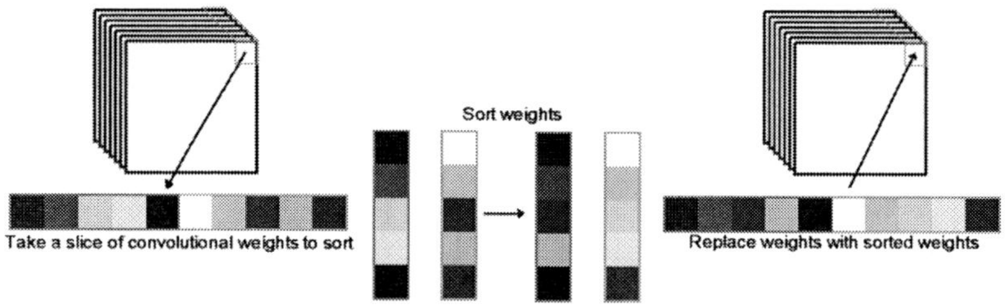

Figure 3. GroupSort with grouping size of 2, weights are sliced, sorted and replaced.

Spectral Norm. Spectral norm is a layer-wise normalization scheme introduced in [10] that enforces the largest singular value of the weight matrix to be less than 1 at each layer. This is achieved by estimating the singular vectors using power iteration and normalizing with respect to the found singular values:

$$\sigma(W) := \max_{h:h\neq 0} \frac{\|Wh\|_2}{\|h\|_2} = \max_{\|h\|_2\leq 1} \|Wh\|_2 \tag{2}$$

Where W are the weights and h is the largest singular value found by power iteration.

Björck convolution. The Björck convolution [2] is an iterative layer that lets us approximate a orthonormal weight matrix which ensures that singular values are exactly 1, which is achieved by iteratively applying the Taylor expansion of the polar decomposition.

Parseval Network. The Parseval regularization requires orthogonal weight matrices, this can be achieved with optimizing on the Stiefel manifold which is too expensive and can therefore be approximated with the layer-wise regularizer to ensure *Parseval Tightness* given by:

$$R_\beta(W) = \frac{\beta}{2}\left\|W^TW - I\right\|_2^2 \tag{3}$$

[1] `https://en.wikipedia.org/wiki/Lipschitz_continuity#/media/File:Lipschitz_Visualisierung.gif`

Optimizing $R_\beta(W)$ to convergence is too expensive, therefore the authors recommend only taking one step of descent, where the gradient is used to update the weight matrices by $W \leftarrow W(1+\beta) - \beta W W^T W$ Where β is the gradient step size.

4. Proposed Approach

The base architecture used is RetinaNet [9] with a ResNet-50 [6], Feature Pyramid Network [8] (FPN) backbone, as shown in figure 4. As discussed earlier, such networks can suffer from instability when exposed to noisy input. To combat the instability of the chosen neural networks, the architecture is modified to be 1-Lipschitz, as explored in [4] and [1]. As discussed above, the work of [4] proposes a regularization of the weights applied to each training loop given orthonormal weight matrices for each layer. On the other hand, [1] introduces the activation GroupSort (a sorting activation) and the Björck [2] convolution.

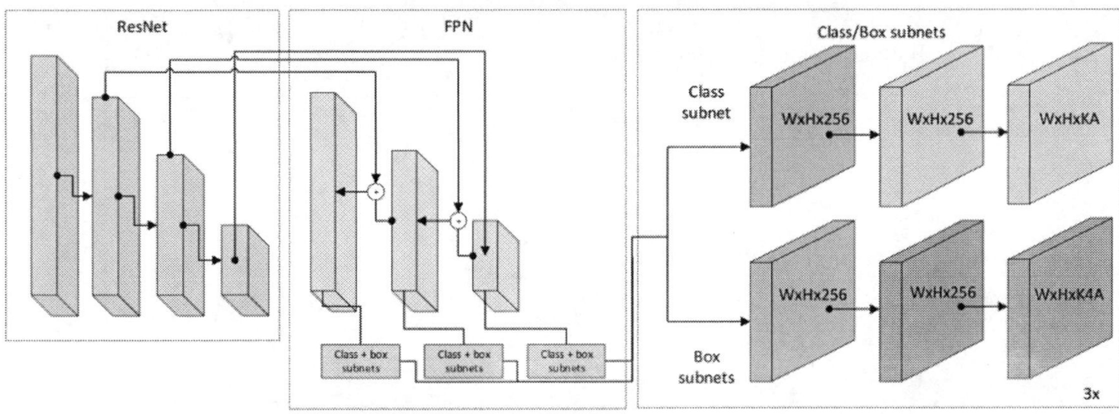

Figure 4. RetinaNet architecture.

The work done in [1] identified GroupSort activation and the Björck Convolution to be the most promising approach. However, using the Björck convolution on the full RetinaNet in its current form, is not feasible. Björck convolution can be applied to the classification/regression networks and use Spectral Normalization on the ResNet base.

Spectral Normalization is used as the regularization of the convolutions. Thus the selected modifications are GroupSort and spectral Normalization to ensure the Lipschitz constraint and thereby improving network robustness. These are selected as they are the least computationally expensive and the fastest to implement. The architectural modifications will be evaluated using similar methods as described by [13]. Several networks will be trained with differing GroupSort group sizes and will be compared to an equivalent network without any modification.

5. Experimental Setup

5.1. Dataset

The effectiveness of the selected networks architectures are evaluated on the dataset, consisting of a total of 9229 images with 20150 annotations divided into two classes namely; boat (6256) and buoy (13894). Each image was captured as a 2560×2048 pixels RGB image and subsequently downsized to 1440×1080 pixels. Data was acquired on-board ferries operating in the Southern Funen Archipelago. Images were captured every second from midday till dusk, ensuring a broad range of scene illumination(s). This yielded a training set containing 8306 images and a validation set containing 923 images. The image in figure 1 shows a cropped section of an annotated image from the dataset.

5.2. Training of the Neural Networks
The RetinaNet is trained with the ADAM optimizer [7], with an initial learning rate of 10^{-5} with a plateau scheduler. Each model is trained for approximately 110 epoch using early stopping, with a batch size of 2. There is a big difference in training time as the operations for 1-Lipschitz networks are computationally demanding.

6. Experimental Evaluation
The mean Average Precision metric (mAP) will be used as the method for comparing the selected networks. Average Precision is the area under the Precision-Recall curve shown in figure 5. Precision is found by $\frac{TP}{FP+TP}$ and recall with $\frac{TP}{FN+TP}$ where TP is true positive, FP is false positive and FN is false negative. Precision and Recall both change with the certainty threshold set for the network, which in these experiments is set to 0.5; this threshold can range from 0 (where everything is accepted) to 1 (where the network has to be extremely certain). The mAP is then found by the mean of the AP of each class in the system.

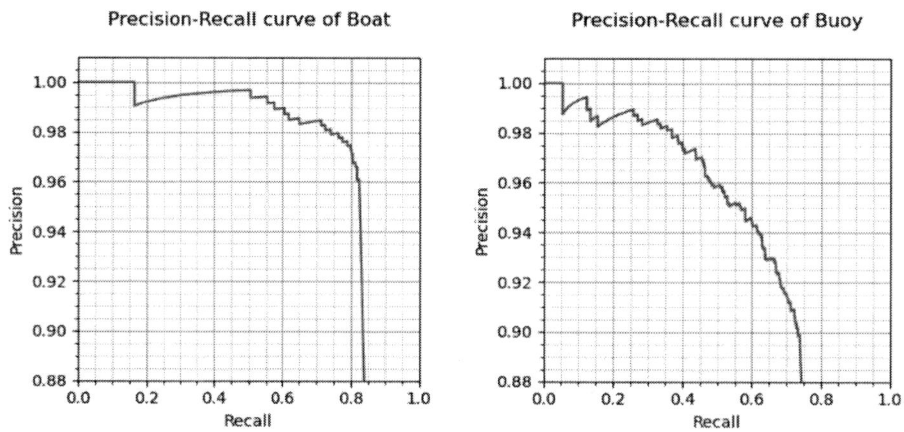

Figure 5. Precision-Recall curves for Boat and Buoy.

The selected networks will be tested on 3 different image corruption techniques; Gaussian noise, Gaussian Blur, and Salt and Pepper noise with increasing levels of intensity. The techniques were chosen as they loosely resemble the outcome of certain problems that could be expected during long-term operation of imaging sensors in harsh marine environments. Furthermore, to gather a deeper insight the precision and recall with respect to the area of each bounding box will be explored; this will provide insight on the size of objects the network has difficulty detecting.

6.1. Gaussian Noise
Corrupting an image with Gaussian noise is achieved by varying the intensities values of each color channel (R,G,B) for each pixel by some value drawn from a normal distribution. An example image, corrupted with various levels of Gaussian noise is shown in figure 6.

Table 1 summarizes the mean Average Precision (mAP) results for various levels of corruption of the images using Gaussian noise, in the following tables, GS is GroupSort and SN is Spectral Normalization. For each RetinaNet model the activation function is indicated, where ReLU corresponds to the implementation of [13], while the 3 remaining entries are our considerations.

Finally, figure 7 and figure 8 show the detection histograms, precision and recall as a function of the bounding box area for boats and buoys respectively. These figures compare the standard

Figure 6. Images corrupted with different levels of Gaussian noise. From left to right: no noise, $\mathcal{N}(0, 0.05)$, $\mathcal{N}(0, 0.075)$.

Table 1. mAP Results for various levels of Gaussian Noise. All results are ±0.02

RetinaNet Model	No Noise	$\mathcal{N}(0, 0.025)$	$\mathcal{N}(0, 0.05)$	$\mathcal{N}(0, 0.075)$
ReLU	0.76	0.65	0.47	0.32
MaxMin, SN	0.75	0.68	0.49	0.29
GS 4, SN	0.76	**0.71**	**0.54**	0.32
GS 8, SN	**0.77**	0.70	0.50	**0.35**

ReLU network (figures on the left) versus the proposed "GroupSort 8, SN" network (figures on the right) for increasing levels of induced Gaussian noise (increasing from top to bottom).

6.2. Gaussian Blur

Corrupting an image with Gaussian blur is achieved by convolving the input images with a 2D Gaussian kernel. Increasing standard deviation of the used kernel results in a stronger blurring effect. An example image, corrupted with various levels of Gaussian noise is shown in figure 9.

Table 2 summarizes the mean Average Precision (mAP) results for various levels of corruption of the images using Gaussian Blur.

Table 2. mAP Results for various levels of Gaussian Blur. All results are ±0.02

RetinaNet Model	No Noise	$\sigma = 1$	$\sigma = 1.5$	$\sigma = 2$
ReLU	0.76	0.69	0.66	**0.58**
MaxMin SN	0.75	0.71	0.66	0.52
GS 4, SN	0.76	0.73	0.69	0.56
GS 8, SN	**0.77**	**0.74**	**0.700**	0.55

6.3. Salt and Pepper Noise

Corrupting an image with Salt and Pepper noise is achieved by randomly selecting a percentage of pixels and assigning the minimum or maximum intensity value to them. An example image, corrupted with various percentages of pixels affected by Salt and Pepper noise is shown in figure 10.

Table 3 summarizes the mean Average Precision (mAP) results for various levels of corruption of the images using Salt and Pepper noise.

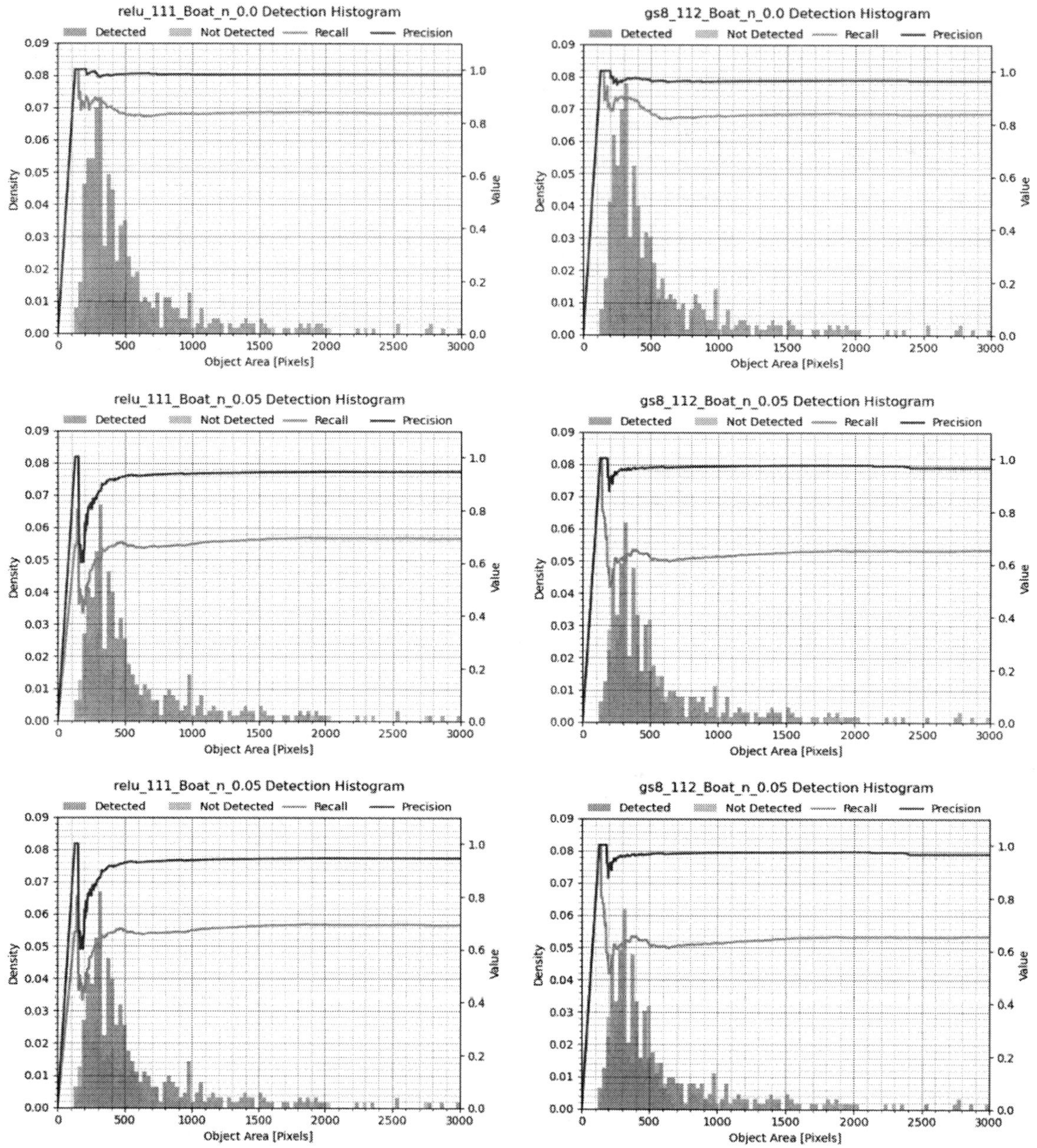

Figure 7. Detection histograms, precision and recall curves as a function of area of boat bounding boxes. The blue histogram is the detected objects where the orange is the objects not detected, dark green curve is the precision and light green is the recall. Comparison of standard ReLU network (figures on the left) versus the "GroupSort 8, SN" network (figures on the right) with no Gaussian noise (top), with $\mathcal{N}(0, 0.05)$ noise (middle) and $\mathcal{N}(0, 0.075)$ (bottom). The left vertical axis show the density of objects of this size in the dataset; the right vertical axis show the percentage of recall/precision and horizontal axis show the bounding box area of each object.

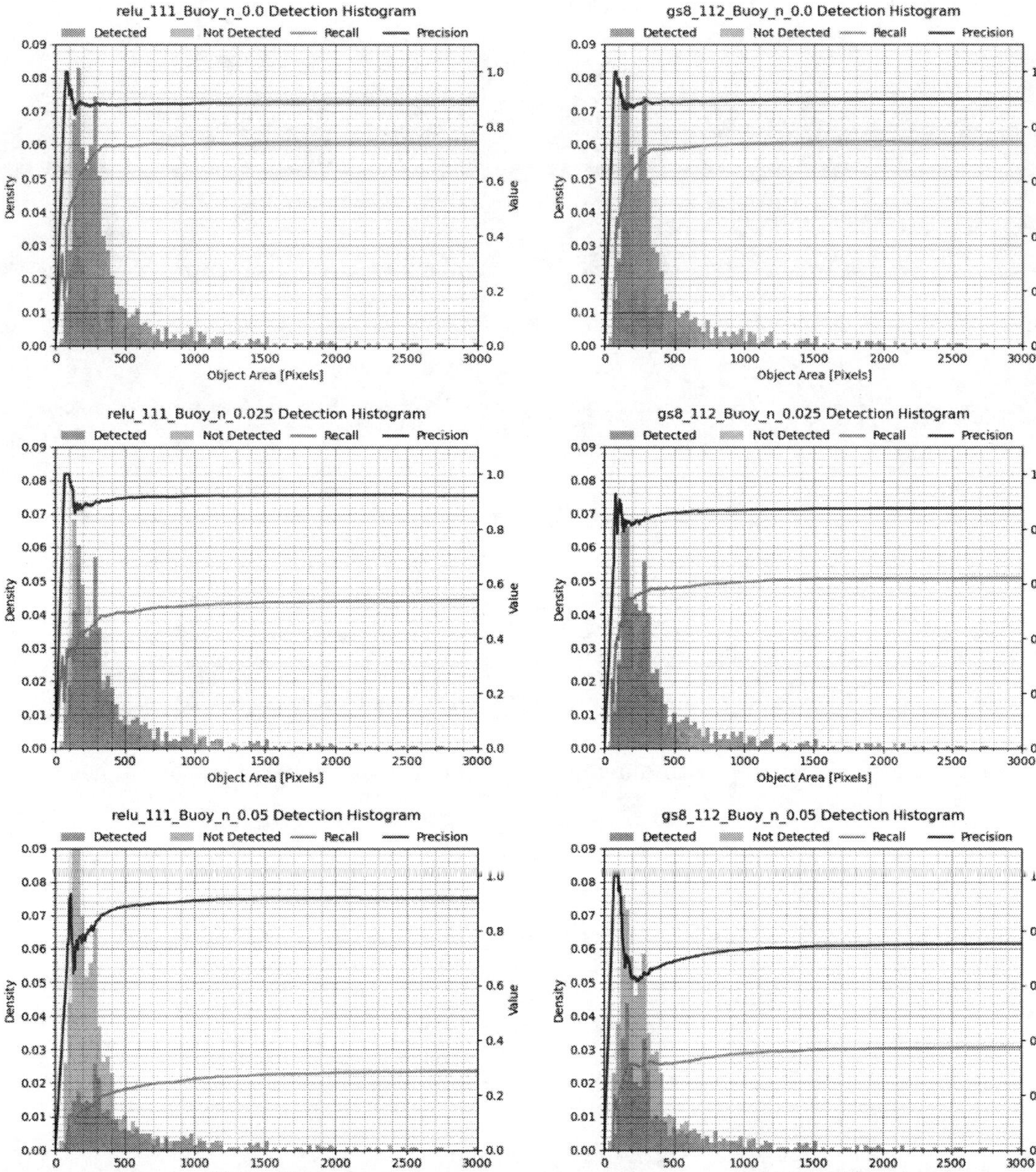

Figure 8. Detection histograms, precision and recall curves as a function of area of buoy bounding boxes. The blue histogram is the detected objects where the orange is the objects not detected, dark green curve is the precision and light green is the recall. Comparison of standard ReLU network (figures on the left) versus the "GroupSort 8, SN" network (figures on the right) with no Gaussian noise (top), with $\mathcal{N}(0, 0.025)$ noise (middle) and $\mathcal{N}(0, 0.05)$ (bottom). The left vertical axis show the density of objects of this size in the dataset; the right vertical axis show the percentage of recall/precision and horizontal axis show the bounding box area of each object.

Figure 9. Images corrupted with different levels of Gaussian blur. From left to right: no noise, $\sigma = 1.5$, $\sigma = 2$.

Figure 10. Images corrupted with different levels of Salt and Pepper noise. From left to right: no noise, 0.5%, 1%.

Table 3. mAP Results for various levels of Salt and Pepper Filter. All results are ±0.02

RetinaNet Model	No Noise	0.1%	0.5%	1%
ReLU	0.76	**0.70**	**0.58**	**0.50**
MaxMin SN	0.75	0.67	0.54	0.48
GS 4, SN	0.76	0.68	0.55	0.50
GS 8, SN	**0.77**	0.69	0.56	0.49

7. Discussion

In this work, Lipschitz constrained network architectures were explored in order to improve general robustness of an object detection network. The architecture used was identical to the one explored in [13], the aforementioned modifications were applied to explore if out of the box modifications could help improve robustness. To get a wider comparison of the modifications early stopping was used and all models were approximately trained to 110 epochs. Training the network to convergence took up to 10 days on a Tesla V100 with 16 GB of memory. In this work, the optimization was not explored to its full extent, as the focus was not the most optimal network but rather how these modifications improve the networks' relative to one another. Given this, it is to be expected that the networks will perform better when trained to convergence.

Most of the tested architecture modifications do not have any negative effect on the zero-noise, uncorrupted data. This fact is very positive, as inferior performance on this base test scenario, would pose questions about the usefulness of the propose modifications.

Furthermore, as shown in table 1 for the case of Gaussian noise corruption, a noticeable improvement appears once noisy data is used for testing. The standard ReLU network shows a fall in mAP as the level of noise increases, whereas each of the Lipschitz constrained networks do better. Similarly, with Gaussian blur, as shown in table 2, the Lipschitz constrained networks performed better overall.

However, applying Salt and Pepper noise to the input seems to cause a decrease in performance for the GroupSort networks see table 3. A reason for this drop in performance can be the proposed Lipschitz constrained networks. They were conceived to be robust against small specific augmentations of the input which is often the case in adversarial attacks. When the induced noise level is radical, e.g. in Salt and Pepper noise pixel values abruptly change to the highest or lowest possible level of the intensity value range–this type of networks cannot handle the change, resulting in inferior results. The same reasoning can explain the worse performance for the highest tested noise level of Gaussian blur.

For boats, in figure 7, the area of the precision curve of the GroupSort networks is larger as noise is introduced. In the recall curve of the same figure, there is a sharp downward trend with smaller vessels which reduces the total area under the curve, which is not as extreme with the GroupSort networks. With respect to Buoys, see figure 8, as noise is introduced the precision curve is slightly higher on larger vessels, but on smaller vessels it shows that the GroupSort network does better. The area under the recall curve in this figure is also noticeably larger for the GroupSort networks, which translates to a higher percentage of detected buoys.

8. Conclusion & Future Work

In this work, Lipschitz constrained neural networks were explored with the hope of increasing the robustness of object detection at sea. Several Lipschitz constrained modifications were explored with the intent of making a pre-selected network architecture, RetinaNet as used in [13], more robust to noisy input data. The selected modifications, e.g. the GroupSort activation function and Spectral Normalization, ensure the network is Lipschitz continuous. These modified networks have an advantage in the case of Gaussian noise and Gaussian blur and provide only slightly worse results for salt and pepper noise. In all cases though, if the noise extends beyond a certain point the Lipschitz constrained networks break down, whereas the standard ReLU network will still function to some degree.

The obtained results are encouraging, even though not entirely conclusive. For future work, an interesting direction is introducing the explored noise as augmentations of the input for the training of the network. The Lipschitz constrained networks break down when the input is too different from what they have been trained on. If adding augmentations to the input of the network allows the Lipschitz constrained networks to handle bigger variation while still maintaining the higher recall/precision, with respect to small changes it would be very positive. Furthermore some of the Lipschitz modifications were not used due to them being computationally inefficient, such as the Björck convolution. Given additional computational resources, these can be explored as it is only demanding in training time and not in inference time.

Acknowledgments

This research is sponsored by the Danish Innovation Fund, The Danish Maritime Fund, Orients Fund and the Lauritzen Foundation through the ShippingLab project.

References

[1] C. Anil, J. Lucas, and R. Grosse. Sorting out Lipschitz function approximation. In *36th International Conference on Machine Learning, ICML 2019*, volume 2019-June, pages 432–452, 2019.

[2] A. Björck and C. Bowie. Iterative algorithm for computing the best evidence of an orthogonal matrix. *SIAM Journal on Numerical Analysis*, 8(2):358–364, 1971.

[3] A. Chernodub and D. Nowicki. Orthogonal permutation linear unit activation function (OPLU). *Lecture Notes in Computer Science (including subseries Lecture Notes in Artificial Intelligence and Lecture Notes in Bioinformatics)*, 9887 LNCS:533–534, 2016.

[4] M. Cisse, P. Bojanowski, E. Grave, Y. Dauphin, and N. Usunier. Parseval networks: Improving robustness to adversarial examples. *34th International Conference on Machine Learning, ICML 2017*, 2:1423–1432, 2017.

[5] R. Dulski, S. Milewski, M. Kastek, P. Trzaskawka, M. Szustakowski, W. Ciurapinski, and M. Zyczkowski. Detection of small surface vessels in near, medium, and far infrared spectral bands. In *Electro-Optical and Infrared Systems: Technology and Applications VIII*, 2011.

[6] K. He, X. Zhang, S. Ren, and J. Sun. Deep residual learning for image recognition. In *Proceedings of the IEEE Computer Society Conference on Computer Vision and Pattern Recognition*, 2016.

[7] D. P. Kingma and J. L. Ba. ADAM: A method for stochastic optimization. *3rd International Conference on Learning Representations, ICLR 2015 - Conference Track Proceedings*, pages 1–15, 2015.

[8] T. Y. Lin, P. Dollár, R. Girshick, K. He, B. Hariharan, and S. Belongie. Feature pyramid networks for object detection. In *Proceedings - 30th IEEE Conference on Computer Vision and Pattern Recognition, CVPR 2017*, 2017.

[9] T. Y. Lin, P. Goyal, R. Girshick, K. He, and P. Dollar. Focal Loss for Dense Object Detection. *Proceedings of the IEEE International Conference on Computer Vision*, pages 2999–3007, 2017.

[10] T. Miyato, T. Kataoka, M. Koyama, and Y. Yoshida. Spectral normalization for generative adversarial networks. In *6th International Conference on Learning Representations, ICLR 2018 - Conference Track Proceedings*. International Conference on Learning Representations, ICLR, 2018.

[11] L. Nalpantidis and A. Gasteratos. Stereo vision for robotic applications in the presence of non-ideal lighting conditions. *Image and Vision Computing*, 28(6):940–951, 2010.

[12] F. E. T. Schöller, M. Blanke, M. K. Plenge-Feidenhans'l, and L. Nalpantidis. Vision-based object tracking in marine environments using features from neural network detections. In *IFAC Worls Congress*, Berlin, Germany, 2020.

[13] J. D. Stets, F. E. T. Schöller, M. K. Plenge-Feidenhans'l, R. H. Andersen, S. Hansen, and M. Blanke. Comparing Spectral Bands for Object Detection at Sea using Convolutional Neural Networks. *Journal of Physics: Conference Series*, 1357:12036, oct 2019.

[14] C. Szegedy, W. Zaremba, I. Sutskever, J. Bruna, D. Erhan, I. Goodfellow, and R. Fergus. Intriguing properties of neural networks. In *2nd International Conference on Learning Representations, ICLR 2014 - Conference Track Proceedings*, 2014.

Additional Test Manoeuvres for Autonomous Inland Vessels

G Peeters[1,2], G Yayla[1], M R Afzal[1], S Van Baelen[1], T Catoor[1], M Vanierschot[1], R Boonen[1] and P Slaets[1]

[1] KU Leuven, Department of Mechanical Engineering, 3000 Leuven, Belgium
[2] Research Foundation - Flanders (FWO), 1000 Brussels, Belgium

E-mail: gerben.peeters@kuleuven.be

Abstract. Standardised test manoeuvres are used to identify the manoeuvrability of a vessel. Currently, most of these tests are unsatisfactory to properly judge the manoeuvring capabilities of an inland vessel which sails in different environments than the seagoing ships for which these tests were initially designed. Moreover, inland vessels tend to have non-conventional propulsion systems and configurations which further decreases the adequacy of these standardised tests. Therefore, this study investigates the inland-applicability of these existing standardised test manoeuvres. In addition, this study suggests three new manoeuvres: (i) the Counter Thrust Test, (ii) the Sine Angle Test, and (iii) the Simultaneous Zigzag Test. In order to validate the suitability of these tests, an autonomous inland cargo vessel performed two of these suggested manoeuvres, (i) and (ii), and one existing manoeuvre, named the Crabbing test. The data of these experiments show fruitful insights in the wide range of manoeuvring capabilities of the examined inland vessel, which the standardised tests would not have uncovered.

1. Introduction

The existing standardised test manoeuvres aim to reveal the capability of a vessel to perform certain manoeuvres. Hence, these tests allow for the identification of certain hydrodynamic manoeuvring characteristics and metrics. These characteristics can provide an operator with crucial information in order to take correct and timely decisions to safely and effectively navigate the vessel. The International Maritime Organization (IMO) issued the most-known manoeuvrability standards which include checking the manoeuvring performance of a vessel in terms of its inherent dynamic stability, course keeping ability, initial turning and course changing ability, yaw checking ability, turning ability, and stopping ability [1]. Note, however, that these IMO standards should be applied to "ships of all rudder and propulsion types, of 100m in length and over, and chemical tankers and gas carriers regardless of the length [1]". In Addition, the manoeuvring committee of the International Towing Tank Conference (ITTC) examined 19 manoeuvring tests proposed by various organizations and recommended procedures and guidelines for 14 of them which are able to measure the above-mentioned characteristics [2]. Currently, the most widely accepted tests for full scale trials consist of the Turning Circle Test, the 10/10 and 20/20 Zigzag Manoeuvre Test, and the Stopping Test.

However, the typical inland cargo vessel differs significantly from its seagoing counterparts. Firstly, inland cargo vessels tend to operate in spatially restricted shallow or confined water without the help of tug boats. Therefore, they are often built with multiple rudder and/or propeller configurations in order to boost their manoeuvring performance [3]. Even the addition of an azimuth thruster, or a transversal bow thruster, is not uncommon to further increase the manoeuvrability of a vessel [4]. Secondly, the vertical restrictions in the inland waterways usually result in speed regulations in order to avoid the touching of the bottom due to the arising squat effect. Hence, inland vessels usually perform their manoeuvres at low speeds and need to keep course under these conditions. Thirdly, due to the horizontal restrictions such as banks, heavy traffic in access areas, and the necessity of extreme close passages in narrow sections of canals, inland vessels also have to change their courses frequently and usually do a manoeuvre similar to cars changing lanes on the road [5]. Therefore, contrary to seagoing vessels where course keeping dominates, initial turning or course changing ability are crucial metrics for inland vessels.

Consequently, smaller vessels which navigate in shallow water with non-conventional propulsion types, do not directly fall under these above-mentioned IMO standards [4]. Therefore, some authorities, like the Central Commission for the Navigation of the Rhine [6] and the European Commission [7], offered regional standards for inland vessels [8]. Nevertheless, these regulations have fewer manoeuvrings tests and criteria and, therefore, the IMO standards are generally used as a guide for inland vessels [5]. To bridge this gap in adequate inland manoeuvres, [5] suggested additional manoeuvres and subsequent metrics for inland vessels. Thereupon, this work expands these additional inland manoeuvres and discusses the outdoor executions of some of these existing and newly added manoeuvres with an unmanned autonomous inland cargo vessel.

This paper continues as follows: firstly section 2 lists some of the existing standardised test manoeuvres, and afterwards it adds three new manoeuvres to this list. Section 3 details the design of the inland cargo vessel that conducted the experiments of section 4, based on which section 5 draws the conclusions of this study.

2. Method

Section 2.1 briefly discusses the existing manoeuvres which focus on non-conventional propulsion systems or inland vessels. Afterwards, section 2.2 introduces additional test manoeuvres tailored for inland vessels with non-conventional propulsion systems.

2.1. Existing Ship Manoeuvres

In the proposed guidelines of the ITTC, 2 of the 14 tests mention the use of a lateral thruster: (i) the *Thruster Test* suggests performing the turning or zigzag manoeuvre with the lateral thruster as a steering input and the rudder at midship, and (ii) the *Crabbing Test* suggests lateral movements at zero forward speed [2]. Furthermore, in the proposed tests for inland vessels of [5], the *Hard Turning*, *T-junction*, and *Lane Changing* tests can use, and benefit from, non-conventional thrusters and their position.

2.2. Suggested Additional Vessel Manoeuvres

Given the fact that inland vessels tend to use non-conventional propulsion configurations and non-conventional actuators, the here-suggested test manoeuvres assume a generic inland vessel which has two propulsion systems: one at the bow and one at the stern. Moreover, both propulsion systems are presumed to have a controllable thrust force magnitude, T, and orientation, θ. Evidently, the maximum thrust force depends on the geometrical design of the system and the engine powering it, whereas the limits of its orientation lie between $[0, 360]°$. Or, more conveniently, one can denote this orientation by measuring the angle between T and the longitudinal axis of the vessel which points to the bow and encapsulate this angle between $[-180, 180]°$ with a positive sign for counter-clock-wise angles. Accordingly, for the remainder of this study, $T \in [0, T_{max}]$, $\theta \in [-180, 180]°$, and the superscripts '*s*' and '*b*' will be used to denote the stern and bow thruster respectively. Finally, it could happen that the internal control angle of the actuation system, θ_i, and the orientation of the resulting output force, θ_o, are not aligned, hence a differentiation will be made by using the subscripts '*i*' or '*o*'. Under these assumptions, the following three additional manoeuvring tests for inland vessels are suggested; 2.2.1: Counter Thrust Test, 2.2.2: Sine Angle Test, and 2.2.3: Simultaneous Zigzag Test.

2.2.1. Counter Thrust Test This test aims to locally rotate the vessel by covering as little space as possible. For this purpose, both the bow and stern thruster should try to exert an equal but opposite force. If the thrust forces do not be equal each other in magnitude, there will be a resulting lateral force which would make the vessel drift during the manoeuvre. Evidently, this manoeuvre could be performed in a clock-wise or counter-clock-wise turning direction, hence:

- $T^b = T^s$
- θ_i^b = 90° (or -90°)
- θ_i^s = -θ^b

2.2.2. Sine Angle Test This test helps to uncover the broad spectrum of achievable turning rates and thruster-angle changes. This test can be performed for both thrusters separately with the other thrusting providing no thrust at all or another desired value. Furthermore, both thrusters could perform the test simultaneously. The test parameters per thruster consist of its thrust magnitude, the maximum angle of its orientation, and the frequency of the sine function:

- T = chosen value
- $\theta = \theta_{max} \sin(2\pi f)$

2.2.3. Simultaneous Zigzag Test In line with the suggested additional zigzag manoeuvre from [2], where the lateral thruster provides the steering, one could also instruct both thrusters to work simultaneously. Towards this end, both thrusters should have an opposite angle to cooperatively perform the zigzag manoeuvre:

- T^b = chosen value
- T^s = chosen value
- θ^b = chosen value (e.g. 30°/-30°)
- θ^s = - θ^b

3. Material

The European Watertruck+ project is constructing a novel fleet of 31 inland cargo vessels of which 16 are self-propelled barges which have an over-actuated propulsion system enveloping a 360-degrees-steerable steering-grid thruster in the bow in combination with a 360-degrees-steerable 4-channel thruster in the stern [9]. The authors believe that these vessels exhibit a high potential for the further automation of their operations which could pave the way for future unmanned or even autonomous vessels. Consequently, to study this automation potential, they constructed a scale model of such a self-propelled barge [10, 11]. Figure 1a shows four real-size barges of European Class type I [12] which are being pushed by a push boat, and in Figure 1b the scale model, named '*Cogge*' can be seen sailing on the Yser river (in Belgium).

(a)

(b)

Figure 1: (a) Four European Class type I vessels from Watertruck^{+} pushed by a push boat [9], (b) The KU Leuven scale model, named Cogge [10].

3.1. Design of the Cogge

The Cogge has a length of 4.81m, a beam of 0.63m, and a maximal draft of 0.35m with a similar geometry as its Watertruck^{+} counterpart. Figure 2a shows the 3-dimensional design drawing of the Cogge with the stern at the left and the bow at the right hand side of the image. The stern and bow thrusters can be seen on these respective locations. Thereupon, Figure 2b portrays an abstract top view of the four-channel stern thruster together with its angle convention, and Figure 2c illustrates the steering-grid thruster and its analogous angle convention. Both thrusters have a propeller which draws water from underneath the hull, but they have a different 360-degrees-rotatable mechanism to control the orientation of their water outflow. The steering grid bends the flow approximately 180° and then exhausts its water flow underneath the hull, whereas the stern thruster bends the flow approximately 90° and then exhausts its water flow through the sides of the hull (except when it is orientated to provide backwards thrust, then the water flow will also be expelled underneath the hull). The magnitude of the thrust force of both thrusters can be controlled by their propeller speed, n. Due to the differences in their steering mechanism design and their propeller size, both thrusters generate different thrust forces at equal propeller speeds, of which more details can be found in [10].

3.2. Onboard Sensors

Two sensors measured the movements of the vessel during the conducted experiments of section 4. A Septentrio AsteRx-U Marine GNSS receiver with two mushroom antennas measured the heading and position of the vessel [13], and a SBG Ekinox-2E IMU/INS provided its yaw-rates [14]. This IMU also measured the reported headings from section 4 based on an internal Kalman filter which received corrections from the GNSS. The main GNSS antenna was placed at the stern of the vessel and reported its own position. The auxiliary antenna was placed 4,44m in front of the main antenna and thus unlocked the access to heading information. The IMU was positioned 95cm in front of the main GNSS antenna.

4. Results and Discussion

This section shows three of the aforementioned manoeuvres for inland vessels which were performed with the Cogge on a lake in Rotselaar (Belgium). The Cogge conducted all these experiments autonomously with no crew on board. The values of the parameters of these tests were empirically chosen. Section 4.1 discusses an ITTC-suggested manoeuvre, namely the Crabbing Test, whereas sections 4.2 and 4.3 respectively handle the here-proposed Counter

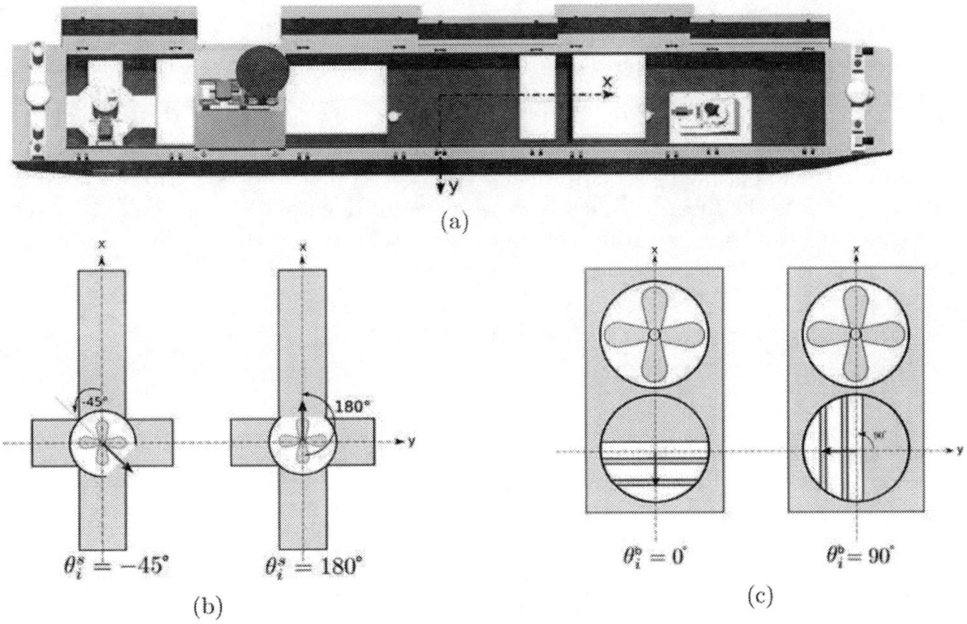

Figure 2: (a) design drawing of the self-propelled scale model barge with its (b) 360-degrees steerable 4-channel thruster at the stern (abstract top view of its bottom section to show the angle convention θ_i^s), and (c) 360-degrees-steerable steering grid thruster in the bow (abstract top view of its bottom section to show the angle convention θ_i^b), adapted from [11, 10].

Thrust and Sine Angle tests. In order to give an idea of the testing conditions, a top view photo of a counter-clockwise Counter Thrust Test can be seen in Figure 3.

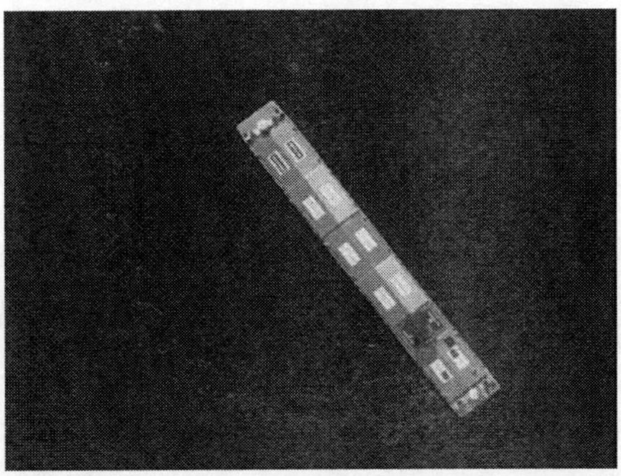

Figure 3: Counter clock-wise manoeuvre with the Cogge on Rotselaar lake, this picture was not taken simultaneously with the shown test data of section 4.2.

4.1. The Crabbing Test

Figure 4a shows all the system states measured during the 40s crabbing manoeuvre, i.e. propeller speeds, n, and internal thruster angles, θ_i. The vessel performed a starboard crabbing manoeuvre with simultaneously changing propeller speeds for the bow and stern thruster. Ideally, both thrust forces exert an equal but opposing torque on the vessel such that the vessel does not rotate. During the experiment, an empirically derived propeller speed ratio was used to achieve this. Figure 4b plots the measured headings (relative to the start heading) and yaw-rates which indicate that the vessel kept a rather straight orientation during the whole manoeuvre. Finally, Figure 4c displays the trajectory the main GNSS mushroom covered.

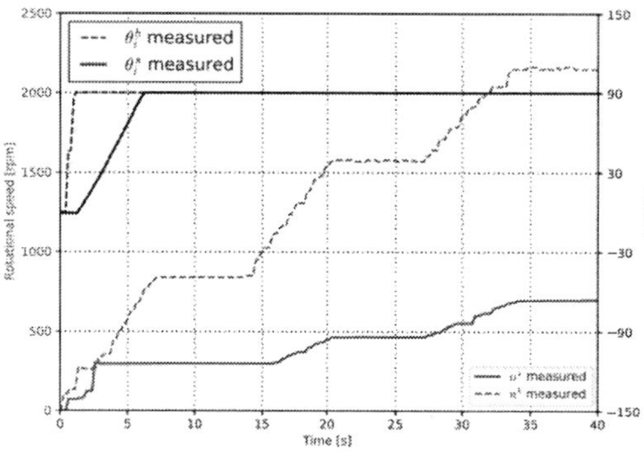

(a)

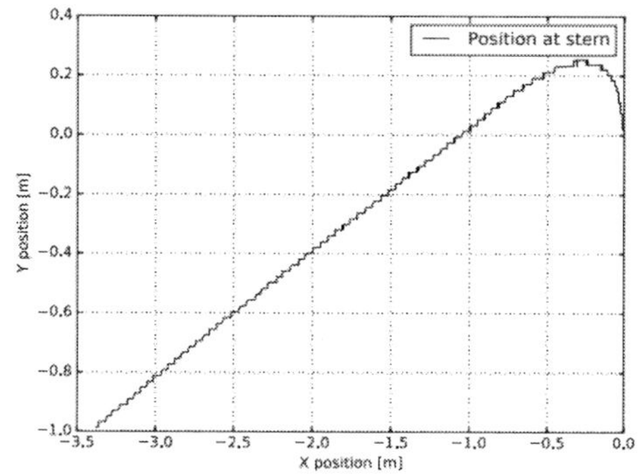

(c)

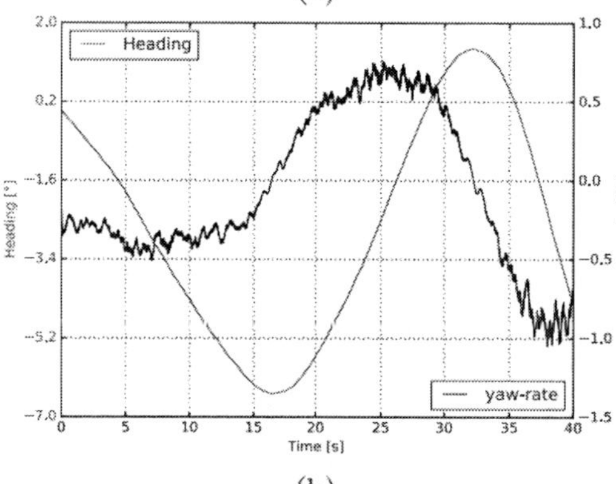

(b)

Figure 4: The Crabbing manoeuvre with (a) $\theta_i^b = \theta_i^s = 90°$, $n^b = 0$ to 2150rpm using a staircase profile, and $n^s = 0$ to 700rpm using a staircase profile, (b) shows the measured headings and yaw-rates, and (c) shows the trajectory of the main GNSS mushroom.

4.2. The Counter Thrust Test

Figure 5a demonstrates the measured system states during the performed counter-clockwise Counter Thrust Test. Here too, a propeller speed ratio was used in order to provide approximately equal thrust magnitudes, similar to section 4.1. As the measured headings of Figure 5b illustrate, the Cogge rotated almost twice over the 60s time span of the manoeuvre. Lastly, Figure 5c exposes that the Cogge did not drift far during the experiment, as the position of the main GNSS mushroom, which is placed at the stern of the 4,8m long vessel, described a spiraled-ellipse shape within a 5m by 5.5m rectangular box. The spiral shape of the GNSS mushroom antenna indicates that there was a small deviation of the starting position which might be caused by: (i) the error on the experimentally tuned propeller speed ratio which could be diminished in the future, and (ii) external disturbances which might have been present.

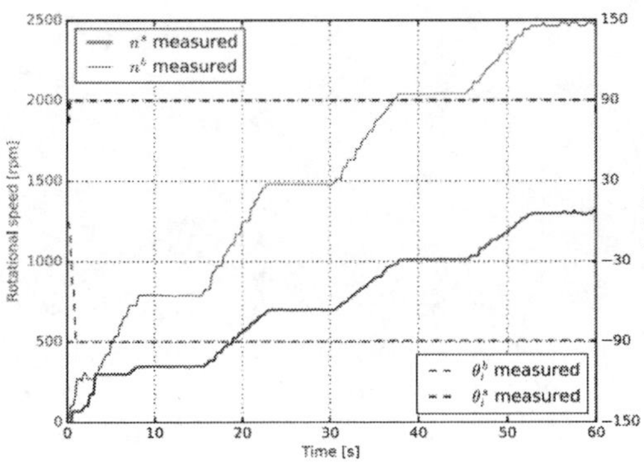

(a)

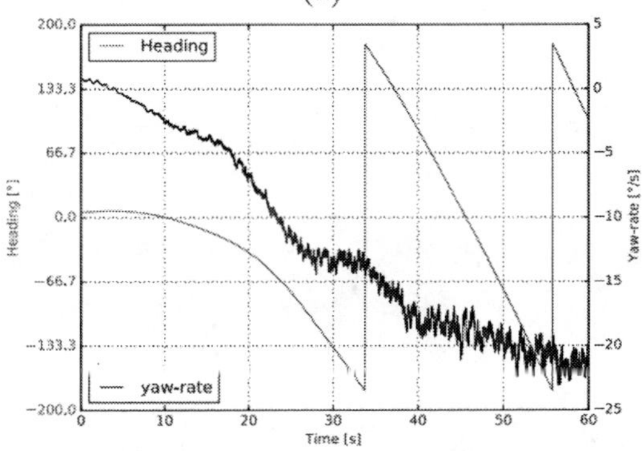

(b)

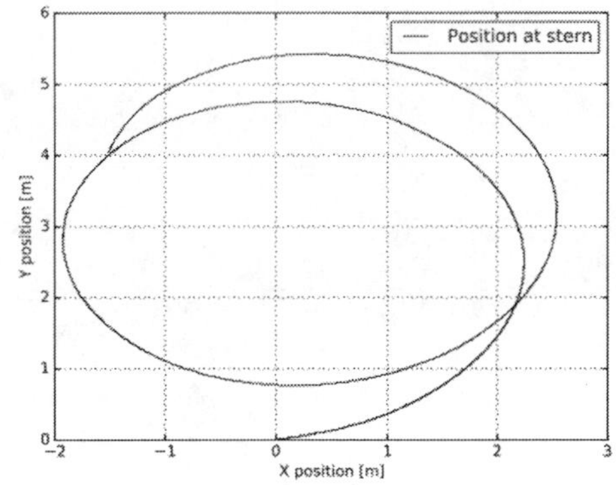

(c)

Figure 5: Counter-Clockwise Counter Thrust manoeuvre with (a) $\theta_i^b = -90°$, $\theta_i^s = 90°$, $n^b = 0$ to 2450rpm using a staircase profile, and $n^s = 0$ to 1300rpm using a staircase profile, (b) shows the measured headings and yaw-rates, and (c) shows the trajectory of the main GNSS mushroom.

4.3. The Sine Angle Test

Figure 6a depicts the measured system states during a Sine Angle Test conducted by the bow thruster, hence the system states for the stern both equalled zero. The bow propeller reached a rotational speed of 2050rpm and the angle of the steering grid oscillated between [-180, 180]° over a period of 40s, which resulted in the measured headings and yaw-rates of Figure 6b. These yaw-rates offer an impression of the achievable turning rates of the vessel at low advance speeds. Evidently, the same manoeuvre could be performed over a longer period, giving the yaw-rates more time to settle, or with higher advance speeds. Finally, Figure 6c portrays the distance covered by the main GNSS mushroom during this experiment.

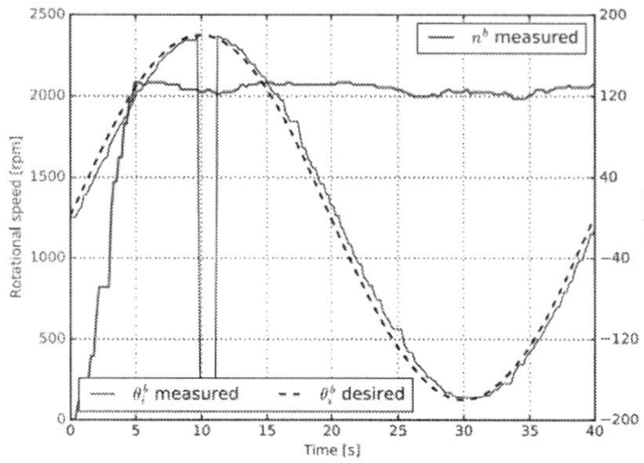

(a)

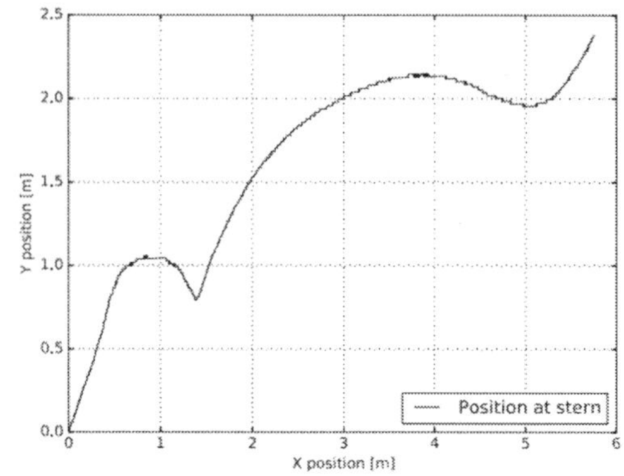

(c)

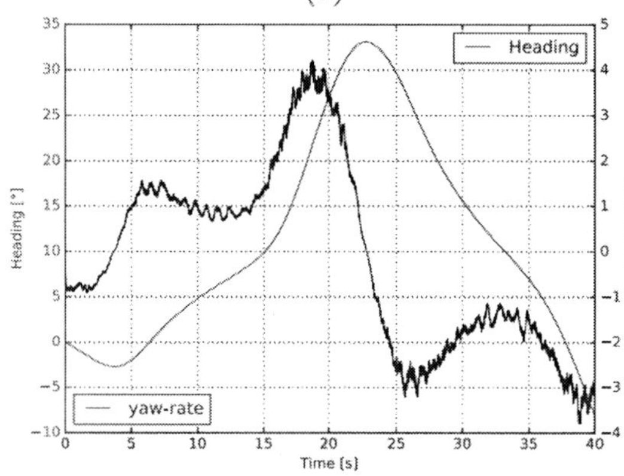

(b)

Figure 6: Sine Angle Test with the bow thruster: (a) θ_i^b oscillating between $= [-180, 180]°$ over 40s, n^b = 2050rpm, θ_i^s = 0°, and n^s = 0rpm, (b) shows the measured headings and yaw-rates, and (c) shows the trajectory of the main GNSS mushroom.

5. Conclusion and Future Work

This paper reviewed the existing standardised test manoeuvres in order to highlight the inland-applicable test manoeuvres as these inland vessels often tend to have non-conventional propulsion systems or an unconventional placement of their actuators. Due to the scarcity of such manoeuvres, this study suggested three additional manoeuvres: (i) the Counter Thrust Test which can judge the turning capabilities of the vessel in spatially restricted areas, (ii) the Sine Angle Test which can uncover a wide range of achievable turning rates, and (iii) the Simultaneous Zigzag Test which can utilise the full capability of the vessel to perform a regular zigzag test. Afterwards, three manoeuvres were conducted with an unmanned autonomous inland cargo vessel: (i) the existing Crabbing Test, (ii) the proposed Counter Thrust Test, and (iii) the suggested Sine Angle Test which was performed with the bow thruster. The successful execution of these manoeuvres provided fruitful insights in the dynamic capabilities of the vessel at hand. For example, the Counter Thrust Test showed how the vessel can locally rotate with little drift. Evidently, the currently performed tests and the list of suggested additional manoeuvres do not exhaust the wide range of possible and useful manoeuvres which are tailored for inland cargo vessels. Consequently, in their future works, the authors aim to define more manoeuvres and their subsequent metrics with an additional focus on the automation potential of the freshly introduced fleet of inland cargo vessels.

Acknowledgments
The Flanders Research Foundation (FWO) funds the doctoral scolarship of Gerben Peeters, 1S12517N. The Cogge was funded by the EFRO-Flanders project Autonoom Varen in de Westhoek.

References
[1] International Maritime Organization 2002 Standards For Ship Manoeuvrability Tech. Rep. 76 International Maritime Organization

[2] International Towing Tank Conference 2017 Full Scale Manoeuvring Trials - ITTC Tech. rep. International Towing Tank Conference

[3] Liu J, Hekkenberg R, Quadvlieg F, Hopman H and Zhao B 2017 *Ocean Eng.* **137** 287–308 ISSN 00298018 URL http://dx.doi.org/10.1016/j.oceaneng.2017.04.008

[4] Liu J, Hekkenberg R, Rotteveel E and Hopman H 2015 *Ocean Eng.* **106** 458–471 ISSN 00298018 URL http://dx.doi.org/10.1016/j.oceaneng.2015.07.021

[5] Liu J 2017 *Impacts of Rudder Configurations on Inland Vessel Manoeuvrability* Phd dissertation TU Delft https://doi.org/10.4233/uuid:bef6c12f-5804-407b-a4c6-f7949541f21c

[6] Central Commission for the Navigation of the Rhine 2018 Reglement onderzoek schepen op de Rijn (in Dutch) Tech. rep. CCNR URL https://wetten.overheid.nl/BWBR0025973/2018-10-07/0/afdrukken+informatie

[7] European Commission 2008 Commission Directive 2008/126/EC amending Directive 2006/87/EC of the European Parliament and of the Council laying down technical requirements for inland waterway vessels Tech. rep. European Commision URL https://eur-lex.europa.eu/LexUriServ/LexUriServ.do?uri=CONSLEG:2006R1881:20100701:EN:PDF

[8] Quadvlieg F, Delefortrie G, Duffy J, Furukawa Y, Guillerm P E, Kim S Y, Simonsen C, Tannuri E and Mao X F 2014 Final Report and Recommendations to the 27th ITTC Tech. rep. International Towing Tank Conference Copenhagen URL https://ittc.info/downloads/proceedings/27th-conference copenhagen-2014/volume-1/ on 19/12/2018

[9] European Commission 2018 Watertruck+ project URL www.watertruckplus.eu

[10] Peeters G, Kotzé M, Afzal M R, Catoor T, Baelen S V, Geenen P, Vanierschot M, Boonen R and Slaets P 2020 *Ocean Engineering* **201** 17 URL https://doi.org/10.1016/j.oceaneng.2020.107056

[11] Peeters G, Catoor T, Afzal M R, Kotze M, Geenen P, Van Baelen S, Vanierschot M, Boonen R and Slaets P 2019 Design and build of a scale model unmanned inland cargo vessel : actuation and control architecture *J. Phys. Conf. Ser.* URL http://dx.doi.org/10.1088/1742-6596/1357/1/012016

[12] CEMT 1992 Resolution No.92/2: On new classification of inland waterways Tech. Rep. 92 European Conference of Ministers of Transport

[13] Septentrio 2017 AsteRx-U User Manual Tech. rep. Septenrio Leuven

[14] SBG 2018 Ekinox Surface Series Tactical Grade MEMS Inertial Sensors Hardware Manual Tech. rep. SBG Systems

IOP Conf. Series: Materials Science and Engineering **929** (2020) 012025 doi:10.1088/1757-899X/929/1/012025

Towards autonomous superyacht navigation

C. Balls.

Government Administration Building, 133 Elgin Avenue, George Town, Grand Cayman, Cayman Islands.

christopher.balls@cishipping.com

Abstract. The paper aims to show possible ways in which increasing automation can be utilised in vessel navigation control especially in relation to Superyachts. It will start with a brief section on how navigation systems have been evolving to where we are now. It will then explore what is currently accepted and what benefits superyachts in particular may gain from further autonomy mainly from the aspect of possible changes to space utilisation. Finally a suggested path of evolution towards autonomy, in order to give enhanced owner and guest experience, will be outlined.

1. Evolution of navigation systems

Navigation is widely agreed to have existed since the early middle ages and in the 15th century the first world explorers utilised very basic instrumentation such as compasses to use landmarks and celestial observation and thus conduct and repeat voyages. By the early 18th century Sextants and chronometers were in use to conduct ever more predictable voyages.

In the United Kingdom regulations regarding navigation originated in the early 19th century and were developed by Trinity House and the Admiralty covering such things as rules for avoiding collision and display of navigation lights. The sinking of Titanic in 1912 was the main catalyst for the first International Convention for the Safety of Life at Sea – SOLAS 1914.[§] This convention has been updated ever since to the latest version; SOLAS 1974, which has been regularly updated to incorporate latest technological developments.

Early in the 20th century there were very basic bridges with lookouts posted on bridge wings and sometimes up a mast in the crows nest. These lookouts were to observe by both sight and hearing as rules existed specifying navigation lights and sound signals such that the lookout could report to the officer of the watch what other vessels in the vicinity were doing as well as observations of shoreline features including lighthouses. Upcoming weather and sea conditions could also usually be observed better from higher observation points.

Along came radar which assisted the lookout function especially in poor visibility. It has long been the understanding that keeping an effective lookout shall be not just by sight but also by hearing with sound signals being a requirement for vessels navigating in restricted visibility. The evolution of fully enclosed bridges effectively meant the lookout coming inside and thus the requirement was made of sound reception facilities on vessels with fully enclosed bridges. Taking of sights is also more difficult as bridges became more enclosed. However electronic navigation systems had meant less need for sights except in event of failure which become increasingly rare as reliability improved.

A helmsman was on duty at all times until autopilots evolved initially just for use deep sea and then as they became more sophisticated, so they became used potentially throughout passages.

Radar systems were developed to allow detection of surface objects in reduced visibility these developed into ARPA (automated radar plotting aid). Echo sounders give details of proximity of the seabed and trends in depth. Paper Charts are being replaced with ECDIS (electronic chart display information systems). Radar and AIS data can be integrated with the ECDIS to provide good situational awareness. Dynamic Position Systems have developed to enable very accurate station keeping of specialist vessels which require special skills especially related to such things as knowing when human intervention into control systems is needed.

2. Current expectations

2.1 General

Vessels bridges are governed by regulation covering layout and visibility primarily in SOLAS V[1], with lookout requirements contained in STCW (International Convention on Standards of Training, Certification and Watchkeeping for Seafarers, 1978) Chapter VIII Regulation VIII/2.

Photo. 1, An example of a yacht bridge with lounge above.

2.2 Bridge layout and visibility

SOLAS V Regulation 22 covers bridge design to ensure good visibility. The specifications are for ships of 55m or more in length built on or after 1 July 1998. Administrations are given discretion for ships of unconventional design which may be a useful factor for development of autonomous vessels.

The length requirement means that many Large Yachts do not strictly fall under these requirements however the REG Yacht Code[2] Section 18.6 requires navigation bridge visibility to comply with SOLAS V/22 with vessels under 55 metres in length required to comply as far as reasonable and practicable. Windows may be inclined from the vertical plane provided that, where necessary, appropriate measures are taken to avoid adverse reflections from within. Where the ship's side cannot be fully visible from the bridge wing, wing station or manoeuvring station, the use of cameras may be specially considered by the Administration giving consideration to image quality, night vision, display screen size and location.

When the ship's bridge is totally enclosed Sound reception facilities are requited under SOLAS V Regulation 19 2.1.8 to enable the officer in charge of the navigational watch to hear sound signals and determine their direction; This feature and the use of cameras for viewing blind spots are examples of technological developments that will help towards less conventional bridges in the future and indeed potentially the eventual moving control off vessels.

2.3 Watchkeeping

STCW Chapter VIII Regulation VIII/2 states that the officer in charge of the navigational watch is, responsible for navigating the ship safely during their periods of duty, when they shall be physically present on the navigating bridge or in a directly associated location such as the chartroom or bridge control room at all times. This is one of the main regulatory hurdles in the move to unmanned bridges.

The regulation also states that an appropriate and effective watch or watches shall be maintained for the purpose of safety at all times, while the ship is at anchor or moored. Also an appropriate and effective watch or watches has to be maintained for the purposes of security.

A dedicated lookout is needed at night in addition to person in charge of navigational watch. One man bridge operation was proposed to IMO several years ago but was rejected at that time.

Casualties resulting from the watchkeepers being asleep has resulted in such developments as comprehensive hours of work and rest regulations under both IMO (International Maritime Organisation) and ILO (International Labour Organisation). Also the fitting of Bridge Navigation Watch Alarm Systems (BNWAS) has become a requirement under SOLAS V, Reg.19.2.2.3. However casualties still occur where the alarm is not switched on.

2.4 Emergency situations

In most superyachts the bridge tends to be the emergency control centre as many safety systems are indicated there and external communication systems are also located there as well. Various developments are occurring related to autonomy in such areas as remotely controlled lifebuoys for assisting with respect to rescuing persons overboard.

As manning levels reduce so more automation of onboard safety systems will be needed as there will be less persons available t assist in such actions as firefighting, casualty handling, and even such areas as emergency towing arrangements.

It will also be necessary to develop suitable means for external assistance to board the vessels but of course this presents problems of the conflict between security and accessibility.

3. Possible benefits of increased automation of navigation systems

With improved visual and audio sensors etc. bridge manning requirements could be reconsidered, there are currently possibilities for safely removing personnel from the bridge, certainly during long quiet ocean passages. Reduced manning of bridge frees up personnel for other duties. Use of optical sensors could easily improve the all round lookout when compared to the current practices with human lookouts. The consequent improvement in overall safety could be seen as beneficial.

It is imperative that the interaction between autonomous systems and humans is fully considered. Even in changes in bridge design over the years new problems have crept in. For example a forward mounted bridge with bridge windows unconventionally orientated was considered a significant contributory factor in a collision investigated by the U.K. Maritime Accident and Incident Investigation Board (MAIB) where relative motion illusion[3] affected the situational awareness of the Pilot. These kinds of factors will be amongst those in need of consideration if significant reliance is to be put on human interpretation of numerous display screens.

Financial benefits are possible from the possibilities of reducing crew numbers and thus the associated operational costs. Further benefit can potentially be gained by changing space utilisation which is investigated in more depth in the following section. Any perceived financial benefit would of course have to consider ongoing maintenance costs as well as costs of any additional verification etc. which would be in addition to the capital cost of installing the necessary systems.

In general it would be expected that the cost of installing systems on vessels will not be directly proportional to the size of vessel with a system on a small vessel probably costing nearly as much as on a larger one although of course larger vessels would probably need more sensors and cabling etc. That said proportionately the benefits of space reutilisation would be greater for smaller vessels as shown in table 1 and section 4.

There are possibilities of developing technology that can be utilised for developing convoy systems for the relocation of superyachts from one cruising area to another provided there is a good level of standardisation in the automated systems being developed, Thus just the lead vessel would be manned with the others following at safe distances behind. Similarly Superyachts that currently tow tenders may benefit from having these tenders operating independently without the need for personnel on board. Thus tender and mother vessel will not be subject to towing forces which not only affect structure but stability as well and of course this would remove the need for monitoring and marinating the tow line. Even of tenders are towed then there are benefits from good monitoring data of their condition being transmitted to the mother vessel.

4. Possible space reallocation within Superyachts by utilizing automated systems for navigation.

4.1 Typical Current space utilisation

Commercial sensitivity considerations make conducting a proper cost benefit analysis difficult, so it was decided to examine the available general arrangement and safety plans of 13 existing superyachts and a comparison of percentage of enclosed space utilized for various functions was made. The names of the yachts chosen have not been included in order to avoid infringing confidentiality, tonnages are approximate Gross Tonnage. From that a very basic design; m.y. Example was developed to demonstrate possible effect on these ratios by rededicating current bridge space and reducing manning (see figure 1). There are of course significant other cost variations especially from an operational cost point of view and in many ways a more autonomous yacht would be best designed as that from scratch, rather than just adapting an existing conventional design.

Figure 1, m.y. Example, rough G.A. plan.

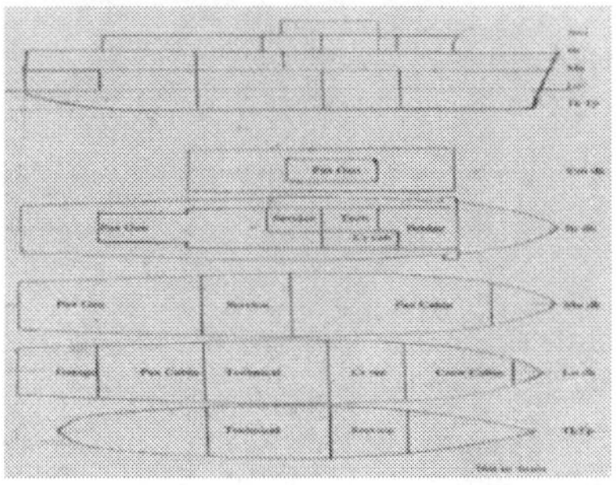

4.1.1.Space Utilisations considered. Guest Cabin, Guest General Use, Crew Cabin, Crew Recreation, Technical spaces, Garage space (space used for storing tenders and "water toys"), Service spaces (Pantries, Bars, Galleys, Storage rooms) Bridge.

4.1.2 Assumptions. Only spaces with reasonable headroom were considered with alleyway and stairway spaces being allocated in accordance with use of the immediately adjacent spaces. Engine Control Rooms are included within Technical Spaces as they represent under 1% of total useable interior space.

4.1.3.. The data. Table 1 below shows the data obtained with a line for each yacht and percentages being given for each space on it. The first 12 entries are certified as Large Yachts and the 13th is a Passenger Yacht. Data used for m.y. Example which uses space allocation similar to the mean values is also shown, as this is used as the basis for the effects comparison in subsequent tables.

Table 1. Comparison of space utilization in various superyachts.

Tonnage	Guest Cabins	Guest General	Crew Cabin	Crew Recreation	Technical Spaces	Garage space	Service spaces	Bridge	Guest / Crew numbers
399	26	23	10	5	12	7	12	5	10/7
496	29	24	15	3	11	0	11	7	12/9
642	25	17	12	3	29	3	7	4	12/13
656	26	21	13	3	19	5	8	5	12/13
769	14	24	16	6	23	0	11	6	12/11
996	30	20	9	6	14	5	13	3	12/16
1102	25	21	11	5	16	2	16	4	12/14
1416	14	25	14	5	29	2	9	2	12/22
1428	16	23	12	2	25	3	15	4	12/14
1592	23	22	12	3	23	6	8	3	12/18
2658	20	28	9	4	19	11	6	3	12/27
2744	15	31	14	2	23	5	8	2	12/30
6862	21	22	13	3	22	10	7	2	31/43
Mean	22	23	12	4	20	5	11	4	
m.y.Example	22	23	14	5	17	5	11	3	12/12

4.2 Effect of reduced human lookout

This section assumes vessel remains much the same but with no requirement for a human lookout at night which could reduce crew numbers to say 11 and reasonable to assume a small reduction in space needed albeit not exactly proportional to total crew space due to consideration of such issues as cabin sharing. This sees guest space increase from 45% to 46% as shown in table 2, m.y. Example 2r.

Table 2. Showing space utilisation by removing human lookout requirement.

	Guest Cabins	Guest General	Crew Cabin	Crew Recreation	Technical Spaces	Garage space	Service spaces	Bridge
m.y. Example	22	23	14	5	17	5	11	3
m.y. Example 2r	22	24	13	5	17	5	11	3

4.3. Effect of rededicating the bridge area usage

This section uses the example vessel and demonstrates possible changes without any reduced manning.

Table 3. Showing space utilization with bridge space rededicated and control remaining on board.

	Guest Cabins	Guest General	Crew Cabin	Crew Recreation	Technical Spaces	Garage space	Service spaces	Bridge
m.y. Example	22	23	14	5	17	5	11	3
m.y. Example 2	22	25	14	5	18	5	11	0
m.y. Example 3	24	23	14	5	18	5	11	0

4.3.1. Bridge used as Guest General Space. Probably an observation lounge which would reasonably practically give possibility of retaining some navigational equipment in the space. It is assumed that part of technical space would be rededicated as the navigational control station. Thus guest space increases from 45% to 47% as shown in table 3, m.y. Example 2. That said yacht bridges already usually have seating areas that can be utilized by selected guests.

4.3.2. Bridge used as Guest Cabin. Unlikely that navigational equipment would be retained in the space. Essentially a similar change of space allocated to guests occurs from 45% to 47% as shown by m.y. Example 3.

4.3.3. Bridge space used for guest space and navigation functions transferred ashore. Essentially this means no increase in technical space needed as shown in table 4, however no manning reduction is considered in this table and guest spaces increase from 45% to 48%. m.y. Example 2a is with bridge space used as general guest space and m.y. Example 3a with use as guest cabins.

Table 4. Showing space utilization with bridge space rededicated and control moved off the vessel.

	Guest Cabins	Guest General	Crew Cabin	Crew Recreation	Technical Spaces	Garage space	Service spaces	Bridge
m.y. Example	22	23	14	5	17	5	11	3
m.y. Example 2a	22	26	14	5	17	5	11	0
m.y. Example 3a	25	23	14	5	17	5	11	0

4.4 Additional effect of potential reduced manning
This section considers changes of space allocation possible by removing say 3 navigational related seafarers as well as removal of the bridge. This shows guest space increasing from 45% to 52% for m.y. Example 2n.

Table 5. Showing space utilization with rededication of bridge and seafarer number reduction.

	Guest Cabins	Guest General	Crew Cabin	Crew Recreation	Technical Spaces	Garage space	Service spaces	Bridge
m.y. Example	22	23	14	5	17	5	11	3
m.y. Example 2n	22	30	10	5	17	5	11	0

4.5 Conclusion with respect to potential financial benefits of space reallocation
The increases in relative guest space allocation is quite modest and at this moment in time it is unlikely that the expense of the necessary sensing and monitoring systems needed would be justified. Unfortunately, there are large variations in construction costs of superyachts with numerous factors and commercial confidentiality does make conduct of full cost benefit analysis difficult.

Of course some owners do just want the latest technology even at a large cost so there are possibilities there.

There main benefits will potentially be for yachts close to the various sizes where regulatory compliance changes, such as 400GT. for MARPOL certification requirements, for SOLAS 500GT is a major changing point. Here if a vessel just under the relevant criteria can have an actually larger guest space than one just over then the benefits of automation may be larger, in other words it becomes a situation where a smaller yacht could have the same guest space as a larger one.

In the examples of bridges effectively being done away there is another drawback which is that it means there is less space being used by crew members to meet up.

It would be reasonable to assume a construction cost of say $25 million for a vessel like m/y example so a 1% change in useable space could be considered worth $250,000 Similarly if a crew member position cost $50000 per year to fill over 20 years that would be a $1million saving by not having to carry such a person. Although very rough these figures give some idea of the additional costs of fitting automated systems that could be financially justified. However very few superyachts have high utilization rates, so not that much time is spent at sea, thus navigational watchkeeping is not required all the time. However, security considerations mean that the technology that is utilized for reducing lookout requirements at sea can similarly be utilized for changing the way security watches are conducted in port, many vessels already utilize security cameras to assist with this.

IOP Conf. Series: Materials Science and Engineering **929** (2020) 012025 doi:10.1088/1757-899X/929/1/012025

Photo. 2, A typical Superyacht Bridge.

Photo. 3, A typical Superyacht Control Room.

5. Possible path of evolution for Superyachts by utilizing automated systems for navigation

5.1 Reduced human lookout requirements

The first stage could be to utilize the greatly improved visual and audio sensors and systems to remove the need for a separate dedicated lookout stationed on or in immediate contact with the bridge, thus enabling single watchkeeper operations in many circumstances.

The overall bridge design will still be expected to conform to requirements for visibility etc. Instead of a Human lookout this function would be undertaken by optical and audio sensors which would be programmed to alarm at pre set criteria and provide similar or enhanced information to that would be given by a typical human lookout.

The main advantage of this would be that the automated lookout system should not get "tired" and fall asleep. Also the automated system should not be afraid to report things, as sometimes occurs with human lookouts where there may be cultural and other clashes between the officer of the watch or Master.

The main drawback with this could be that excessive alarms so the appropriate programming of the sensors is important and elements of "machine learning" need to be carefully considered.

A further development of this would be along the lines of the current thinking on a B0 (B0 – a conditionally and periodically unmanned bridge) type scenario which is essentially for a tradition wheelhouse type arrangement but with it being periodically unattended in much the same way as engine control rooms often are. Here the main savings would be by reduced manning requirements. Without the need for continuous manning savings would very much depend upon operational areas and the residual demand for attendance on the bridge.

5.2 Partial rededication of bridge space

The next stage could be to allow for operation from an operating station still in the traditional wheelhouse location but where that space is used as a guest lounge and the navigational equipment has a smaller footprint and is only used during navigation or possibly just for maneuvering. The cost benefit is also difficult to determine here due to the numerous variables involved.

5.3 Effective Relocation of bridge space

This stage would see the moving away from the traditional bridge location and maybe sharing with engine control room in a location where space cannot really be utilized by guests or is certainly less attractive for guest use. There would still be a remote operation capability from a position suitable for direct observation of operations by a human.

It should be noted that the views contained in this paper are those of the author and not necessarily the organisation for whom he works.

References

[1] Various references have been made to publicly available International Maritime Organization documents throughout the paper for example; STCW including 2010 Manila Amendments, 2017 Edition - SOLAS Consolidated Edition, 2020 - MARPOL, Consolidated Edition, 2017 - Current full listing at; http://www.imo.org/en/Publications/Pages/CurrentPublications.aspx

[2] Red Ensign Group Yacht Code January 2019 Edition; https://www.redensigngroup.org/media/1094/reg-yacht-code-january-2019-edition-part-a.pdf

[3] MAIB investigation report 3-2017: City of Rotterdam/Primula Seaways; https://assets.publishing.service.gov.uk/media/58984f60ed915d06e1000025/MAIBInvReport3-2017.pdf

IOP Conf. Series: Materials Science and Engineering **929** (2020) 012026 doi:10.1088/1757-899X/929/1/012026

Standardized navigational data for situational awareness during simultaneous maritime operations

M. Hagaseth[1], S. P. Berge[1]

[1]SINTEF Ocean, Marinteknisk senter, Otto Nielsens vei 10, 7052 Trondheim, Norway

Marianne.Hagaseth@sintef.no

Abstract. The objective of this paper is to describe a protocol for exchange of navigational data between ships operating in close proximity. The protocol supports the exchange of data needed for relative positioning and includes geometry data of the objects involved in the operation in addition to inertial and positional data. The protocol takes into account the variable and limited communication capabilities that may be available for maritime operations. The paper describes one use case where two ships perform a simultaneous operation and another use case where a ship broadcasts its intended route. A S-100 Application Schema covering the required data is presented. Examples of the message exchange and bandwidth requirements are given. The resulting protocol will be useful for autonomous systems and ships that need standardized exchange of data to support situational awareness during their passage. The paper is written as part of the H2H project [1] which is developing an open system architecture to ensure interoperability between different vendors of autonomous systems when it comes to the data exchange related to position and geometry data.

1. Introduction

This paper describes the S-100 Application schema [2] for the H2H Framework, Figure 1. The H2H Framework is an open system architecture providing standardized information exchange between ships to help and assist both mariners and lately unmanned autonomous vessels to take the correct navigational decision when navigating in close proximity of other objects. The H2H Framework consists of two parts, the information model and the service description. The information model (domain model) covers the minimum set of data needed to compute the desired information for safe autonomous navigation including relative distances, velocities and the uncertainty zone. The service description includes services to establish the connection between two ships, to request data from a ship and to receive data that has been requested. The services are grouped as follows:

- Services to check the availability of the H2H system, its services and sensor data, and to establish connections to H2H Engines and Applications.
- Services to request data from a H2H Engine. This can be both sensor data but also calculated values as for instance uncertainty zones, relative distance and speed, in addition to object geometry.
- Services to receive data that has been requested.

In this paper, we will focus on the information model of the H2H Framework. The framework describes the information needed internally in the H2H Engine in addition to information that needs to be exchanged between two H2H engines on board different objects/ships (own and target) (see red arrows in Figure 1) and also the information exchanged between a H2H Application and the H2H Engine onboard the same object/ship (see red arrows in Figure 1). E2A means Engine to Application, while E2E stands for Engine to Engine. The operation of each H2H Engine, both on own object/ship and on target object/ship are transparent, meaning that the internal calculations done in the engines will not be part of the framework. Thus, the framework allows many H2H Applications and H2H Engines to work together in a total H2H System.

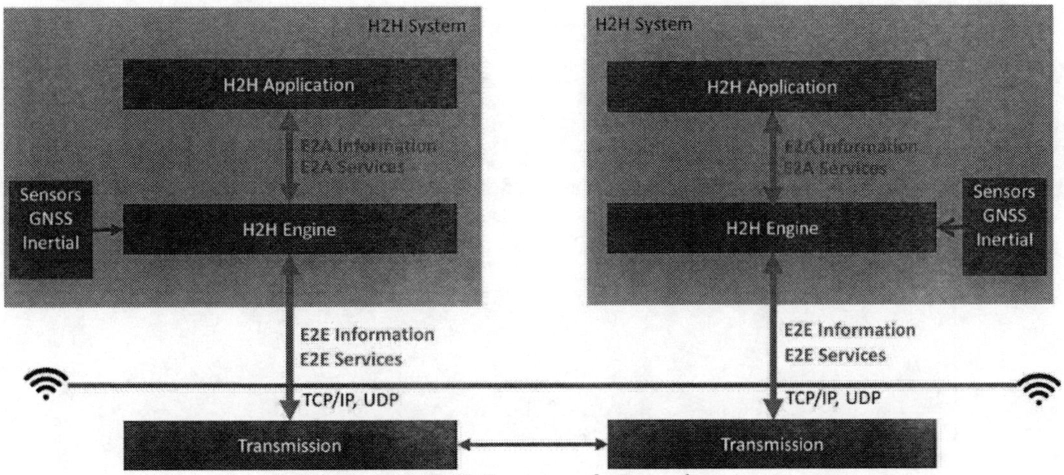

Figure 1. H2H Framework Overview.

2. Simultaneous Operation (SimOps) Use Case

Figure 2 gives an outline of a simultaneous operation where the own ship is leading an operation involving another (following ship). This use case covers in principle any scenario where two or more ships perform a simultaneous operation in close proximity of each other, and where the goal is to operate as close as necessary while still maintaining safe distance.

The ship that is leading the operation starts the operation by broadcasting a message to possible participants. The own ship also defines a safety zone which shall not be entered by other vessels, unless approved by the leading ship. They also define an escape zone where the following ship should enter if the leading vessel need to abort the side-by-side-operation.

Figure 3 gives an overview of the processes, sub-processes and events that may be involved in a simultaneous operation. The main processes are Approach, Operation, and Finish. During the approach, the initial information is exchanged among the ships inside the Area of Interest (AOI), and end-to-end communications are set up. Also, alternative communication means including their priorities are agreed among the participants. A simple geometry model including the length and breadth of the ship is exchanged at the start of the approach. Later, before the operation starts, more detailed geometry models are exchanged. The level of detail of the models are decided during the exchange of the models. Also, all ships participating in the operation must exchange information about the available sensors to others, including the latency of the data and the position of the sensor relative to the ship's physical origin. Also, the requested update rates on the data are agreed among the ships.

During the operation, data from the sensors are exchanged among the participating ships, including position data, raw GNSS data, inertial data and range sensor data. Based on this data from its own sensors and also similar data received from the other participants, a ship can calculate both its own uncertainty zone (UZ) and the other ship's uncertainty zone (UZ). This uncertainty zone represents the uncertainty in the hull's position, specified with a certain probability, often 95%, Figure 4. This is thus a well-suited tool for estimation of the relative distance, speed and bearing between two ships which operates in close proximity, see the red arrows in Figure 4.

3. S-100 Application Schema for H2H

Figure 6 and Figure 7 show the S-100 Application Schema for the information that is needed to compute the position of each ship in a simultaneous operation with a certain probability. It also includes classes to describe the intended route as a list of waypoints and prediction zones that may be sent to others to be used during autonomous navigation. The rest of this section gives a brief description of the most important feature types and information types in the H2H S-100 Application Schema.

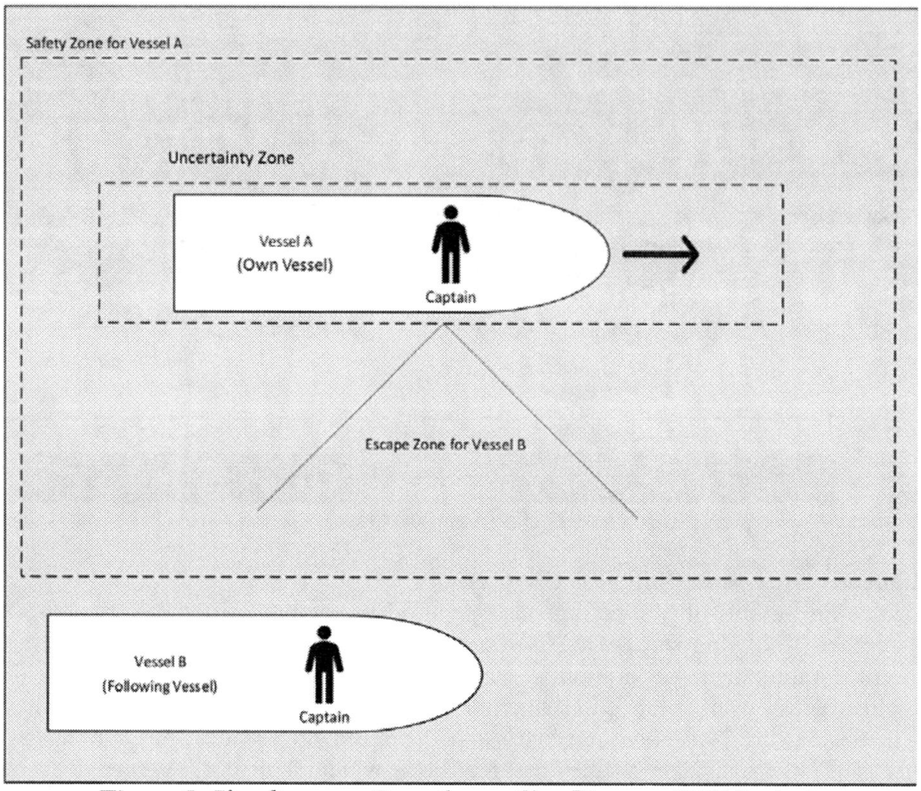

Figure 2. Simultaneous Operation outline [Source: H2H Project, https://www.sintef.no/projectweb/hull-to-hull/].

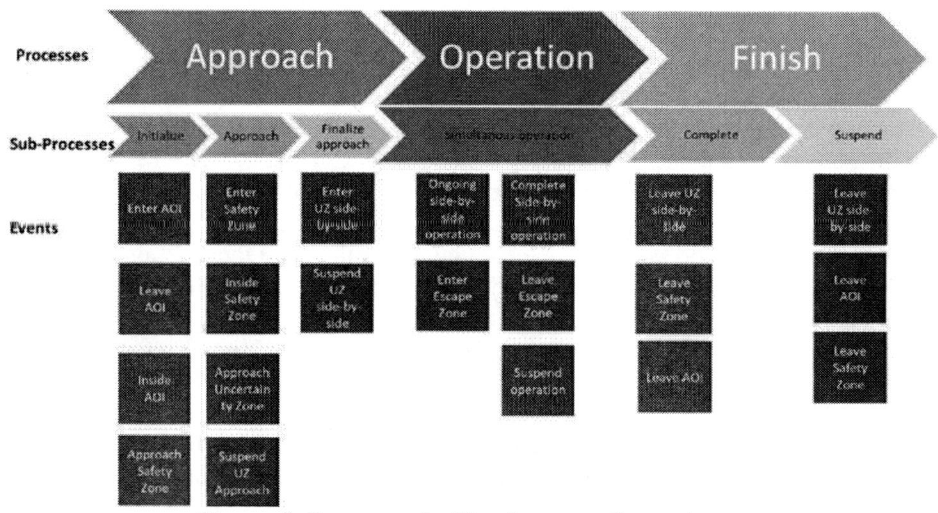

Figure 3. Processes in Simultaneous Operations.

3.1. *ObjectPoint*

This is the class to represent a point related to the physical object/ship. The fromAft, fromCenter and fromKeel coordinates specifies the position relative to the ship's physical origin in three dimensions. The accuracy is represented by a number for each of the dimensions. *ObjectPoint* can represent the position of a real sensor or a virtual sensor. Also, a set of instances of *ObjectPoint* is used represent the object geometry (*ObjectGeometry*). The *ObjectPoint* class is used to define a vector between points on two different objects, and further to do the calculations of relative distance, velocity and bearing between these two points. The *ObjectPoint* class is also used to describe an uncertainty zone or an operational zone.

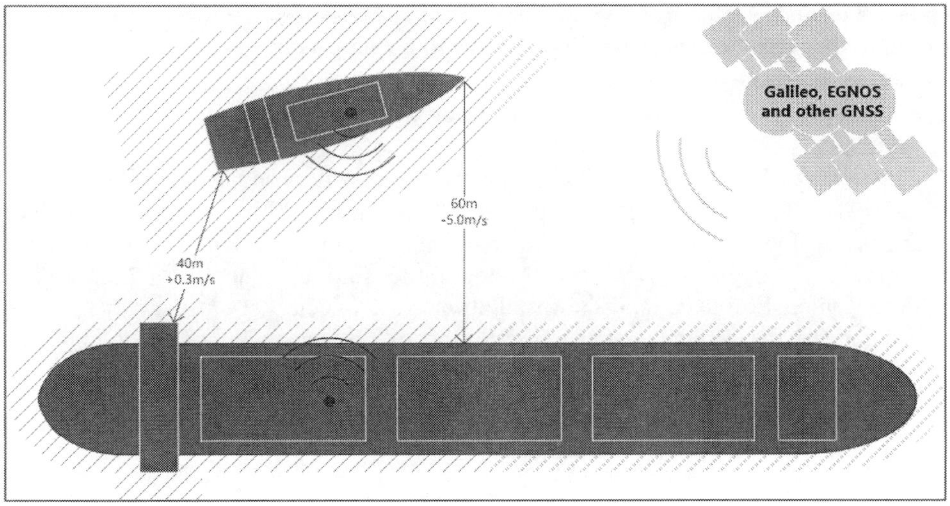

Figure 4. Uncertainty Zone (in red) [Source: Kongsberg Seatex].

3.2. *ObjectGeometry*
This is the list of points that represents the detailed geometry of a ship's hull, or more generally, it describes the geometry of an object. The geometry has a certain accuracy. An *ObjectGeometry* consists of a list of *ObjectPoints*.

3.3. *RequestedObjectGeometry*
For each operation, a ship/object can request an object geometry from another ship/object with certain accuracy and size, dependent on the communication means available and dependent on the calculations that is to be done.

3.4. *ShipDynamics*
This is the position, speed, course and heading for a selected point on the ship, among other motion information related to a ship. Also, the intention of the voyage or leg can be given as a textual description. A more formal way of describing the intention, is to add a set of waypoints as an intended route.

3.5. *Communication*
This class contains information about the communication technologies that are available for this ship/object to be used in the calculations. Each technology is listed with their bandwidth, latency and the type of communication.

3.6. *Ship*
This class contains both the static and dynamic information about a ship, including the simple geometry model (length and breadth). It also contains a description of the position of the ship's physical origin, described as the distance from bow and aft, starboard and port side, and also in the z direction.

3.7. *ObjectData*

This class contains information that is common for both fixed and moveable objects, that is, for both ships, quays, bouys, locks etc.

3.8. Sensor
This class contains the information that is needed to describe a sensor onboard a ship, including the type, update rate, and whether it is available or not. It also contains the lever arm for the sensor, that is, the position of the sensor onboard the ship relative to the ship's physical origin.

- *Name*: The name of the sensor
- *IsAvailable:* This is a Boolean indicating whether the sensor is available for use by the calculations.
- *UpdateRate:* This is the update rate that the sensor will produce data at.
- *LeverArm*: This is the location of the sensor onboard the ship or on the fixed object, relative to the ship's/object's origin. This is shown as the "Sensor A modelled location" (a green circle) in Figure 5. The physical location onboard the ship or object is represented by FromAft, FromCenter, FromKeel and the installation error (AccuracyX, AccuracyY, AccuracyZ). This accuracy is the accuracy of the installation of the sensor onboard the ship or on the fixed object. This corresponds to the distance shown as "Installation error" in Figure 5.
- *SensorType*: This is the type of the sensor.

Each sensor produces a set of *SensorData*.

3.9. SensorData
The *SensorData* class and its specializations contain the actual measurements from the sensors. The *SensorData* contains the properties that are common for all different types of sensor data. This includes the key identifier of the sensor data and the Timestamp for the sensor data. The concepts linked to *SensorData* includes *PositionData*, *GNSSRawData*[1], *InertialSensorData*, *AngleSensorData* and *RangeSensorData*. This means that all these concepts inherit the SensorDataId and Timestamp from the SensorData concept. Each instance of the SensorData concept represents data generated by a certain sensor. The *Accuracy* property indicates the accuracy of the data produced by the sensor, that is, the uncertainty shown as "Sensor Error" in Figure 5.

3.10. VirtualSensor
This class represents a set of calculated values in a position related to the ship's/object's physical origin. This means that this concept represents a virtual sensor, since no physical sensor is located at this position, but instead, a set of values for this position is calculated based on the sensor values otherwise available for the ship/object. The virtual sensor data is used to calculate relative distance, velocity and bearing between two points for two different ships/objects.

3.11. RelativeDistanceAndVelocity
The *RelativeDistanceAndVelocity* class contains information about relative distance and velocity from a given start point to a given end point. The start and end points normally refer to points on own and target geometry objects, respectively. It also contains the bearing from start to end reference point. This is shown in Figure 4 as the red arrows.

3.12. Uncertainty Zone
The *UncertaintyZone* class contains information about an uncertainty zone that has been calculated by a H2H Engine. Since both sensor installation accuracy, sensor data accuracy and object geometry accuracy are used in the calculations of the uncertainty zone, the actual uncertainty zone geometry is more complex than just the object geometry adjusted with a specific sigma value. Therefore, a

[1] Data in RTCM format

separate property *UncertaintyZone* is needed to contain the actual geometry of the uncertainty zone. One object can have several uncertainty zones, both because several calculations based on different sensors and sensor data will result in different uncertainty zones, but also because one calculation of an uncertainty zone can be viewed as several slices of the uncertainty zone geometry. Examples of these are that several 2D slices of the 3D uncertainty zone can be defined, and thus result in several uncertainty zones. Examples are a 2D slice at the water level or at the level where the geometry has the largest areal or based on some other definition.

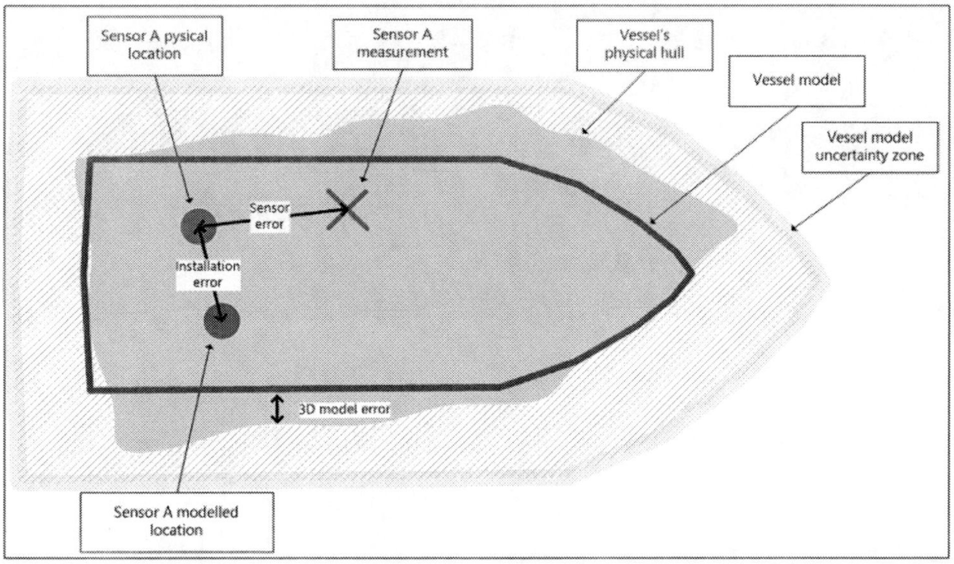

Figure 5. Various Sensor Uncertainties [Source: H2H Project].

3.13. Operation
This class contains the properties to describe an operation, which can be SimOps, Docking, Lock or Sailing. An operation can involve several objects, while each object can participate in several operations.

3.14. OperationalZone
This class contains properties to describe operational zones and is closely linked to the applications. Notice that an operational zone can be defined independent of sensor values and object geometries, for instance in the case of escape zones. However, in other cases, it can be related to sensor values and geometries, for instance when covering regions related to trajectories. Area of interest and safety zone are two types of operational zones needed in the simultaneous operation case.

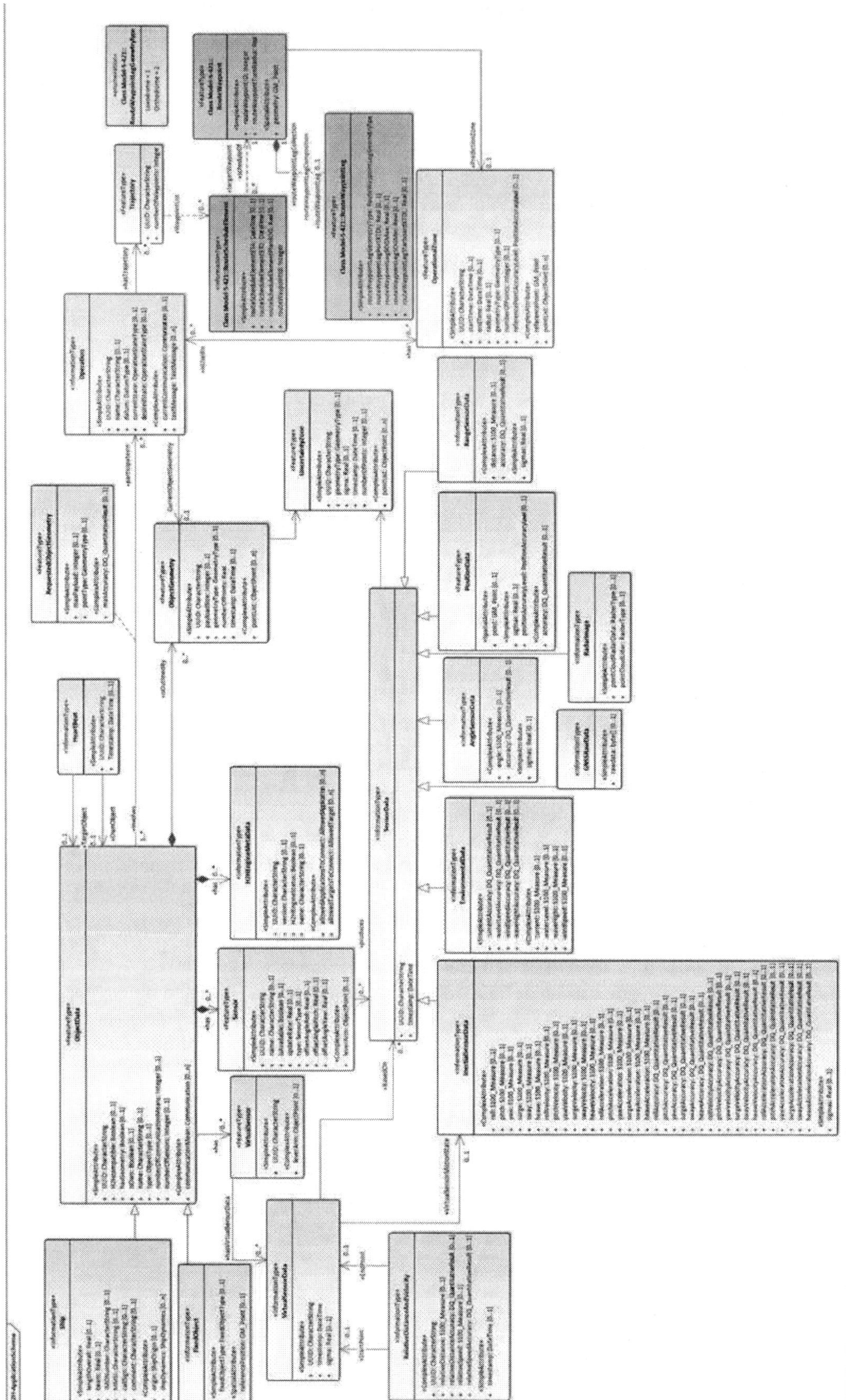

Figure 6. S-100 Application Schema for H2H information.

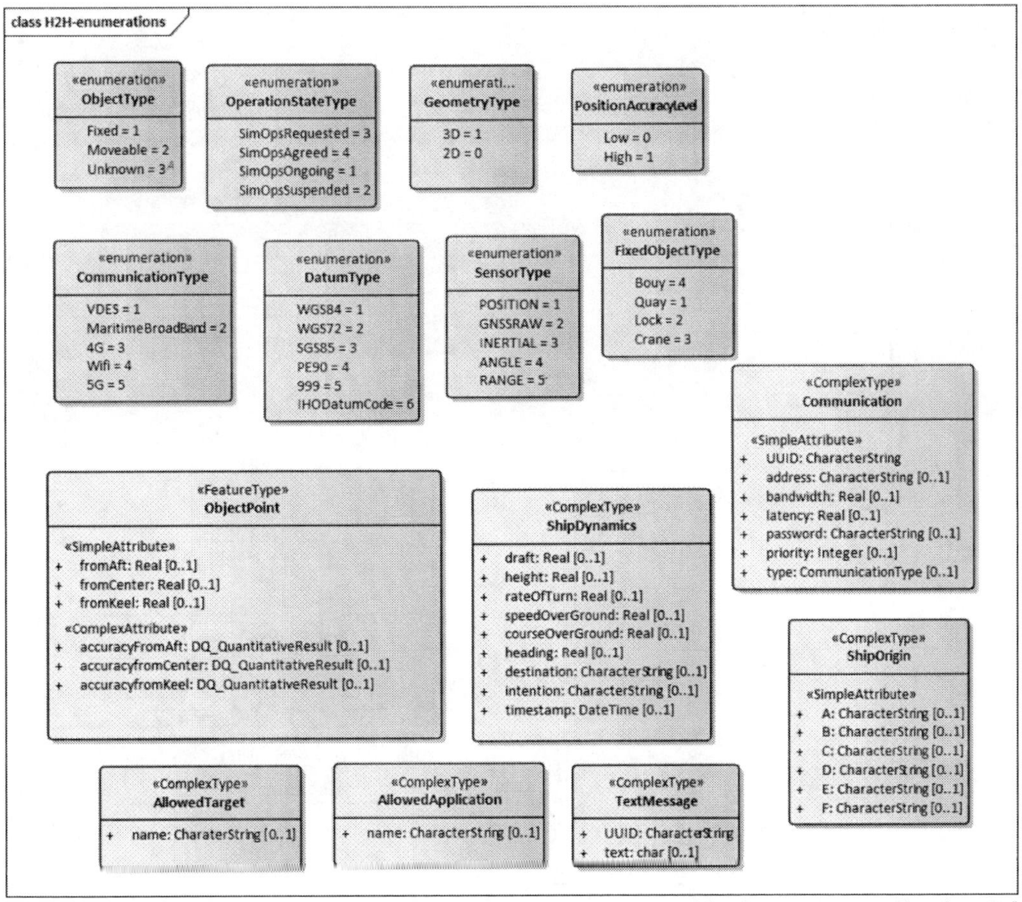

Figure 7. H2H enumerations, complex types and feature types used in the S-100 Application Schema.

3.15. Heartbeat
This class contains information saying that the H2H Engine is alive.

3.16. H2HEngineMetaData
This class contains properties needed to keep track of the status of a H2H Engine. Each object can have more than one H2H Engine.

4. Message Sequencing and VDES
Figure 8 shows a sequence diagram for the SimOps use case. Figure 9 shows the S-100 message exchange model representing the response geometry message. The content of each of message type is defined based on the S-100 Application schema presented in this paper. An example message is shown in Table 1 for the ResponseGeometryMessage. This message contains a list of points that defines the geometry of an object in 2D or 3D including the accuracy of each of the points. The level of detail of the geometry model is decided by the number of points in the point list. When it comes to the VDES coding of the geometry, 10 points in the object geometry will give a message size of 1135 bits which

fits into one VDES packet having 3 slots. Larger geometries can be sent in a message spanning more VDES packets[2].

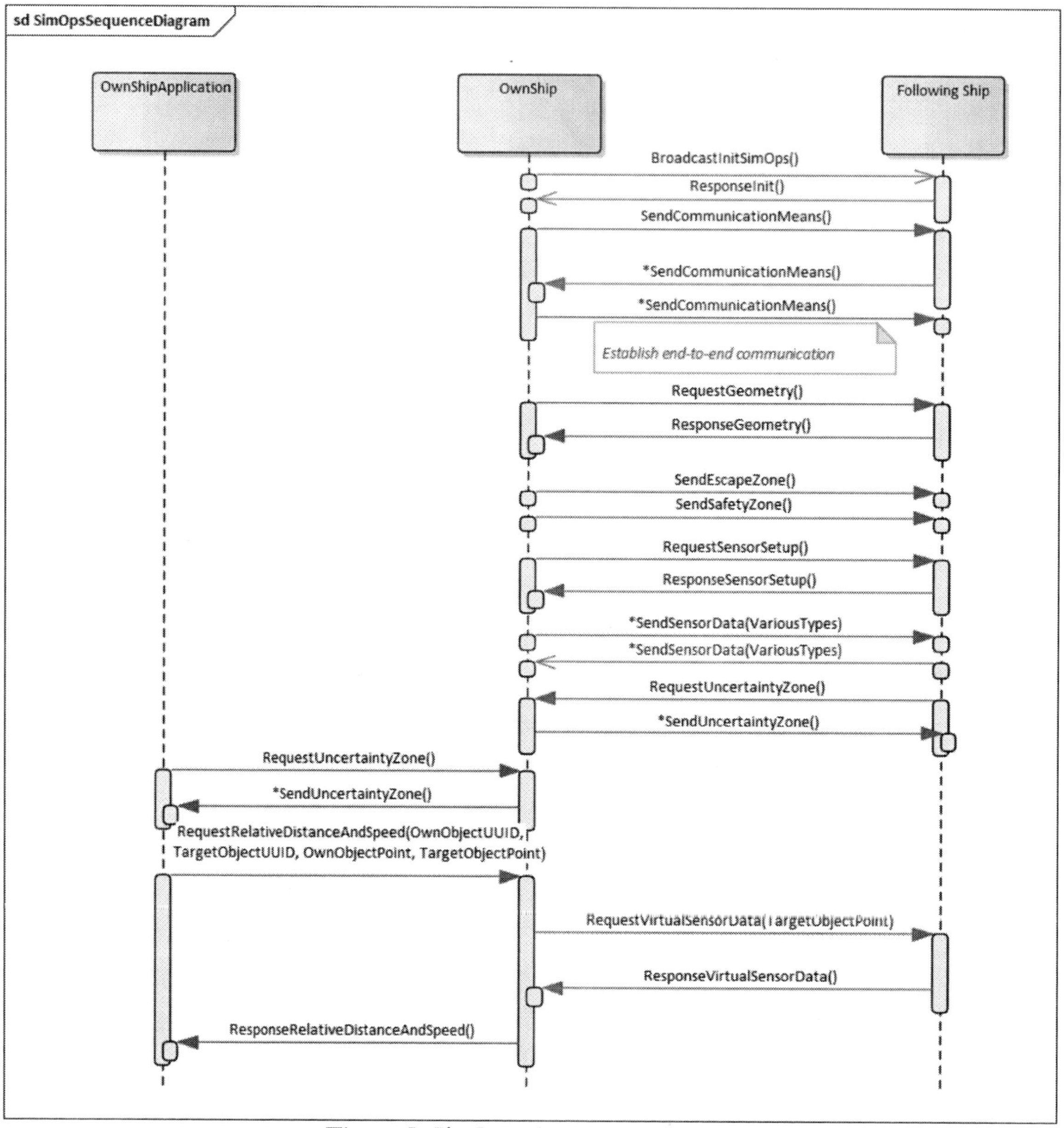

Figure 8. SimOps Message Sequences.

5. Intended Route

An important functionality for autonomous ships that is related to the SimOps case is the possibility to broadcast the intended route to others. This is covered in the H2H Framework by amending the model with classes from the S-421 Route exchange product specification and selecting those parameters needed.

---

[2] VDES is a newly specified digital communication technology utilizing the VHF band to provide free ship-ship communication with more bandwidth (ca 300 kbps for each channel (transmit and receive) in full duplex mode).

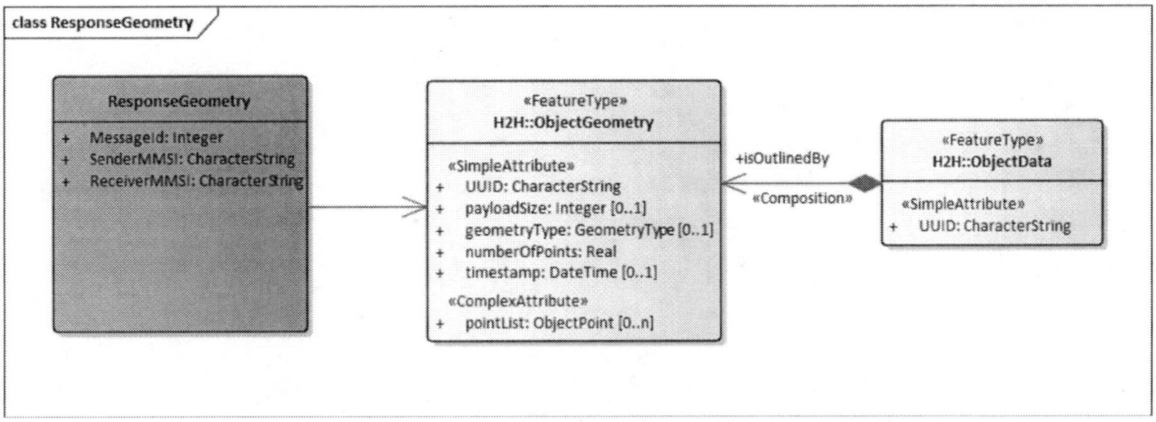

Figure 9. S-100 Exchange Message for sending geometry data.

Table 1. Message example: ResponseGeometryMessage.

Parameter	Number of bit	Description
MessageId	6	**ResponseGeometryMessage.** The message type is used to send a response on an object geometry request.
SenderMMSI	30	The sender ship MMSI
ReceiverMMSI	30	The receiver ship MMSI
ObjectId	128	The object id (UUID) for the sender MMSI
ObjectGeometryUUID	128	UUID of the sender's object geometry
payloadSize	16	Size of the geometry in kBytes
geometryType	1	2D or 3D points?
NumberOfPoints	16	Number of points in the PointList
timestamp	6	Timestamp of the geometry data
fromAft	16	**pointList:** Length in meters, including 2 decimals, from the ship's origin or from another reference point in the X direction.
fromCenter	16	**pointList:** Length in meters, including 2 decimals, from the ship's origin or from another reference point in the Y direction.
fromKeel	16	**pointList:** Length in meters, including 2 decimals, from the ship's origin or from another reference point in the Z direction.
AccuracyFromAft	12	**pointList:** Accuracy in X direction, up to 40.96 meter
AccuracyFromCenter	12	**pointList:** Accuracy in Y direction, up to 40.96 meter
AccuracyFromKeel	12	**pointList:** Accuracy in Z direction, up to 40.96 meter
TOTAL	295 + 84x NumberOfPoints bits	The geometry size is dependent of the number of points: • A simple "shoe box" representation of the ship geometry with 4 points gives 697 bits => 88 bytes • A simple representation with 5 points gives 715 bits • A geometry with 100 points gives 8695 bits • With a VDES addressed message with 3 slots having a total of 1272 bit payload, **11 points** can be transmitted in one VDES

Parameter	Number of bit	Description
		message coded as a ResponseGeometry message.

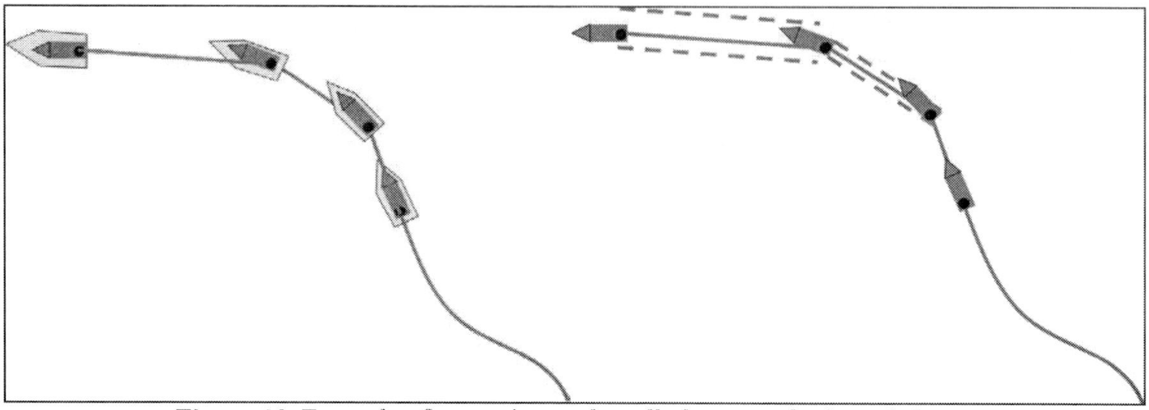

Figure 10. Example of waypoints and prediction zone for intended route.

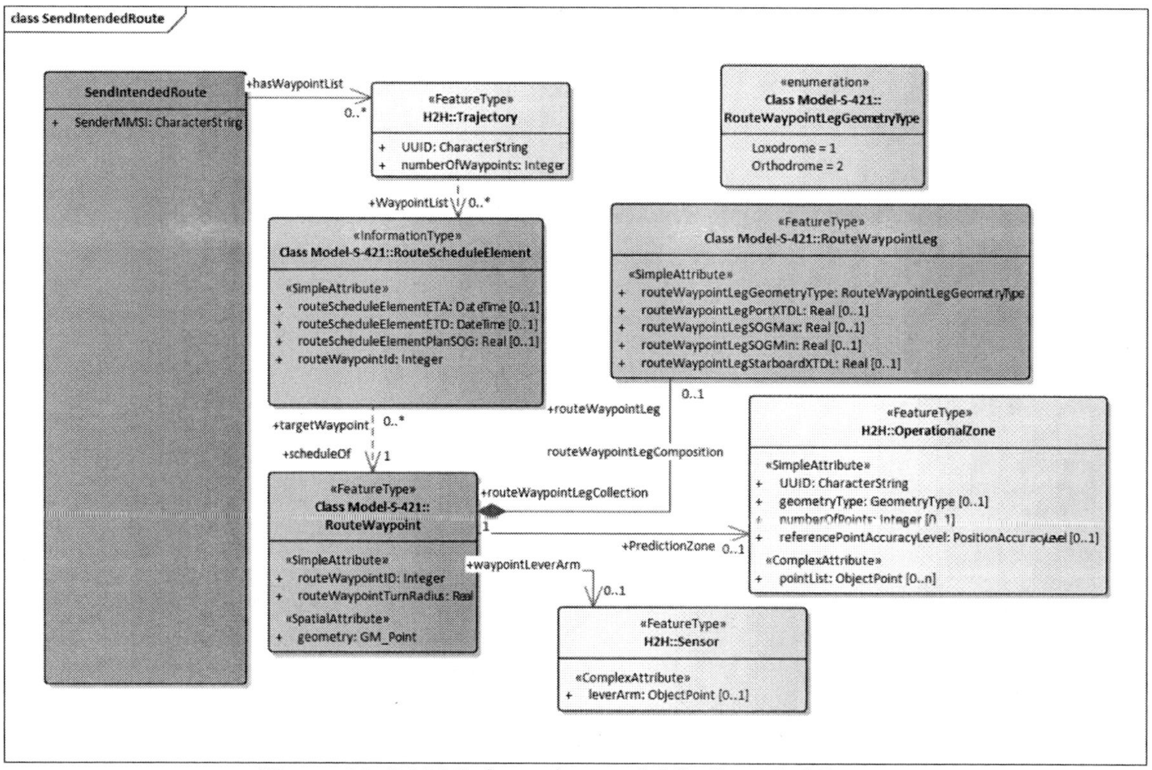

Figure 11. S-100 Information model for Intended Route.

The route can be specified as a set of waypoints (black dots) defined as Latitude and Longitude and possibly with the addition of the accuracy of the Latitude and Longitude values, left part of Figure 10. In addition, a prediction zone (operational zone) that represents the uncertainty of the position of the ship's geometry can be specified for each waypoint. This is an optional information for the route intension. In addition, the uncertainty of the leg between each waypoint can be specified, right part of Figure 10. This includes a description of the legs between the waypoints including the size of the expected deviation from the intended course both to starboard and port side.

The S-100 information model for the intended route is shown in Figure 11. By proposing VDES messages for the exchange of intended routes and analysing this with respect to the bandwidth requirement, we find that a route consisting of 10 waypoints will take about 9.7 kbit. This can be reduced to about 3.5 kbit if the prediction zones for each waypoint are omitted, and only the waypoint and legs with uncertainties are transferred.

6. Conclusions

The proposal for a S-100 Application Schema covering position data, sensor data, geometry information, uncertainty zones and operational zones are defined. Based on this information, services to calculate the relative distance, speed and bearing between two points on two ships in close proximity can be defined. The message exchange needed to support calculation of uncertainty zones, relative distances, speeds and bearings can be specified using the same coding as for AIS data according to Recommendation ITU-R M.1371-2 [3]. When using VDES as the carrier, the message exchange will be free for the ships, and the bandwidth will be sufficient for simple geometry models and sensor data. When communication means with higher bandwidth are available, more detailed geometry models and more sensor data can be exchanged according to this Application Schema. Calculations of the total bandwidth requirement for the message exchange has been done, for different situations:

Two ships participate in the SimOps, one own ship, and one following ship, each ship has 5 sensors: For a geometry with 5 points for each ship, ca 84 kbit of data must be exchanged. With each ship geometry consisting of 100 points, 103 kbit must be exchanged, while 1000 points gives about 280 kbit of data for the initial setup. With a maximum capacity of 307.2 kb/s for VDES, this shows that the initialization should be feasible also in cases where the communication conditions are significantly worse that in the optimal case.

Six ships participate in the SimOps, one own ship, and five following ships, each ship has 5 sensors: In this case, an object geometry with 5 points gives a data exchange of 288 kbits, 100 points gives 580 kbits, while 1000 points gives 3344 kbits. This is still possible given the typical bandwidth of VDES communications.

In addition of proposing a new S-100 standard, we have made use of the S-421 route exchange standard to handle intended routes for autonomous ships. We have also recognize that **a common S-100 feature catalogue for the operational domain** is needed: Although a feature catalogue with concepts exists, no common set of definitions that can be reused, exists. For instance, the properties describing various ship particulars as for instance ship ids (MMSI, call sign) or ship size (length, breadth) are found in several product specifications. It is not clear which of these definitions should be reused when specifying a new application schema. This means that a common reference model consisting of a minimum common set of attributes, features, information types and associations should be defined and included to the S-100 feature catalogue.

Acknowledgement

This work has been done as part of the H2H project which has received funding from the European GNSS Agency under the European Union's Horizon 2020 research and innovation programme grant agreement No 775998.

References

[1] https://www.sintef.no/projectweb/hull-to-hull/
[2] http://s100.iho.int/S100/home/s100-introduction
[3] ITU, Rec. ITU-R M.1371-2, Technical characteristics for a universal shipborne automatic identification system using time division multiple access in the VHF maritime mobile band ANNEX 1

IOP Conf. Series: Materials Science and Engineering **929** (2020) 012027 doi:10.1088/1757-899X/929/1/012027

Voyage performance evaluation based on a digital twin model

M Liu[1], Q Zhou[1], X Wang[2], C Yu[3], and M Kang[3]
[1]Seastel Marine System (Canada) Inc., 1250 René-Lévesque Blvd W Suite 2200, Montreal, QC H3B 4W8, Canada
[2]Newcastle University in Singapore, SIT Building @ Ngee Ann Polytechnic, 537 Clementi Road #06-01, 599493, Singapore
[3]Seastel Marine System (Shanghai) Co. Ltd., Room 7-102, 555 Dongchuan Road, Shanghai, 200241, China

Abstract. Using digital twins in voyage performance evaluation is becoming critical for ocean vessels to reduce GHG emissions. A novel GBM approach is proposed in this paper to establish a digital twin model for voyage performance prediction. The weather hindcast data are introduced to enrich noon reports (NR) and automatic identification system (AIS) datasets, which are split into training and validation sets to develop GBM. The NR and AIS datasets collected from a 57000DWT bulk carrier are used to demonstrate the fidelity and capability of the proposed GBM. The voyage performance prediction from the GBM shows better accuracy than those from pure WBM or pure BBMs. An arrival time forecast and a weather routing showcase are also presented to demonstrate the application effects of GBM. The proposed GBM provides a satisfying prediction of ship speed and fuel consumption without mandatory sensor-collected data, thus applicable for a varity of vessels. In those cases where more sensors are available onboard, the proposed approach can incorporate sensor data to improve the model accuracy further.

1. Introduction

In recent years, voyage performance evaluation and operational optimization based on digital twin models are becoming increasingly critical to reduce Greenhouse Gas (GHG) emissions in shipping industry. There are three mainstream modeling techniques to build the digital twins, namely white, black, and grey-box models[1].

A white-box model (WBM) expresses the system mechanism involving physical principles and numerical/experimental results. Many attempts have been made to improve the model accuracy by modifying empirical formulas or simulation methods. Moreno-Gutiérrez compared four existing methods for calculating energy consumption and emissions[2]. The advantages of those methods were discussed and combined into a new approach, which demonstrated better accuracy in predicting the energy consumption and emissions for a particular scenario. Nielson developed several WBMs to simulate the required power and fuel consumption of an academic hull in the time domain[3]. Comparisons were conducted to investigate the effects of operational and environmental factors on voyage performance.

A black-box model (BBM) describes input-output relation based on measured data. Bialystocki used a statistical approach to analyze a dataset from noon reports (NR)[4]. The impact of the ship's draft, displacement, hull roughness, and weather condition on ship's fuel consumption was investigated. A simple prediction model with acceptable accuracy was proposed. Safaei established a prediction model of fuel consumption based on multiple linear regression[5]. The NR and AIS data of four VLCCs were used in the model establishment. Gkerekos compared the performance of different machine learning models on predicting ship fuel consumption[6]. The various methods such as support vector machines, random forest regressors, extra trees regressors, artificial neural networks, and ensemble methods were analysed in the study with the data from NR and Automated Data Logging & Monotoring (ADLM) system. The comparison results revealed that th artificial neural networks (ANN) model is preferred and high-quality training data from ADLM could significantly improve model performance. Pedersen and Larsen established a nonlinear neural network model to predict ship propulsion performance[7]. The sensor dataset collected from a tanker with high data frequency contained ship speed, relative wind speed, and the temperature of air and seawater. Liang compared the performance between the traditional machine learning method and deep learning method on predicting ship propulsion power[8]. AIS data, ship propulsion power measurements, and weather data were employed in the study. Farag developed a ship performance model based on ANN and multi-regression techniques to predict the main engine power and fuel consumption in varying operating conditions[9]. The sensor datasets collected from a VLCC with high data frequency were deployed with AIS data for model establishment. Coraddu created a ship digital twin to estimate the speed loss caused by fouling[10]. A data-driven approach based on deep extreme learning machines was proposed with sensor datasets from two tankers.

A grey-box model (GBM) combines the physical principles with data-driven modeling results. Leifsson established GBMs with both serial and parallel approaches to estimate the speed and fuel consumption of a container vessel[11]. The results showed that GBM could significantly improve the prediction of fuel consumption compared with pure WBM and pure BBM. The real sensor data is necessary for evaluating voyage performance and optimizing ship operations.

Since high-quality sensor data are not complete in most cases, the commonly recorded noon reports (NR) and automatic identification system (AIS) data are valuable for digital twin modelling. However, NR data is not regarded as high-quality data due to the possible human errors and low frequency of collection. Some attempts have been made to overcome these problems. Safaei employed the statistical methods to correct and enrich the NR data of four VLCC[12]. Th raw NR data were refined, and the newly generated data were statistically in harmony with the original data. Based on the refined NR and AIS data, the fuel consumption prediction model was successfully established[5].

In this paper, a novel GBM approach is proposed to establish a digital twin model for voyage performance evaluation. The weather hindcast data are introduced to enrich the NR and AIS data, which are split into training and validation sets to develop GBM. An arrival time forecast and a weather routing showcase of a 57000DWT bulk carrier are presented to demonstrate the fidelity and capability of the proposed GBM. The paper proceeds as follows. Section 2 describes the structure of the GBM model and the enrichment of the NR and AIS datasets. Section 3 establishes the GBM of a 57000DWT bulk carrier for prediction of ship speed and fuel consumption. Section 4 presents both validation and application results of the GBM. Section 5 gives the conclusions and discussions.

2. Modelling

2.1 GBM Structure

The GBM combines both WBM and BBM to utilize the prior knowledge, physical principles, and data-driven results to form the digital twin model. Both WBM and BBM are treated as sub-models in a GBM. There are mainly two types of GBM structure, namely serial and parallel[11]. In a GBM with serial structure, two or more sub-models are connected in serial order, and the output of each sub-model can be used as the input of the following sub-models. The output of a serial GBM is the output of its last sub-model. In a GBM with a parallel structure, two or more sub-models are distributed parallel with the independent input data. The output of a parallel GBM is the combination of all sub-model outputs.

A novel GBM with the mixed structure, as shown in figure 1, is proposed to build a digital twin model for voyage performance prediction. The input variables of the GBM are weather information and propeller revolution rate, while the output variables are corresponding ship speed and the accumulative fuel consumption over a certain period. The sub-model WBM_Voyage serves as the GBM backbone to calculate the ship speed and fuel consumption with calm water effect, current effect, and wind effect. The sub-model BBM_Speed predicts the wave-caused speed loss *Speed_BBM*, while sub-model

IOP Conf. Series: Materials Science and Engineering **929** (2020) 012027 doi:10.1088/1757-899X/929/1/012027

BBM_Oil predicts the wave-caused fuel consumption increase *Oil_BBM*. The outputs from the WBM and BBMs are added together to form the GBM outputs *Ship Speed* and *Fuel Consumption*, just like a normal parallel GBM. Meanwhile, the outputs from the WBM are also directed into the BBMs as part of inputs, just like a normal serial GBM.

The sub-model WBM_Voyage also serves as a measure to enrich the NR and AIS datasets via weather hindcast data, which is critical to recover the weather change process and ship sailing process during voyage. The enriched datasets are split into training and validation sets to develop the sub-model BBMs.

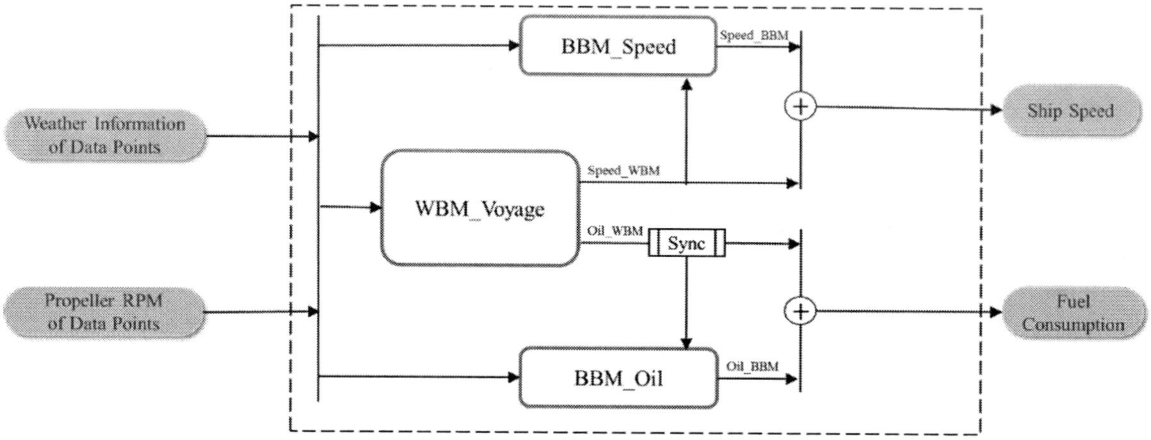

Figure 1. Illustration of GBM Structure.

2.2 *WBM_Voyage Sub-model*

The sub-model WBM_Voyage serves as the GBM backbone to calculate the ship speed and fuel consumption with calm water effect, current effect, and wind effect. Among the total ship resistance components, only the wave added resistance is excluded from resistance calculation in WBM[13].

The clam water resistance R_{calm} could be calculated following Holtrop and Mennen[14]. The current effect is included by replacing the absolute ship speed with the relative ship speed.

The wind resistance F_{wind} could be calculated by equation (2.1). A_f is the frontal area of the superstructure exposed in the wind, and V_{wind} indicates the wind speed relative to the ship. The wind load coefficient C_x could be obtained from Blendermann's method[15].

$$F_{wind} = \frac{1}{2} C_x \cdot A_f \cdot \rho_{air} \cdot V_{wind}^2 \tag{2.1}$$

The propeller thrust T and propeller torque Q under open water condition are expressed as equations (2.2) and (2.3). The interaction between the propeller and ship hull is represented by the wake coefficient ω_p and thrust deduction coefficient t_p[13].

$$T = k_T \rho n^2 D^4 \tag{2.2}$$

$$Q = k_Q \rho n^2 D^5 \tag{2.3}$$

For a specific propeller revolution rate n retrieved from NR dataset, the corresponding steady speed could be easily determined from the balance between the propeller thrust T and ship resistance[15]. This steady speed *Speed_WBM* is one output of the sub-model WBM_Voyage.

IOP Conf. Series: Materials Science and Engineering **929** (2020) 012027 doi:10.1088/1757-899X/929/1/012027

At the steady speed, the propeller torque Q is counterbalanced by the driving torque from the main engine. The engine brake power P_B is determined by the propeller delivered power and the shaft efficiency η_s. Once the Specific Fuel Oil Consumption (SFOC) of the engine is known, the engine fuel consumption rate (FCR) could be calculated by equation (2.4). Integrating FCR over the sailing period as equation (2.5) gives the accumulative fuel oil consumption (FOC), which is another output *Oil_WBM* of the sub-model WBM_Voyage.

$$FCR = SFOC \cdot P_B = SFOC \cdot \frac{Q \cdot 2\pi \cdot n}{\eta_s} \tag{2.4}$$

$$FOC = \sum_{i=1}^{n} FCR_i \cdot \Delta t_i \tag{2.5}$$

2.3 BBM_Speed Sub-model

The sub-model BBM_Speed predicts the wave-caused speed loss. The multilayer perceptron (MLP) is adopted here for BBM modelling. There are three layers of nodes as illustrated in figure 2. The output values are expressed in equations (2.7) and (2.8). w_{ij} is the connection weight from the ith neuron in input layer to the jth neuron in hidden layer. v_{jt} is the connection weight from the jth neuron in hidden layer to the tth neuron in output layer. θ_j is the bias of the jth neuron in hidden layer. γ_t is the bias of the tth neuron in output layer. f is the activation function of hidden layer neurons. g is the activation function of output layer neurons. The activation function of the hidden layer is commonly selected as a continuous, smooth, and monotone increasing function with upper and lower bounds.

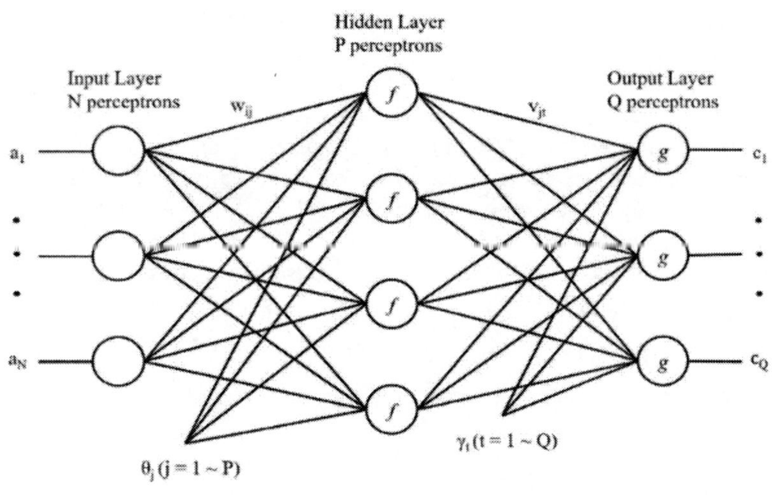

Figure 2. Illustration of MLP network.

$$b_j = f(\sum_{i=1}^{n} w_{ij} a_i + \theta_j) \tag{2.6}$$

$$c_t = g(\sum_{j=1}^{p} v_{jt} b_j + \gamma_t) \tag{2.7}$$

The output variable of the BBM_Speed is the wave-caused speed loss *Speed_BBM*. The input variables are related to wave information, propeller revolution rate, and WBM output *Speed_WBM*.

2.4 BBM_Oil Sub-model

The sub-model BBM_Oil predicts the wave-caused fuel consumption increase. The polynomial regression is adopted here for BBM modeling[9]. It is a form of regression analysis in which the relationship between the output variable and the input variable is modeled as an *n*th degree polynomial. The output variable of the BBM_Oil is the wave-caused fuel consumption increase *Oil_BBM*. The input variables are related to wave information, propeller revolution rate, and WBM output *Oil_WBM*.

2.5 Data Enrichment

The NR dataset records the noontime weather information, noontime vessel motion, noontime operation state such as propeller revolution rate, and daily fuel consumption for the past 24 hours. The AIS dataset supplements relatively high-frequency motion state at each moment of signal transmission. The typical NR and AIS datasets are plotted together in figure 3. Between two adjacent NR data points, the ship speed fluctuated with the time, which means the weather conditions could not be constant within that period. Since the daily fuel consumption represents the accumulative amount for the past 24 hours, it must be associated with the weather change process and ship sailing process within the same period, which could not be simply represented by the noontime instant information in NR dataset. Obviously, only NR and AIS datasets could not serve well in revealing the correlation among fuel consumption, weather condition, and ship operation states.

The weather hindcast data are introduced here to enrich the NR and AIS datasets. For a certain AIS data point, the wind and current information are determined by querying from the weather hindcast dataset with the timestamp, the latitude, and the longitude of the AIS data point. The propeller revolution rate is retrieved from the neighbouring NR data point. Through the sub-model WBM_Voyage, the steady speed without wave effect *Speed_WBM* is determined and assumed to be constant till the next AIS data point. The fuel consumption over this period without wave effect *Oil_WBM* is also determined. Repeating the above process could supplement every AIS data point with weather information, steady speed value *Speed_WBM*, and fuel consumption value *Oil_WBM*. The wave-caused speed loss *Speed_BBM* and wave-caused fuel consumption increase *Oil_BBM* on each NR data point could be derived from AIS data points.

The weather change process and ship sailing process between two adjacent NR data points are recovered mostly from the enriched NR and AIS datasets. The characteristcis of those stochastic variables are better represented by the statistics such as mean value and variance value over the same period. Therefore, the input variables of the sub-model BBMs are set to be statistics of weather information, propeller revolution rate, and WBM outputs. The output variables of the sub-model BBMs are statistics of the wave-caused speed loss and wave-caused fuel consumption increase, respectively. The training data and validation data for the sub-model BBMs could be obtained from the enriched NR and AIS datasets.

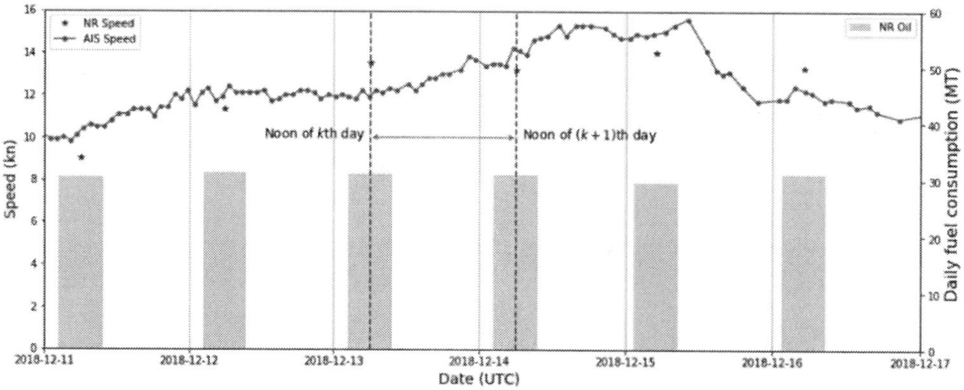

Figure 3. Typical Example of NR and AIS data.

3. Case Study

3.1 Datasets

The NR and AIS data collected from a 57000DWT bulk carrier are used to demonstrate the fidelity and capability of GBM model. The vessel particulars are listed in table 1. The weather hindcast data are retrieved from NOAA and ECMWF[16]. The variables available in the datasets are listed in table 2. The sampling period for the AIS dataset is 1 hour, and updating period of the weather data is 3 hours.

The datasets cover the period from August 15, 2018 to May 10, 2019, which contains 138 days in sailing. The 138 data points in NR dataset are crosschecked with AIS data and weather hindcast data to remove the incorrect NR data caused by human errors. After data pre-processing, 51 effective data points are left in the NR dataset for further training and validation.

Table 1. Vessel particulars.

Vessel type	Bulk carrier
Year built	2012
Gross Tonnage	33044
summer DWT	56741t
Length overall	189.99m
Breadth moulded	32.26m
Depth moulded	18.0m
Draft Scantling moulded	12.8m
Main Engine	MAN B&W. 6S50MCC
M.C.R	9480kW × 127r/min

Table 2. Variable List in Dataset.

Variable	Source	Variable	Source
NR timestamp	NR	Draft	NR
NR ship location	NR	Trim	NR
Propeller speed	NR	AIS timestamp	AIS
Daily fuel consumption	NR	AIS location	AIS
Daily traveled distance	NR	AIS speed	AIS
Daily steamed time	NR	AIS heading	AIS
Daily average speed	NR	Hindcast Wind speed	ECMWF
Daily wind direction	NR	Hindcast Wind direction	ECMWF
Daily wind grade	NR	Hindcast Current speed	NOAA
Daily current direction	NR	Hindcast Current direction	NOAA
Daily current grade	NR		

3.2 Parameters in WBM

The open water characteristics of the propeller, as shown in figure 4, are obtained from CFD simulation. The values of the propeller-hull interaction coefficients ω_p and t_p are set to 0.2 and 0.1, respectively. The SFOC for the engine is shown in figure 5.

IOP Conf. Series: Materials Science and Engineering **929** (2020) 012027 doi:10.1088/1757-899X/929/1/012027

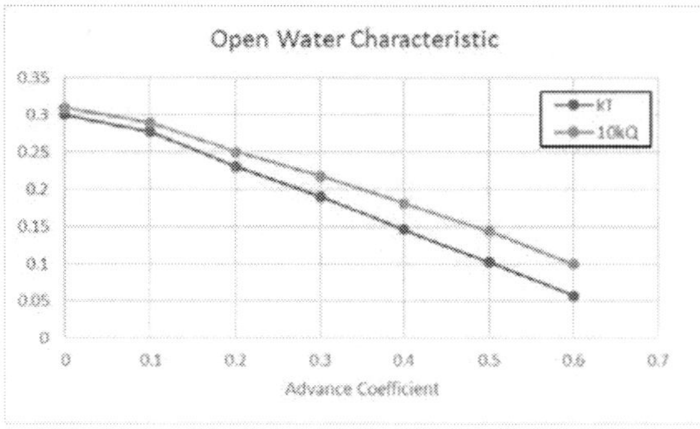

Figure 4. Propeller open water performance.

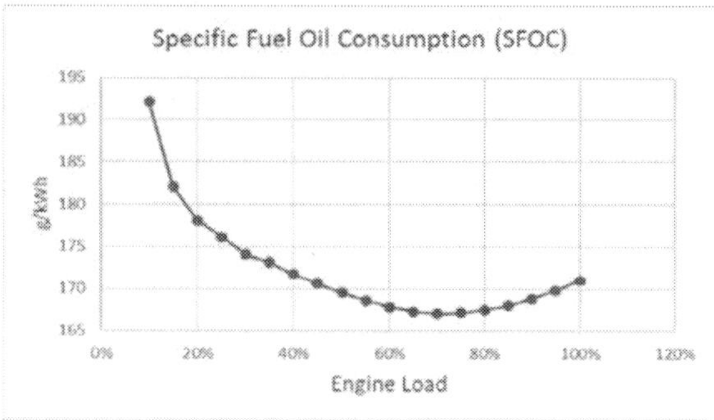

Figure 5. Engine Specific Fuel Oil Consumption.

3.3 Features and Hyperparameters of BBMs
The multilayer perceptron (MLP) is adopted in sub-model BBM_Speed to predict the wave-caused speed loss. The polynomial regression is applied in sub-model BBM_Oil to predict the wave-caused fuel consumption increase. The features of the BBMs are determined by the correlation analysis and presented in table 3. The hyperparameters for the BBMs are listed in table 4.

Table 3. Correlation analysis.

Parameters	Corr. with speed	Corr. with fuel consumption
longitudinal relative current speed	0.903	0.612
longitudinal relative wind speed	0.416	0.351
transverse relative current speed	0.099	0.137
transverse relative wind speed	0.150	0.093
propeller revolution rate	0.053	0.496

Table 4. Hyperparameters of BBMs.

Model	Hyperparameter	Value
BBM_Oil	Polynomial Order	6
BBM_Speed	Number of hidden layers	1
	Hidden layer size	8
	Activation function	hyperbolic tan function
	Solver	lbfgs
	L2 penalty	1
	Tolerance	10^{-3}

4. Results

4.1 Validation
The enriched NR and AIS datasets are split into training set and validation set by K-Folds cross validator. The total number of the folds is set to 4. The training set is used to train the GBM and the pure BBMs, respectively. A pure WBM is also adopted for comparison[1]. The validation set is used to test the performance of above three models. From the validation results presented in table 5, the GBM could predict the fuel consumption with the relative error 4.8%, which is much better than 8.5% of the BBM and 28.5% of WBM.

Table 5. Validation results.

		GBM	BBMs	WBM
Speed prediction	Error mean[a]	3.6%	4.0%	2.9%
	Error standard deviation[a]	4.9%	5.2%	2.2%
Fuel consumption prediction	Error mean	4.8%	8.5%	28.5%
	Error standard deviation	17.9%	18.8%	5.0%

[a] An error of a data point is calculated by $error = |estimation - true| \cdot (true)^{-1} \cdot 100\%$. "Mean" represents the average of all errors, and "standard deviation" represents the standard devation of all errors. The true value of speed is from the AIS dataset and the true value of fuel consumption is from the NR dataset.

4.2 Application

The established GBM could be applied to predict the estimated time arrival (ETA) and the total fuel consumption of any voyage. The prediction results for a past voyage are presented in table 6. The case vessel departed from Port Gibraltar at 20:53UTC, Oct. 10, 2018, and arrived at Port Lagos at 01:29UTC, Oct. 22, 2018. Accroding to the NR dataset, the actual fuel consumption of this voyage is 339.4MT, while the GBM prediction is 337.5MT. The relative error of prediction is only 0.6%.

Table 6. GBM application results.

Voyage	Source	Departure time	Arrival time	Voyage time (h)	Distance traveled (nm)	Total oil (MT)
Gibraltar to Lagos	GBM	10/10/2018 20:53	10/22/2018 01:13	290.1	3368	337.5
	NR	10/10/2018 22:30	10/22/2018 04:00	295.3	3419	339.4
	AIS	10/10/2018 20:53	10/22/2018 01:29	292.8	3379	NA

Another important application of the proposed GBM is weather routing. Figure 6 presents a 12-day voyage optimization result from Port Louis, Mauritius to Port Singapore, Singapore. The total fuel consumption of the actual voyage is 382.3MT, while the optimized route in red line could save 28MT fuel oil (7.3%) with the same arrival time.

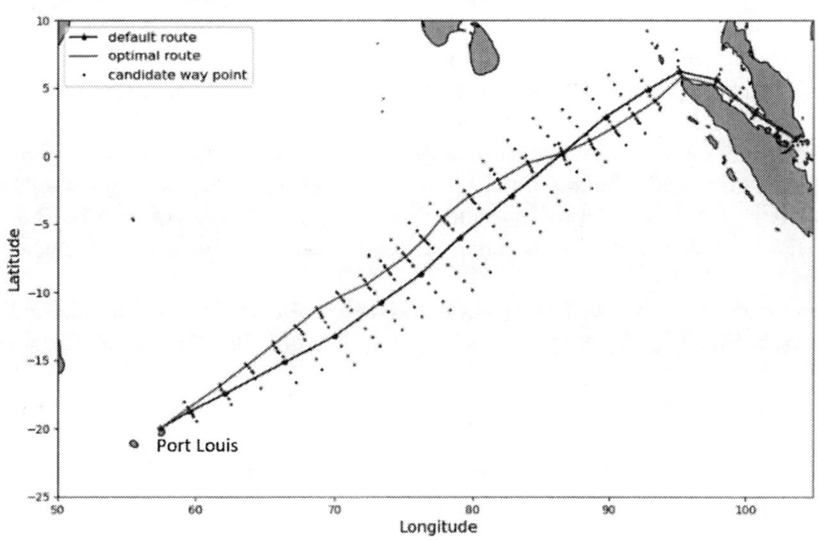

Figure 6. Weather Routing Optimization.

5. Conclusion

In this paper, a novel GBM approach is proposed to establish a digital twin model for voyage performance prediction. The NR and AIS datasets collected from a 57000DWT bulk carrier are used to demonstrate the fidelity and capability of the proposed GBM. The weather hindcast data are introduced to enrich the NR and AIS datasets, which are split into training and validation sets to develop GBM. The voyge performance prediction from the GBM shows better accuracy than those from pure WBM or pure BBMs. An arrival time forecast and a weather routing showcase are also presented to demonstrate the application effects of GBM.

The proposed GBM provides a satisfying prediction of ship speed and fuel consumption without mandatory sensor-collected data, thus applicable for a varity of vessels. In those cases where more sensors are available onboard, the proposed approach can incorporate sensor data to improve the model accuracy further.

6. Reference

[1] Haranen M, Pakkanen P, Kariranta R and Salo J 2016 White, Grey and Black-Box Modelling in Ship Performance Evaluation *1st Hull Performence Insight Conf.* pp 115–27

[2] Moreno-Gutiérrez J, Pájaro-Velázquez E, Amado-Sánchez Y, Rodríguez-Moreno R, Calderay-Cayetano F and Durán-Grados V 2019 Comparative analysis between different methods for calculating on-board ship's emissions and energy consumption based on operational data *Sci. Total Environ.* **650** pp 575–84

[3] Nielsen J B, Sandvik E, Pedersen E, Asbjørnslett B E and Fagerholt K 2019 Impact of simulation model fidelity and simulation method on ship operational performance evaluation in sea passage scenarios *Ocean Eng.* **188** p 106268

[4] Bialystocki N and Konovessis D 2016 On the estimation of ship's fuel consumption and speed curve: A statistical approach *J. Ocean Eng. Sci.* **1** pp 157–66

[5] Safaei A A, Ghassemi H and Ghiasi M 2019 VLCC's fuel consumption prediction modeling based on noon report and automatic identification system *Cogent Eng.* **6** p 1595292

[6] Gkerekos C, Lazakis I and Theotokatos G 2019 Machine learning models for predicting ship main engine Fuel Oil Consumption: A comparative study *Ocean Eng.* **188** p 106282

[7] Pedersen B P and Larsen J 2009 Modeling of Ship Propulsion Performance *World Marit. Technol. Conf.* pp 1–10

[8] Liang Q, Tvete H A and Brinks H W 2019 Prediction of vessel propulsion power using machine learning on AIS data, ship performance measurements and weather data *J. Phys. Conf. Ser.* **1357** p 012038

[9] Farag Y B A and Ölçer A I 2020 The development of a ship performance model in varying operating conditions based on ANN and regression techniques *Ocean Eng.* **198** p 106972

[10] Coraddu A, Oneto L, Baldi F, Cipollini F, Atlar M and Savio S 2019 Data-driven ship digital twin for estimating the speed loss caused by the marine fouling *Ocean Eng.* **186** p 106063

[11] Leifsson L T, Sævarsdóttir H, Sigurdsson S T and Vésteinsson A 2008 Grey-box modeling of an ocean vessel for operational optimization *Simul. Model. Pract. Theory* **16** pp 923–32

[12] Safaei A A, Ghassemi H and Ghiasi M 2018 Correcting and enriching vessel's noon report data using statistical and data mining methods *Eur. Transp. - Trasp. Eur.* pp 1–13

[13] Lewis E V 1988 *Principles of naval architecture: Volume II: Resistance, propulsion and vibration*

[14] Holtrop J and Mennen G G J 1982 An Approximate Power Prediction Method *Netherlands Sh. Model Basin, NSMB, Wageningen, Publ. No. 689, Publ. Int. Shipbuild. Progress, ISP, Vol. 29, Nr 335* pp 166-70

[15] Blendermann W 1994 Parameter identification of wind loads on ships *J. Wind Eng. Ind. Aerodyn.* **51** pp 339–51

[16] Zyczkowski M, Szłapczyńska J and Szłapczyński R 2020 Review of weather forecast services for ship routing purposes *Polish Marit. Res.* **26** pp 80–9

IOP Conf. Series: Materials Science and Engineering **929** (2020) 012028 doi:10.1088/1757-899X/929/1/012028

From Sensors to MASS: Digital Representation of the Perceived Environment enabling Ship Navigation

H-C Burmeister[1]; M Constapel[1], C Ugé[1], C Jahn[1]

[1] Fraunhofer Center for Maritime Logistics and Service, Am Schwarzenberg-Campus 4, 21073, Hamburg, Germany

E-Mail: hans-christoph.burmeister@cml.fraunhofer.de

Abstract. Maritime Autonomous Surface Ships (MASS) do require shifting from human information processing to automation functions. While supporting the sensory processing of human navigation by automated information acquisition from different sensors is an already established (and even in many cases mandatory) technology with a set of framing standards and information exchanges, defining a common data set for a perceived environment model is currently neglected which hinders modularization of MASS technology development. Beginning with an introduction into the navigational process as well as information processing, this paper subsequently proposes an initial minimum data set for different aspects of ship navigation.

1. Introduction

Already today, automation is present in maritime navigation, especially in the area of information acquisition. Despite human sight and hearing, a variety of sensor equipment is nowadays (mandatory) state of the art on ship bridges, like e.g. Radar, Global Navigation Satellite Systems as well as the Automated Identification Systems. They all perform automated information acquisition functions for the officer of the watch, meaning the sensing and registration of (raw) input data relevant for the navigation process. Existing standardization even allows the exchange and integration of this raw data between different systems, e.g. by the common NMEA format IEC 61162 [1].

In contrast, information analyses on board do still mainly rely on human perception and are less supported by automated information analyses functions, meaning prediction and integration of the (raw) data sets, as stand-alone ECDIS or radar system do primarily only allow overlaying and displaying of raw (NMEA) data sets, as well as some low-level predictions like CPA calculations. Recent Integrated Navigational Systems (INS) provide some more capabilities, as they should monitor the availability, validity, and integrity of connected sensors [2]. However, even in the case of an INS, the system's intention is to hand over its information to a human for decision-making, resulting primarily in visual output.

Digitalization and further automation are ongoing developments on the way to Smart or even Maritime Autonomous Surface Ships (MASS). MASS are vessels, that can «operate to a varying degree independently of human interaction» up to the stage, where «the operating system of the ship is able to make decisions and determine actions by itself» [3].

Recent MASS feasibility studies as well as published class guidelines on MASS have in common, that MASS with the highest degree of autonomy does conceptually require an Automated Sensor System (information analyses function - also often referred to as 'situational awareness systems') as well as an Autonomous Navigation System (decision and action selection function) on board as proxies for the human lookout and officer of the watch[1] [4] [5] [6] [7]. Nevertheless, what is lacking is the definition of a specific information transfer between these two systems. While practical function developments for the Automated Sensor System can at least still easily link by the existing sensor standards, no common definition exists regarding the digital description of the perceived environment of a vessel. However, a precise definition of this perceived environment and its related data elements is crucial for further MASS development, as it first enables modular development and connectivity within MASS technology.

Based on an introduction into maritime information processing in the following and MASS sensor system concepts in the subsequent chapter, chapter 4 proposes a minimum data set to be delivered by any to information analyses to allow for a proper maritime decision-making within MASS.

2. Information Processing on board

2.1. *Information Processing Theory*

Human and automation information processing can be described by a four-stage model, which is also used in some of the recent MASS guidelines [6] [7]. On a conceptual level, this model distinguished four stages of human information processing [8]:

(i) sensor processing,
(ii) perception,
(iii) decision-making, and
(iv) response selection.

In an automation context, these must be replaced by corresponding automation functions:

(i) information acquisition,
(ii) information analyses,
(iii) decision and action selection, and
(iv) action implementation.[2]

While *information acquisition* is limited to *"sensing and registration of input data"*, *information analyses* include predictions based on the raw data *"augmenting human operator perception and cognition"* up to a level to *"provide context-dependent summaries"* [8].

2.2. *Maritime Information Acquisition*

Today's ships are obliged to fulfill the carriage requirements for navigational equipment according to SOLAS Regulation 19 and 20 [9]. These include (besides others) depending on ship size and sailing area the following sensory equipment:

- Magnetic Compass
- GNSS receiver
- Echo sounder

[1] a shore-side unit is normally also foreseen during information processing of MASS, but as long as no direct remote control is executed, this unit does normally act in a 'supervising' function for the on board system, which is why it is not further investigated in this paper, as in principle an operation independent from a direct shore connection is aspired for the highest degree of autonomy

[2] called condition detection, condition analyses, action planning and action control in [6]

- Radar
- AIS
- Gyro compass

While this sensory equipment is available from a variety of manufacturers, common maritime standards are in place ensuring connectivity with third-party equipment, like e.g. the established NMEA-standard for most sensor data [1], common de- and encoding for AIS data [10] and newer developments like the Asterix-Format for radar data [11]. Thus, for automation in the information acquisition context, a technical framework, that MASS can build upon, is existing, even though further sensory equipment might be necessary.

2.3. *Maritime Information Analyses*

This sensory equipment normally provides their data to further mandatory navigational equipment for further information analyses [9]:

- Bridge Navigational watch alarm system (BNWAS)
- Automatics Radar Plotting Aid (ARPA)
- Electronic Chart Display Information System (ECDIS)
- Voyage Data Recorders (VDR)

However, in the context of maritime traffic, information processing with regards to *navigation* is the officer of the watch's (OOW) responsibility on an operational level, which includes tasks like e.g. to plan and conduct the voyage and to maintain a safe navigational watch. Hereby, the OOW can be supported by a lookout who reports e.g. sound and light signals as well as detected other objects as well as relative bearings [12]. Even though, the OOW is already supported by the aforementioned systems, those serve primarily as a data display during the information analyses stage as they only conduct data integration or prediction on a very low level of automation.

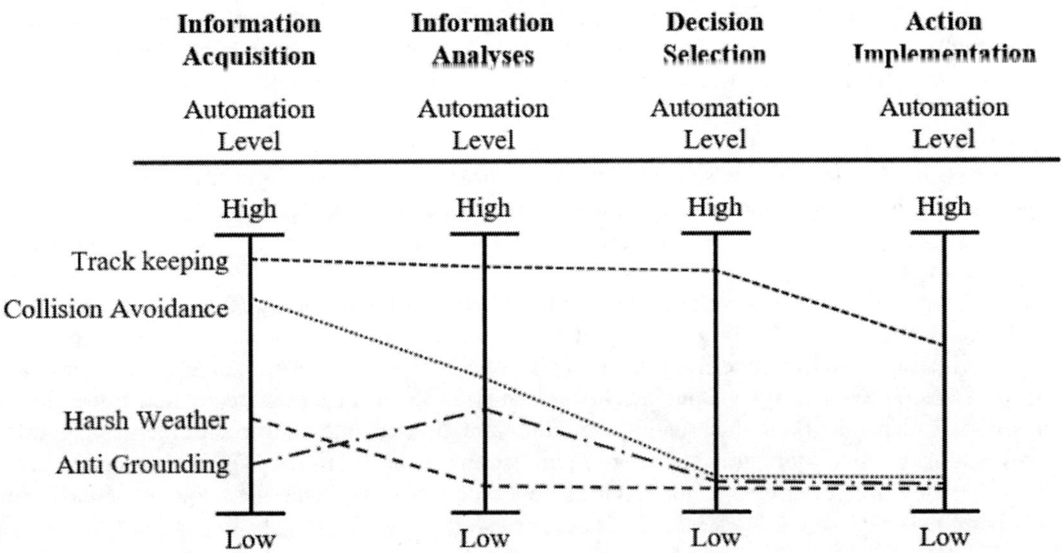

Figure 1. Level of automation for different functions of maritime navigation (own drawing based on [8]).

As indicated in figure 1, several nautical sub-tasks are supported by automated information acquisition, as e.g. the task of collision avoidance, which is automatically getting object information by Radar as well as the Automated Identification System (AIS) and not only by sight and hearing. Thereby, high automation means, that the system decides and executes automatically with or even without information to the human [8]. For the remaining information processing stages, automation is however comparatively low besides some CPA calculations, even within modern INS, meaning that only limited assistance or decision alternatives are provided. Thus, creating the full navigational situation picture by using all available information and check it for plausibility as required by COLREG [13] is still a primarily non-automated task. Also, for further nautical sub-tasks like e.g. anti-grounding or harsh weather operation, a modern INS might get a very limited set of environmental data as sensory inputs, but predicting based on this data as well as all further decision steps rely solely on the OOW. In terms of marine navigation, the only exception with regards to state-of-the-art bridge system is track control, where fully automated systems are available [14].

2.4. *Maritime Decision-Making*

With the exception of the aforementioned track control systems, decision-making on the bridge is fully up to the OOW. Besides alarms, no concrete recommendations for e.g. collision avoidance, anti-grounding, or harsh weather are normally given. As the OOW fulfills a double function as an information analyst and decision-making unit, no clear data set definition exists, describing the perceived environment which an OOW is using to base its decisions upon, as this is normally hidden within the nautical officer's mental model.

3. MASS Situational awareness

On a MASS with the highest degree of autonomy, this still human-centered functions of the lookout and the OOW are going to be replaced by automation. Thus, the relevant artifacts of the Maritime Perceived Environment (e.g. own ship's state, approaching vessels, navigational aids) need to be collected and represented internally through a sensor platform. Generically, this sensor platform relies on integrated onboard sensors (e.g. radar and AIS) and additional inertial and short-range sensors (e.g. cameras and radars in the millimeter frequency range). A sufficient perception of the maritime environment from this sensor system is needed to carry out safe nautical decision-making by any navigation system during transit on a given route mainly for collision avoidance and grounding avoidance over a longer period of time and under consideration of existing regulatory, internal and environmental restrictions.

As seen in figure 2, the sensor system captures sensory input via the Sensor Data Access and processes (i.e. filters, smoothes, associates, estimates and transforms) the gathered sensor data via its core component: Multi Sensor Data Fusion. Subsequently, it transfers the transformed and associated measurements and estimates (i.e. High Level Features, which yield pieces of information with navigational importance or relevance) from the sensor fusion component via three logical interfaces (Objects, Environment, Body) to the navigation system. These interfaces serve the information needs of the navigation system concerning the perceived objects, the environmental parameters, and the own ship's state. The sensor system decides how often High Level Features are created since the information content is dependent on the quality, amount, and update rate of measurements. Nevertheless, the navigation system may request the creation and submission of High Level Features for a given point in time as they are required for the establishment of situation awareness and as a basis for nautical decision making.

However, a common generic description of the maritime perceived environment regarding the three logical interfaces is necessary to enable independent development and implementation of Situational Awareness and Autonomous Navigation Systems.

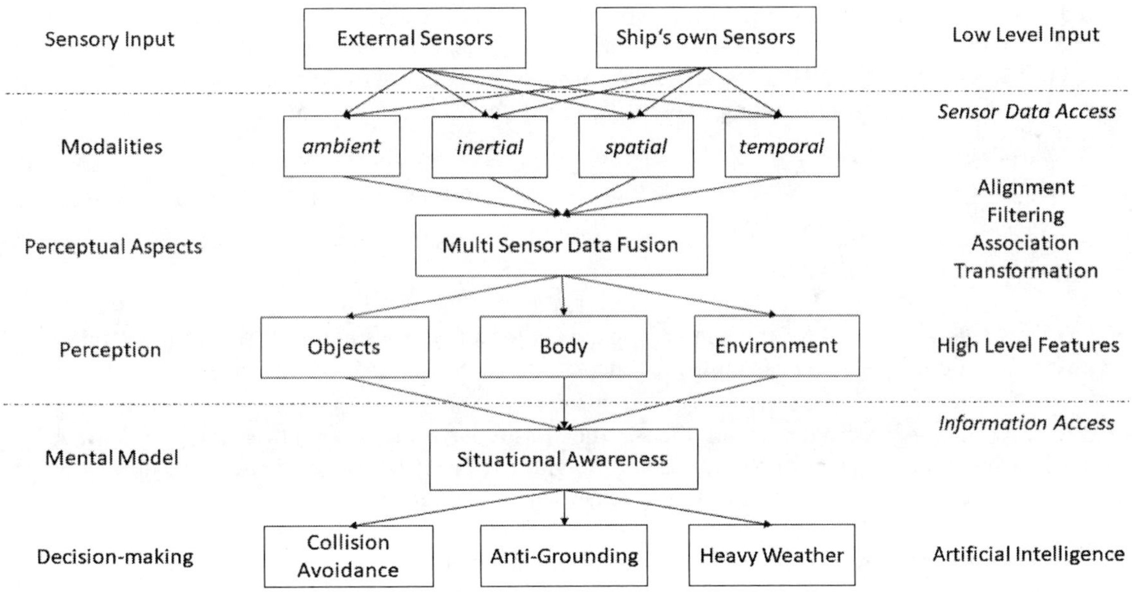

Figure 2. Generic overview about sensors, perception, and decision-making on MASS

4. Maritime Perceived Environment Proposal

The idea of the digital representation of the maritime perceived environment is to define a (minimum) data set to be provided by any sensor system enabling the decision-making capabilities of the MASS' navigation system. Thus, it is necessary to stress that this perceived environment is sensor-independent, as it shall not define how this data shall be gathered, but solely which data points are necessary to allow for connectivity between different MASS modules. Conceptually, the Maritime Perceived Environment is established and refined over time through the High Level Features created by the sensor system. Any High Level Feature is temporarily aligned corresponding to the position of the own ship at a given timestamp since locations are always represented relative to the own ship's position. This scheme applies also to spatial and temporal context data, e.g. tide, or chart related data. Ideally, an update frequency of 1 Hertz is assumed to be sufficient for time-dependent navigational data.

4.1. *Objects*

The perceived environment data set must consist of the detected objects in the vicinity of the own ship recording at least (1) system timestamp, (2) unique object identifier, (3) the classification or type of the object°, (4) relative bearing*, and (5) distance* corresponding to the position of the own ship at the system's timestamp. Relevant object types to be distinguished shall be vessels, aids to navigation, small unlit boats, floating logs, oil drums, containers, buoys, or ice [6].

Moreover, the data set contains additional information for an object and thus further augment the record with information for (1) speed over ground*, (2) course over ground*, (3) heading*, (4) dimensions* (length and width), and in case of vessels (5) navigational status°. For larger objects (i.e. length greater than 20 m), that augmentation is considered mandatory.

In case multiple sensors are used by the sensor system to define one data set, interval values shall be given for discrete values (marked with *) and the number of sensor confirmations compared with the number of available sensors types for type assignments (marked with °).[3] It shall be noted, that the authors assume, that information analyses in the context of collision avoidance does not include determining COLREG statuses and obligations, which rely not just on the object itself but on the given traffic situation. COLREG assessments are indeed considered to be a decision and thus part of the next step of information processing, which is why it shall not be part of the perceived environment description.

4.2. *Environment*

Ambient data of the own ship's surroundings is subdivided into disjoint records describing either the wind state, sea state, or visual state.

4.2.1. Wind state A record for the wind comprises at least (1) timestamp, (2) wind force, and (3) wind direction. Again, in case multiple sensors interval values shall be given.

4.2.2. Sea state A record describing the sea state has to have at least (1) timestamp, (2) wave height and, (3) wave direction. Optionally, (4) swell height, (5) swell period, and (6) swell direction can be added to the record. Again, in case multiple sensors interval values shall be given.

4.2.3. Visual state A record on visibility conditions needs to be composed of (1) timestamp, (2) visibility (i.e. range), and (3) an occurring sector designated by two relative bearings.

4.3. *Body*

The data of the own ship's state are stored in a record with at least (1) timestamp, (2) actual position in a global reference frame, (3) position accuracy, (4) speed over ground, (5) course over ground, (6) heading, (7) speed through water.[4] Additional information about the own ship's state may extend the record: (1) torsion forces, momenta, and force vectors attached to the hull,
(2) rotational motion, (3) attitude (i.e. pose).

5. Conclusion

This paper is proposing an approach for a maritime perceived environment data set as an interface from any Situational Awareness to any Autonomous Navigation System, which is decoupled from the specific sensor set-up. Such a harmonization or even a standard is a necessary prerequisite, to improve parallel development of MASS algorithms and technologies for information analyses and decision making, respectively. Besides the data definition, a common, specified, and non-proprietary exchange format would be a necessary future work to ensure modular design and thus flexibility for MASS bridge equipment.

Acknowledgment

The B ZERO project has received funding by the German Federal Ministry of Economic Affairs and Energy under support code 03SX500A.

[3] if a sensor system is e.g. capable to detect the navigational status by image recognition of lights and shapes, but also uses AIS information, it shall indicate if the current type value is determined by only one or both possible inputs.

[4] besides the navigational status of the MASS itself being important, it shall not be part of the perceived environment data definition, as it is presumably an input, like e.g. the voyage plan, and nothing that will be given from a sensor system

IOP Conf. Series: Materials Science and Engineering **929** (2020) 012028 doi:10.1088/1757-899X/929/1/012028

References

[1] IEC 2016 *IEC 61162: Digital interfaces for navigational equipment within a ship* (IEC)

[2] IEC 2012 *IEC 61924-2: Maritime navigation and radiocommunication equipment and systems - Integrated navigation systems - Part 2: Modular structure for INS - Operational and performance requirements, methods of testing and required test results* (IEC)

[3] IMO 2018 *MSC 99 Report on the Maritime Safety Committee on its ninety-ninth session* (London: IMO)

[4] Burmeister H C, Bruhn W, Rødseth nd Porathe T 2014 *International Journal of E-Navigation and Maritime Economy* **1** 1–13

[5] Poikonen J, Hyvönen M, Kolu A, Jokela T and Paasio A UN aawa position paper URL https://www.rolls- royce.com/ /media/Files/R/Rolls-Royce/documents/customers/marine/ship-intel/aawa- whitepaper-210616.pdf

[6] GL D Class guideline: Autonomous and remotely operated ships (dnvgl-cg-0264) URL https://rules.dnvgl.com/docs/pdf/DNVGL/CG/2018-09/DNVGL-CG-0264.pdf

[7] Veritas B Guidelines for autonomous shipping (ni641) URL https://marine-offshore.bureauveritas.com/ni641-guidelines-autonomous-shipping

[8] Parasuraman R, Sheridan T B and Wickens C B 2000 *IEEE Transactions on Systems Man and Cybernetics - Part A Systems and Humans* **30** 286–297

[9] IMO 2019 *SOLAS - Safety of Life at sea and Amendments in 2020* (London: IMO)

[10] ITUTechnicalcharacteristicsforanautomaticidentificationsystemusingtimedivisionmultipleaccessinthe vhf maritime mobile frequency band URL https://www.itu.int/dms$_p$*ubrec/itu* − *r/rec/m/R* − *REC* − *M*.1371 − 5 − 201402 − *I*!!*PDF* − *E.pdf*

[11] EUROCONTROL Specification for surveillance data exchange - part 1 all purpose structured eurocontrol surveillance information exchange (as- terix) URL https://www.eurocontrol.int/sites/default/files/2019- 06/part$_1$ $_e$*urocontrol$_s$pecification$_a$sterix$_s$pec* 149$_e$*d*$_2$.4*.pdf* −

[12] IMO 2011 *STCW Convention and STCW Code including 2010 Manila Amendments* (London: IMO)

[13] IMO 2019 *COLREG - Collision Regulations 1972* (London: IMO)

[14] Burmeister H C, Bruhn W and Walther L 2015 *TransNav, the International Journal on Marine Navigation and Safety of Sea Transportation* **9** 31–40

IOP Conf. Series: Materials Science and Engineering **929** (2020) 012029 doi:10.1088/1757-899X/929/1/012029

Research on ontology-based situation understanding and decision-making approach for MASS

AN Lan-xuan[1], ZHANG Xin-yu *,[1], DING Zhi-guo[1], WANG Cheng-bo[2]
[1] Maritime intelligent Transportation Research Group, Dalian Maritime University, China
[2] School of marine electrical engineering, Dalian Maritime University, China a1120180001@dlmu.edu.cn,*zhang.xinyu@sohu.com

Abstract. Aiming at the problems of miscellaneous navigation safety information and the difficulty of multi-source heterogeneous information fusion, a novel ontology-based collision avoidance decision making method is proposed in this paper. first of all, the structured ontology model of navigation situation is built on the basis of analysing the hierarchy and interactivity of various scene elements in driving scene and constructing the semantic model of entity class and binary attribute of scene elements, implement effective modelling of navigation scenarios. Secondly, the knowledge base of collision avoidance decision and online reasoning system are constructed based on the semantic expression of typical scene elements to realize the efficient utilization of navigation prior knowledge. Finally, the simulation experiment is carried out for the typical ship encounter scenarios in open water. The results show that the ontology method can effectively improve the cognition ability of Maritime Autonomous Surface Ships (MASS), and meet the requirements for real-time, safety and rationality of collision avoidance decision.

1. Introduction

The concept of scene understanding was first proposed by Endsley, and was used to describe the process of pilots' perception, interpretation, and prediction of aircraft operations [1]. The perception and understanding of the ship's driving scene are the primary prerequisite for the safe navigation of an autonomous ship. A true autonomous ship should be able to understand the semantic meaning of sensor data, so as to make safe driving decisions such as steady course, steering, and emergency stop. Most of the existing driving scene analysis and applications are mainly focused on the perception of the scene. The driving scene data is obtained through various sensors such as AIS, radar, electronic charts, etc. Amit Sharma determined the specific information needs at three different levels of perception, comprehension, and projection by analyzing the factors that affect the understanding of the scene in the actual goal directed task [2], Mogens Blanke discretized the elements of the navigation scene and proposed an autonomous situation awareness framework based on the DES theory, which realized the demonstration of the collision risk of three ships [3], X.Y. Zhou used probability theory to quantify the scene perception process to achieve the goal of guiding the navigational safety of autonomous ships [4], O.V. Smirnova analyzed the current existing situational awareness model, neither used the ontology-based method to perceive the navigational situation nor gave out of the implementation method [5], Wanqi Wei et al. constructed a multi-entity Bayesian network for the uncertainty of navigation information and expressed the fuzzy navigation information in a probabilistic form[6]. These methods focus on the process of environment perception, and process the acquired data information to realize the scene understanding based on rules and driving environment. However, they are not capable of combining the scene understanding and collision avoidance decision, and this makes it difficult for these methods to cope with the navigation needs of ships in different sailing areas and different encounter scenarios. Therefore, it is necessary to construct a knowledge representation method capable of handling the fusion of complex and multi-source safety information, and able to imitate the process of experienced ship pilots cognizing the current ship's navigation environment through lookout, so as to assist autonomous ships generate reasonable driving decisions.

The ultimate goal of perception and understanding of navigation scenarios is to support the decision-making of autonomous ships' driving behavior. The autonomous decision-making of ships is an important manifestation of the highly intelligent ship's autonomous navigation. There are many methods of research on autonomous ship collision avoidance decision-making, which are mainly divided into the following four categories: The first is to build a collision avoidance expert system based on driving knowledge and collision avoidance rules, and to realize anthropomorphic collision avoidance decisions by matching the ship's encounter eigenvalues [7-9], the second is the artificial intelligence algorithm, which uses reinforcement learning, deep learning and other methods to set rules as reward functions to train agents to achieve the goal of safe navigation of ships [10-12]. The third is a deterministic algorithm, typically there are artificial potential field [13], velocity obstacle algorithm [14], graph search algorithm [15], etc. In order to plan the navigable area of the ship to achieve the goal of collision avoidance decision, these methods handle obstacles reasonably, such as constructing a repulsion field, calculating the speed obstacle interval and soon. The fourth is intelligent optimization algorithms, such as A* algorithm [16-17], evolutionary algorithm [18], ant colony algorithm [19], etc. using search strategies, information exchange between groups to find an approximate optimal path, and calculate collision avoidance decision information. In summary, most collision avoidance algorithms convert collision avoidance rules into navigational constraints and give rule-based collision avoidance decisions. However, this process failed to truly combine the ship's navigation safety information with collision avoidance decisions, also lacked a comprehensive understanding of the driving environment.

This paper proposes an ontology model containing navigation scenes based on ontology methods and a reasoning rule library based on driving experience. Compared with the previous algorithm the model based on ontology for scene understanding and decision-making has two advantages: At the algorithm level, the scene understanding and collision avoidance methods based on the ontology model have good interpretability. The reason is that the existing algorithms lack of the support of the driving experience, the method proposed in this paper is based on the COLREGS and good seamanship. In the rule base, the steering action is based on the rules of the decision-making, which are in line with actual maneuvering requirements. In terms of environment modelling, compared with the widely used grid map method obtained entity attributes and relationships by searching row by row, the environment representation method based on ontology model can directly restore the attributes and relationships of scene elements from the acquired information, and semantic-based navigation scenarios can improve the reasoning ability of rules and provide decision support for safe navigation of ships.

In this study, it is assumed that the navigation area is open water, and that radar data and decoded AIS data are known. Using AIS data containing rich ship information to construct the ship navigation environment information, including wind, wave and flow properties. In addition, the use of onboard radar information to construct the properties of the target ship, including the target ship's course, velocity, the range between target ship and own ship as well as TCPA and DCPA. When building the rule base, the main consideration is the COLREGS (include R13, R14, R15, R16) of the two ships encounter situations, and the good seamanship of the ship's decision-making behavior.

This paper mainly considers the restrictions of rules and maneuvering experience on the decision-making behavior of ships under the situation of two ships' encounter. Furthermore, this work has not deal with the problem of multi-ship collision avoidance, the multi-ship scene understanding and collision avoidance issues will be further discussed in the future work.

2. Navigation scene ontology model

In order to express the relationship between the autonomous ship and the navigation environment more clearly and accurately, the entity and corresponding entity attributes are constructed by analysing the composition of the entity model and the scene elements of the real navigation environment. An instantiated scene expression is based on the two ships' encounter situation in the open water is shown.

2.1. Ontology

Ontology first originated in the field of philosophy, and recently developed into fields such as computational science and information science. It has been widely used in the semantic web, biomedicine and other fields. Ontology contains different types of knowledge such as facts, rules, definitions, etc. It has no inference function itself, and needs to be combined with an inference engine to form a knowledge system to solve the problem of artificial intelligence. The composition of the ontology model is as Figure1.

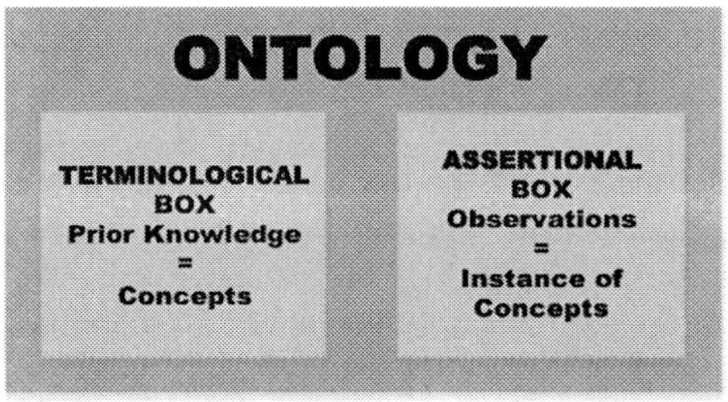

Figure 1. Composition of ontology model

A complete ontology knowledge model consists of Terminological Box (TBox) and Assertional Box (ABox), where TBox is used to construct a conceptual model of objectively existing objects, which consists of classes, attributes, and relationships, and is stored in the knowledge base in the form of prior knowledge. ABox is a materialized representation of the concepts in TBox, which instantiates the individuals actually observed in the scene and stores it in the knowledge base as the scene knowledge in the scene. For example, in the actual navigation scenario, the class "obstacle" is defined in the TBox. When there is a target ship in front of the own ship, MMSI= 412701000, it is instantiated as a specific " obstacle " instance and stored in the ABox.

2.2. Elements of a navigation scene

Determining the scene elements is the first step in constructing the ontology model. By combining the rules and the constraints of ship navigation restrictions, the ship's navigation scene elements are divided into autonomous ships (own ship), environmental elements, obstacle elements, and chart elements (navigation area elements). The composition of scene elements is shown in Figure 2.

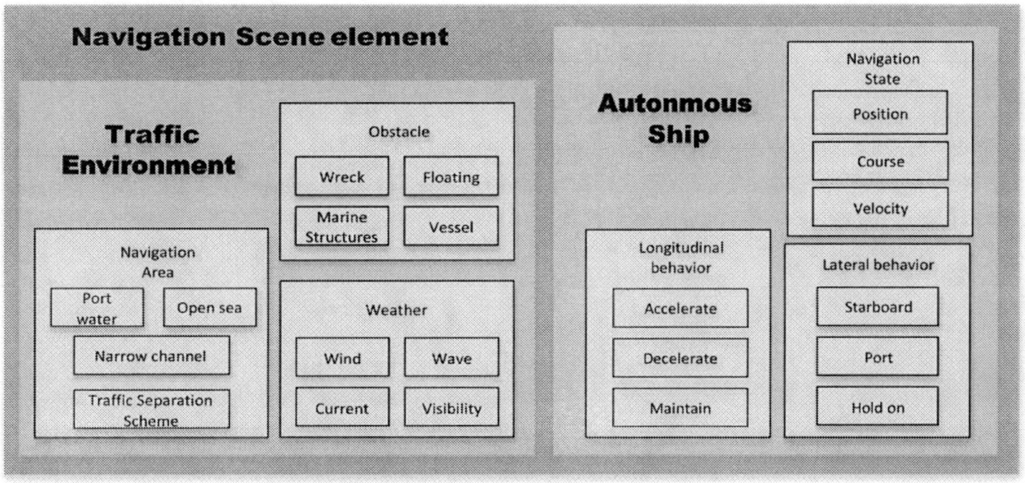

Figure 2. Navigation scene element

From the perspective of autonomous ship collision avoidance decision, this figure divides the navigation scene elements into two parts. One part is the information related to the autonomous ship itself, including position and collision avoidance decision information. The decision information includes lateral steering decision and longitudinal speed change decision. The other part is related information of traffic environment, mainly including weather conditions, obstacle conditions and navigation area. The specific composition and interpretation of the entities of various elements are elaborated in Section 2.3.

2.3. Navigation scene ontology model

After the hierarchical division of the scene elements, these elements need to be conceptualized to construct the TBox to complete the construction of the navigation scene in ontology model. The autonomous ship (own ship), environmental elements, obstacle elements and chart elements are transformed into entities, and the attributes of the entities are constructed to describe the relationship between the autonomous ship and the navigation environment. The conceptual model of the navigation scene ontology based on figure 2 is organized by 4 parts:

- Chart entity: Chart is an important source of knowledge of ship's navigation scene, The ship's navigation scene is constrained by constructing the type pairs of different navigation areas to facilitate the construction of navigation rules for different navigation areas.
- Environmental entity: Environmental entities are used to describe current weather conditions and sea conditions, mainly including wind, waves, currents and visibility, wind, waves, currents, water depth, etc.
- Autonomous ship entity: this entity is used to describe autonomous ship's navigation state, which include position, velocity, course. possible driving decision include the change of velocity and course.
- Obstacle entity: Obstacle entities are divided into two categories, one is dynamic obstacles including various types of ships and marine animal, and the other is static obstacles including various types of navigation aids and marine structures;

After the construction of the entity class is completed, the interactivity and characteristics of the scene elements need to be converted into properties of related entities. The properties mainly have three parts including position attributes, relationship attributes and data attributes. The properties listed in the relevant part of this article are shown in Table 1.

Table 1. Navigation ontology model property

Name	Property	Domain	Range
Position property	hasLongitude	autonomous ship /target ship	double
	hasLatitude	autonomous ship /target ship	double
Relationship property	hasForntObstancle	autonomous ship	obstacle(dynamic/static)
	hasLeftObstancle	autonomous ship	obstacle(dynamic/static)
	hasRightObstancle	autonomous ship	obstacle(dynamic/static)
	hasBehindObstancle	autonomous ship	obstacle(dynamic/static)
Data property	hasTCPA	target ship	double
	hasDCPA	autonomous ship /target ship	double
	hasVelocity	autonomous ship /target ship	double
	hasRange	autonomous ship /target ship	double
	hasCourse	autonomous ship /target ship	double
	hasVelocity	autonomous ship /target ship	double
	hasWind	Navigation Area	int
	hasCurrent	Navigation Area	double
	hasVisibility	Navigation Area	int

By constructing the entity and the properties of the entity, the dynamic and static characteristics of the autonomous ship and the navigation environment in the actual navigation scene can be clearly expressed. the reconstructed navigation safety information scene is realized in the "brain" of own autonomous ship's driving system, this process can be regarded as the polit's cognitive process of looking at the surrounding environment in the brain.

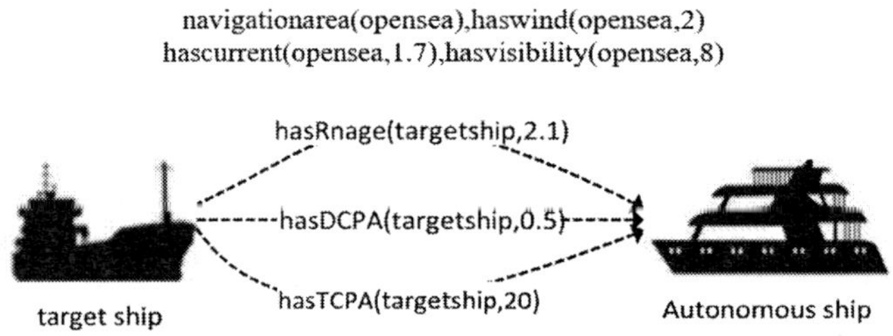

Figure 3. Instantiation of collision avoidance scenarios in open sea

Figure 3 shows the instantiation of ABox in an open sea navigation scenario.

Assuming that the autonomous ship's navigation area is open sea, there are no obstacles other than a target ship, the current navigation environment is instantiated as *navigationarea (opensea)*, the wind level is instantiated as *haswind (opensea,2)*, and the current rate is instantiated as *hascurrent (opensea,1.7)*, the visibility level is instantiated as *hasvisibility(opensea,8)*, the obstacle information is instantiated as *obstancle(targetship)*, the course and velocity of the target ship can be instantiated as hasCourse(targetship,180), *hasVelocity(targetship ,7.5)* ,Instantiate the calculated distance of the target ship, TCPA, DCPA into *hasRange (targetship,2.1)*, *hasTCPA (targetship,20)*, *hasDCPA (targetship, 0.5)*. the course and speed related to the autonomous *ship (ownship)* can be turned into *hasCourse(ownship,000)*, *hasVelocity(ownship,8.6)*, At this point, the instantiation process of the entire navigation element is completed, and that can be regard as the process of ABox.

IOP Conf. Series: Materials Science and Engineering **929** (2020) 012029 doi:10.1088/1757-899X/929/1/012029

3. Navigation Decision Rule Base

The construction of the navigation scene ontology model is equivalent to the pilot's prior knowledge of the environment, and the navigation decision rule base is equivalent to the acquired and accumulated driving experience of the ship's pilot, including seamanship and navigation rules. By combining the ontology model and the rule base, the process of simulating experienced pilots to make collision avoidance decisions can be realized.

Considering the huge workload of building a complete rule base for the entire navigation area, this paper uses SWI-Prolog software to build a navigation rule base for open sea to store and query rules.

The rule base lays the foundation for building a comprehensive rule base and solves the problem of collision avoidance decision based on scene understanding. The rule base is mainly composed of two parts, The first part uses the results of the instantiation of the ontology model to reason about the encounter situation to come up with the specific encounter situation, the second part further refines the constraints under the current situation and matches matching collision avoidance decision information with driving experience. The following explains the two parts of the rules.

3.1. Rules for situation classification

On the basis of following the spirit of COLREGS, the encounter situation between the two ships is divided into five situations: heading-on, overtaking, overtaken, left crossing situation, right crossing situation. Consider specifying in the provisions of the collision avoidance rules'When a vessel is in any doubt as to whether such a situation exists she shall assume that it does exist and act accordingly' And may involve a situation where uncoordinated collision avoidance exists, this paper sets the angle of overtaking as 107.5 , and 351 as the boundary between the heading-on and crossing encounter situation. Examples and explanations of some prolog rules in the prolog rule base are shown in Table 2.

Table 2. The definition encounter situations in the rule base

Encounter Situation	Instantiation information	Reasoning condition
Overtaking	ts_hasBT(ts,BT), ts_hasTCPA(ts,TCPA), ts_hasRange(ts,R),ts_hasVelocity(ts,VT), os_hasVelocity(os,VO),ts_hasCT(ts,CT), os_hasCO(os,CO).	107.5+CT≤CO+BT+180≤CT+252.5 VO>VT, R≤3, TCPA>0.
Heading-on	ts_hasBT(ts,BT), ts_hasCT(ts,CT), ts_hasCR(ts,CR), os_hasCO(os,CO),	BT>=351(or BT≤5), 174≤|CT-CO|≤186, CR>0.
Right crossing	ts_hasBT(ts,BT), ts_hasCT(ts,CT), ts_hasCR(ts,CR),os_hasCO(os,CO).	0≤BT≤117.5,0<|CT-CO|<180, CR>0.

In Table 2, the velocity and course attributes of the target ship will be dynamically loaded according to the system settings. It should be noted that the variables in the rule base are expressed in capital letters (the same settings are used in Table 3), such as ts_hasTCPA(ts,TCPA), it indicates that the target ship (ts) has the attribute value TCPA at the current moment, After the decision is made, the data will be automatically deleted, a new round of data will be instantiated to express navigation information about own ship and target ship.

3.2. Rules for avoidance decision

In nautical practice, an experienced ship pilot does not give a specific steering angle, but instead steers the ship by order such as 'port pass astern' or 'starboard pass ahead'. Therefore, on the basis of meeting the rules for encounter situation, the rule base combines the common practices of seafarers and the influence of seamanship on the steering collision avoidance method, and simplifies the ship collision avoidance actions into 16 action rules. The base includes encounter situation (ES), decision condition (DC) and execution action (EA), Some rules are shown in Table 3.

Table 3. Reasoning and conditions for driving decisions

Encounter situation	Prolog rules	
	Decision conditions	Action rules
heading on	encounter_situation_TS(ts,heading_on)	starboard
right cross	180≤CT-CO≤270, CT>=BT+180	starboard
(small angle)	180≤CT-CO≤270, CT<BT+180	port
right cross	270≤CT-CO≤360, CT>=BT+180	port
(large angle)	270≤CT-CO≤360, CT<BT+180	starboard

4. Experiment

4.1. Simulation environment

In order to verify the usability and effectiveness of the proposed model, a two-dimensional simulation environment based on python was built, and the connection between prolog and python was realized through the plug-in pyswip. By converting various types of navigation data in python into facts in prolog to implement rule-based query, python implements collision avoidance decision by turning decisions returned by the rule base. In order to realize collision avoidance decision based on prior knowledge simulation experiments are carried out on the encounter situations that may be occurred in navigation practice such as heading on, overtaking, and right crossing in simulation environment. In open sea, the initial navigation states information of own ship and target ship is shown in Table 4.

Table 4. Navigation states information of ship

Encounter type	Ship in situation	Initial position of the ship	Direction of approach of the obstacle	Velocity of the ship (nm)
Heading on	own ship	(118.30°,38.82°)	—	(+6.8, -4.2)
	targetship1	(118.26°,38.87°)	Bow	(-6.5, +3.8)
Overtaking	own ship	(118.30°,38.82°)	—	(+6.8, -4.1)
	targetship2	(118.29°,38.84°)	Bow	(+6.3, -4.0)
Crossing1 (small angle)	own ship	(118.25°,38.82°)	—	(+5.3 +6.2)
	targetship3	(118.32°,38.84°)	Starboard	(-2.1, -11.8)
Crossing2 (large angle)	own ship	(118.25°,38.83°)	—	(+4.6, +5.9)
	targetship4	(118.27°,38.82°)	Starboard	(+10.5, -0.9)

4.2. Heading-on

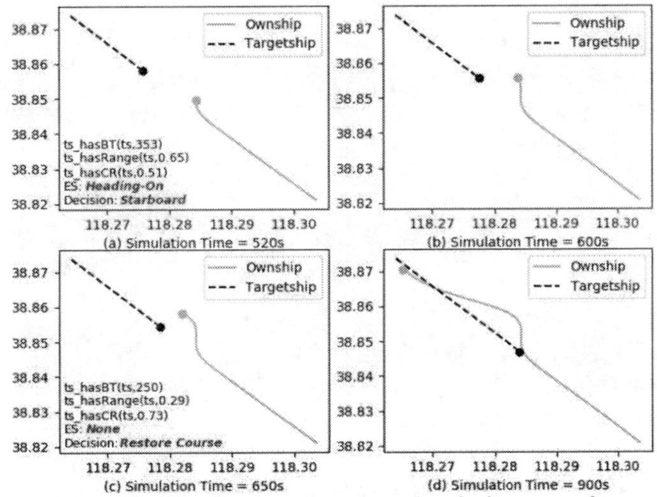

Figure 4. Heading-on situation simulation experiment

In Figure 4(a), own ship takes 520s from the initial position to the current position, At this time, the relative bearing and distance of the target ship are instantiated as *ts_hasBT(ts,353)* and *ts_hasRange(ts,0.65)* in the prolog rule base, which satisfies the reasoning system's conditions. the system output the situation is heading-on, and the total collision risk reaches 0.5, which meets the collision avoidance decision requirements for starboard under the encounter situation, and feeds the decision back to own ship to perform the right turn avoidance action such as Figure 4(b) shows. At t = 650s, own ship and target ship have not formed encounter situation, the reasoning system output meeting situation is None, and own ship arrived at CPA (closest point of approach), it is confine with navigation practice 'when the other ship passed through own ship's abeam, own ship start restore course.' The condition of starting restore course is satisfied, the reasoning system outputs the decision of restore course of flight as shown in Figure 4(c), after the restore course is executed, own ship has returned to the original course, as shown in Figure 4(d).

4.3. Overtaking

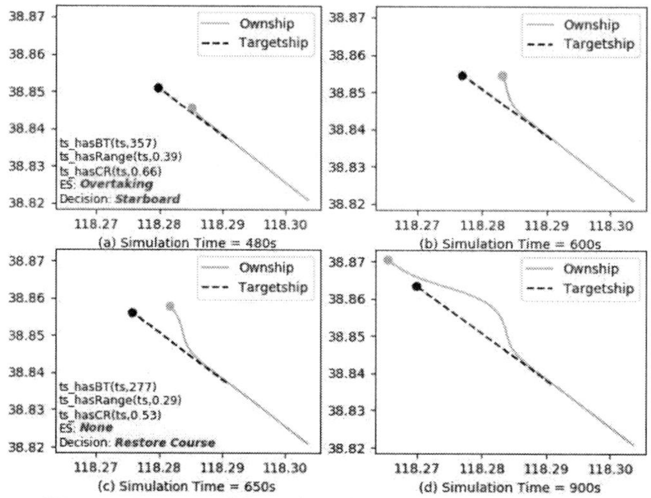

Figure 5. Overtaking situation simulation experiment

In this scene, when own ship sails to the position shown in Figure 5(a), the relative bearing and distance of the target ship are instantiated as *ts_hasBT(ts,357)* and *ts_hasRange(ts,0.39)* in the prolog rule base. The condition of overtaking from the right behind of the target ship is satisfied in the reasoning system, and the reasoning result will be executed. The effect after turning is as shown in Figure 5(b). When own ship and target ship reach the position in Figure 5(c), own ship and target ship no longer constitute an encounter situation, so the output of the reasoning system is None, and own ship will pass the CPA, the reasoning system output the restore course decision. Figure 5(d) shows the effect after the decision is completed.

4.4. Crossing 1

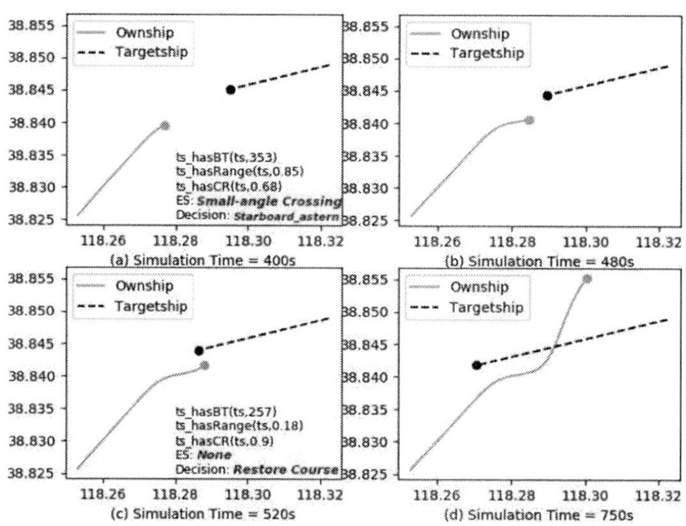

Figure 6. Crossing1 situation simulation experiment

Figure 6 shows the execution process of the reasoning system when small angle crossing occurs.When own ship navigates to the position shown in Figure 6(a), the relative bearing and distance of the target ship are instantiated as ts_hasBT(ts,353) and ts_hasRange(ts,0.85) in the prolog rule base. At this time, the judgment condition of small angle crossing is activated, and the system deduced that the encounter situation is small angle crossing. At the same time, according to the judgment basis of risk degree proposed in Section 3.2, the reasoning system give the decision of starboard stern pass, which conforms to the nautical practice experience. Figure 6(b) shows the results of the collision avoidance decision. When own ship sails to the CPA in Figure 6(c), own ship starts to make the decision to restore course. After the decision to restore course is made, own ship has returned to the original course, as shown in Figure 6(d).

4.5. Crossing 2

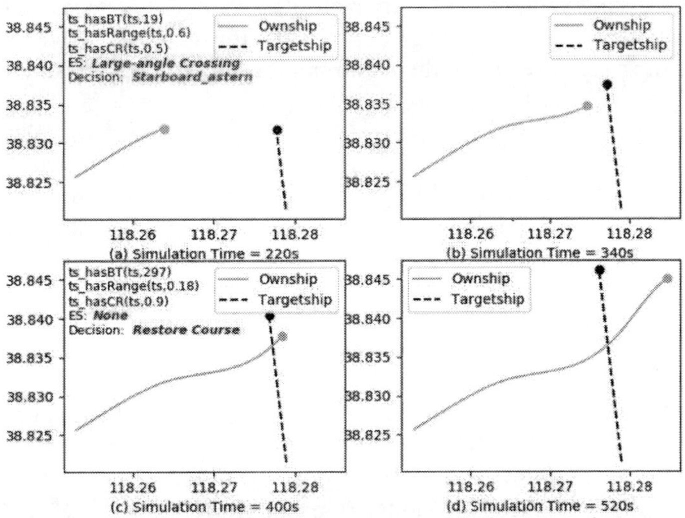

Figure 7. Crossing2 situation simulation experiment

Figure 7 shows a complete right large angle crossing situation and the reasoning process of own ship. Figure 7 (a) shows that the target ship will pass by starboard of own ship in the direction of own ship's beam, and own ship's responsibility is give-way. The relative bearing and distance of the target ship are instantiated as *ts_hasBT(ts,19)* and *ts_hasRange(ts,0.6)* in the prolog rule base, these information together with own ship's course and speed is transmitted to the reasoning system, and it can be concluded that the situation is right large angle crossing. Meantime, the conclusion the avoidance action deduced is starboard astern pass, Figure 7(b) shows the results of the collision avoidance decision. When own ship sails to the CPA in Figure 7(c), own ship starts to make the decision to restore course. own ship has returned to the original course, as shown in Figure 4(d).

4.6. Result analysis

The results of simulation experiments show that the method based on ontology model proposed in this paper can deduce the steering and collision avoidance action conforming to the actual pilot experience, which provides support for the safe navigation of ships. However, there are still a few problems in the experiment, such as crossing situation, restore course given by reasoning system may contradict with collision risk degree, this phenomenon may be caused by the collision of time or space collision risk accidentally become bigger when own ship is steering, nevertheless, collision risk degree start to decrease sharply after this moment, so this method still guarantees the safety of navigation of ship, and this measure is allowed in the navigation practice.

5. Conclusion

Aiming at the problems of complex navigation safety information and difficulties in the fusion of multi-source heterogeneous information in the process of collision avoidance, an entity and entity attributes are constructed to semantically express the elements of the navigation scene, moreover, a completed ontology model of the conceptual navigation scene is constructed for the entire navigation area. At the same time, the navigation rule base based on a priori knowledge are designed at open sea for realizing the reasoning and query of the ship's encounter situation and collision avoidance actions. Furthermore, a simulation experiment is designed based on the velocity obstacle method to verify the proposed model. However, the rule base designed in this paper does not contain the rules for collision avoidance in restricted waters. It is necessary to further increase the completeness of the rule base in accordance with the detailed provisions of the collision avoidance rules for the sake of realizing the anthropomorphic driving decision of ships in different navigation areas.

Acknowledgments

This work was supported by the National Key Research and Development Program of China (Grant no. 2018YFB1601502), and the National Natural Science Foundation of China (Grant no. 51779028).

References

[1] Endsley, M. R. 1995. Toward a theory of situation awareness in dynamic systems. *Human factors.* **37** 32-64

[2] Sharma, A., Nazir, S., Ernstsen, J. 2019. Situation awareness information requirements for maritime navigation: A goal directed task analysis. *Safety Science.* **120** 745-752

[3] Hansen, Peter, Papageorgiou, Dimitrios, Blanke, Mogens, Galeazzi, Roberto, Lützen, Marie, Mogensen, John, Bennedsen, Mette, Hansen, Dorte. 2020. COLREGs-based Situation Awareness for Marine Vessels -a Discrete Event Systems Approach

[4] Zhou, X. Y., Liu, Z. J., Wu, Z. L., Wang, F. W. 2019. Quantitative processing of situation awareness for autonomous ships navigation. *TransNav: International Journal on Marine Navigation and Safety of Sea Transportation.* **13** 25-31

[5] Smirnova, O. V. 2018. Situation Awareness for Navigation Safety Control. *TransNav: International Journal on Marine Navigation and Safety of Sea Transportation* **12 383-388**

[6] Wei, W., Gao, S., Chu, X., SidibÃ©, A. 2018, May. SNS-MEBN Based Method for Situational Awareness of Ship Navigation. In *2018 3rd International Conference on Information Systems Engineering* (China: Shanghai) pp 73-77

[7] Li, L., Chen, G., Li, G. 2015. Study on Auto Decision-Making and Its Simulation Control for Vessel Collision Avoidance. In *Proceedings of the 2015 Chinese Intelligent Automation Conference* (Berlin: Heidelberg) pp 265-275

[8] Chen, G., Yin, Y., Li, L. 2010, October. Mechanism and simulation of personifying intelligent decision-making for vessel collision avoidance. In *2010 International Conference on Computer Application and System Modeling* (China: Taiyuan) chapter 4 pp V4-681

[9] Li, L., Yang, S., Suo, Y., Chen, G., Wang, J. 2008, Automation method for personifying intelligent decision-making for vessel collision avoidance. In *2008 IEEE International Conference on Automation and Logistics* (China: Qingdao) pp 1876-1881

[10] Zhao, L., Roh, M. I. 2019. COLREGs-compliant multiship collision avoidance based on deep reinforcement learning. *Ocean Engineering.* **191** 106436

[11] Ma, Y., Zhao, Y., Wang, Y., Gan, L., Zheng, Y. 2020. Collision-avoidance under COLREGS

for unmanned surface vehicles via deep reinforcement learning. *Maritime Policy Management*.**47** 665-686

[12] Guo, S., Zhang, X., Zheng, Y., Du, Y. 2020. An Autonomous Path Planning Model for Unmanned Ships Based on Deep Reinforcement Learning. *Sensors*. **20** 426

[13] Lyu, H., Yin, Y. 2019. COLREGS-constrained real-time path planning for autonomous ships using modified artificial potential fields. *The Journal of Navigation*. **72** 588-608

[14] Huang, Y., Chen, L., van Gelder, P. H. A. J. M. 2019. Generalized velocity obstacle algorithm for preventing ship collisions at sea. *Ocean Engineering*. **173** 142-156.

[15] Niu, H., Savvaris, A., Tsourdos, A., Ji, Z. 2019. Voronoi-visibility roadmap-based path planning algorithm for unmanned surface vehicles. *Journal of Navigation*. **72** 850-874.

[16] Singh, Y., Sharma, S., Sutton, R., Hatton, D., Khan, A. 2018. A constrained A* approach towards optimal path planning for an unmanned surface vehicle in a maritime environment containing dynamic obstacles and ocean currents. *Ocean Engineering*. **169** 187-201.

[17] Gu, S., Zhou, C., Wen, Y., Zhong, X., Zhu, M., Xiao, C., Du, Z. 2020. A motion planning method for unmanned surface vehicle in restricted waters. *Proceedings of the Institution of Mechanical Engineer*.M **234** 332-345.

[18] Zhang, W., Xu, Y., Xie, J. 2019, April. Path Planning of USV Based on Improved Hybrid Genetic Algorithm. In *2019 European Navigation Conference ENC* (Poland:Warsaw) pp 1-7.

[19] Wang, H., Guo, F., Yao, H., He, S., Xu, X. 2019. Collision avoidance planning method of USV based on improved ant colony optimization algorithm. *IEEE Access*. **7** 52964-75.

AEGIS: Advanced, efficient and green intermodal systems

Ørnulf Jan Rødseth[1,*], Harilaos N. Psaraftis[2], Stefan Krause[3], Jesper Raakjær[4], Nelson F. Coelho[4]

[1]SINTEF Ocean, Trondheim, Norway
[2]Technical University of Denmark
[3]Institut fur Strukturleichtbau und Energieeffizienz, Germany
[4]University of Aalborg

*Corresponding author, e-mail: Ornulfjan.Rodseth@sintef.no

Abstract. The European maritime transport policy recognizes the importance of the waterborne transport systems as key elements for sustainable growth in Europe. A major goal is to transfer more than 50% of road transport to rail or waterways within 2050. However, waterways are at a disadvantage as they normally depend on transhipment and land transport to and from final destination. To meet this challenge we need a completely new approach to short sea and inland waterways shipping in Europe. This needs to include ships as well as ports and the digital information exchanges between them. A key element in this is automation of ships, ports and administrative tasks. The AEGIS project has been funded by the EU Commission to develop new knowledge and technology to address this challenge.

1. Introduction

The European maritime transport policy [1] recognizes the importance of the waterborne transport systems as key elements for sustainable growth in Europe: In EU, 30% of road freight over 300 km should shift to rail or waterborne transport by 2030, and more than 50 % by 2050. So far, this ambition has failed [2]: In the period 1995 to 2016, the share of road transport in intra-EU ton-km increased from 45% to 49%, whereas the share of short sea shipping (SSS) dropped from 33% to 32%. Inland had a decrease from 4.3% to 4.0%. Rail was reduced from 13.6% to 11.2%. Total freight, on the other hand, increased with 29% (ton-km).

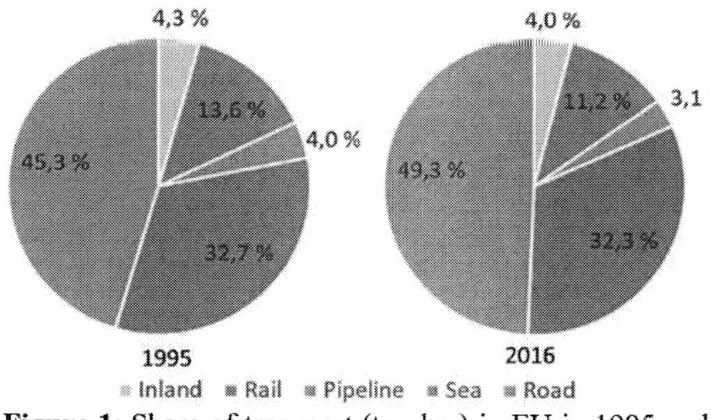

Figure 1: Share of transport (ton-km) in EU in 1995 and 2016 [3].

Waterborne transport has a great potential to reduce road congestion as well as pollution from the transport sector. When used correctly, it is very energy efficient, and new ship types can now operate with a combination of electric batteries, fuel cells and, when necessary, highly efficient combustion engines, e.g. powered by biogas or LNG. This can make them virtually emission, dust and noise free. However, ships have for a long time grown larger to reduce energy and operations cost. This restricts the terminals they can call on, increases costs and sizes of these terminals and reduces service flexibility by reducing frequency and fixing one speed for all cargo on one ship. Figure 2 summarizes the conclusions on why short sea waterborne transport today cannot compete efficiently with trucks, from the Auditor General in Norway [3]. Ship transport is generally cheaper and friendlier to the environment, but it loses on offered service quality. Fewer and larger ships are one part of this problem.

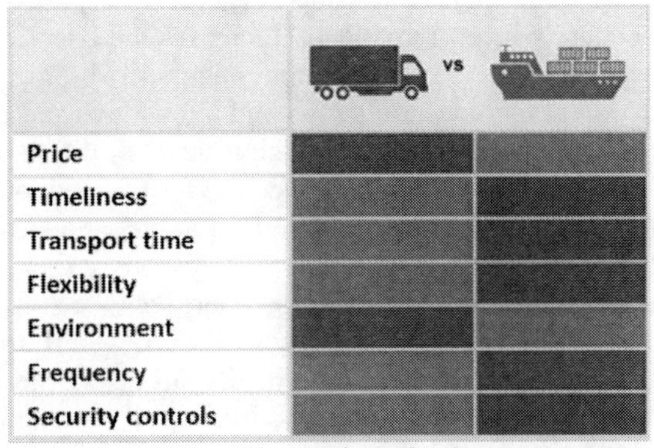

Figure 2: Why Norwegian short sea transport fails to compete with trucks [3].

AEGIS proposes a concept with more and smaller ships that can be used to increase frequency, differentiate speeds, reduce terminal costs and reduce time in port for the larger ships as they will not need to call smaller ports. Several smaller ships also increase reliability and resilience in the transport systems. Breakdown of one vessel has little impact on transport capacities. Automated cargo handling and standardized cargo units will further reduce problems and transhipment costs in ports and on ships. In addition, ships are most efficient when the cargo holds are full, so AEGIS also addresses how to attract new cargo, inbound as outbound, to waterborne transport. This requires new types of services, new business models, better logistics and more streamlined digital solutions.

This paper will first explain why autonomy and automation are critical factors in tomorrow's waterborne transport systems (section 2). It will then give an overview of the AEGIS use cases and explain how autonomy and automation is used in these (section 3). We will then briefly describe the main technologies and operations developed in the project (section 4), how we plan to measure economic and societal benefits (section 5), before we summarize and conclude in section 6.

2. Autonomy and automation are critical factors

2.1. Autonomy can defeat economy of scale

Autonomy can defeat economy of scale simply by reducing personal costs, increasing efficiency, resilience, speed and frequency and thus the competitiveness compared to other modes of transport. Small automated shuttles can operate more flexible and are much more economic because the crew costs will be much lower. By reducing the crew, the personnel and related auxiliary costs can be decreased significantly. That is very important due to the fact that crew-related expenses are approximately between 10 to 44 % of a ship owner’s operating expenditure. In addition to the described cost savings, autonomous ships offer additional benefits due to changes in ship design and possibilities for more efficient use of fuel. Data from the MUNIN projects predict savings of USD 7 Million over a 25-year period for an autonomous merchant ship, due to improved fuel consumption and reduced crew supplies and salaries [4].

It is expected that autonomy and automation in ships can improve safety, increase efficiency and relieve humans from unsafe and repetitive tasks. A study by the insurance company Allianz shows that between 75% and 96% of maritime accidents are caused by human errors. Autonomous ships also offer a possibility to solve the growing problem of lack of qualified personnel onboard. In addition new types of jobs will be created on land by automation such as high-skilled route operators, new kinds of pilots and technical maintenance. Some claim that autonomous hips are more likely to alter jobs rather than eliminate jobs. Combined with the creation of new types of jobs, a greater prosperity is expected [5].

2.2. Automated cargo handling and more standardized cargo units
Cargo handling is a considerable cost driver in transport. In waterborne logistics, transhipments are often needed, which dramatically increases the costs. Terminal handling charges (THC) are typically on the order of around EUR 100 per container [6]. Ship to ship transhipment is even more expensive and THC does not include storage beyond a few days and not transport to and from the terminal gate. For shorter distances transport, this is a significant addition to the basic freight rate. As an example, a typical short sea route on the Norwegian costs is estimated to cost about EUR 284 for 236 km by road while a small autonomous feeder ship can do it at a cost as low as EUR 133 (194 km by sea) [7]. However, adding two THC to the by sea cost will obviously create problems for the short sea alternative. In addition to that, one has to add the various port fees the ship may have to pay.

To be able to develop competitive waterborne door-to-door logistics systems, more automation in cargo handling is required. Ideally, all discharge and loading operations should be automated. This can be used to minimize labour costs which are one important factor. Operation around the clock will also reduce costs and increase service quality for the users but will also require full automation to avoid problems with working hours and additional shift costs.

In addition to personnel costs, capital costs will also be a significant factor in THC. Thus, it is necessary to find more reasonably priced equipment for automated cargo handling. One possibility for cost reduction is improved standardization of cargo units. This applies to physical sizes as well as to maximum weights. One element in this may be to provide automated handling only for lower weight, e.g. max 25 ton 20 feet containers and charge a higher price for manually handling larger containers with more powerful cranes. This could also provide positive feedback by defining a more competitive business model for using the standard container sizes wherever possible. This issue is a part of the investigations in the AEGIS project.

2.3. Automation of administrative processes
Physical automation is an important part of AEGIS, but cargo information management must also be automated. The World Bank [8] has analysed 190 countries and shows that in certain cases, the administrative paperwork related to export through seaports can incur direct costs of more than USD 2000 and delays of more than 14 days per container. In addition to import and export, come all logistic related information exchanges. This is particularly important for seamless integration with automated cargo handling and efficient transport to and from storage in the terminal.

Another central issue in this work is to enable the seamless transhipment between the larger international cargo ports with their ISPS (International Ship and Port Security code) compliant and customs guarded terminals, and smaller rural or urban terminals without such facilities. This means that customs clearance and ISPS security checks must be done in a way that do not hinder efficient movement of goods (see case A in section 3.1), but that it neither interferes with the correct tariff payment nor the assurance of the security and safety of the ship and port.

AEGIS will investigate these issues and develop new concepts for port community systems (PCS) in small ports. These systems will be based on full digitalization of all relevant information messages and an improved logistics management system. It will make use of new emerging message standards based on the new IMO Reference Data Model [9] and extensions to that, developed through the AEGIS project. Inherent in this concept is also use of electronic signatures to ensure sufficient trust in the information received to enable automated processing of it [10]. The public key infrastructure will not be developed in AEGIS but will be reused from related and ongoing projects.

3. Three AEGIS use cases

3.1. Case A: Short sea, urban and rural terminals in Norway
This case is led by North Sea Container Line in cooperation with the inter-municipal port of Trondheim. It covers transport from large ports in Europe (Rotterdam) to smaller destinations along a less populated coast of Europe (Norway). This line is a load-on, load-off (LOLO), and mainly container-based transport system. The idea in this use case is to use fewer terminals closer to the main fairway in Norway to

increase liner service speeds and then shuttle cargo to its final destination with small unmanned and preferably autonomous and electric vessels. Figure 3 shows the relevant coastal liner route.

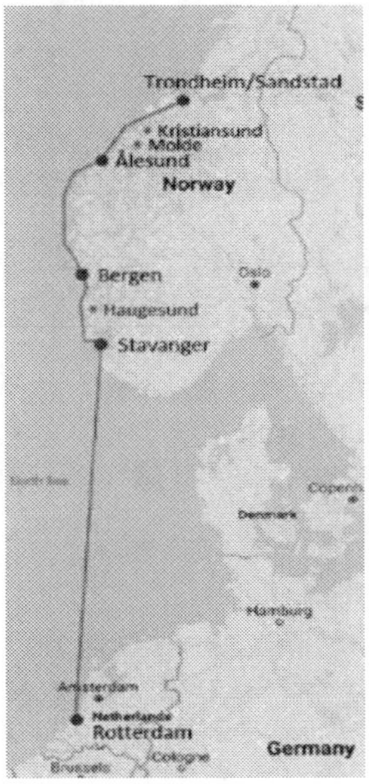

Figure 3: Short sea connection to Norway

Figure 4 shows the regional distribution system. The new fairway terminal is indicated as the large grey circle at Sandstad. By loading and discharging at this terminal instead of going to the commonly used ports inside the fjord (red circles), the liner ship will save between 5 and 10 hours. The rest of the system consists of small international urban ports (red), non-international rural terminals (yellow) and one suggested new terminal with possibilities for transhipment to rail and air (green). The main route northward is indicated with a dashed line as it is not part of the use case today, but will be considered if the coastal route should be extended in that direction. The case will also look at local shuttling to the south-west of the new terminal. It is possible to operate in sheltered waters all the way down to the next major city, Kristiansund (orange terminals).

The use-case will also investigate the complete container transport pattern along the Norwegian coast to find possibilities for better cooperation between transporters to provide more user-centric services. At the time of writing, there are eight operators that provide different types of scheduled services (LOLO, RORO, Heavy lift) on the coast, but with different routes and schedules. A better coordination, possibly also with transhipments between main lines, could drastically improve the user's service quality on seaborne transport in this area.

The project will also investigate a future small and non-storage terminal in cities where cargo can be transferred directly to and from the shuttle without intermediate storage.

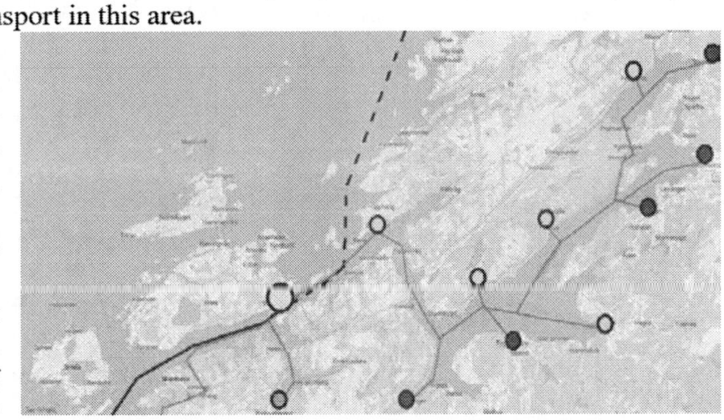

Figure 4: Local distribution in Trondheimsfjorden

The key element in this case is the small container shuttles, perhaps as small as 10 to 20 TEU that provide low cost and flexible transport to rural as well as urban terminals. If roll-on/roll-off (RORO) solutions are used, one may significantly reduce the noise related to the cargo handling process. In rural areas load-on/load-off (LOLO) may be considered, but RORO may also here be a preferred solution as one may be able to manage with a relatively simple ramp into the water for the shuttle to dock to.

3.2. Case B: Short sea and inland shipping interface in Belgium and Netherlands

This use-case is led by DFDS and covers the interface between DFDS' RORO transport from several North European ports to Rotterdam (Vlaardingen – with an existing link to Antwerp), Ghent and Zeebrugge and further on to smaller inland destinations in Flanders with waterway connections. The Rotterdam, Ghent and Zeebrugge roads are heavily congested, but it is currently difficult to get efficient handling of RORO type cargo on barges in the area. The idea is to bring cargo as close to the end destination as possible with small vessels with zero emission propulsion (battery, fuel cells, etc.) and in the future autonomous vessels. The main inland waterway connections between the case ports are shown in Figure 5.

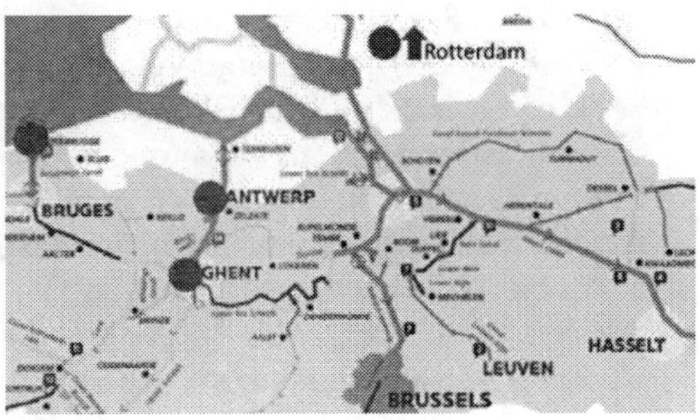

Figure 5: The inland interface case

The use case will consider the use of existing RORO inland vessels, but these are few and difficult to get hold of. Thus, the use case will also need to look at alternative vessel designs to simplify transhipments and cargo handling in general. It may also be relevant to look at AGV use, but this will depend on the characteristics of the different ports. It is also interesting to look at delivery closer to final destination with RORO vessels.

The main emphasis will be on the logistics perspective. A detailed analysis of transport needs between different areas and cities will be carried out followed by an analysis of time lost in transhipment and administrative tasks and other obstacles (mapping according to economic and environmental criteria). AEGIS will develop solutions to minimize such obstacles and will involve vessel size and design, intelligent routing (emissions, time, cost, etc.), cargo handling (ramps for trailers/vehicles) and study potential impact on congestion and other relevant KPIs. The project will take into consideration vessel sizes in relation to the inland network with CEMT (Conference of European Ministers of Transport) categories, specifically to find optimal size to reach most destinations. An ICT solution will also be included, with a proof of concept being developed, in the form of a digital portal, to suggest to the customer and logistics companies the best way to transport the cargo to destination (time, environmental impact, cost, etc.).

3.3. Case C: Revitalizing regional ports and city centre terminals in Denmark

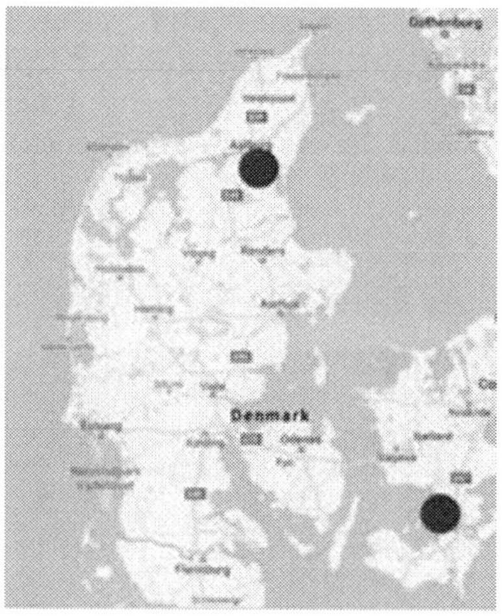

Figure 6: The regional port case

The purpose of this use case is to consider how concepts and solutions developed in AEGIS can be applied to increase efficiency to the value chain and thus increase competitiveness towards road transport in small and medium sized ports. It will be using the two Danish ports Aalborg and Vordingborg as pilot sites. The case will estimate the potential gross volume that can be shifted from road transport to short-sea-shipping in Denmark and analyses of the price structure for transportation of different types of goods by both road transport and short-sea shipping and hereby unfold the potential for a shift towards short-sea-shipping in Denmark. Based on analyses the project will identify potentials and constraints for enhanced short sea transport through the two ports.

In order to examine the challenges and opportunities for increased short-sea shipments in the above mentioned and between neighbouring ports separate and confidential SWOT analyses in the respective ports, including an evaluation on how implementation of AEGIS concepts being may affect the shift towards waterborne transport solutions, and to what degree autonomous solutions may be implemented, including considering if the concept of satellite and autonomy ports within a group of ports can be used to promote fast ship handling and frequent calls. Explore the potential for integrating RORO calls to the existing route schedules giving priority to, but not limited to project partners.

Based on the concepts developed in AEGIS, a new terminal layout will be proposed for the Port of Vordingborg comprising: A multipurpose port area for multiple users on short lease and/or slot booking basis; A multipurpose storage and parking areas with multiple users on short lease and/or slot booking basis; A set-up for autonomous ship handling and self-services packages for both conventional and future autonomous ships. Whereas, for the Port of Aalborg it will be demonstrate how an Intermodal Automatic Green Terminal in medium ports like Aalborg can be designed and technologically equipped based on AEGIS concepts and solutions: A proposal will be made for a new fully automated terminal in the Port of Aalborg, with interface to handle intermodal transport - green rail and short-sea shipping covering RORO, containers and bulk operations.

4. Critical Technologies and Operations

The technical activities in AEGIS are concentrated around seven "Critical Technologies and Operations" (CTO) that are described in the following.

4.1. Logistics system redesign

Automated cargo handling, new terminal types and use of lightweight shuttle vessels to complement and increase overall speed of the main liner ships will enable significant changes in the logistics systems that can address several weak spots. It also makes it possible to stretch waterborne supply chains further into the coastal areas and inland waterways, further increasing flexibility and attract new customers to sea transport. Use of roll-on/roll-off (RORO) solutions combined with automated guided vehicles for the small shuttles can also reduce use and cost of transhipments, for certain cargos, both in cities and in urban areas. Improved standardization of unit cargo will be an important factor in efficient realization of several of these improvements. Also simplification of the digital systems handling safety and security as well as import and exports regulations is important. Finally, we will also need to ensure that the new system is as least as robust and resilient as the old. AEGIS will develop user-centric logistics systems with new components, better service quality and much lower impact on environment and society.

The logistic system redesign will be based on the three use-cases in AEGIS, all representing existing logistics chains operated by ship owners, terminal operators and ports in the consortium, and break them down into the various elements of the complete logistics chain. This will be the basis for further studies to identify bottlenecks and cost drivers by means of qualitative and quantitative methods (optimization and simulation). The outputs of this analysis will show where and how implementation of new business models, technology and systems can further improve the waterborne logistics systems to be a better alternative for a shift from road to sea.

4.2. New terminal concepts

Large terminals already have very high degrees of automation, as e.g. in Singapore and Rotterdam. There are also many publications on various optimization in such ports, e.g. on container storage, intra-port movements, ship arrival optimization etc. [11]. The target of AEGIS is smaller ports with a yearly capacity of typically in the range 20 000 to 100 000 TEU and its links to even smaller terminals with yearly capacities in the 100's of TEU. This means that similar concepts as is used in the large ports have to be scaled down and made applicable to smaller and more cost-sensitive environments.

The concept proposed by AEGIS in case A has similarities to "dry ports" [12], which is an inland intermodal terminal that acts as a centre for transfer of cargo to inland destinations. Here, the emphasis is on efficient truck or rail last mile transport, but AEGIS will extend this idea to the use of more remote seaport terminals, linked to automated water shuttles with a corresponding change in automated cargo handling systems, on the liner ships, in the new remote terminal, and on the shuttles. The concept will in a certain sense look like a large transhipment port like Singapore, where most of the cargo never leaves the port area, but on a much smaller scale. The urban and rural terminals will in most cases rely on self-loading and unloading shuttles, either using RORO or LOLO technology. Capacity for these terminals is on the order of some hundred to a few thousand TEU per year, and large investments in infrastructure cannot be made. The complete system must be digitally integrated to allow paperless transfer of cargo between the terminals.

Case B looks at better utilization of inland waterways (IWW) and in particular the possibility to scale down barge sizes to reach longer into the hinterland and provide more flexible services, similarly to what is proposed in case A. However, case B will mainly interface with an international RORO shipping line, which typically has a large part of their cargo on trailers. These are not normally shipped on IWW vessels and one of the challenges here is to find a good solution to how this type of cargo can be handled in the terminal and on the old and new IWW vessels.

Case C represents an upgrade of medium sized terminals in terms of increased physical and digital automation. The goal is to give better service to the immediate hinterland and to waterborne transport in general, as well as investigating how more of the current truck transport can be encouraged to use the sea. It is necessary to investigate how layout and design have an impact on the efficiency of terminal operations and thus also the efficiency of waterborne logistics systems. Qualitative and quantitative analyses based on simulation will be carried out. Automation in cargo handling is decisive for competitiveness. Concepts will be developed and validated by means of numeric simulation.

4.3. Automatic cargo handling

Cargo handling by state of the art handling systems is a considerable cost driver in waterborne transport. In waterborne logistics, transhipments are often needed, which dramatically increases the costs. THC is around EUR 100 per direction (section 2.2). To be able to develop competitive waterborne door-to-door logistics systems, more automation in cargo handling is required. This may also affect the need for novel and more standardized cargo units.

The new logistics system relies on more flexible exchanges between larger liner ships with LOLO or RORO capabilities, various forms of inland or short sea shuttles and last mile transport on land. This requires cargo transfers and AEGIS needs to automate all or most of these forms for cargo handling. This may require more standardized cargo units to reduce complexity of the equipment, both in terms of unit sizes and maximum weights. The trade-off between automated cargo handling equipment on ships and in terminals will also be investigated. This trade-off will obviously be different for RORO versus LOLO and in small versus large terminals.

New equipment may also include use of automated trucks for certain types of short distance transport from terminals to end users close to the terminals, e.g. if shuttles use RORO solutions. This is most relevant for the rural terminals, but could also be considered in urban areas.

In order to enable an automatic cargo handling of standardized cargo units in terminals and on-board innovative concepts will be developed and validated in simulators, and the impact of novel solutions will be benchmarked concerning profitability and externalities in cost benefit analysis and environment assessments. Therefore the necessary cargo handling requirements and terminal designs will be developed with AEGIS.

4.4. Autonomous shuttle

Small, lightweight and fully automatic cargo shuttles will be used to link waterborne transport from the large ports and terminals to rural or inner-city areas. As stated in Global Marine technology trends 2030 report [13], we need to extend the reach of waterborne further into the logistics chain, including new solutions for last mile transport.

Small autonomous shuttles can significantly reduce transit and port time for the larger ships on their main lines while providing efficient transport to and from end users located further from the main ports and terminals. This can also reduce infrastructure costs in smaller ports and terminals and allow new terminal concepts with less negative impacts on society.

To realize this ambition, the shuttles must also facilitate automatic berthing and cargo operations as well as shore support in the form of electricity or other necessary services. The logistics capabilities of the ship in the selected use cases are the main focus in AEGIS, e.g. related to door-to-door logistics represented by a set of use-cases; - smaller autonomous shuttle vessels from bigger ports along the main fairway to e.g. destinations inside fjords; - how autonomous vessels can be used for last mile deliveries into city centres to reduce road transport and also to reduce the need for storage capacities inside city centres; - how autonomous inland shuttles can be linked to bigger RORO-operations calling the Benelux area to be a competitive alternative to road transport in this region; - and finally how short distance last mile transport on land by means of AGVs can be used to "close the loop" towards the consignee.

The advances beyond state of art in AEGIS is to look at autonomous shuttle as an integral part of the transport system and optimize cargo handling on board with facilities in non-city terminals and at private and public quays. The main case is expected to be LOLO, but in Norway it is, e.g. possible to use car ferry ramps during night time for such vessels, if they carry RORO cargo. Also, the RORO liner case will have to consider the use of RORO solutions, also on the small shuttles.

4.5. Digital connectivity

An addition to physical hindrances, international ship traffic and local port logistics are also hindered by digital barriers. A lack of digitalization and automation of the clearance and planning processes between ship, port, logistics operators and authorities adds significant to the cost of transhipment. In the digital connectivity field, some improvements have been made, but these must be leveraged by supporting new digital cargo and ship documents as well as automation of work processes in shipping and within authorities. Digital connectivity to ships is still much inferior to what land-based transport has. This applies to communication systems, protocols and to integrated management tools.

An important part of this research is communication efficiency and cyber security. The project will make an inventory of the most relevant ship to shore communication systems and their characteristics [14], [15]. The characteristics is mainly the quality of service (e.g. bandwidth, reliability), but special emphasis will be put on properties related to cyber security, e.g. how easy is it to jam or spoof messages when using the communication system. The latter will be used in a criticality analysis of the communication between the different ship types, particularly the autonomous small shuttles, and relevant shore or waterborne assets.

Another important area for research is digital interoperability. Based on work already done in the IMO's Facilitation committee (FAL) [9] and in ISO 28005 [16], the project will extend the scope of IMO reference data model and the implementation of it in the ISO 28005 standard. This is partly a traditional data modelling exercise, but a challenge is that the developments need to be aligned with international work undertaken in various organizations, such as IMO, ISO and IALA [17].

The third activity is an improved port community system (PCS) for small ports. PCS is common in large ports and is used to automate interactions between all commercial parties in the port as well as with authorities' systems [18]. The challenge is to develop a new type of PCS that fits smaller ports, and which can be operated by the port itself. This PCS should also make more extensive use of AIS signals from approaching ships to automate logistics and planning processes.

4.6. Policy measures

The concept designed by AEGIS will require high initial investments and new business models. Deploying this concept without considering the pre-existing socioeconomic scenario may result in less efficiency being brought to the logistics chain rather than more, or to the concept being altogether disregarded by market actors. Policy support will provide an important level of legal certainty and predictability for the necessary investments and business model readjustments to be made.

The transition to short-sea shipping automation bears an impact at various levels of the logistics chain. Automated vessels will at first coexist with unmanned and manned vessels, as well as with non-automated terminals. An overall efficiency gain in that chain produces externalities, namely those associated with the substitution of seafarers with programmers and remote pilots. This transition requires that actors indirectly involved in waterway and coastal navigation are aware of and educated on shipping automation. A preliminary research task involves assessing who was involved in the process so far and what other actors may have a role to play in the governance arrangement renegotiation.

The political ambition to ensure a transition towards short-sea shipping will possibly require public authorities to intervene at the national level and at the international level (EU/IMO). This intervention translates into two types of approaches: the removal of existing obstacles and the creation of new incentives. The need to exempt, either temporarily or permanently, automated vessel operators and terminal handlers from fiscal and administrative burdens applicable to their competitors requires sound evidence. Such evidence can for example be determined by reference to qualitative KPI (section 5) and quantitative assessments of use cases specific issues. The policy design may also consist of new incentives designed specifically for the AEGIS concept. A positive discrimination between alternative but equally lawful transportation solutions may however be considered an anticompetitive behaviour. Fine-tuning the possible nudges to ensure the AEGIS concept remains attractive all the while not impairing fair competition is a key activity in this regard. For that purpose, it is necessary to establish a dialogue between all the parties present in the AEGIS consortium, together with external stakeholders and policymakers, on perceived hurdles and alternative policy solutions.

The successful delivery of policy support to the implementation of the AEGIS concept depends on the willingness of all the actors consulted to participate and contribute to producing the knowledge associated with the governance of AEGIS concept, and to reflect on it. The outcome will expectedly be a revision of existing norms governing short-sea shipping in a way that accommodates the new AEGIS concept and steers all actors participating in the logistics chain towards shifting freight from land to sea.

4.7. Safety, security and resilience

The whole idea of AEGIS is to develop more integrated transport systems, both physically and digitally. On one hand, this creates new threats as the system may become more integrated and reliant on individual components, but it also creates new possibilities in that the system is more diverse, has smaller units and may be easier reconfigured in case of major problems or attacks. AEGIS aims both to reduce threats to individual components and increase resilience by offering more diversity and easier reconfiguration. Thus, a focus both on the possibilities for increased resilience as well as the more complex threat picture is an important part of AEGIS.

Safety, security and resilience are not in a separate work item in AEGIS. It is an embedded component of most activities. This reflects the ubiquitous nature of these factors. Research in AEGIS on safety, security and resilience will mostly follow the lines of previous research, but more focused on the cargo handling, smaller last mile vessels and how these are integrated in the whole transport system. A particularly important issue is establishment of digital trust in digitalized and automated work processes. This requires electronic signatures, integrity checks and non-repudiation of electronic exchanges. This is an issue similar to cyber security, but with implications much beyond that. AEGIS will also support the development of a related work item in the IMO Facilitation Committee where ISO have provided some suggestions for how this can be implemented [19].

Resilience is a special theme in the logistics systems redesign where it is linked to the implementation of the more distributed and automated transport system. In general, one may expect that more complex technology and higher reliance on wireless communication reduces resilience, but a system with several smaller shuttle vessels may also turn out to be more robust against disruptions. We will follow traditional research approaches but will need new metrics on the robustness of the automated features of the transport system. Also, the risk of cyber security attacks needs to be included in the analysis.

Safety of autonomous ships is a relatively well published area although one cannot say that any of the theories have been verified or tried. However, much research has been done and AEGIS will make use of already published results. One area of special interest is the danger of "stowaways" on small cargo shuttles. When operating close to land without crew on board, it can be an interesting target for youths that are out in small boats in the area the shuttles operates. We will also need to consider similar "attacks" where the vessel is forced off-course by deliberately running small boats on collision course at its intended route.

5. Measuring the economic and societal effects

Whatever solutions are contemplated in AEGIS, it is imperative to assess them in a holistic fashion, so as to capture the effects of all conceivable cross-linkages and interdependencies and hopefully obtain what we call "win-win" solutions. The objective here is to perform a comprehensive Cost Benefit Analysis and environmental assessment including also social aspects, as is indicatively depicted in the figure in the context of sustainable maritime transport. One can see from Figure 7 that a great number of factors can be at play. In the context of the AEGIS project, the distinct perspective is the interaction among the *environmental,* the *economic* and the *social* dimensions [20]. Achieving a balance among these is important as it would make no sense for the vessels, the shipping operations or the maritime supply chain systems under study in AEGIS to be performing well economically but be non-viable from an environmental or social perspective. So, it is imperative that all trade-offs be evaluated and achieving what we call "win-win" solutions is an indispensable prerequisite. The study on these impacts will link to the case studies examined in AEGIS and also to several other activities of the project.

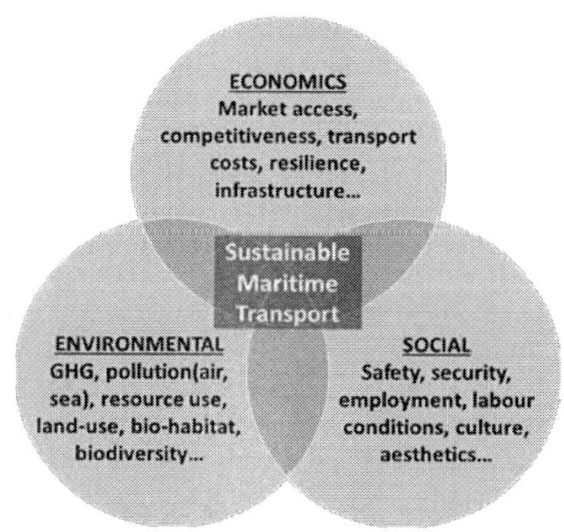

Figure 7: Sustainable Maritime Transport, based on [20].

To perform this analysis, our work will consist of the following tasks:

Identification of KPIs: This task will develop a comprehensive set of Key Performance Indicators (KPIs). KPIs represent the criteria under which the set of solutions developed under AEGIS will be evaluated. They include criteria grouped under the following classes (refer also to figure above):

- Economic KPIs including cost, profit, logistical efficiency and others
- Environmental KPIs including GHG and other emissions
- Social KPIs including safety, security, externalities and others.

Economic analysis: This task will evaluate the solutions developed under AEGIS in terms of the economic KPIs developed as per above. The full range of AEGIS case studies will be investigated. Economic data will be collected for all these case studies including their variants.

Environmental analysis: Similar to the economic analysis task, this task will evaluate the solutions developed under AEGIS in terms of the environmental KPIs identified in the first task.

Social analysis: Also running in parallel to the previous two tasks, this task will evaluate the solutions developed under AEGIS in terms of the social KPIs,

Identification of win-win solutions. For each of the user case studies studied in the AEGIS project, and the analyses performed as per above, this task will identify "win-win" solutions, as well as the conditions for these solutions to be realized. A "win-win" solution is defined in terms of being acceptable in terms of most of the KPIs that have been identified as important. These solutions will also be compared to the solutions in common use today.

6. Conclusions

The central objective of AEGIS is to develop a new waterborne transport system for Europe that leverages the benefits of ships and barges while overcoming the conventional problems like dependence on large terminals, high transhipment costs, low speed and frequency and low automation in information processing.

To achieve this objective, AEGIS will leverage a multidisciplinary team to integrate new innovations from the area of Connected and Automated Transport (CAT), including more diverse sizes of ships and more flexible ship systems, automated cargo handling, ports and short sea shuttles, standardized cargo units and new digital technologies to design the next generation sustainable and highly competitive waterborne transport system in Europe.

AEGIS will also integrate technological advances on ship and cargo systems, development of improved digital connectivity and automation, and specific logistics use cases with studies on how this can be realized in terms of necessary policy measures and how the effects can be measured by economical, environmental and societal key performance indicators.

Many of the activities in AEGIS will be of a pre-competitive nature and the project plan includes provisions for extensive publication both of project reports and public papers. These will be made available on the project web site [21] as well as in other channels.

Acknowledgements

The AEGIS project has received funding from the European Union's Horizon 2020 research and innovation program under Grant Agreement N°859992. Partners in AEGIS are Port of Aalborg (DK), Aalborg University (DK), Kalmar (FI), DFDS (DK), DTU (DK), Grieg Connect (NO), Institut für Strukturleichtbau und Energieeffizienz (DE), MacGregor (FI), NCL (NO), SINTEF Ocean (NO), Port of Trondheim (NO) and Port of Vordingborg (DK).

References

[1] Strategic goals and recommendations for EU's maritime transport policy until 2018 COM/2009/0008 final/Brussels, 21 January 2009.
[2] EU 2018, *Statistical pocketbook 2018*, European Commission, Mobility and Transport.
[3] Auditor General Norway, 2018, *Transport of cargo*, Retrieved June 2020 from: https://www.riksrevisjonen.no/rapporter/Sider/Godstransport.aspx (In Norwegian).
[4] The MUNIN project, https://cordis.europa.eu/project/id/314286/reporting (Retrieved July 2020)

[5] The Manufacturer, https://www.themanufacturer.com/articles/autonomous-shipping-future-seafaring/ (Retrieved July 2020).

[6] Hackett, B., 2009. Terminal handling charges during and after the liner conference era. Competition Report, Raven Trading Limited for EC Directorate Competition, EU's Publications office.

[7] Gribkovskaia, V., Borgen, H., Holte, E.A., Lindstad, E. and Nordahl, H., 2019, October. Autonomous ships for coastal and short-sea shipping. In *SNAME Maritime Convention*. The Society of Naval Architects and Marine Engineers.

[8] World Bank Group, 2020, *Doing Business 2020 - Comparing Business Regulation in 190 Economies*, The World Bank, 2020. ISBN: 978-1-4648-1440-2.

[9] IMO, 2020, *The IMO Compendium on Facilitation and Electronic Business*, On-line version of the compendium and the reference data model, retrieved June 2010 from: http://www.imo.org/en/OurWork/Facilitation/Pages/IMOCompendium.aspx

[10] Rødseth Ø.J., Frøystad C., Meland P.H., Bernsmed K., Nesheim D.A., 2020, The Need for a Public Key Infrastructure for Automated and Autonomous ships, in *Proceedings of the International Conference on MASS*, Ulsan, Korea, November 2020.

[11] Paddeu, D., Calvert T., Clark B., Parkhurst G., 2018, *New Technology and Automation in Freight Transport and Handling Systems,* Government Office for Science, London UK.

[12] Roso, V., Woxenius J., Lumsden K., 2009, The dry port concept: connecting container seaports with the hinterland., *Journal of Transport Geography* 17.5 (2009): 338-345.

[13] Global Maritime Trends: https://www.lr.org/en/insights/global-marine-trends-2030/global-marine-technology-trends-2030/ (retrieved July 2020).

[14] Rødseth, Ø.J. and Kvamstad, B., 2009. Communication Bandwidth Requirements for Future e-Navigation Services. *European Journal of Navigation*, *7*(1), p.10.

[15] Rødseth, Ø.J. and Lee, K., 2015. Secure communication for e-navigation and remote control of unmanned ships. In *Proc. of the 14th Conference on Computer and IT Applications in the Maritime Industries-COMPIT* (Vol. 15).

[16] ISO, 2013, ISO 28005 Security management systems for the supply chain — Electronic port clearance (EPC), International Organization for Standardization.

[17] Rødseth Ø. J., Lee, K., 2017, Supporting Operational Data Exchanges in Shipping with the Common Maritime Data Structure, In *Proceedings of the 5th International Port-Maritime Technology Conference* (MTEC) 2017, Singapore, April 2017.

[18] Rødseth, Ø. J., and Kapidani. N., 2017, A Taxonomy for Single Window Environments in Seaports. In *Proceedings of the 5th International Port-Maritime Technology Conference* (MTEC) 2017, Singapore, April 2017.

[19] ISO, 2018, FAL 43/8, 20 December 2018, *Developing Guidance for Authentication, Integrity and Confidentiality of Content for the Purpose of Exchange of Electronic Information*, Input document to IMO FAL Committee.

[20] Benamara, H., Hoffmann, J., Youssef, F., 2019, Maritime Transport: the Sustainability Imperative. In Psaraftis, H.N. (ed.) 2019, *Sustainable Shipping: a Cross-Disciplinary View*, Springer.

[21] The AEGIS Web site: http://aegis.autonomous-ship.org/ (Retrieved July 2020).

Towards an investigation of a MASS-assisted anti-grounding service through simulated nautical scenarios in a ship handling simulator

V E Schneider[1], M Constapel[1], J Riordan [2,3], F Flannery[2], H-C Burmeister[1], C Jahn[1]

[1] Fraunhofer Center for Maritime Logistics and Service, Am Schwarzenberg-Campus 4, 21073, Hamburg, Germany
[2] SonarSim Ltd., Limerick, Ireland
[3] School of Computing, Engineering, and Physical Sciences, University of the West of Scotland, Hamilton International Technology Park, South Lanarkshire, G72 0LH, Scotland

E-mail: vincent.schneider@cml.fraunhofer.de

Abstract. Vessel groundings pose a major risk for maritime safety, constituting 20 percent of all incidents in the last decade. Frequent dredging and bottom mapping are resource intensive solutions currently employed, but these services cannot be maintained with the necessary frequency in all critical areas. While ships are carrying echo sounders to acquire precise and current under keel clearance data, it does not allow the vessel to react to possible deviations from the map data. Within the RoboVaaS project, a service is designed, implemented and tested in simulation. This service consists of one or more small MASS travelling ahead of a merchant vessel, collecting bathymetric data with enough lead time for the merchant vessel to react to possible threats, e.g. by course correction. This service is tested in compliance with the IMO HCD guideline in a quick approach and safe environment using a ship handling simulator. The simulator is augmented by a display system based on an ECDIS map and displaying the bathymetric data in three different scenarios. These scenarios are tested with nautical officers to collect feedback for service design and implementation with trained personnel and show the effectiveness of the chosen human machine interface.

1. Introduction

One of the major risks for maritime safety is vessel groundings. Within the last decade, approximately 20 percent of all total losses were related to grounding incidents [1]. In European waters, over 2000 accidents that involved groundings have been registered between 2011 and 2018 [2]. Areas affected by shallow water effects and tidal bore are especially prone to grounding incidents. These circumstances can be found in port areas and approaches, where recurring sedimentation can restrict large ships. To combat these problems, the need for dredging and frequent bottom mapping arises, leading to resource-intensive missions with manned vessels and high-tech equipment. Moreover, waterborne is the largest international transport sector, carrying 90 percent of global trade, with ports and harbours serving as the maritime epicenter of the arterial inter-modal network. Port transit delays on intercontinental routes can quickly eliminate the single-digit profit margins of major carriers. In the European context, the majority of about 58 percent of shipping is internal short-sea, with a further 13 percent through inland waterways, rendering it particularly vulnerable to incidents of shallow water groundings [3]. Currently, anti-grounding measures in place are mainly bathymetric data in electronic navigational charts, marked fairways with ensured minimum depth as well as pilot services, which all rely on historical data (in case of chart data, they might even rely on measurements from up to several decades ago). Additionally, they might not be available especially in areas of lower traffic density due to high costs. While each merchant vessel is equipped with echo sounders, which provide online under keel clearance (UKC) data, those are only of limited applicability for anti-grounding decisions. In transit or approach in confined water, large vessels require maneuvering decisions up to 20 minutes ahead, thus the risk of shallow waters up to that point must be known. Even though forward-looking sonars, that can be mounted on a vessel, exist, none of them is currently capable to safely detect shallow waters in the required range due to the unfavorable horizontal measurement angle. A safer vertical measurement would however require a second sensor carrier unit in due distance ahead of the main vessel. This demand can be met by the recent technology developments in the area of Maritime Autonomous Surface Ships (MASS). Thus, real-time bathymetric data including the current tide offset shall be provided. In contrast to purely adding tide gauge measurements to an ENC, dynamic changes in the seabed are thus immediately detected and exact measurements between the depth contour lines are also possible by this concept providing a more accurate safety contour for merchant vessels in shallow waters

In this paper, a MASS-based anti grounding service concept along with a design and simulation implementation will be presented serving three purposes: 1) development and design of a service that provides bathymetric data on request as a foundation for anti-grounding decisions, 2) implementation of the service design into a ship handling simulator for evaluation purposes with certified nautical officers, and 3) building a deploy-able on-demand service without significant data transmission delays for real applications.

2. Service Concept

The RoboVaaS (Robotic Vessels as a Service) project investigates different use cases, how small MASS can support maritime transport and port operations. Despite the the small size the term MASS is chosen instead of the also frequently used term Autonomous Surface Vehicle ASV, as ASV is not defined in the context of the IMO and this vessel category will most likely be one derivative from MASS, as e.g. in [4], which also defines vessels below 7m length as *ultra-light MASS*. One of the envisioned use cases is the anti-grounding service. In the scheme of this service, a fleet of small MASS, is envisioned to be deploy-able on short notice. Those MASS sail ahead of a merchant vessel to collect bathymetric data of the seafloor with more favorable vertical measurement. Based on the spatial data of the MASS and the information of the ship requesting the service, the UKC is determined and displayed in an Electronic Chart Display and Information System (ECDIS). On the vessel's bridge, the information is color-encoded and displayed in an overlay of the ECDIS to visualize shallow waters quickly and precisely. The service concept is illustrated in Figure 1.

A MASS is equipped with a standard AIS transponder, a side-beam sonar, and computational resources for collecting and preprocessing the bathymetric data from the sonar while surveying during the mission. The preprocessed data of the MASS is shown to the officer of watch on the merchant ship indicating shallow areas in three different representations: 1) visualization of AIS safety-related messages in the event the MASS recognizes a shallow area violating the ship's under keel clearance (UKC), 2) addressed AIS messages sent periodically yielding a route for safe passage, and 3) heat maps of the bathymetric data showing the actual state of the seafloor, which is illustrated in Figure 1.

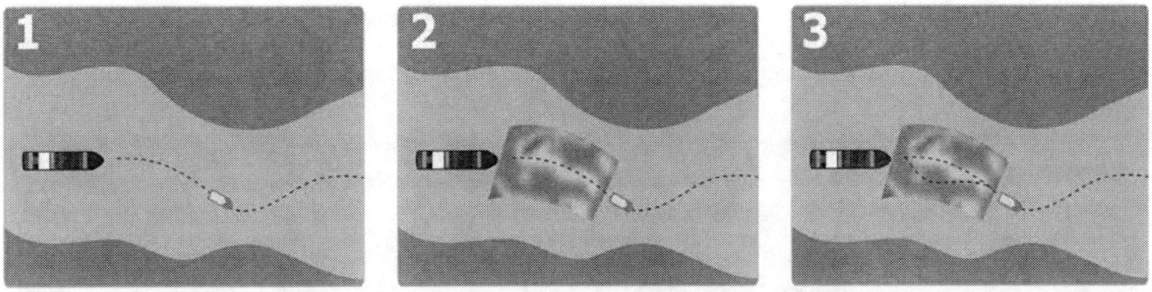

Figure 1. The application of the anti-grounding service is composed of three subsequent steps: 1) an ASV is sailing ahead and well in advance of a merchant vessel in demand of the anti-grounding service, 2) the ASV collects bathymetric data while sailing on a route requested by the merchant vessel, 3) the merchant vessel can react accordingly to the actual bathymetric data collected by the ASV. The bathymetric raw data is pre-processed on the ASV and either transmitted to the merchant vessel via VDES or coastal 5G networks for display of high density heat maps or via addressed AIS messages for indicating spots violating the UKC requested in an ECDIS.

3. Service Backend

Mission planning is a crucial part of the anti-grounding service and the deployment of MASS, thus the basic parameters of mission planning are briefly described. The aforementioned MASS platform is following a preplanned route. The route is initially based upon the route of the approaching vessel, which transfers its monitored route including its schedule, the merchant's vessel velocity (MVV), and allowed cross-track errors by the RTZ-format to the MASS mission planning. This gives the lower boundary for the MASS' velocity. The upper-velocity limit is given by maximum survey velocity (MSV) of the sonar. If the MVV exceeds the MSV, a lead time of the MASS for each way point has to be computed, so that safety restrictions (e.g. stopping distance of the merchant vessel) are not exceeded. In the long run, a system is aspired, where MSV is greater than the expected MVV. For the high-level control loop of the MASS, the Guidance Navigation and Control (GNC) paradigm is used, where the guidance component data is used as a reference, such as a set point, path, or trajectory [5]. Furthermore, environmental effects such as tide and sea state are considered [6], which can affect the effective velocity of the MASS. With the effective velocity of the MASS and the specific restrictions of the merchant vessel the bathymetric data can be collected to be effective for the anti-grounding service. These constraints are considered for the simulations as well to create a realistic scenario.

4. Experimental Setup

To avoid financial and logistical endeavors for the investigation of such a service in the real world, suitable bathymetric data is generated for several scenarios by simulation and deployed to a virtual full-scale simulation (VFSS). From port studies, it is known that real-time ship handling simulation (SHS) "is the only way to ensure that technical ship handling and the important human factors, are sufficiently incorporated", which is why this set-up is also used as methodology for assessing and reviewing the service in an early development phase [7]. However, this requires a realistic implementation within the VFSS' SHS set-up. This set-up is illustrated in Figure 2. The simulation environment is bound to the port of Hamburg as shown in Figure 3. It consists of a supervised merchant ship and a guiding MASS sailing in advance on the merchant ship's route. However, transmitting heat maps to the merchant ship raises technical requirements on the telemetry communication and data link since, dependent on the size and speed of the guided merchant ship, the distance between ship and MASS units has to be at least a few nautical miles to permit maneuvers to avoid a potential grounding given reaction delays, ships inertia, and constraints in maneuverability.

Thus, for the heat map representation as shown in Figure 3, any MASS is assumed to have advanced telemetry capabilities besides a standard AIS transponder. Assuming such advanced communication technology will be available off-the-shelf in the future (e.g. by VDES or coastal 5G networks), the heat map representation is integrated as an overlay to the ECDIS for the service implementation.

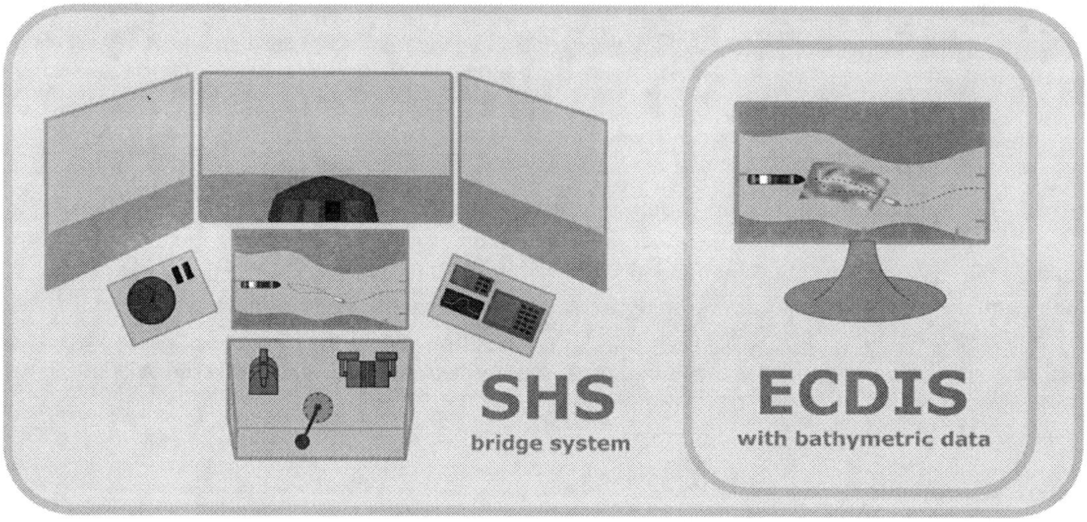

Figure 2. The bathymetric data captured and transmitted from the MASS fleet is shown at the Ship Handling Simulator (SHS) - or on the merchant vessel's bridge - to the officer of watch in an overlay in the Electronic Chart Display and Information System (ECDIS). The merchant vessel is responsible for requesting the MAAS service well in advance for a passage of an area or channel at an estimated time period as well as providing appropriate safety settings, i.e. draught and UKC, for surveying, and additional information, e.g. estimated maximum speed and ships dimensions for mission planning.

5. Service Evaluation

The MASS-assisted anti-grounding service introduces a new concept and its own ways of data representation into the bridge environment, which has to be tested carefully and thoroughly. With choosing a SHS as evaluation environment, the IMO HCD guideline to ensure quick testing in a safe environment can be fulfilled while having the flexibility to introduce the new system [8]. The experimental setup for the evaluation in the ship handling simulator is separated into three independent scenarios for anti-grounding given a merchant ship approaching to Hamburg port within a predefined and fixed nautical environment regarding weather, tide, seafloor, route, and own-ship dynamics. The three scenarios are: 1) approach without the aid of the MASS service just using the ship's own echo sounder, 2) approach with safety-related AIS messages given from one or more MASS to the merchant ship when detecting shallow waters or areas of caution, 3) approach with color-encoded ECDIS overlays indicating a channel of safe passage. The three scenarios will be executed several times with changing depth profiles to simulate runs with and without deviations from the existing ECDIS and record the nautical officers' reactions in form of control inputs such as speed and rudder. This will give insight into how quickly decisions are taken based on the additional information. Furthermore, feedback of the nautical officers will be collected on how the information is displayed. The focus lies on differences in perceived benefit between the three scenarios and the overall evaluation of the service. The evaluation of those approaches and actions taken while perceiving the augmented data in the ECDIS is investigated from a human-computer interface (HCI) point of view. By using the service an increase in the situational awareness for the nautical officers is expected to allow them to react faster and more effectively to situations that could risk the safety of the ship and crew with regards to groundings. One of the main goals of the evaluation is to determine if the designed service will be received positively in its current form and to incorporate the vital feedback of experiences officers.

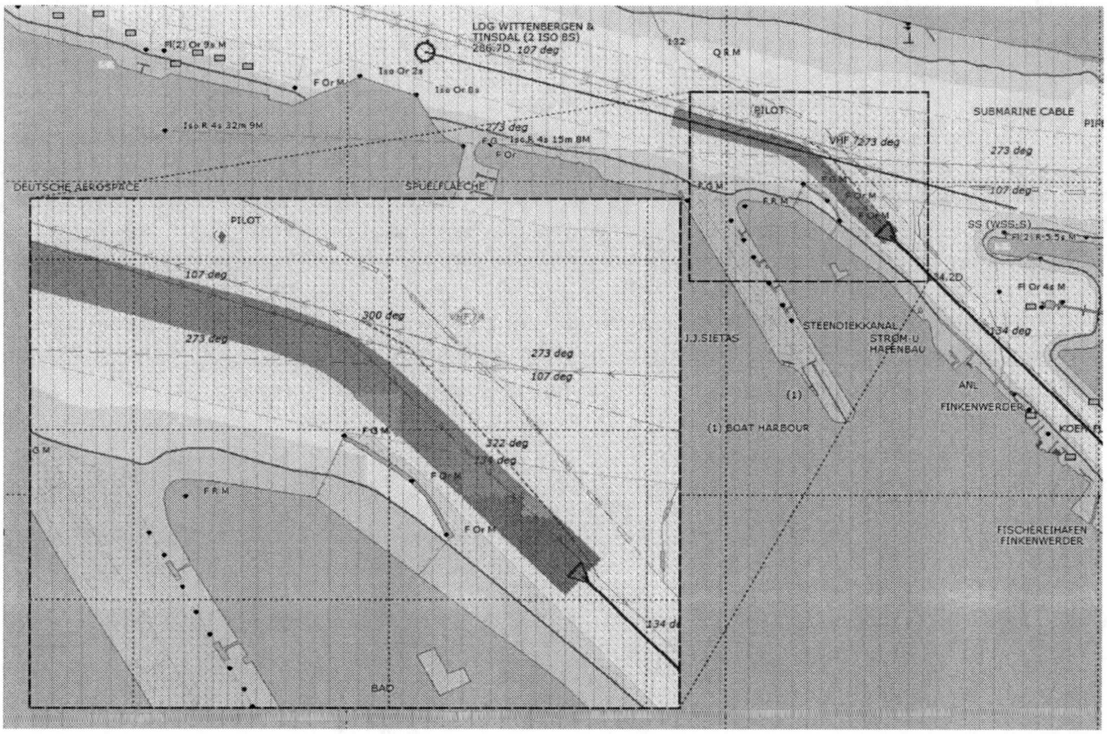

Figure 3. Screen capture of the ECDIS overlay showing a heat map (based on generated bathymetric data) covering the path of a surveying MASS with a beam width of 50 meters: 1) green shades indicate a channel of safe passage given the draught and requested UKC (received via AIS or set during the mission planning) of the supervised merchant ship, 2) yellow shades are areas of caution, which highlight bathymetric measurements where the depth is smaller than the sum of draught and requested UKC plus 1 meter, 3) red shades are shallow waters where the actual depth is smaller than the requested depth of the supervised merchant ship

6. Results and Future Work

The envisioned anti-grounding service has been successfully designed. Its findings have been used to implement three data visualisation scenarios with the SHS to create the test bed capable of evaluating the service design with certified nautical officers. Unfortunately, the envisioned testing with certified nautical officers could not be performed due to the regulations for personal contact active at the time of writing. Therefore the evaluation was limited to functional testing in-house at CML. The situation is monitored closely and invitations will be send to nautical officers as soon as the situation allows to continue the testing.

Acknowledgments

Acknowledgement The RoboVaaS project has received funding by the German Federal Ministry of Economic Affairs and Energy under support code 03SX463A, Universities and Research (MIUR), and ERA-NET CofundMarTERA (contract 728053). SonarSim Ltd. received funding support from Marine Institute Ireland (Grant-Aid Agreement No. PBA/AT/18/02) under the Marine Research Programme by the Irish Government and co-funded by the European Commission under the framework of MarTERA (Maritime and Marine Technologies for a new Era) ERA-NET Cofund in conjunction with JPI Oceans.

References

[1] Allianz safety and shipping review 2019 URL https://www.agcs.allianz.com/content/dam/onemarketing/ agcs/agcs/reports/AGCS-Safety-Shipping-Review-2019.pdf
[2] EMSA annual overview of marine casualties and incidents 2019 URL http://www.emsa.europa.eu/ news-a-press-centre/external-news/download/5854/3734/23.html
[3] UN review of maritime transport 2018 URL https://ec.europa.eu/eurostat/statistics-explained/ index.php/Maritime_ports_freight_and_passenger_statistics
[4] Maritime UK 2019 *Maritime Autonomous Surface Ships (MASS) UK Industry Conduct Principles and Code of Practice* URL https://www.maritimeuk.org/documents/478/code_of_practice_V3_2019_8Bshu5D.pdf
[5] Fossen T I 2011 Handbook of marine craft hydrodynamics and motion control (Chichester: Wiley)
[6] Lee T, Kim H, Chung H, Bang Y and Myung H 2015 *Ocean Engineering* 107 118–131
[7] PIANC 2014 *Harbour Approach Channels Design Guidelines: Report Nr.* 121 (Brussels: PIANC Secr´etariat G´en´eral)
[8] IMO 2015 *Guideline on Software Quality Assurance and Human Centred Design for e-navigation* (London: International Maritime Organization)

IOP Conf. Series: Materials Science and Engineering **929** (2020) 012032 doi:10.1088/1757-899X/929/1/012032

Autoferry Gemini: a real-time simulation platform for electromagnetic radiation sensors on autonomous ships

K Vasstein, E F Brekke, R Mester and E Eide

Norwegian University of Science and Technology, Høgskoleringen 1, 7491 Trondheim, Norway

E-mail: kjetv@stud.ntnu.no, edmund.brekke@ntnu.no, rudolf.mester@ntnu.no, egil.eide@ntnu.no

Abstract. Testing that ships are compliant to specified safety requirements have traditionally relied on real world data, which is not scalable and limited to testable scenarios due to financial and ethical reasons. Low fidelity simulations have been used to counteract some of these problems, which is sufficient for emulating simpler systems such as radar detectors, but not for testing complex systems as found in computer vision. In the automotive industry the use of game engines have shown to be a great testing platform due to their customizability, and combination of real-time physics with computer graphics to create large volumes of high fidelity images. In the work presented here, development of an open-source maritime platform named **Autoferry Gemini** based on the Unity game engine is used to simulate sensors in real-time. Utilizing simulated optics and general purpose GPU programs, the render pipeline is capable of modeling lidar, radar, visible-light and infrared camera sensors simultaneously. Results from visible-light cameras and lidar have already proven to satisfy other research activities on sensor fusion for autonomous ship technology. Infrared cameras motivates further research in gathering empirical data, while GPU algorithms have made it possible to simulate 3D radar models and multiple lidar types in real-time.

1. Introduction

Traditionally, testing control and navigational systems for ships have to a large extend been done through real world testings. In addition, such systems have relied on being operated or supervised by human pilots, thus simple systems have been sufficient. Due to this simplicity, the use of idealized or analytical mathematical models for simulating control and navigation systems, have been a good substitution for real testings. With the development towards autonomous ships, these systems have in contrary become more complex due to the introduction of probabilistic models and other methods from the field of artificial intelligence (AI), such that traditional simulation techniques is less capable of determining if the control and navigation is safe, stable and predictable. In addition, advancement in fields such as machine learning, computer vision and the integration of these with conventional model based control and navigation, such as optimal-control, target tracking, localisation and mapping, requires data from other sensors than just IMU, GPS, and simple radar detectors. This includes sensors with high data rates such as lidars, but also more fidelity sensitive sensors such as visible-light (VL) and infrared (IR) cameras. This have led to an increasing demand for sensor data with high quality and volume. As real world sensor data from both cars and ships have proven to be costly, moreover difficult to receive for research outside companies, there is a demand for high fidelity simulations.

High fidelity video games, which goes by the nickname *AAA-Games* have traditionally used game engines developed in-house. However, in the last decade multiple third party game engines have enabled several simulators such as AirSim created with the Unreal Engine [1] and LGSVL created with Unity [2] to simulate sensors for both drones and cars. Such game engines have also been used in the maritime industry, primarily for training and educating human operators in a real-time environment [3]. Among these, the *Offshore simulation center* (OSC) stationed in Ålesund are the best known high fidelity maritime simulator to the author. In addition for being used in education and training, it have been used for research on human behaviour in offshore operations [4, 5], control designs related to cranes operating in difficult environments [6] and testing multi-sensor systems for enhancing situational awareness (SITAW) [7]. Unfortunately OSC's simulator is not open source, and since it focuses on VL cameras, other sensors of maritime interest such as IR, lidar and radar have received little attention.

Simulators for IR images have primarily focused on modeling IR materials on specific objects such as aircraft [8] and oceans [9], while various video games have in contrast relied on artistic implementations to emulate IR textures. More recent studies have shown results of generalizing IR for various scene objects with the use of GPU shaders [10] and examples can be seen in game engines [11] that have previously only used shaders for VL simulations.

Simulating VL cameras, have been the primary reason for the use of game engines in simulators such as Carla [12], LGSVL [2] and AirSim [1]. LG's automotive simulator [2] was early to adapt Unity's new high definition render pipeline (HDRP) which demonstrated its high fidelity capabilities for cameras on autonomous vehicles, and the Unreal engine have been well known for its fidelity in the AAA-game marked for over a decade. The use of these simulators for autonomous control have been vast, but mainly focused on computer vision techniques such as Visual Odometry [13].

Lidar simulations using GPU was also demonstrated in LGSVL at the end of 2018, using depth-buffers from Unity's older built in render pipeline (BRP) to simulate dense point clouds from virtual lidars running in real-time. Previous research on virtual lidars [14] have also been known to use depth-buffers, which are among the two most common depth techniques used in render pipelines, dating back to 1974 [15]. Ongoing research focuses on enhancing the lidars fidelity by introducing effects such as reflection, refraction, clutter and raydrops with conventional modeling techniques [16] and AI [17].

Radar simulations using game engines are to the author not known, except for simple collision sensors for cars seen in various simulators [1, 12, 2]. The vast selection of different radar types have probably not made the task any easier, such as the need for dedicated hardware to simulate synthetic aperture radars [18]. For conventional long range radars for ships, there are however non real-time solutions such as Carpet [19], that functions as a separate radar program capable of handling atmospheric disturbances, Doppler and polarization effects, antenna radiation pattern etc.

In a student project at NTNU during the spring of 2019, implementation of several EMR sensors was done using a CPU based lidar with raycasts from Unity's physics engine, and virtual VL camera from Unity's BRP [20]. This was done for the autonomous ferry Milliampere, which gave a demonstration of the possibility of simulating sensors in a virtual environment. The study showed the possibility to run virtual sensors simultaneously, with relatively high fidelity, in a maritime environment considering hydrodynamic forces, vessel control and human interaction. This have since been used in research regarding situational awareness, such as feature-based pose estimation of ships in monocular camera images [21], which demonstrated the use of data from game engines for testing computer vision on ships.

The main contribution of this paper is to present the open source maritime simulation platform *Autoferry Gemini*, with the sensor models for IR, radar, VL and lidar. The paper builds upon the student projects [20, 22], which can be obtained from contacting the first author. By utilizing the Unity game engine, HDRP is used to model IR and VL sensors, while depth-buffers and GP-GPU algorithms are used to simulate lidar and radar. We believe from the mentioned use cases [13, 7, 6, 21] that this can give researchers a new platform for obtaining EMR sensor data from virtual marine vessels that have previously been spread across multiple platforms. As a summary, this entails:

- Creating an open-source maritime simulator emphasising autonomous ships.
- Concurrent simulation of radar, lidar, IR and VL sensors.
- Real-time simulation by utilizing the GPU for sensor modelling.

2. Sensor simulation methods

The core idea behind our method is to utilize Unity's real-time render pipeline to generate data. By configuring it with the use of *command buffers* [22], creating several separate GPU data pipelines based on virtual cameras graphic-buffers (G-buffers), which contains information of surface normals, materials and depths from the deferred shading technique [23], one guarantees that all of the sensor data comes from the same state. This also makes it possible to run all of the sensors concurrently, as Unity schedules the pipelines running on the GPU.

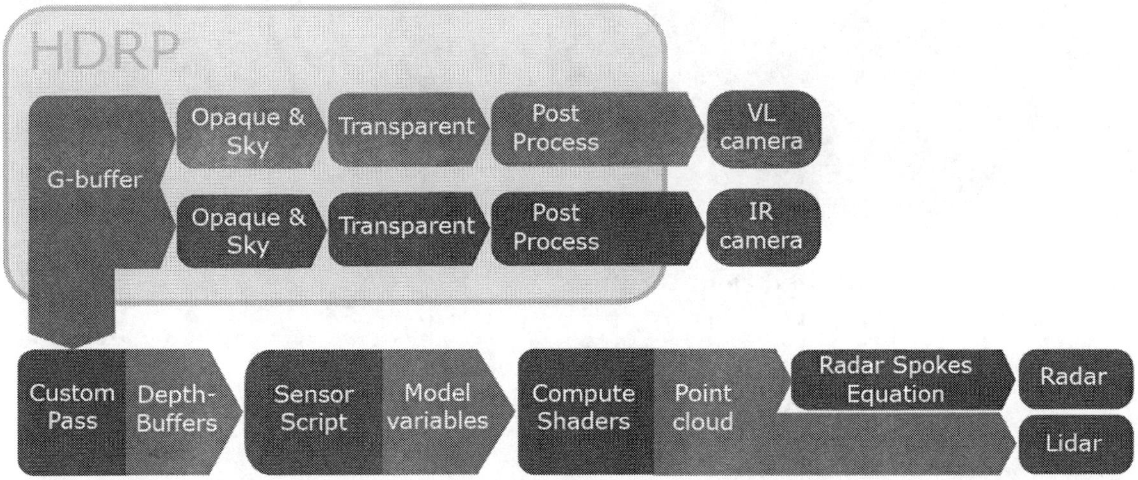

Figure 1. GPU data pipeline using HDRP to simulate VL and IR, and a custom pipeline to simulate radar and lidar. Blue boxes cover the implemented work, red the resulting EMR sensors, while purple is implemented by Unity.

2.1. HDRP

Unity's HDRP are used to implement the VL and IR camera sensors. Unity's *Lit* shader are used to display most of the surfaces for the VL sensor by utilizing physically based rendering (PBR) materials from third party sources. Custom shaders are used for the ocean created with HDRP's *Shader Graph* tool, while clouds are created using particles from the *Visual Effects Graph* (VFX).

A culling mask is created by using Unity's layer system, in order to create the separate pipeline for the IR camera shown in Figure 1. In addition, Shader Graph is used to implement the thermal radiation from Stefan-Boltzmann law [24] with an emissivity texture ϵ which gives the render pipeline the emissivity at different surface locations on a 3D model:

$$W(T) = \epsilon\sigma T^4. \qquad [\mathrm{W\,m^{-2}}] \quad (1)$$

Due to the lack of empirical data, artistic emissivity textures are used. Assuming that the 3D models are close to the ambient temperature $T \approx 300K$, one would get from Wiens's displacement law [24] that the radiations peak wavelength $\lambda_{peak} = \frac{b}{T} \approx 9.7\mu m$ is in the infrared spectrum. Due to this, Stefan-Boltzmann law becomes a good first order approximation, since close to ambient temperature ranges, infrared imaging devices are classified as total-radiation radiometers [25], where 98% of the total black body radiation described by Planck's law are in the infrared spectrum.

IOP Conf. Series: Materials Science and Engineering **929** (2020) 012032 doi:10.1088/1757-899X/929/1/012032

2.2. Depth-Buffer to Point Cloud

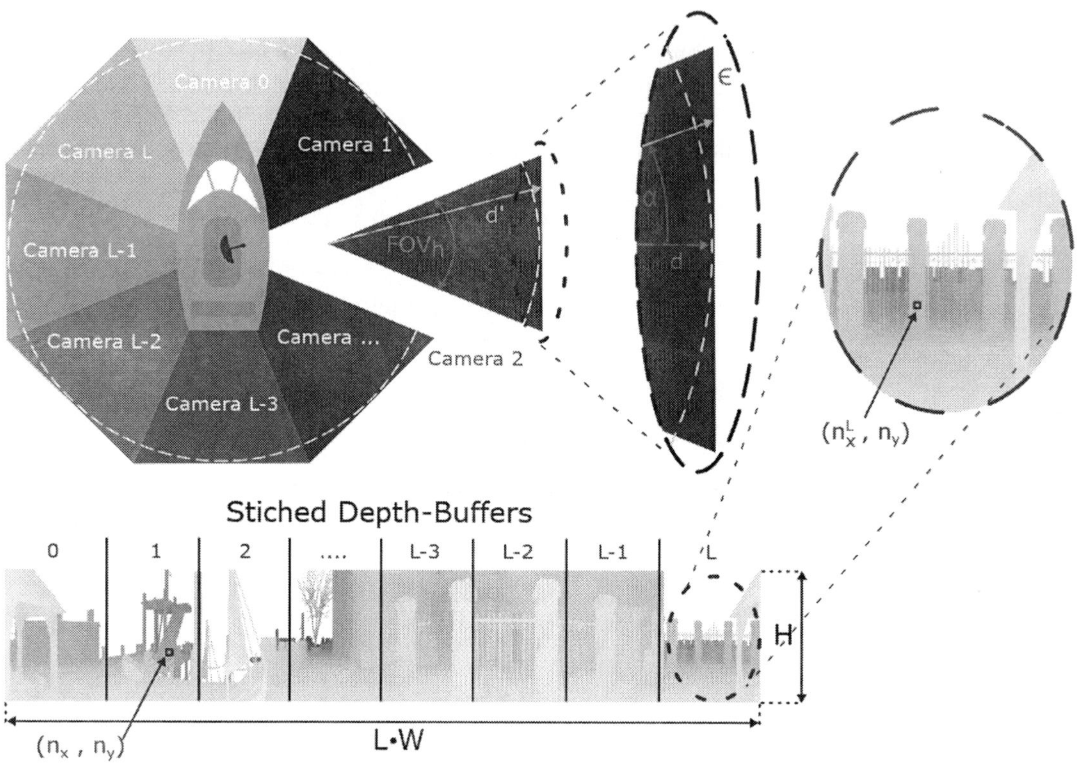

Figure 2. Depth-buffers from virtual cameras used to approximate cylindrical beams with a convex regular polygon. Stitching the depth-buffers together, forms a 2D depth array.

A camera's depth-buffer is fetched from the G-buffer with the use of a *command buffer* from a *custom pass* class in Unity seen in Figure 1. This is sent to a sensor script that dispatches the depth-buffer along with other sensor parameters such as the sensor's max target distance, resolutions, etc, to a compute shader that creates a point cloud using the GPU.

The depth-buffer for camera l is defined as a 2D pixel array $Z^l(n_x^l, n_y) \in \mathbb{R}$ of normalized values $0 \leq Z^l \leq 1$, and pixel positions $(n_x^l, n_y) \in \mathbb{N}^{H \times W}$ where H and W are the depth-buffer's height and width. From this, homogeneous coordinates in *image space* can be written as:

$$\bar{\boldsymbol{X}}^{i,l} = [\bar{n}_x^l, \bar{n}_y, Z^l(n_x^l, n_y), 1]^T \tag{2}$$

where $\bar{n}_x^l = \frac{2n_x^l}{W} - 1$, $\bar{n}_y = \frac{2n_y}{H} - 1$ places origo at the depth-buffers center and remaps the pixel coordinates to be values between $(-1, 1)$. To convert from *image space* coordinates $\bar{\boldsymbol{X}}^{i,l}$ to the camera's *view space* coordinates $\boldsymbol{X}^{v,l}$, a projection matrix $\boldsymbol{P}_i^v$ with parameters for the virtual cameras far plane f, near plane n and horizontal field of view FOV_h seen in Figure 2 is used:

$$\boldsymbol{P}_i^v := \begin{bmatrix} \cot(\frac{FOV_h}{2}) & 0 & 0 & 0 \\ 0 & \cot(\frac{FOV_h}{2}) & 0 & 0 \\ 0 & 0 & -\frac{f+n}{f-n} & -2\frac{f*n}{f-n} \\ 0 & 0 & -1 & 0 \end{bmatrix}^{-1} \tag{3}$$

$$\bar{\boldsymbol{X}}^{v,l} = \boldsymbol{P}_i^v \bar{\boldsymbol{X}}^{i,l} \tag{4}$$

$$\boldsymbol{X}^{v,l} = \frac{1}{\bar{w}^{v,l}} \bar{\boldsymbol{X}}^{v,l}. \tag{5}$$

The last equation divides all elements by $\bar{\boldsymbol{X}}^{v,l}$ last element such that it becomes 1 as in (2). From this, a range $R^l(n_x^l, n_y)$ defined as the distance between camera l's center to a objects rasterized surface in *view space* with respect to the depth-buffers pixel positions, can be created from (2)–(5):

$$R^l(n_x^l, n_y) := ||\boldsymbol{X}^{v,l}||_2 = \left\| \frac{1}{\bar{w}^{v,l}} \boldsymbol{P}_{\boldsymbol{i}}^{\boldsymbol{v}} \left[\frac{2n_x^l}{W} - 1, \frac{2n_y}{H} - 1, Z^l(n_x^l, n_y), 1 \right]^T \right\|_2 . \tag{6}$$

In order to approximate a circular beam, L cameras is spaced around the sensor's azimuth-axis with vertical field of view $FOV_h = \frac{2\pi}{L}$ as seen in Figure 2. This creates L depth-buffers $Z^l(n_x^l, n_y)$ that can be stitched to a 2D depth array $Z(n_x, n_y)$ seen in Figure 2:

$$l := n_x \setminus W \tag{7}$$

$$Z(n_x, n_y) := Z^l(n_x^l, n_y) \tag{8}$$

where $0 \leq n_x < L \cdot W$ substitutes the camera dependent pixel coordinate n_x^l seen in Figure 2, and "\" is defined as the integer division. Since each camera is spaced around the azimuth-axis, they have a rotation matrix $\boldsymbol{R}^{\boldsymbol{l}}$ relative to the first camera which transforms each cameras *view space* coordinates $\boldsymbol{X}^{v,l}$ to the first camera $\boldsymbol{X}^{v} := \boldsymbol{X}^{v,0}$:

$$\boldsymbol{R}^{\boldsymbol{l}} := \boldsymbol{R}_{\boldsymbol{y}}(FOV_H \cdot l) := \begin{bmatrix} \cos(\frac{2\pi l}{L}) & 0 & \sin(\frac{2\pi l}{L}) & 0 \\ 0 & 1 & 0 & 0 \\ -\sin(\frac{2\pi l}{L}) & 0 & \cos(\frac{2\pi l}{L}) & 0 \\ 0 & 0 & 0 & 0 \end{bmatrix} \tag{9}$$

$$\boldsymbol{X}^{v} = \boldsymbol{R}^{\boldsymbol{l}} \boldsymbol{X}^{v,l}. \tag{10}$$

This results in a circular range equation:

$$\begin{aligned} R(n_x, n_y) &= ||\boldsymbol{X}^{v}||_2 = ||\boldsymbol{R}^{\boldsymbol{l}} \boldsymbol{X}^{v,l}||_2 = ||\boldsymbol{R}^{\boldsymbol{l}}||_2 ||\boldsymbol{X}^{v,l}||_2 = ||\boldsymbol{R}^{\boldsymbol{l}}||_2 R^l(n_x^l, n_y) \\ &= ||\boldsymbol{R}^{\boldsymbol{l}}||_2 \left\| \frac{1}{\bar{w}^{v,l}} \boldsymbol{P}_{\boldsymbol{i}}^{\boldsymbol{v}} \left[\frac{2n_x^l}{W} - 1, \frac{2n_y}{H} - 1, Z^l(n_x^l, n_y), 1 \right]^T \right\|_2 \\ &= \left\| \frac{1}{\bar{w}^{v,l}} \boldsymbol{R}^{\boldsymbol{l}} \boldsymbol{P}_{\boldsymbol{i}}^{\boldsymbol{v}} \left[\frac{2n_x}{W} - 1, \frac{2n_y}{H} - 1, Z(n_x, n_y), 1 \right]^T \right\|_2 \end{aligned}$$

$$R(n_x, n_y) = \left\| \frac{1}{\bar{w}^{v, n_x \setminus W}} \boldsymbol{R}^{\boldsymbol{n_x \setminus W}} \boldsymbol{P}_{\boldsymbol{i}}^{\boldsymbol{v}} \left[\frac{2n_x}{W} - 1, \frac{2n_y}{H} - 1, Z(n_x, n_y), 1 \right]^T \right\|_2 . \tag{11}$$

Spherical coordinates can be transformed from pixel coordinates in the following manner:

$$\theta = \frac{2\pi n_x}{L \cdot W}, \quad \phi = FOV_v \left(\frac{2n_y}{H} - 1 \right) \tag{12}$$

θ is the azimuth angle: $0 \leq \theta < 2\pi$,
ϕ is the elevation angle: $-\frac{FOV_v}{2} \leq \phi < \frac{FOV_v}{2}$,
FOV_v is the sensors vertical field of view.

2.3. Radar Spokes Equation

A single radar spoke, which contains cells of power at different target ranges, can be described with the radar range equation for an antenna used for both transmitting and receiving [26]:

$$P_r = \frac{P_t G^2 \lambda^2 \sigma}{(4\pi)^3 R^4} \tag{13}$$

with radar wavelength λ, antenna gain G, power P_t, target range R, and radar cross section area defined as $\sigma = \lim_{x \to \infty} 4\pi R^2 \frac{E_s^2}{E_i^2}$ from Maxwell's equations [26].

To simplify the problem, we assume $\sigma = 1$ and use the range $R(n_x, n_y)$ from (11). The antenna gain is created by introducing a normalized radiation pattern $G(n_x, n_y)$ with an amplitude $|G_{max}|$ such that $G = |G_{max}|G(n_x, n_y)$, where we create a single main lobe as an ellipse with the height spanning FOV_v as defined from 12 with a beam width of g, defined to be where the radiation pattern is 3dB lower than the maximum of the main lobe. With these representations (13) becomes:

$$P_r(n_x, n_y) = \frac{P_t \lambda^2 |G_{max}|^2 G(n_x, n_y)^2}{(4\pi)^3 R(n_x, n_y)^4}. \tag{14}$$

To create a spoke, we bin this power by introducing the spoke range $r \in \mathbb{N}$ with a resolution of r_{max} cells:

$$r = \frac{R(n_x, n_y)}{f} r_{max} \tag{15}$$

where f is the far plane distance from (3) and r rounds to the closest integer. Any range larger than r_{max} due to the beam-shape error is dropped. With this, we marginalize (14) to form a single spoke equation that takes into account the 3D surroundings:

$$P_{spoke}(r) = \frac{P_t \lambda^2 |G_{max}|^2}{(4\pi)^3} \sum_{n_x=0}^{L \cdot W} \sum_{n_y=0}^{H} \frac{G(n_x, n_y)^2}{R(n_x, n_y)^4}. \tag{16}$$

As typical ship radars have multiple spokes arranged around the azimuth-axis, a finite number of spokes h is sampled from the total sweep resolution $h = \frac{L \cdot W}{2^n}$, $n \in \mathbb{N}$, with a spoke index $0 \leq k < h \in \mathbb{N}$, resulting in h spokes displaced around the radar by offsetting the radiation pattern:

$$P_{spokes}(r, k) = \frac{P_t \lambda^2 |G_{max}|^2}{(4\pi)^3} \sum_{n_x=0}^{L \cdot W} \sum_{n_y=0}^{H} \frac{G(n_x + k \cdot \frac{L \cdot W}{h}, n_y)^2}{R(n_x, n_y)^4}. \tag{17}$$

A first order approximation of the radar range resolution is done by taking the radar bandwidth B into account. This is done by convolving a discrete Gaussian function $f(x)$ with $P_{spokes}(r, k)$, creating a finite summation with window size $2M$, where $M = 50$ turned out to be sufficient for experiments:

$$f(x) = \frac{1}{\sqrt{2\pi s^2}} e^{-\frac{x^2}{2s^2}}, \quad s^2 = \frac{c}{2B} \tag{18}$$

$$\bar{P}_{spokes}(r, k) = (P_{spokes} * f)(r) = \sum_{m=-M}^{M} f(r - m) P_{spokes}(r, k) \tag{19}$$

where c is the speed of light in a vacuum. Combining (17) - (19) the *radar spokes equation* with radar range uncertainty becomes:

$$\bar{P}_{spokes}(r, k) = \frac{P_t \lambda^2 |G_{max}|^2}{(4\pi)^3 \sqrt{\pi \frac{c}{B}}} \sum_{m=-M}^{M} e^{-\frac{(r-m)^2}{(\frac{c}{B})^2}} \sum_{n_x=0}^{L \cdot W} \sum_{n_y=0}^{H} \frac{G(n_x + k \cdot \frac{L \cdot W}{h}, n_y)^2}{R(n_x, n_y)^4}. \tag{20}$$

For implementation, (15) and (17) is used to create the first *compute shader kernel*, while a second kernel uses (19) to compute the final radar spokes results from (20).

IOP Conf. Series: Materials Science and Engineering **929** (2020) 012032 doi:10.1088/1757-899X/929/1/012032

2.4. Beam-Shape Error

The use of L depth-buffers creates a convex regular polygon beam that approximates a circle as seen in Figure 2. This creates an error as a function of both distance and angle relative to the depth-buffers center for a camera l:

$$\epsilon(\alpha, d) = d' - d = \frac{d}{\cos(\alpha)} - d = d\left(\frac{1}{\cos(\alpha)} - 1\right) \tag{21}$$

From Figure 2, it can be seen that the maximum error is located where the depth-buffers meet: $\alpha_{max} = \frac{FOV_h}{2} = \frac{\pi}{L}$. Distance d's largest value is the virtual cameras far plane f, giving a *maximum beam shape error*:

$$\epsilon_{max} = f\left(\frac{1}{\cos(\frac{\pi}{L})} - 1\right). \tag{22}$$

Further, an average beam shape error with respect to distance can be calculated as:

$$\begin{aligned}\epsilon_{avg}(d) &= \frac{1}{\alpha_{max}}\int_0^{\alpha_{max}} \epsilon(\alpha, d)\partial\alpha = \frac{L}{\pi}\int_0^{\frac{\pi}{L}} d\left(\frac{1}{\cos(\alpha)} - 1\right)\partial\alpha \\ &= \frac{L \cdot d}{\pi}\left[\ln\left(\frac{1+\sin(\alpha)}{\cos(\alpha)}\right) - 1\right]_0^{\frac{\pi}{L}} \\ &= d\left[\frac{L}{\pi}\ln\left(\frac{1+\sin(\frac{\pi}{L})}{\cos(\frac{\pi}{L})}\right) - 1\right]\end{aligned}$$

and the final *average beam shape error* becomes:

$$\epsilon_{avg} = \frac{1}{f}\int_0^f \epsilon(d)\partial d = \frac{1}{f}\left[\frac{L}{\pi}\ln\left(\frac{1+\sin(\frac{\pi}{L})}{\cos(\frac{\pi}{L})}\right) - 1\right]\left[\frac{d^2}{2}\right]_0^f \tag{23}$$

$$= \frac{f}{2}\left[\frac{L}{\pi}\ln\left(\frac{1+\sin(\frac{\pi}{L})}{\cos(\frac{\pi}{L})}\right) - 1\right] \tag{24}$$

3. Evaluation

Sensor parameters used to obtain the results are based on specifications from a *Simrad 4G* for radar, *Velodyne VLP-16* for lidar, while IR and VL are chosen arbitrarily, if not other is specified. As studying how the sensor outputs could be transferable to the real-world was beyond the scope of the paper, only a qualitative evaluation is made.

The results came from a laptop with the following specifications:

- *OS*: Windows 10 Home, 64-bit
- *CPU*: Intel(R) Core(TM) i7-8565U, X64 Quad Core CPU
- *Memory*: 16 GB RAM, 512 GB SSD
- *GPU*: NVIDIA GeForce GTX 1050, 4 GB GDDR5 VRAM

Table 1. Virtual camera parameters.

Sensors	W [pixels]	H [pixels]	L	FOV_h [°]	FOV_v [°]	f [m]	n [m]
VL	1920	1080	1	50	90	10^5	10^{-1}
IR	1920	1080	1	34	60	10^5	10^{-1}
Lidar	64	16	32	11.25	30	100	1
Radar	64	128	16	22.5	20	100	1

Table 2. Additional radar parameters.

P_t [W]	λ [m]	B [MHz]	G_{max} [dBi]	M	h	r_{max}	g [°]
20	λ	50	1	50	512	1024	5

3.1. Results compared to real sensor data

Figure 3. Simulated VL camera of Milliampere 2.0 outside Trondheim to the left, and real VL image of Milliampere 1.0 at Ravnkloa in Trondheim.

High fidelity results for the VL camera with physical correct lightning such as ocean reflection, sub surface scattering in the clouds and colour temperature relative to the time of day, can be seen from Figure 3. Comparing the real and synthetic data for such high quality built scenes, there is not much to remark. A more noticeable effect was discovered by changing the scene content with low resolution models and materials, which resulted in lower fidelity. This could be seen at Milliampere's operational area at Ravnkloa, where to little details resulted in an unrealistic look, even though the city model of Trondheim seemed realistic at far distances.

Similar observations was made for the IR camera, which had no empirical based emissivity textures to emulate real IR images, which can clearly be seen by comparing the human faces in Figure 4. Lightning effects such as glow and blurriness did however reassemble real IR photographs.

As for lidar, experiments showed the ability to simulate data rates up to 30 million points per second while still having 50 fps as from Table 3. Because of this, it was possible to emulate multiple lidar types between 16-128 lasers, one of which is demonstrated in Figure 5. Using depth-buffers did however introduce a *beam shape error* which created a trade-off between accuracy and performance seen in Figure 7. Synchronisation errors was also discovered while experimenting with multi-sensor simulations, but no attempts where made to solve it.

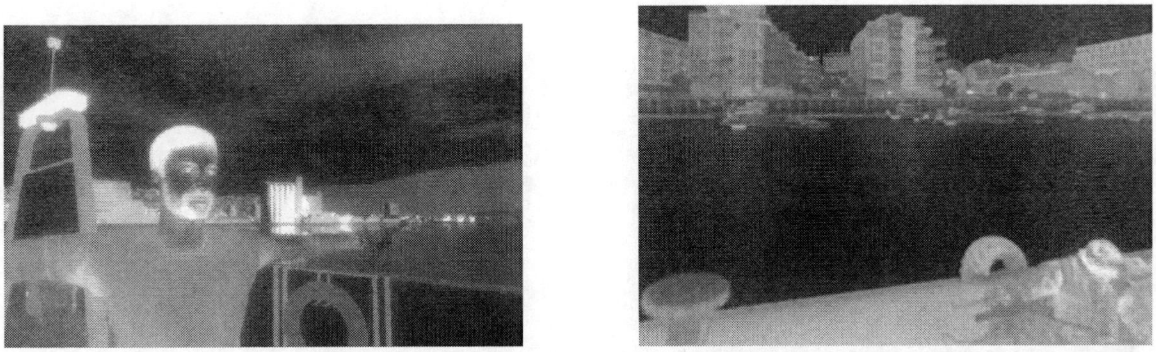

Figure 4. Simulated IR camera of a human outside Trondheim using artistic textures to the left, and real IR image from Milliampere 1.0 with a human in the bottom right corner.

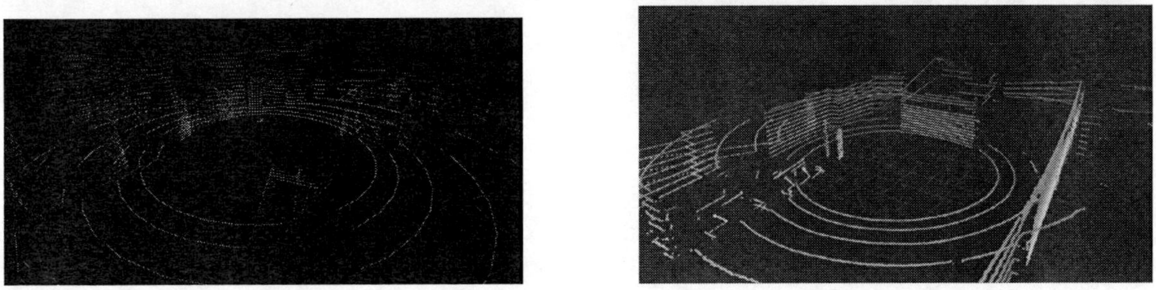

Figure 5. Simulated results from the Old Seaport model of a VLP-16 to the left, and real data from [27] to the right.

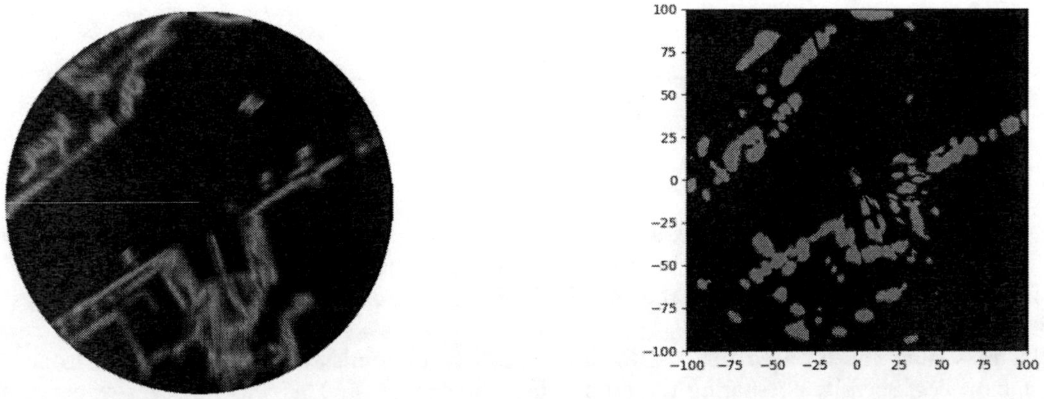

Figure 6. Simulated radar plot for Simrad 4G at Ravnkloa to the left, with corresponding real radar data to the right. All plots uses a 100m radar range.

Modeling radars using depth-buffers, showed that 3D information from both the radiation pattern and the area of operation could be considered. However, results such as from Figure 6 showed more details in the area than in the real world. Constructing the virtual operational area with sufficient content as in reality, was also an issue. Testing side and back lobes in the radiation pattern created mirrored patterns as in real life, but no further observations towards its fidelity was be made.

3.2. Errors and real-time performance

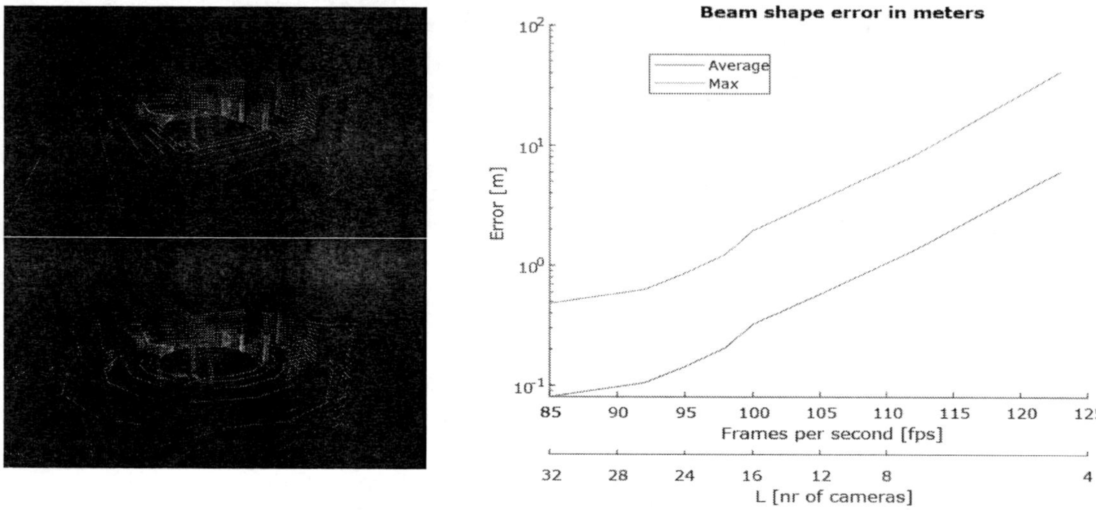

Figure 7. Observable beam shape error in the Old Seaport model for lidar in the left images, *top: L = 4* and *bottom: L = 16* cameras. Graph to the right depicts the trade-off between error and real-time performance sampled from using different camera numbers.

Table 3. Real-time results measured in frames per second (fps or Hz) in the Trondheim city model for individual sensors and results from a multi-sensor simulation.

VL camera	IR camera	Lidar	Radar	All sensors
45	80	50	35	15

4. Discussion

Results showed high fidelity VL images seen in Figure 3, close to what can be observed in existing automotive simulators [1, 12, 2]. However, the dependency on quality and quantity of scene content such as materials and 3D models, shows the current limitation and challenges of using game engines as simulation platforms.

Similar reasoning can be used for the IR camera, where the emissivity texture proved to be the limiting factor, best seen by comparing the human faces in Figure 4. This was however as expected, since artistic textures where used when no empirical data was available, moreover 3D models with both IR and VL textures. Lightning did however reassemble real IR photographs, indicating a potential for further work.

Lidar was capable of generating high volumes of data in real time, while being able to maintain the simulations frame rate, very similar to LGSVL's GPU accelerated lidar [2]. As expected, a beam shape error from stitching multiple depth-buffers did occur as seen in Figure 7, though the unforeseen trade-off between performance and accuracy in increasing the number of cameras, did limit the proposed lidar model. This shortcoming could be explained by the increased *draw calls* the render pipeline needs to handle as the number of virtual cameras increases.

A proposed 3D sensor model for radar showed potential of running in real-time on a GPU seen in Table 3. Results carried little resemblance to existing stand alone solutions such as Carpet [19], moreover with real world data as seen in Figure 6, some of which can be explained by the lack of virtual content relative to the real world. However, the main contributing factor that limits the fidelity is probably the negligence of RCS.

Table 3 shows acceptable real-time performances for all sensors, especially considering it was running on a laptop. Also the 15Hz performance for multi-sensor simulations are believed to be sufficient for testing virtual vessels using control systems with similar update frequency, such as the autonomous ship Milliampere. As real-time rendering through G-buffers compromises on fidelity versus performance, sensor models is constrained by the poly count in 3D meshes and lightning fidelity in comparison to other rendering techniques such as ray tracing. Because of this, simulators based on other game engines, e.g Unreal [1], that utilizes real-time global illumination based on ray tracing, might be able to improve the sensor models further.

5. Conclusion

In this paper, we showed simulations of EMR sensors with high fidelity VL camera similar to automotive simulators, a customizable lidar capable of generating high volumes of data, a IR camera using lightning based on black body radiation and artistic textures, and a radar taking into account 3D information, all running on models implemented on the GPU. This demonstrated that game engines are capable of simulating multiple EMR sensors of maritime interest, running concurrently in real-time given the proper hardware. From these results in context of related work stated in the introduction, it can be concluded that *Autoferry Gemini* is the first known simulator that combines VL, IR, lidar and radar simulations in such a way for autonomous ships. Further work will need to emphasize data gathering techniques and modelling to address the lack of EMR content, such as emmissivity and 3D models. Sensitivity analyses could also be carried out to account for real-world imperfections that may occur from actual sensors.

Acknowledgments

A special thanks to Master student's at NTNU's Department of Design, for donation of 3D models such as Milliampere 2.0 by Petter Mustvedt [28] and the ferry's docking stations designed by Peter Glesaaen and Henrik Ellingsen. Many thanks to Rambøll for providing the city model of Trondheim, Renderpeople for providing 3D human models, and Chritopher Fantauzzo for his Old Seaport Unity asset. The last thanks goes to Erik Wilthil and Øystein Kaarstad Helgesen that provided real VL and IR images in addition to radar data from Milliampere 1.0.

References

[1] Shah S, Dey D, Lovett C and Kapoor A 2018 AirSim: High-fidelity visual and physical simulation for autonomous vehicles *Proc. of Field and Service Robotics* ed Hutter M and Siegwart R (Cham: Springer International Publishing) pp 621–635

[2] LG 2018 LGSVL simulator `https://www.lgsvlsimulator.com/`

[3] Li G, Mao R, Hildre H P and Zhang H 2020 Visual attention assessment for expert-in-the-loop training in a maritime operation simulator *IEEE Transactions on Industrial Informatics* **16** 522–31

[4] Mao R, Li G, Hildre H P and Zhang H 2019 Analysis and evaluation of eye behavior for marine operation training - a pilot study *Journal of Eye Movement Research* **12** 3

[5] Ana M, Li G and Zhang H 2019 A decentralized sensor fusion approach to human fatigue monitoring in maritime operations *Proc. of ICCA* vol 15 (Edinburgh: IEEE) pp 1569–74

[6] Sanfilippo F, Hatledal L I, Styve A, Pettersen K Y and Zhang H 2016 Integrated Flexible Maritime Crane Architecture for the Offshore Simulation Centre AS (OSC): A Flexible Framework for Alternative Maritime Crane Control Algorithms *IEEE Journal of Oceanic Engineering* **41** 450–61

[7] Sanfilippo F 2016 A multi-sensor system for enhancing situational awareness in offshore training *Proc. of CyberSA* (London: IEEE) pp 1–6

[8] Li N, Su Z, Chen Z and Han D 2013 A real-time aircraft infrared imaging simulation platform *Optik* **124** 17
[9] Chen X, Zhou X, Qian W, Chen Q, Wan M and Gu G 2018 Infrared simulation and implementation of virtual ocean scene *Proc. of SPIE* vol 10765 ed Strojnik M and Kirk M S (Bellingham: SPIE) pp 165–172
[10] Dai Y, Chen B and Ming D 2013 Research on real-time infrared simulation of ground scene *Proc. of SPIE* vol 8917 ed Zhang X and Liu J (Bellingham: SPIE) pp 100 – 106
[11] Zhang F, Wang T, Huang K, Mao J, Li M and Wang Z 2018 One improved real-time infrared simulation system based on unity3d *Journal of Computer-Aided Design and Computer Graphics* **30** 1177
[12] Dosovitskiy A, Ros G, Codevilla F, Lopez A and Koltun V 2017 CARLA: An open urban driving simulator *Proc. of Machine Learning Research* vol 78 ed Levine S and et al (Zürich: PMLR) pp 1–16
[13] Baldini F, Anandkumar A and Murray R M 2019 Learning pose estimation for uav autonomous navigation and landing using visual-inertial sensor data *ArXiv* abs/1912.04527
[14] Su H, Wang R, Chen K and Chen Y 2019 A simulation method for lidar of autonomous cars *IOP Conf. Series: Earth and Environmental Science* **234** 012055
[15] Sutherland I E, Sproull R F and Schumacker R A 1974 A characterization of ten hidden-surface algorithms *ACM Computing Surveys* **6** 1–55
[16] Muckenhuber S, Holzer H and Bockaj Z 2020 Automotive lidar modelling approach based on material properties and lidar capabilities *Sensors* **20** 3309
[17] Manivasagam S, Wang S, Wong K, Zeng W, Sazanovich M, Tan S, Yang B, Ma W C and Urtasun R 2020 Lidarsim: Realistic lidar simulation by leveraging the real world *Proc. of CVPR* pp 11167–76
[18] Li W, Zhang H, Hildre H P and Wang J 2014 An FPGA-based real-time UAV SAR raw signal simulator *IEICE Electronics Express* **11** 20140168
[19] Huizing A G and Theil A 1991 CARPET: A versatile program for radar performance analysis Final Report Physics and Electronics Lab. TNO
[20] Hem A G, Leinebø D, Opheim H V, Vasstein K and Skarshaug T 2019 Digital tvilling Project report from a group in the village Autoferge in Experts in Team at NTNU
[21] Lauvsnes M 2020 *Feature-based pose estimation of ships in monocular camera images* Master's thesis NTNU
[22] Vasstein K 2020 *Real time simulation of EMR sensors for autonomous ferries* Project thesis NTNU
[23] Deering M, Winner S, Schediwy B, Duffy C and Hunt N 1988 The triangle processor and normal vector shader: a VLSI system for high performance graphics *SIGGRAPH Comput. Graph.* **22** 21–30
[24] Young H D and Freedman R A 2016 *Sears and Zemansky's University Physics with Modern Physics* (Edinburgh: Pearson Education Limited) pp 1328 – 31
[25] Meola C, Boccardi S and Carlomagno G M 2017 *Infrared Thermography in the Evaluation of Aerospace Composite Materials* ed Meola C and et al (Cambridge: Woodhead Publishing) pp 57 – 83
[26] Richards M A, Scheer J A and Holm W A 2010 *Basic Principles* (*Principles of Modern Radar* vol 1) ed Kay D R and Janelle K (New Jersey: SciTech Publishing) pp 64, 211–19
[27] Morales J, Plaza V, Mandow A, Gomez J A, Serón J and García A 2018 Analysis of 3d scan measurement distribution with application to a multi-beam lidar on a rotating platform *Sensors* **18** 2
[28] Mustvedt P 2019 *Autonom ferge designet for å frakte 12 passasjerer trygt over Nidelven* Master's thesis NTNU

The 3rd International Conference on Maritime Autonomous Surface Ship (ICMASS 2020) IOP Publishing
IOP Conf. Series: Materials Science and Engineering **929** (2020) 012033 doi:10.1088/1757-899X/929/1/012033

Global linked simulations - a key to the evaluation of future transport concepts and navigation

R. Grundmann[1] and D. Kang[2]

[1] Research Associate, Fraunhofer CML, Hamburg, Germany
[2] Junior Researcher, KRISO, Daejeon, Korea

E-mail: Robert.Grundmann@cml.fraunhofer.de

Abstract. This paper is intended to reflect the achievements of the European Maritime Simulator Network (EMSN Connect) and the Asian Pacific Maritime Simulation Network (APMSN established in 2019), and the achievements of this project, to create a global digital testbed for the entire shipping industry.

1. Introduction

The digitization of transport technologies is developing rapidly. While the focus has so far been primarily on digitized road, rail or in some cases air transport systems, the maritime industry has recently been moving more into focus.

The future design of this digital testbed structure can be used for testing further supporting software and hardware, for evaluating new navigation systems and for large-scale scenarios between (partly) autonomous ships, remotely controlled ships and manned ships, but also for completely different parts of the maritime transport chain.

The European Maritime Simulator Network (EMSN) has set itself the task of increasingly highlighting and exploiting the advantages of collaborative simulation in a digitized world in order to validate exemplary developments of route exchange formats or other services. In addition to testing technological developments, there is a great deal of interest in the ever-growing field of networked simulations in the education and training of seafarers.

Since this network is still relatively young and the possibilities have not yet been extensively researched, some of the currently possible, but also future domains shall be mentioned here, as well as the technical implementation and realization in the different disciplines.

2. History

Till now, Ship Handling Simulator (SHS) experiments have been limited to a small number of simulated vessels in a joint exercise, thus organizational affects or large-scale changes to waterborne processes, procedures and technologies could barely be assessed. According to [1], large-scale virtual test facilities are required to bridge "major gaps with regards to development of safe waterborne connected and automated transport".

IOP Conf. Series: Materials Science and Engineering **929** (2020) 012033 doi:10.1088/1757-899X/929/1/012033

Such a large-scale virtual test facility exists: The European Maritime Simulator Network (EMSN). Recently, the EMSN has been developed to provide a large-scale test environment for maritime safety. This network evolved from as a proof of concept and was operationalized during the EU-funded MONALISA 2.0 and Sea Traffic Management (STM) Validation project and can be regarded today as a baseline technology/testbed for interdisciplinary research in shipping [2].

The EMSN has already been successfully used several times. First, simulations of the current initial situation were carried out, which were then compared with follow-up test runs involving the newly developed STM applications. The exercises were based on navigational demanding traffic situations in the Kattegat between the island of Anholt and the Great Belt Bridge, in the English Channel and various scenarios were also implemented in the Strait of Gibraltar.

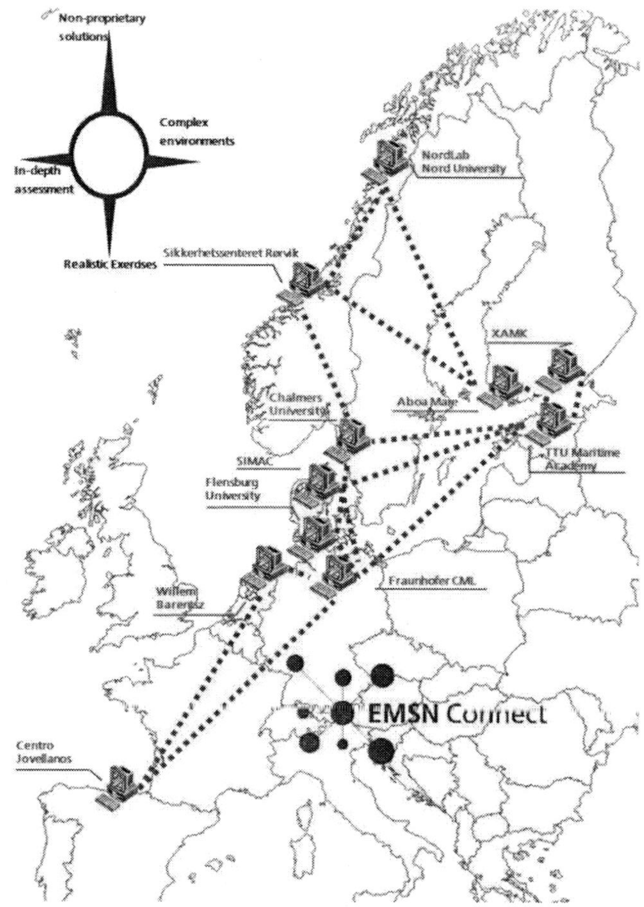

Figure 1: Representation of the simulator centers connected and participating in the EMSN Connect 06/2020 [3]

3. What is EMSN Connect?

EMSN Connect is a network of several Ship Handling Simulators (SHS) from different manufacturers, distributed over European locations.

As depicted in [2], the network has a generally applicable network architecture. Starting from the central node, the hub, each center has its own spokes, which in turn send the simulator data to the other centers via the spokes and the hub.

The network sees itself in its basic infrastructure as a platform for the exchange of simulation data and the exchange of voice communication data. These basic services are implemented in IP (sub)networks and enable participants in EMSN Connect to join a networked simulation exercise. The service networks are implemented in the form of Virtual Private Networks (VPN).

The EMSN Connect structure is based on so-called VPN tunnels, which provide secure connections via the Internet. Virtual private networks can be implemented with different topologies. The topology shown for EMSN Connect is a hub-and-spoke topology. It is a variant of so-called site-to-site VPNs that connects multiple sites with intranets to a common virtual private intranet.

The location of the Fraunhofer CML in Hamburg forms the hub for this. All other locations set up VPN tunnels only with the hub in Hamburg. There are no tunnels between the individual spokes.

The entire data is transmitted from a spoke to a hub and from there to all spokes in the network via multicast.

These VPN tunnels are designed with IPSec, data integrity, data confidentiality, authentication of data origin and confidentiality of traffic flow to ensure maximum network security.

To make the simulator data available to different simulators and different simulator manufacturers the DIS standard was used.

"Distributed Interactive Simulation (DIS) is a government and industry initiative to define an infrastructure for linking simulations of various types at multiple locations into realistic, complex, virtual worlds for the simulation of highly interactive activities. In order for DIS to take advantage of currently installed and future simulations developed by different companies, a means had to be found to ensure interoperability between dissimilar simulations" [4].

Since EMSN Connect does not support a common ship model database, an agreement was reached on a number of ship types that are used for visualization throughout the EMSN, i.e. DIS units.

Each simulator manufacturer has its own model database. A distinction is made between so-called OS models (OwnShip) and TG models (TargetShip). A selected list of OS and TS models was assigned a DIS entity number and made uniformly accessible to all participating centers in the DIS application.

The simulator sites which receive the own ship of another simulator in the network map this ship as a foreign ship (TG model). Based on the definitions in the DIS applications, the simulator maps these foreign ships from its model database. Due to the different simulator manufacturers in the network, there may be slight deviations in shape, color and size. However, this can be neglected if the DIS applications in the network are considered in a uniform way, since similar dimensions of the ships in the network were defined for the respective DIS numbers.

To exchange simulation data between the ship simulators, e.g. ship position and motion parameters and AIS messages, the DIS protocol [4] is used. Each ship is identified by a unique entity ID to ensure that own and target ships are correctly connected. In addition, according to DIS, each Entity ID is assigned a DIS entity number indicating the entity type (e.g. ship type and size). The DIS standard offers many advantages, there is an international standard protocol for simulation networks (IEEE 1278), this standard is open and there are no license costs, the infrastructure requirements are easy to implement and simple to handle.

4. Data evaluation and storage

The consideration of the simulations and the implementation of a functioning infrastructure are realized. For the meaningful use of the network in the context of research and development projects, it must now be possible to record the data. Simulation and voice communication data are exchanged within the VPN. Each simulation centre is connected to the Internet via a VPN router, which creates the virtual EMSN subnets between hubs and spokes.

IOP Conf. Series: Materials Science and Engineering **929** (2020) 012033 doi:10.1088/1757-899X/929/1/012033

Voice over IP (VoIP) tools are used to establish the radiotelephony connection in the simulator, which is an important part of the training of future navigators.

A specially developed software, the EMSN Data Tracker, is used to map the entire situation of the network and the centres and ships in the exercise based on the DIS Data PDU [5]. In addition to the detailed presentation of the respective status of each centre in the exercise, this also serves to record the data.

Once the data tracker receives the encrypted PDU packets, it first decrypts them and then decrypts the PDU packet based on the different PDU types. After decoding each PDU packet, data trackers visualize detailed information on the data tracker. This way the information is also stored in the database. The database is stored locally on the computer and is available for further analysis after the recording is finished. The movement data of each ship that participated in the exercise can be evaluated in detail. Thus, navigator's behavior can be more objectively analyses in contrast to pure expert opinion interviews [6]

5. APMSN

The EMSN Connect network has been strengthened during 2019 by the Asian Pacific Maritime Simulator Network (APMSN).

By creating a counterpart of the European network EMSN Connect it was possible to connect two separately working networks in a joint exercise in February 2020. The two networks offer the possibility to hold exercises only in the respective network, but can also be easily interconnected for exercises between APMSN and EMSN Connect.

Due to the same structural setup as EMSN Connect with its own hub, it can operate separately from EMSN Connect. This means that the APMSN has a similar architecture to the EMSN Network and can operate as a stand-alone network. The two independent networks are then connected via different VPN tunnels with different IP security features. To send information between the two independent networks, we use the static protocols EIGRP and NHRP so that the individual networks are stable and secure according to the network guidelines and the encapsulation is done according to the OSI model

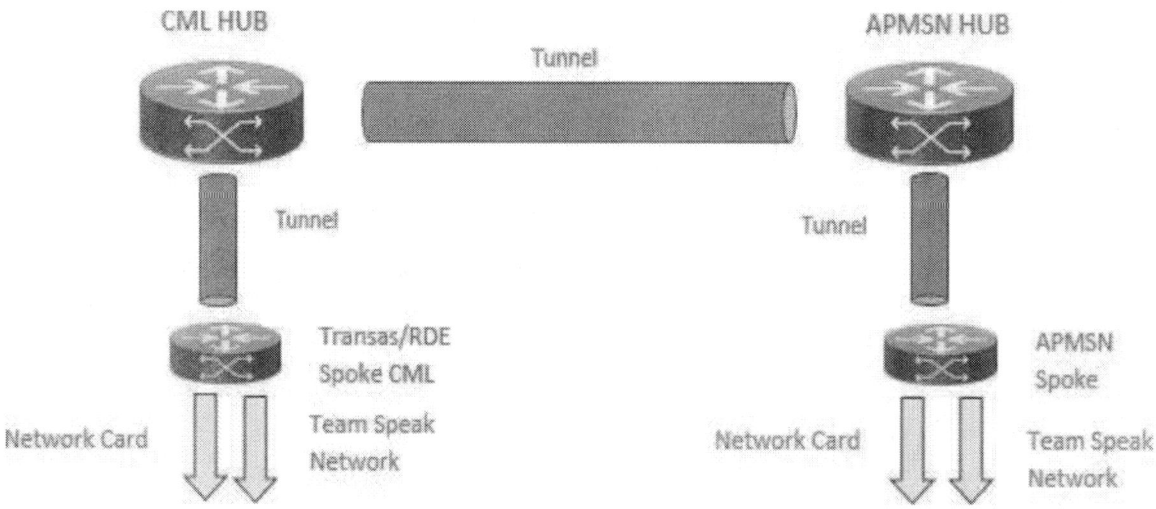

(Open System Interconnection).

Figure 2: simplified presentation of the infrastructure between EMSN Connect and APMSN

6. Joint Exercise

The presentation of scenarios in simulation networks is not fundamentally different from the runs in a local centre.

On the one hand, the respective objectives of the scenario have to be evaluated, on the other hand the preferred personnel, sailing area and the possible ship types. Once the scenario description has been created, these are implemented in the various simulation locations. This includes standardized local tests as well as short technical test setups which can also be carried out before the simulation itself.

During the run the interactions of the ships can be recorded in the EMSN Data Tracker. These recordings can be enhanced with VoIP recordings to get a complete impression of the run.

An example is a run in the combined network EMSN Connect and APMSN, which reflects a port arrival and berthing scenario in the port of Busan with a container ship planned by KRISO (Korea Research Institute Ships & Ocean Engineering) and instructed by a pilot at KRISO, and two tugs controlled by Chalmers University of Technology / Department of Mechanics and Maritime Sciences (Chalmers M2).

The University of Chalmers used Transas simulators and KRISO used STR simulators respectively.

This was the first time a globally networked simulation was run. Further innovations included a port arrival and berthing scenario in the port of Busan using a Chalmers ship planned by KRISO and instructed by a pilot at KRISO, as well as two Chalmers tugs. The pilot's instructions were passed on via the chain of command with repetition and confirmation of the manoeuvres, similar to the real port entrance. However, the manoeuvres were carried out by the experienced crew in Sweden. Particularly noteworthy here is the use of the tugs during the traversing of the ship to the pier.

The objective was to test the implementation of new software and hardware to support pilots in addition to the successful berthing of vessels.

Further implementations of new software are planned for future runs in 2020. Among other things, the ship data provided by the Data Tracker Software will be collected and made available to the pilot for better coordination of the vehicles in the area and timely assessment of potential hazards.

7. A glimpse into the future

The possibilities presented here are only the beginning of the development of a globally operating maritime simulation network. Certainly, with the advancement of automation and autonomation of maritime transport participants in the maritime transport chain, the network will be faced with completely new and diverse tasks in the future. The simulator network can be used in many different ways. The possibilities to take the education and training of seafarers to a new level are as much a topic as the development and testing of new equipment and generally new concepts in maritime transport.

One of the futures and most desirable integrations or innovations in the network may be the integration of (partially) autonomous, autonomous and remote-controlled units [5]. Integrating and evaluating these systems in a networked simulation could be considered within the framework of the two simulation networks. Through the direct connection to traffic control centres and the combination with human simulation participants, possible complex real-world scenarios could be mapped and examined in the networks.

The further development of navigation systems and the integration of these into the bridge is increasingly considered in terms of user interface and user experience factors. The integration of such systems into ship operation requires fundamental tests under as real conditions as possible. Even completely new systems, which are currently used only marginally or not at all on ships, should facilitate the work of navigators in the coming years. To provide important data for the refinement and optimization of such systems, a broadly-based simulator network is suitable.

In a further step, Korea plans to implement a system that will allow simulators to implement the same information available to ports by linking Korean e-navigation services with APMSN technology.

If this is implemented, it is expected that the simulator experiments can also be performed in the real port traffic environment.

In addition to the above-mentioned goals and possible fields of activity of the network, the maritime simulation network is a digital testbed which is prepared for the tasks to come. Beyond the digital infrastructure the network has a large number of maritime experts who are naturally interested in discussing current problems and open questions of the maritime industry and maritime traffic.

In June 2020, the partners of the simulation network were able to show their work to the public in a webinar and let them participate interactively in a run. This first-time public relations effort is to be repeated continuously for different runs.

The goals of this project go beyond a simple online presence. In the future, the society's attention will be focused on ships with different degrees of autonomous systems and their interaction with their human counterparts on other simulator bridges and in the shore stations.

Certainly, this will contribute to an increased public awareness and possibly to a higher acceptance of such systems in society.

The possibilities associated with networked simulations are certainly still in their infancy at the moment, but in view of the current difficult situation with regard to COVID 19, the need for such contactless tests is evident. By building up the network, research and testing can continue on a larger scale, even in times of contact ban.

Acknowledgments

This work received funding from the European MONALISA 2.0 and the STM Validation project as well as from Korean Implementation of Core Technology for Korean e-Navigation Services project. Further, the author would like to thank all participating institutions of the EMSN-Connect initiative named under https://emsn.connect.fraunhofer.de/ for their trust and contribution to this network.

References

[1] European Commission 2017 Connected and Automated Transport - Studies and Reports Available at: http://ec.europa.eu/newsroom/horizon2020/document.cfm?doc_id=46276 [Accessed 10 04 2019].

[2] Rizvanolli A, Burmeister H-C and John O 2015 The Role of the European Maritime Simulator Network in Assessing Dynamic Sea Traffic Management Principles TransNav, the International Journal on Marine Navigation and Safety of Sea Transportation **9**, 559-564

[3] https://emsn.connect.fraunhofer.de/

[4] IEEE Std 1278.1-1995 1995 IEEE Standard for Distributed Interactive Simulation Application Protocols

[5] Burmeister H-C, Scheidweiler T, Reimann M and Jahn C 2020 Assessing Safety Effects of Digitization with the European Maritime Simulator Network EMSN: The Sea Traffic Management Case TransNav, the International Journal on Marine Navigation and Safety of Sea Transportation **14** 91-96Olindersson F, Bruhn W, Scheidweiler T, Andersson A 2017 Developing a Maritime Safety Index using Fuzzy Logics TransNav, the International Journal on Marine Navigation and Safety of Sea Transportation **11** 469-475

The 3rd International Conference on Maritime Autonomous Surface Ship (ICMASS 2020) IOP Publishing
IOP Conf. Series: Materials Science and Engineering **929** (2020) 012034 doi:10.1088/1757-899X/929/1/012034

Analysis of manoeuvring characteristics through sea trials and simulations

SeungGi Gug[1, *], Dimantha Harshapriya[2], Haesang Jeong[1], JuneHo Yun[1], DongJin Kim[3] and YeonGyu Kim[3]

[1] Department of coast guard studies, Korea maritime & Ocean University, 727 Taejong-ro, Yeongdo-Gu, Busan 49112, Republic of Korea.
[2] Department of Naval architecture & ocean systems engineering, Korea maritime & Ocean University, 727 Taejong-ro, Yeongdo-Gu, Busan 49112, Republic of Korea.
[3] Korea Research Institute of Ships and Ocean Engineering (KRISO), Daejeon 34103, Republic of Korea.

* E-mail: cooksg@kmou.ac.kr

Abstract. In recent years, Autonomous Surface Vehicles have secured the focus of major research and industrial initiatives around the world. Following this trend in the wake of fourth industrial revolution of oceans, Marine Autonomous Surface Ships (MASS) have become the focus of numerous research. In order to shift from conventional ships to autonomous transportation, manoeuvring data collected through real-ship trials have to be available, and analysed. In this study, manoeuvring characteristics of a real ship were studied over a period of three years, data were collected and analysed to identify changes of manoeuvring characteristics over time. Real-ship data were then compared with simulation data to validate a manoeuvring simulator program. Through the real-ship experiments, not only the simulation results were validated, but also the possibility of predicting ship manoeuvrability using simulations in controlled environments was identified. In addition, the importance of accumulation of manoeuvring data in helping future designs of ships with similar characteristics is identified.

1. Introduction

In recent years, Autonomous Surface Vehicles have become a significant topic for major research and initiatives throughout academia, industries and governments. Following this viable trend in the wake of fourth industrial revolution of oceans, Maritime Autonomous Surface Ships (MASS) have become the center of numerous maritime-related research. Although the concepts of autonomous transportation systems were introduced decades ago, successful models and applications have only recently become a reality. [1] However, the research and development of MASS have acquired momentum over the years due to the technological advancements in maritime as well as in other academic domains. Even though there are numerous aspirations for developing autonomous ships, outcome focuses on greener shipping, optimum energy consumption and safe, as well as, efficient maneuvering. [2-5] Even though the concept of autonomous vessels seems promising, practical approach to designing an actual autonomous ship is a complex task. Hence, the postulated shift from conventional ships to autonomous ships are likely to follow a circumspect transition, rather than a steep development. [6] Although this has been discussed extensively, current progress of MASS research has been rapid due to positive advancements in technology.

As identified and recommended by recent research [7], the minimum requirement for an autonomous ship to be operational is that it should at least be as safe as a manned, conventional ship. In addition, the safety of operation as well as personnel is identified to be the most important factor in developing MASS [8]. In order to achieve safe maneuvering in an autonomous operation, the control system is required to collect and process information. Based on the processed data, decisions on strategies are made. In a dynamic maritime environment, there are two main phases [9] [10] of motion planning. Namely, Long-term path planning (global path planning), where path or sailing route optimization is considered, and Short-term path motion planning (local path planning), where moving speed, heading and other maneuvering characteristics are considered for optimized sailing performances. Hence, prior accumulated maneuvering data is evidently important for predicting ship motion, as well as for simulating ship motion in a virtual environment. [11]

Ship simulating methods can vary depending on the approach and available data, although the fundamentals used are the same. A previous research done by Barr [12] was able to provide a comprehensive set of ways that can be utilized to predict maneuvering in calm water. Vessel motion prediction is mostly based on steady speed and constant heading, although methods adapting frequency domain approaches are more commonly utilized due to the higher computational efficiency when compared to time domain approaches. Moreover, strip theory is commonly adapted when solving hydrodynamic forces and they are summed over 2D cross sections along the ship length. [13] With the advancements in computational powers, motions in waves now can be predicted accurately. In addition, three-dimensional components of hydrodynamic forces can be simulated with improved precision. [14] However, there are limits to the exact application of vessel's maneuverability because the ship's performance and maneuverability deteriorate with time in sea. Therefore, maneuvering characteristics at design stage has to be tested and compared after the ship is commissioned.

The present study is threefold, structured to adhere further development of future research and a validation of a computer simulator program. First, the study is focused on identifying the benchmark results and maneuvering data for the subject ship, in order to carry out real-ship experiments in an open sea environment. Once the initial conditions are identified and discussed, second part of the paper is focused on standard real-ship experiments, which were carried out over a period of three years. Special equipment used, methodology and results are discussed. Data were accumulated and analyzed in detail to identify deviations and subsequently, to compare with initial benchmark values. Final part of the study is ongoing with a greater scope, where a specific model ship simulator is constructed to predict and analyze the standard maneuvering characteristics in open-sea environments. This particular article includes the simulation data and results of the aforementioned program for clarification and the results from both real-ship trials and simulations are compared to draw conclusions, and to validate the simulator program.

2. Training ship status

The subject vessel is registered under the South Korean flag and is categorized as a research and a training vessel. Personnel capacity of the vessel is 246 and the ship particulars are shown in Table 1.

Table 1. Ship particulars of subject ship.

Ship particular parameter	Value
Length Overall (LOA)	: ~ 117.20 m
Length Between Perpendiculars (LBP)	: 104.00 m
Breadth, Molded	: 17.80 m
Depth, Molded	
Main Deck	: 8.15 m
Upper Deck	: 10.80 m
Designed load draft, molded	: 5.90 m
Displacement	: ~ 6418.00 ton
Deadweight	: 2868.00 ton
Gross tonnage	: ~ 6,700 ton
Scantling draft	: 6.2 m

Initial maneuvering trials were conducted by the ship building company after commissioning the vessel in 2005. (Training ship particulars, Korea Maritime & Ocean University) These tests included standard turning circle test and 20°/20° zig-zag test.

Tactical diameter, transfer at 90° change of heading, maximum advance, times to change heading 90° and 180°, and transfer loss of steady speed were recorded during the turning circle test. Moreover, initial turning time, overshoot angle and yaw checking time were recorded through the 20°/20° zig-zag test.

Test environment and the information related to external factors in turning tests are shown in Table 2.

Table 2. External factors and test environment during turning circle test.

		PORT Turning	STBD Turning
Initial Heading (deg.)		201.1	201.5
Sea	Condition	Fresh Breeze	
	Depth	110	110
T. Wind	Direction	55.0	86.0
	Velocity (m/s)	10.3	9.9
M/E RPM (Initial/End)		176.1/108.8	181.7/129.6
Speed (kts) (Initial/End)		17.6 / 6.0	17.5 / 6.8
Ruder Angle (deg.)		-34.1	32.5
Advance (m)		329.939 (3.2LBP)	311.941 (3.0LBP)
Transfer (m)		195.238 (1.9LBP)	193.897 (1.9LBP)
Tactical Diameter (m)		374.561 (3.6LBP)	407.006 (3.9LBP)

Information of the test environment and external parameters in initial 20°/20° zig-zag test are shown in Table 3, where the initial ship-maneuvering criterion related to ship performance prior to, and during the test are tabulated.

Table 3. External factors and test environment during 20⁰/20⁰ zig-zag test.

Parameter		Value
Initial Heading (deg.)		351.1
Sea	Condition	Strong Breeze
	Depth (m)	110
R. Wind	Direction (deg.)	334.0
	Velocity (kts)	39.7
M/E R.P.M. (Initial/End)		177.5 / 177.4
Speed (kts) (Initial/End)		18.8 / 14.6
1st Overshoot Angle (deg.)		17.6
2nd Overshoot Angle (deg.)		16.2

Initial results of turning circle maneuver and 20⁰/20⁰ zig-zag tests are graphically represented in Figure 1. When the initial results of turning circle diameter test are considered, it was noted that the designed parameters are within the IMO accepted ranges, [15] given that, advance, transfer and tactical diameter are 3.2-3.0 LBP, 1.9-1.9 LBP, and 3.6-3.9 LBP on both port and starboard sides, respectively. Similar, 20⁰/20⁰ zig-zag test results were noted to show that the first overshoot angle is less than 25 degrees and agrees with the IMO standards related to maneuvering performance characteristics. [15]

IOP Conf. Series: Materials Science and Engineering **929** (2020) 012034 doi:10.1088/1757-899X/929/1/012034

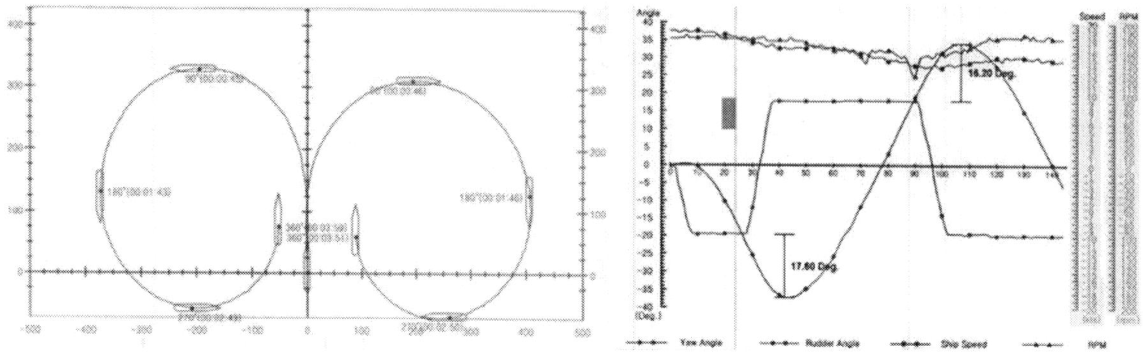

Figure 1: Graphical representation of initial turning circle test (Left) and $20^0/20^0$ zig-zag test (Right).

3. Sea trials and analysis

Main purpose of carrying out the sea trials was twofold: first, to compare current maneuvering characteristics with initial test results to analyze and identify possible deviations; and secondly, to acquire precise maneuvering data to use as the benchmark to validate the simulator program. Thus, special equipment were used and standard turning circle test and 20°/20° zig-zag test were carried out multiple times with different loading and weather conditions. Parameters related to the standard maneuvering tests, as well as additional parameters such as, rolling, trim, pitching and acceleration were recorded for further analysis.

3.1 Methodology of sea trials

For measuring the real-ship maneuvering performance, an Inertial Measurement Unit (IMU) consisted of a 3-axis accelerometer with a built in gyrocompass, two Global Positioning Systems (GPS) receivers and a laptop equipped with dedicated software were used. IMU was installed along the centerline of the vessel, on the deck in par with vessel's center of gravity. Two receivers (main and secondary) were installed on either sides, parallel to the centerline. The IMU used in trials, which is a GPS aided inertial navigation system named "Spatial Fog Dual", is showed in Figure 2. Accurate position, velocity, and acceleration of the vessel during the sea trials were measured using the spatial fog dual. Apparatus was consisted of fibre optic gyroscopes, magnetometers, accelerometers and a pressure sensor with a dual antenna Real-time Kinematic Global Navigation Satellite System (RTK GNSS) receiver.

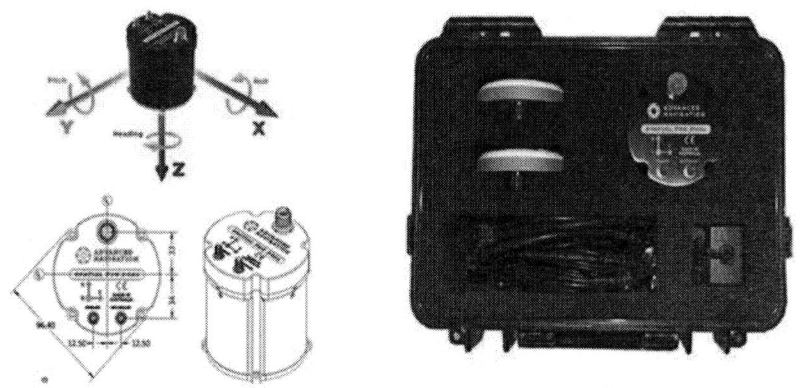

Figure 2. Inertial Measurement Unit (IMU) used in the real-ship trials.

Integration of the GPS receivers, IMU and dedicated software used in real-ship experiments, and the graphical user interface (GUI) of the software are shown Figure 3.

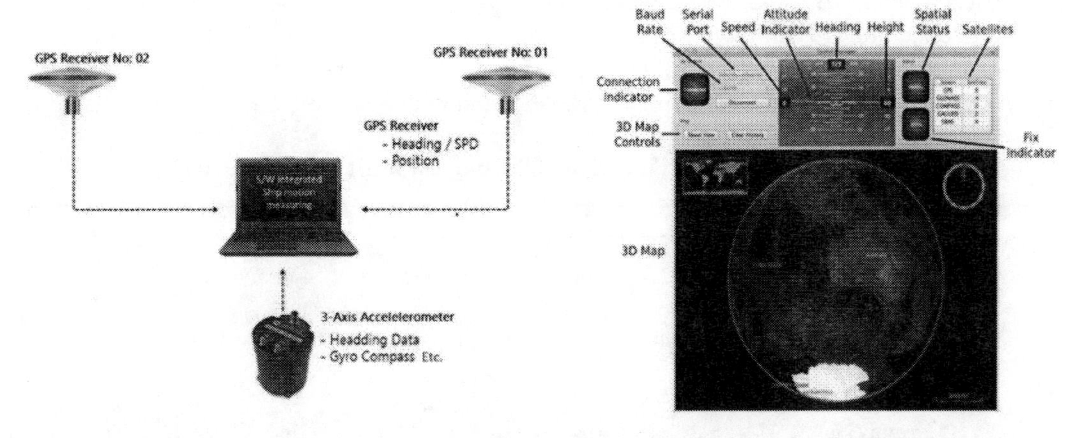

Figure 3. Integration of GPS receivers and IMU (Left) and GUI (Right).

Once the equipment were installed, the ship was maneuvered following the standard turning circle and 20°/20° zig-zag test procedures and data were recorded. Same experiment was carried out over a period of three years; from 2017 to 2019. Each test was carried twice a year to ensure accuracy. This data was then tabulated and represented graphically to identify the changes in maneuvering characteristics of the training ship.

3.2 Comparison of initial and ongoing sea trials data

Importance of using specific equipment for the trials is reemphasized, and data was collected from the turning circle test and 20°/20° zig-zag test over the course of three years, from 2017 to 2019.

Time and location of ship trials are included for precision and transparency. Information such as advance, transfer and turning circle diameter (TCD) related to turning test, and overshoot angles related to 20°/20° zig-zag test are tabulated. For further clarification, initial test data is included in study.

It should be noted that the data recorded using the Navigational Decision Support System (NDSS), together with IMU and other apparatuses, provide an un-calibrated result. In order to identify parameters such as TCD and advance, equipment were calibrated accordingly, and was an important part of the study. Difference between pre and post-calibrated results are shown in Figure 4 for clarification.

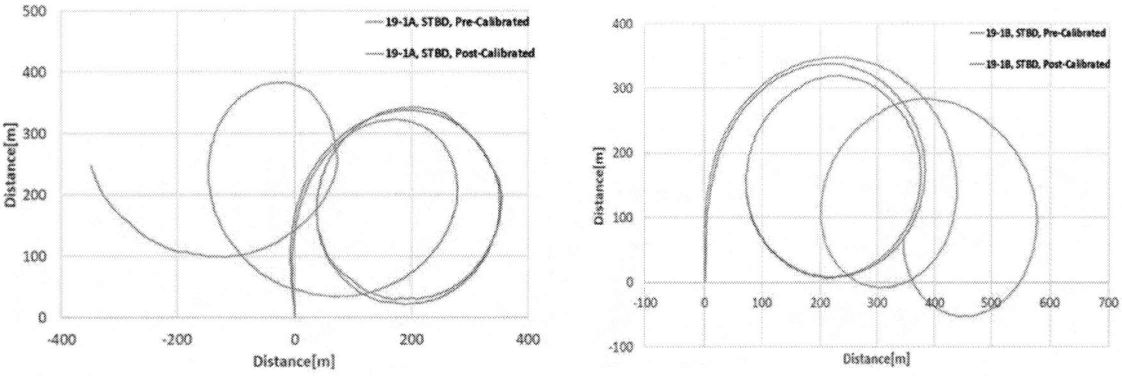

Figure 4. Graphical representation of pre and post-calibrated data of two starboard-side turning circle tests with different heading angles, 90^0 and 270^0.

Each real-ship test included a turning circle test and a $20^0/20^0$ zig-zag test and all parameters were recorded accordingly. Moreover, environmental conditions such as wind, current, tide direction and wave height etc. were measured to evaluate external factors during the tests.

Collected data is summarized in Table 4. 17-1, 17-2, 18-1 and 18-2 denote the first and second tests done in each year in 2017 and 2018, respectively. 19-1A, 19-1B, 19-2A, 19-2B denote the first and second tests done in 2019 with two different heading angles, A and B (90^0 and 270^0, respectively), in order to expand the data range for accuracy, starting from 2019. (20°/20° zig-zag and starboard turning test) Advance, transfer and tactical diameter are tabulated with respect to LBP, which is 104 (m).

Table 4. Comparison of initial test in 2005 and test trails data from 2017 to 2019.

		Initial	17-1	17-2	18-1	18-2	19-1A	19-1B	19-2A	19-2B
Date		2005. 11. 18	2017. 09. 08	2017. 10. 28	2018. 05. 19	2018. 10. 12	2019. 04. 08	2019. 04. 08	2019. 04. 22	2019. 04. 22
Sea Area		Busan	Busan	South China Sea	Vietnam	Jeju	Ulsan	Ulsan	East sea	East sea
Environment Condition	Wind (m/s)	10.3	2.3	5.3	1.9	6	9.1	9.1	12.4	12.4
	Current (kts)	-	1.3	0.4	0.3	0.7	1	1	0.9	0.9
	Tide Direction (rel. deg.)	-	100	220	345	305	90	270	90	270
	wave height (m)	-	0.3	0.3	0.2	0.6	0.3-0.5	0.3-0.6	0.3-0.7	0.3-0.8
Loading Condition	Disp. (ton)	5529	5945	6462	6759	6864	6124	6124	5983	5983
	Trim (m)	-1.137	-1.49	-0.98	-0.54	-0.03	-0.82	-0.82	-0.92	-0.92
	LCG (m)	-	-2.14	-1.77	-1.25	0.48	-1.16	-1.16	-1.22	-1.22
Turning (P)	Init. SPD (kts)	17.6	11.7	11.2	11.5	10.3	10.6		10.2	
	Rudder (deg.)	-34.1	-35.7	33.7	-35.1	-34.3	-34.4		-34.3	
	Advance (m(LBP))	3.2	2.68	3.44	3.16	2.75	3.25		2.82	
	Transfer (m(LBP))	1.9	1.68	1.38	1.62	1.73	2.26		1.56	
	Tac. Dia. (m(LBP))	3.6	3.37	3.36	3.24	2.1	3.51		3.26	
	Drift Angle (deg.)	-	2.5	-2.4	-1.6	0.1	5.1		-1.7	
Turning (S)	Init. SPD (kts)	17.5	10.2	11.2	10.9	10.1	10.4	10.4	10.2	10.2
	Rudder (deg.)	32.5	35.3	30.7	34.2	34.6	34.7	34.7	34.7	34.7
	Advance (m(LBP))	3	2.99	3.33	2.78	3.01	3.25	3.17	3.05	3.06
	Transfer (m(LBP))	1.9	1.57	1.79	1.97	1.56	1.71	2.56	1.41	1.45
	Tac. Dia. (m(LBP))	3.9	3.3	3.94	3.65	3.43	3.38	3.52	3.47	3.42
	Drift Angle (deg.)	-	12.3	-1.2	-0.5	-0.2	5.2	-4.5	3.2	1.6
Zig-Zag 20	Init. SPD (kts)	18.8 (177.5 rpm)	12.5 (120 rpm)	11.2	12.7 (120 rpm)	10	10.3	10.1	9.8	9.9
	1st OA	17.6	8.3	10.5	14.1	9.1	7.8	8.9	7	5.1
	Drift Angle (deg.)	-	5.3	8.4	10.8	10.6	6		6.7	

IOP Conf. Series: Materials Science and Engineering **929** (2020) 012034 doi:10.1088/1757-899X/929/1/012034

Calibrated turning circle data (port and starboard) over the time span from 2017 to 2019 are graphically represented in Figure 5. Changes and differences in parameters in each turning circle test are evaluated through the analysis and the primary benchmark for simulations were set using the aforementioned data. Results of real-ship turning circle tests are summarized in following representation.

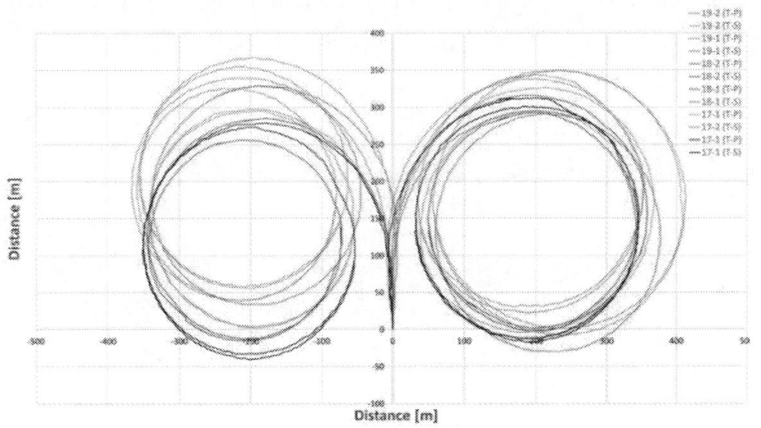

Figure 5. Comparison of calibrated turning circle data.

Similarly, 20°/20° zig-zag tests were carried out twice a year, over the time span of three years. (Data is tabulated in Table 4) In this particular paper, only the graphical representation of the last two zig-zag tests, which were carried out in 2019 are included. First test was carried out (19-1A) with a relative angle of 90 degrees, and the second 20°/20° zig-zag test (19-1B) with a relative angle of 270 degrees. Figure 6 represents the said results; however, the complete data analysis is included in the study.

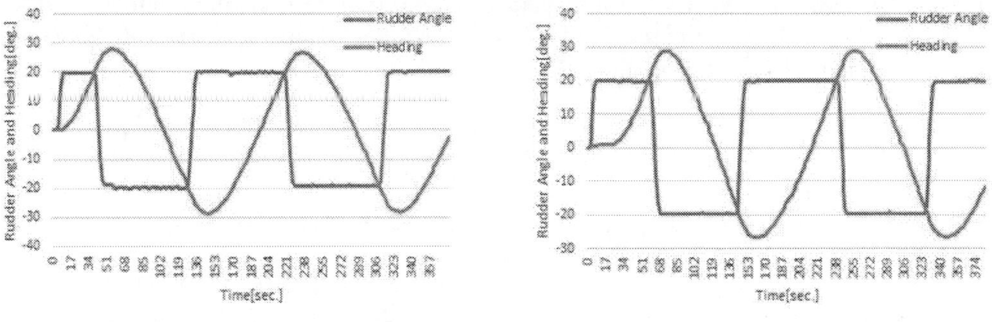

Figure 6. Comparison of $20^0/20^0$ zig-zag tests carried out in 2019.

Similarly, data was collected for all 20°/20° zig-zag tests, which were carried out from 2017 to 2019. Analysis was primarily focused on turning circle data and it was noted that, with the increment of initial speed, advance increases and consequently, advance decreases with the increment of rudder angle. Moreover, increment of turning circle diameter was noted with the increasing values of displacement. However, the need of having a controlled environment for simulations more than the real-ship tests were clarified since external parameters evidently cause an unpredictable variation in maneuvering results. Thus, the importance of accumulating real-ship test results in MASS and decision-making system design was established.

4. Simulations

Two vital parts were focused under the simulations. First, the mathematical model used in simulation is discussed, and second, the process of constructing the simulation program for the subject vessel, using a model approach to predict maneuvering characteristics and validating it with real-ship data were carried out.

Mathematical modeling of seakeeping and simulations have proven to be vital in utilizing automatic control in marine vessels for several years. [16] However, there is a limited number of robust and versatile applications available due to the complexity of the models. A model was proposed by Son and Nomoto [17], combining test data obtained for lateral motion through planar motion mechanism (PMM), integrating different vales of static heel with roll motion tests. Moreover, multiple models were proposed using experimental data in a unique 4 degree of freedom roll planar motion mechanism (RPMM). [18] In this study, however, a standard mathematical model is followed to approach a simpler model to simulate maneuvering procedures of the subject vessel.

4.1 Mathematical model

A bi-dimensional approach is considered in this study, with two reference frames; one of them fixed and the other one moving with the vessel, as shown in Figure 7(a).

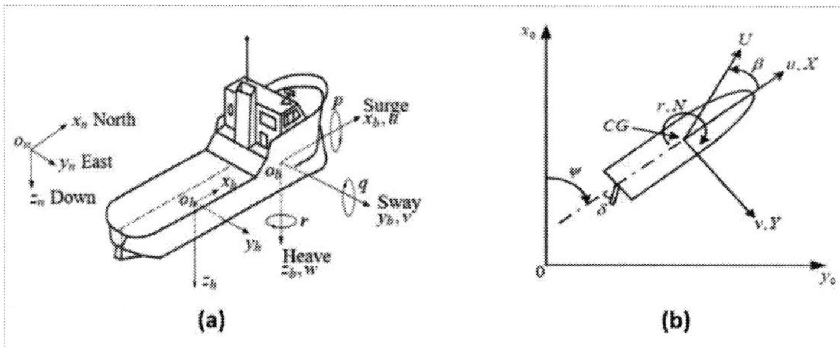

Figure 7. (a) Reference frames and motion variables (Left), **(b)** Motion coordination system (Right).

4.2 Equations of motion

When defining the equations of motions, moving coordination system in Figure 7(b), was used. Actual velocity (U) of vessel is resolved as transversal velocity (v) and advance velocity (u). (N) is the rotational velocity, r is the turning rate, and (β) is the drift angle, which is the angle between U and x-axis. ψ and δ are the ship heading angle and rudder angle, respectively.

Vessel is considered a solid object with three degrees of freedom; surge, sway and yaw. Dynamical model of the ship motion can be written as follows.

$$\left\{(m' + m'_x)\left(\frac{L}{U}\right)\left(\frac{\dot{U}}{U}\cos\beta - \dot{\beta}\sin\beta\right) + (m' + m'_y)r'\sin\beta\right\} = X' \tag{1}$$

$$\left\{(m' + m'_y)\left(\frac{L}{U}\right)\left(\frac{\dot{U}}{U}\sin\beta - \dot{\beta}\cos\beta\right) + (m' + m'_x)r'\cos\beta\right\} = Y' \tag{2}$$

$$\left\{(I'_{zz} + J'_{zz})\left(\frac{L}{u}\right)^2\left(\frac{\dot{U}}{L}r' + \frac{U}{L}\dot{r}'\right)\right\} = N' \tag{3}$$

Superscript (') in equations denotes the non-dimensional quantities and thus, X', Y' and N' define the dimensionless surge, sway forces and yaw moment acting on ship, respectively. m', m'_x, and m'_y denote the dimensionless mass of vessel and added masses in x and y directions. Ship length and draft are denoted as L and T.

The forces and moments noted on the right side of the equations can be described separately, following the concept of maneuvering modeling group (MMG) model, as follows.

$$X' = X'_H + X'_R + X'_P \tag{4}$$

$$Y' = Y'_H + Y'_R \tag{5}$$

$$N' = N'_H + N'_R \tag{6}$$

Where, subscripts H, P and R denote hull, propeller and rudder, respectively. Subsequently, the forces are separately defined as follows.

4.2.1 Forces and moments acting on hull

$$X'_H = \{X'_{\beta r} r' \sin\beta + X'_{uu} \cos^2\beta\} \tag{7}$$

$$Y'_H = \{Y'_\beta \beta + Y'_r r' + Y'_{\beta\beta}\beta|\beta| + Y'_{rr} r'|r'| + (Y'_{\beta\beta r}\beta + Y'_{\beta rr} r')\beta r'\} \tag{8}$$

$$N'_H = \{N'_\beta \beta + N'_r r' + N'_{\beta\beta}\beta|\beta| + N'_{rr} r'|r'| + (N'_{\beta\beta r}\beta + N'_{\beta rr} r')\beta r'\} \tag{9}$$

Longitudinal component of hydrodynamic force (X'_H), lateral force (Y'_H) and yaw moment (N'_H) are approximated at amidships by above polynomials. X'_{uu} in longitudinal component of hydrodynamic force denotes the ship resistance in forward straight motion. Y'_β, …., $N'_{\beta\beta r}$ are non-linear hydrodynamic derivatives, derived from regression on various hull experimental tests. [19]

4.2.2 Forces and moments induced by propeller

Hydrodynamic forces created by the propeller are as follows, as proposed by Kajima et al. [20]

$$X'_P = \frac{[C_{rp}(1 - t_p)n^2 D_P^4 K_T(J_P)]}{0.5\rho L d U^2} \tag{10}$$

$$Y'_P = 0 \tag{11}$$

$$N'_P = 0 \tag{12}$$

$$K_T(J_P) = C_1 + C_2 J_P + C_3 J_P^2 \tag{13}$$

$$J_P = \frac{U \cos\beta\,(1 - w_P)}{n D_P} \tag{14}$$

Where, t_P is the thrust deduction coefficient, N is propeller revolutions, D_P is propeller diameter, and w_p is wake fraction coefficient. K_T: thrust coefficient, C_{rp}, C_1, C_2, C_3 are constant for propeller open water characteristics, and J_P is the advance coefficient. In this section, lateral force and yaw moment are neglected.

4.2.3 Forces and moments induced by rudder

Rudder induced forces and moments are defined following a proposed model. [20]

$$X'_R = -(1 - t_R)F'_N \sin\delta \tag{15}$$

$$Y'_R = -(1 + a_H)F'_N \cos\delta \tag{16}$$

$$N'_R = -(x_R + a_H x_H)F'_N \cos\delta \tag{17}$$

Where, F'_N: dimensionless rudder normal force, δ: rudder angle, and a_H (ratio of additional lateral force), t_R (coefficient for additional drag), x_H and x_R denote the rudder-to-hull interaction coefficients. Total sway and surge forces and yaw moment were calculated following the MMG model as derived in equation 1 to equation 17, and consequently, the distance parameters (X and Y) were obtained. Simulation results of 35^0 starboard and port turning tests are graphically illustrated in Figure 8, and the simulation conditions are tabulated in Table 8.

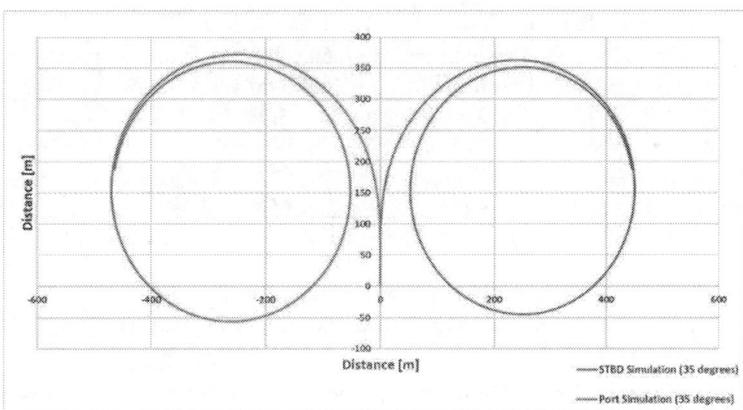

Figure 8. Turning circle (35^0) simulation test results.

Table 5. Simulation conditions for Port and STBD 35^0 turns.

	Port turning test	STB turning test
Draft (m)	5.9	5.9
Trim (m)	0.0	0.0
Initial speed (kts)	10.5	10.5
Advance (LBP) [a]	3.578	3.48
Turning circle diameter (LBP)	4.5	4.32

[a] LBP = 104 (m)

Simulations were carried out to obtain $20^0/20^0$ zig-zag parameters. The simulation conditions are tabulated in Table 6, and the respective results are illustrated in Figure 9.

Table 6. Simulation conditions for $20^0/20^0$ zig-zag

	$20^0/20^0$ Zig-zag test
Draft (m)	5.9
Trim (m)	0.0
Initial speed (kts)	10.5
First overshoot angle (deg.)	11.21
Second overshoot angle (deg.)	9.52

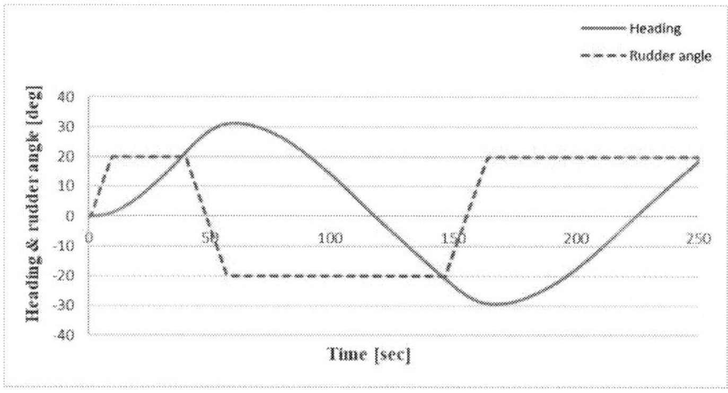

Figure 9. $20^0/20^0$ zig-zag simulation test results.

5. Result analysis

Analysis of preliminary results related to turning circle (port and starboard) and $20^0/20^0$ zig-zag test are tabulated in Table 7. All test results were compared and analyzed with IMO criterion and initial test values of the subject vessel. (Average result of each test represents the average value calculated using the series of respective real-ship test results, tabulated in Table 4.)

Table 7. Turning circle test results analysis.

	Turning circle diameter (LBP) [a]		$20^0/20^0$ zig-zag test
	Portside	Starboard side	First overshoot angle (deg.)
Initial test	3.6	3.9	17.6
Average value of real-ship tests	3.14	3.53	8.85
IMO criterion	5	5	25
Simulation value	3.57	3.49	11.21

[a] LBP = 104 (m)

When test results of turning circle maneuver are considered, it is clear that the simulation results are in par with both average and initial values of turning circle test. Based on the results, the simulation results obtained from the TCD simulations were validated and the vessel is recognized to be operating under maneuvering criterion recommended by the IMO. Comparison of aforementioned data is illustrated in Figure 10.

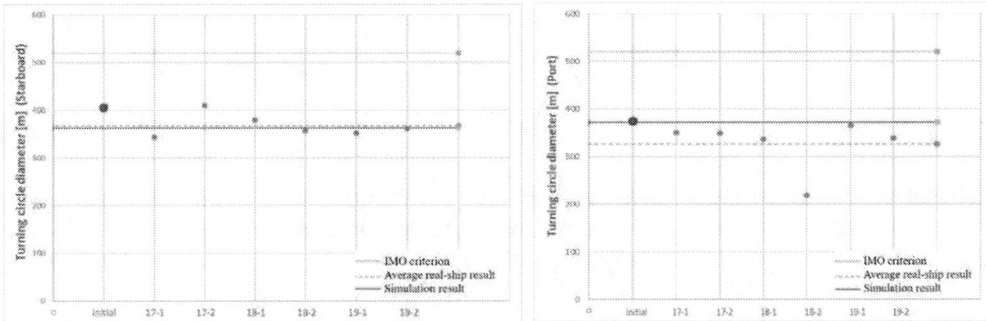

Figure 10. Turning circle test results analysis.

Similarly, $20^0/20^0$ zig-zag test results were analyzed, focusing on the first overshoot angle. As per the recommendation of IMO [15], value of the first overshoot angle in 20°/20° zig-zag test should not exceed 25 degrees. Average value of overshoot angle from each real-ship test was calculated and compared with the simulation results for validation. IMO criterion, first overshoot angles of each test, initial value of first overshoot angle and simulation results of 20°/20° zig-zag test are tabulated in Table 7 and graphically represented in Figure 11.

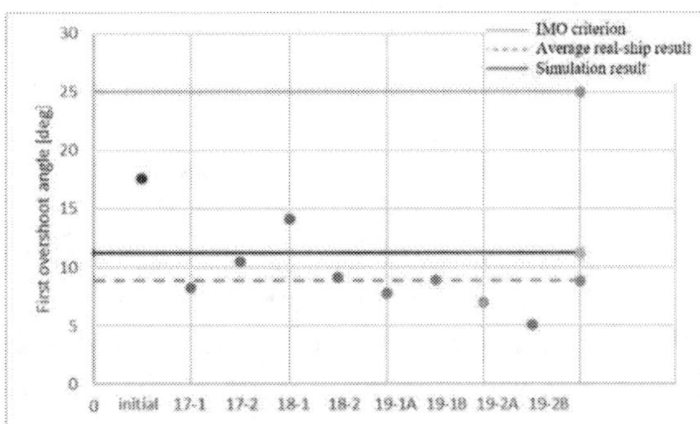

Figure 11. 20°/20° zig-zag test results analysis.

6. Conclusions

Purpose of carrying out the study was primarily to accumulate data through standard maneuvers, in order to analyze maneuvering characteristics of a training vessel with a gross tonnage of 6,700 ton. Secondary focus of the study was to develop and validate a comprehensive yet versatile simulation program for the specific vessel, to simulate and predict its vital maneuvers. Once adequate data were collected, maneuvering characteristics of the ship over a period of three years were analyzed to identify changes, which occurred over time as well as deviations due to different variables, such as displacement, and subsequently LCG and trim.

Analysis suggested that the advance and turning circle diameter of subject vessel tend to vary around base values, which were obtained initially after commissioning. Moreover, all the test results were evidently rested well within the limitations defined by the IMO, concluding the safety of operation in maneuvering of training vessel. Simulation program was used to obtain data for standard maneuvering tests and the data was then compared with values obtained from real-ship experiments. After successful comparisons, the simulation results were validated. Results obtained from simulations were identified to be within the safe limitations recommended by the IMO. Consequently, the developed simulation program was validated. In addition, the importance of accumulating real-ship data in different conditions with varying external factors in constructing simulation environments is emphasized.

Defining and using the external conditions and parameters within a simulated environment for the training vessel is identified as the future work of the study. Moreover, it is planned to carry out the standard maneuvering tests using the subject vessel to accumulate more data, in order to rectify and further validate the results obtained through simulations. It is evident that more variations in parameters and, deviations in experimental values are caused by external factors. Therefore, data that is more continuous are identified to be useful in making an accurate simulator program, which can aid in autonomous decision making when standard maneuvers are needed to be carried out.

7. References

[1] Wahlstrom M, Hakulinen J, Karvonen H and Lindborg I, 2015 Human Factors Challenges in Unmanned Ship Operations - Insights from Other Domains. *Proc. Manuf.* **3**, 1038–45.

[2] Porathe T, 2014 Remote Monitoring and Control of Unmanned Vessels – The MUNIN Shore Control Centre. *13th Int. Conf. on Computer and IT Applications in the Maritime Industries.*

[3] Porathe T, 2016 A Navigating Navigator Onboard or a Monitoring Operator Ashore? Towards Safe, Effective, and Sustainable Maritime Transportation: Findings from Five Recent EU Projects. *Transport. Res. Proc.* 14, 233–42.

[4] Kretschmann L, Burmeister H C and Jahn C, 2017 Analyzing the economic benefit of unmanned autonomous ships: An exploratory cost-comparison between an autonomous and a conventional bulk carrier. *Res. Transport. Bus. Manag.* **25**, 76–86.

[5] Ramos M A, Utne I B, Vinnem J E and Mosleh A, 2018a Accounting for human failure in autonomous ship operations. *European Safety and Reliability Conf.*, pp. 355–63.

[6] Utne I B, Sørensen A J and Schjølberg I, 2017 Risk Management of Autonomous Marine Systems and Operations. In: *Proc. 36th Int. Conf. Ocean. Offshore Arct. Eng.* pp. 1–10.

[7] Thieme C A, Utne I B and Haugen S, 2018 Assessing ship risk model applicability to Marine Autonomous Surface Ships. *Ocean Eng.* **165**, 140–154.

[8] Wrobel K, Montewka J and Kujala P, 2018b Towards the development of a system theoretic model for safety assessment of autonomous merchant vessels. *Reliab. Eng. Syst. Saf.* **178**, 209–224.

[9] Psaraftis H N and Kontovas C A, 2014 Ship speed optimization: Concepts, models and combined speed-routing scenarios. *Transp. Res. Part C Emerg. Technol.* **44**, 52–69.

[10] Tang P, Zhang R, Liu D, Huang L, Liu G and Deng T, 2015 Local reactive obstacle avoidance approach for high-speed unmanned surface vehicle. *Ocean Eng.* **106**, 128–140.

[11] Wang C, Zhang X, Li R and Dong P. 2019 Path Planning of Maritime Autonomous Surface Ships in Unknown Environment with Reinforcement Learning. *ICCSIP, CCIS. Vol 1006.* ed Sun F, Liu H and Hu D. (Springer, Singapore) 127-37.

[12] Barr R. 1993 A review and comparison of ship maneuvering simulation methods. *Trans. SNAME, vol.* **101**, 609-35.

[13] Salvesen N, Tuck E O and Faltinsen O, 1970 Ship motions and sea loads. *Trans. SNAME, vol.* **78**. 250-87.

[14] Papanikolaou A D and Schellin T E, 1992 A Three-Dimensional Panel Method for Motions and Loads of Ships with Forward Speed. *Ship Technol. Res.* **39**, 145–56.

[15] International Maritime Organization 2002 Standards for Ship manoeuvrability, *Report of the Maritime Safety Committee on its Seventy-Sixth Session-Annex 6 Res. MSC.* **137** (76). 1-6.

[16] Fossen T I, 1994 *Guidance and control of ocean vehicles* (Trondheim: British Library) p 1–494.

[17] Son K H and Nomoto K. 1982 On the Coupled Motion of Steering and Rolling of a High-Speed Container Ship. *J. Naval Arch. and Ocean Eng.* **20**.73-83.

[18] Blanke M and Jensen A G, 1997. Dynamic Properties of Container Vessel with Low Metacentric Height. *Tech. Report doc.* **R**-1997-4173.

[19] Inoue S, Hirano M and Kijima K, 1981 Hydrodynamic derivatives on ship manoeuvring. *Int. Shipbuilding Prog.* **28** (321). 112–25.

[20] Kijima K, Katsuno T, Nikiri Y and Furukawa Y, 1991 On the maneuvering performance of a ship with the parameter of loading condition. *Journal of the Society of Naval Architects of Japan.* 168(2) 9-26.

Evaluation of Navigation System Performance Requirements for Safe Autonomous Navigation.

S P Berge[1], J S Dæhlen[1] and M Hagaseth[1]

[1] SINTEF Ocean AS, Svein.Berge@sintef.no

Abstract. *This paper describes how different motion sensors can be integrated and fused to obtain necessary navigational requirements needed for autonomous operations. Both autonomous coastal navigation and automatic docking operations are analysed with respect to trade-offs in position and velocity accuracies, integrity, availability and continuity. By using AIS measurements from typical docking operations for the Norwegian coastal express vessels it is possible to define feasible safe trajectories for the docking operations. Justified requirements can be obtained using closed-loop analysis for the specific vessel and its operation. Both vessel and passenger safety can be related to vessels horizontal velocity and the safe velocity concept has been introduced to derive requirements for integrity, availability, alarm limit and continuity in case of unreported/unhandled errors in the navigation system. The outcome of this study is a design tool, or a generic methodology for selecting the most promising combination of the integrated sensor components to obtain necessary navigation system performance for safe autonomous navigation.*

1. Introduction

This paper describes a methodology for deriving justified technical requirements for the position, velocity and timing (PVT) sensor solution to be used for autonomous navigation. Two different autonomous operations have been investigated: autonomous coastal operation and automatic docking. Based on existing AIS data from the Norwegian coastal express vessels, feasible trajectories with corresponding accuracies for both positions and heading are obtained from several docking operations. When vessel is moored at the quay, also the accuracies for both position and heading are analysed and will be used as input to a noise model when analysing the closed-loop control system. Based on the IMO requirement [1] for position accuracies for automatic docking operation, we propose to justify the requirements to also include requirement for the heading measurements and also the velocities (surge, sway and yaw motion) to obtain required accuracies for closed loop control. Since the vessel is docked manually today, we propose to use the measured trajectories as a requirement for the closed loop control as well. Closed loop control requires that the vessel will follow a desired feasible trajectory which again depends on the actual vessel propulsion configuration and desired closed-loop behaviour. The IMO requirement of 0.1 m horizontal accuracy which is related to accuracy of the position measurement, is not necessarily needed in closed loop control to obtain a **motion accuracy** of 0.1m in for a selected position on the real physical vessel due to the nature of a closed-loop control system.

2. Selected autonomous operations

Typical concept of operations for the coastal express vessels is presented in Figure 1. During undocking and docking operations, GNSS augmentation service (SBAS, GBAS) will give better position accuracies than during the coastal operation. For the coastal express vessels travelling along the Norwegian coastline, the GNSS augmentation system is available for most distances. Since the navigator and captains on the coastal vessel have permission to dock the vessel themselves, there is no need for external piloting. Typical position accuracy requirement for the navigation in coastal areas should be better than 10 m and for the docking and undocking case, the requirement should be much lesser than 10 m. These requirements are rather generic and must be further analysed and adopted to case by case.

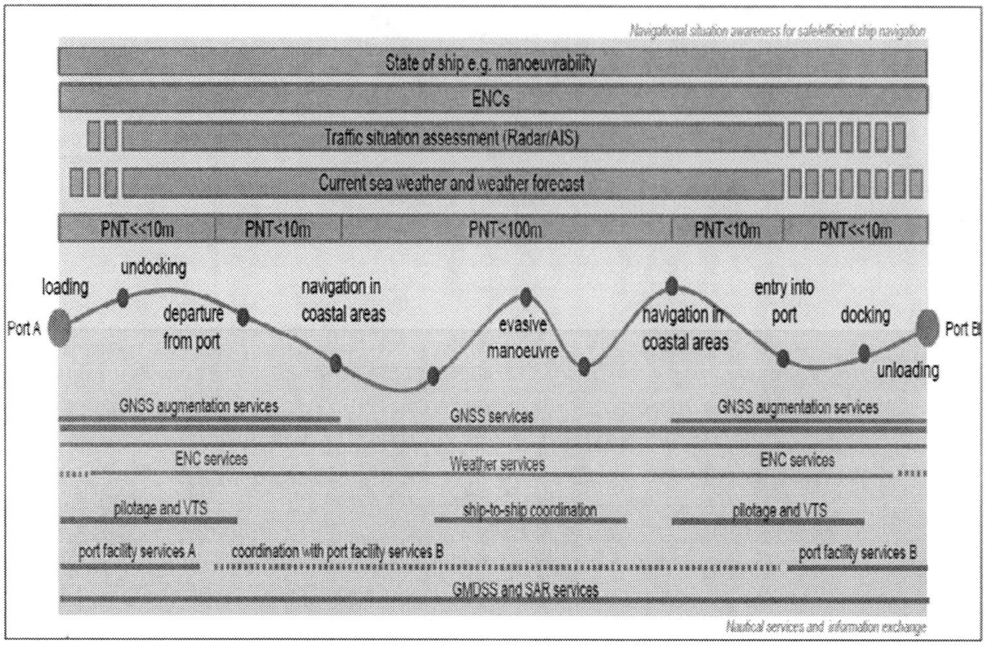

Figure 1*. Internal/external information demand and support by services for ship navigation, [2].*

2.1. Autonomous coastal navigation

During coastal operation, only the main propulsions units are used, not the bow thrusters. The navigator uses the waypoint tracking control function to follow a desired track from the ECDIS system by use of rudder. Typical, feedback signals to the controller are position measurements from GNSS and heading/rate of turn. The vessel speed is still controlled manually. If there is a need for an evasive manoeuvre, the navigator needs to do this manually. For an unmanned, autonomous coastal operation, both own vessel's position (localisation) and the surrounding objects must be identified and used as input to the control system. By fusing cameras and radar images or other sensors it is possible to detect and identify critical objects with properties like speed and geometry that has to be digitized in same manner as human situational awareness (perception, comprehension and projection). Maritime situational awareness is still an uncovered and new research field. Replacing human on bridge with computers is difficult, but it is possible to use available navigation data to **assist navigator** performing safe operation as a first step. In the EU project Hull-to-hull [3], the concept of uncertainty zones is introduced. The uncertainty zone is the outer boundary of the vessel based on GNSS, heading and simple geometry description of the vessel, see Figure 3. Since the AIS data contains a specific position, course over ground and heading, it is possible to compute vessel's own outer boundary with probability. In addition, by transmitting GNSS raw data between several vessels, it is possible to derive relative distances and velocities for neighbouring objects with high precision and high integrity.

Figure 2*. The Norwegian coastal express vessel MS Polarlys.*

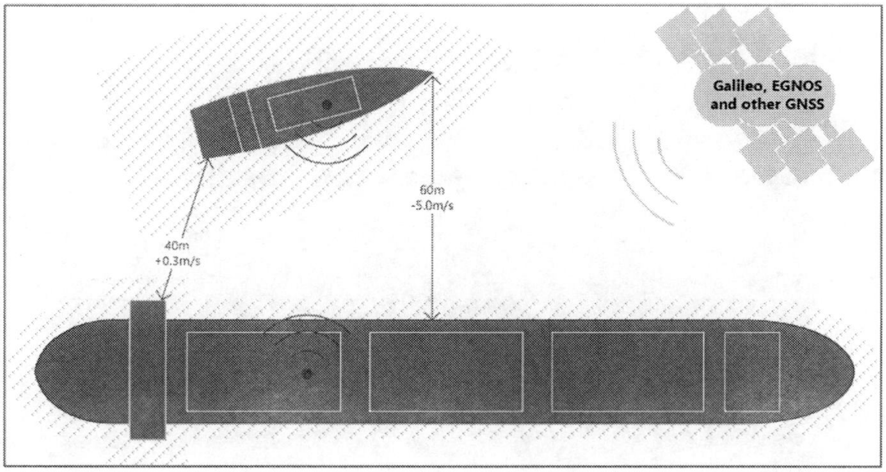

Figure 3*. Uncertainty zone (in red). Courtesy: Kongsberg Seatex AS, [3]*

2.2. Automatic docking

<u>*Analysis of AIS data from the docking operation:*</u>

As a part of the study, analysis of AIS data that has been performed during this task to be able to calculate the relation between the distance to the quay and the ship's speed and heading during a port approach. AIS data from one year for the Port of Trondheim covering the docking of the coastal expresses MS Polarlys (Figure 2). MS Nordnorge and MS Trollfjord has been used. MS Polarlys and MS Nordnorge are both RoPax ships with length 123 meters, width 19,5 meters and gross tonnage 11300, while MS Trollfjord has a length of 136m, width 21m and gross tonnage 16100. These vessels operate on the coastal express route between Bergen and Kirkenes, including port visits to Trondheim both on north- and south-bound voyages. The reported AIS position is referred to a specific location on the vessel and is obtained from one of the AIS messages. It is reported as 4 distances in Table 1.

Coastal express	**To bow (m)**	**To stern (m)**	**To starboard (m)**	**To port (m)**
MS Polarlys	33	90	3	15
MS Nordnorge	47	76	14	10
MS Trollfjord	14	122	11	11

Table 1*. Location of the reported AIS position on the vessel.*

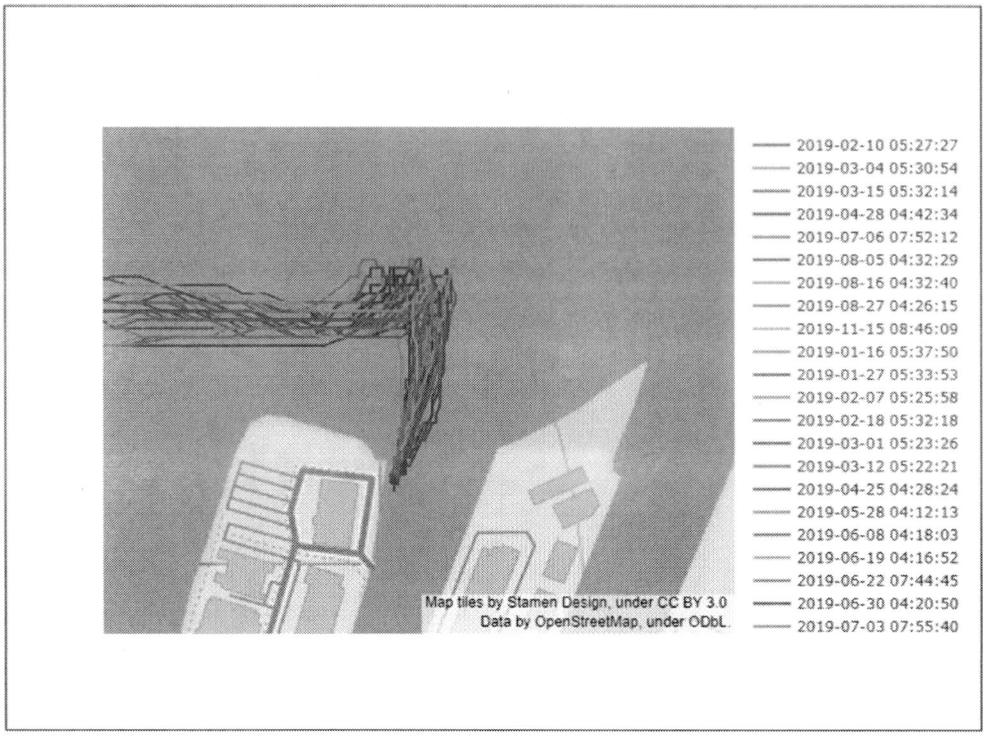

Figure 4. *Approach trajectories of MS Polarlys, MS Nordnorge and MS Trollfjord during 2019.*

Having established a common set of approach trajectories, the speed-over-ground variable of each AIS-message is analysed with respect to the distance to the dock. Please note that as all AIS-messages with speed-over-ground reported to less than 0.3 knots are rejected. The distance from each AIS-message to the final position of each trajectory is calculated, and grouped into intervals of 0-10, 10-20, 20-30, 30-40, 40-50, 50-75 and 75-100 meters. The mean and standard deviation are calculated for the speed-over-ground and heading variable from the AIS-messages for each bin. The result can be seen in Figure 5, Figure 6 and Table 2.

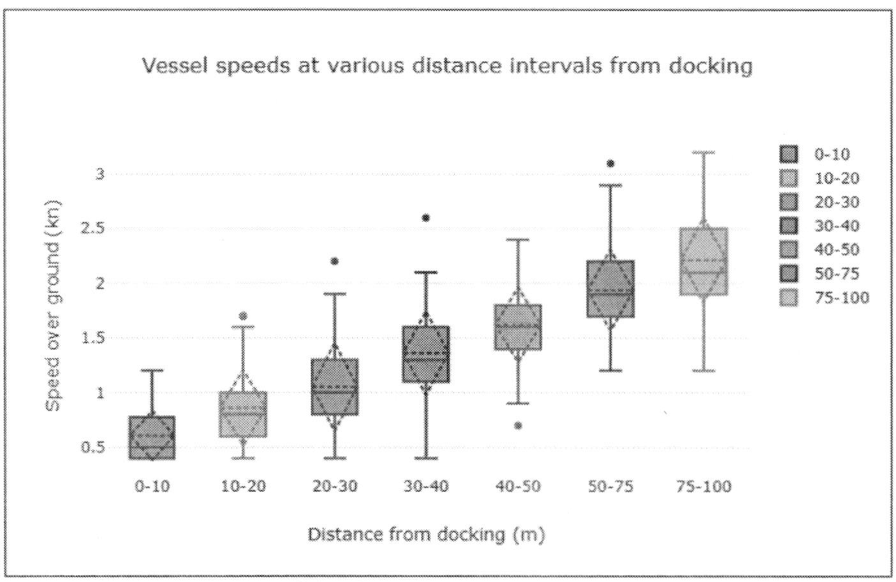

Figure 5. *Speed over ground for the trajectories at various distances from the dock.*

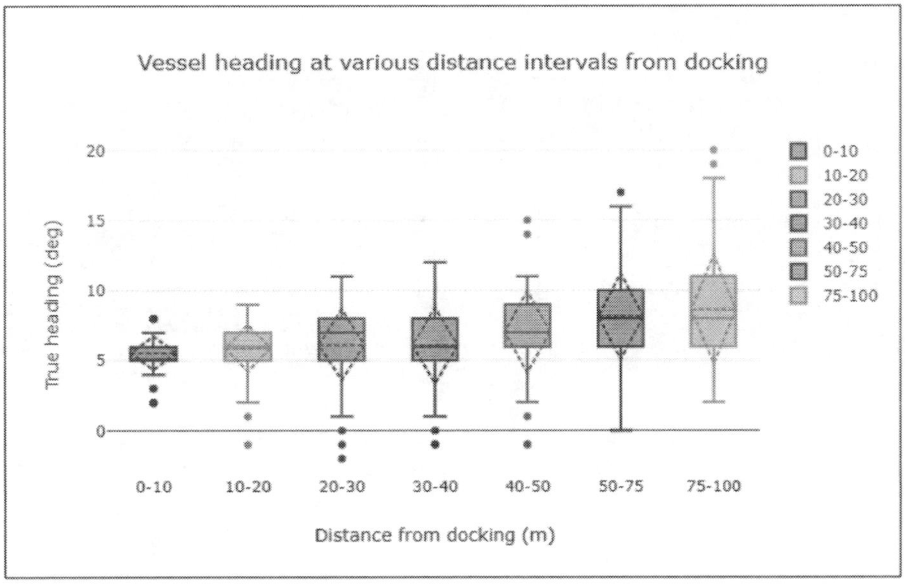

Figure 6. *True heading for the trajectories at various distances from the dock.*

The trend for the heading during the approach is that the variance has more of a decline as the vessel approaches the dock compared to the speed-over-ground, converging to around 6 degrees at dock. The mean value and standard deviations shown in Table 2.

Interval	Speed-over-ground (knots / $\frac{m}{s}$)		Heading (deg)	
	Mean (knots / $\frac{m}{s}$)	Std. dev. (knots / $\frac{m}{s}$)	Mean (deg)	Std. dev (deg)
0-10m	0.61 / 0.31	0.22 / 0.11	5.55	1.21
10-20m	0.86 / 0.44	0.34 / 0.17	5.90	1.66
20-30m	1.05 / 0.54	0.40 / 0.21	6.13	2.53
30-40m	1.36 / 0.70	0.38 / 0.20	6.07	2.71
40-50m	1.62 / 0.83	0.34 / 0.17	7.03	2.92
50-75m	1.94 / 1.00	0.37 / 0.19	8.14	3.08
75-100m	2.22 / 1.14	0.39 / 0.20	8.64	3.89

Table 2. *Summary of mean and standard deviation for speed over ground and heading for various distance intervals.*

Analysis of anchored operation:

When vessel is moored, we assume that the positions are fixed, and corresponding position measurements and heading is sensor noise. The standard deviation for both vessels is similar, indicating that the majority of points are indeed from an actually moored vessel. The results are summarized in Table 3.

Position and heading	MS Trollfjord	MS Nordnorge
Latitude (m)	0.889	0.534
Longitude (m)	0.832	0.497
Heading (deg)	0.446	0.443

Table 3. *Standard deviation for latitude, longitude and heading measurements while moored.*

A typical plot showing the different accuracies are computed for selected points around the hull. The accuracy model is based on 1000 pseudorandom numbers drawn from the standard normal distribution function ***randn*** in MATLAB. As an example, position accuracies for different location around the hull is shown in Figure 7. For illustration, the heading accuracy (one sigma) is increased from 0.44 to 2.0 (deg) to see the effect of position uncertainties around the hull. Position accuracies on latitude and longitude are unchanged for MS Nordnorge which is used in the further study.

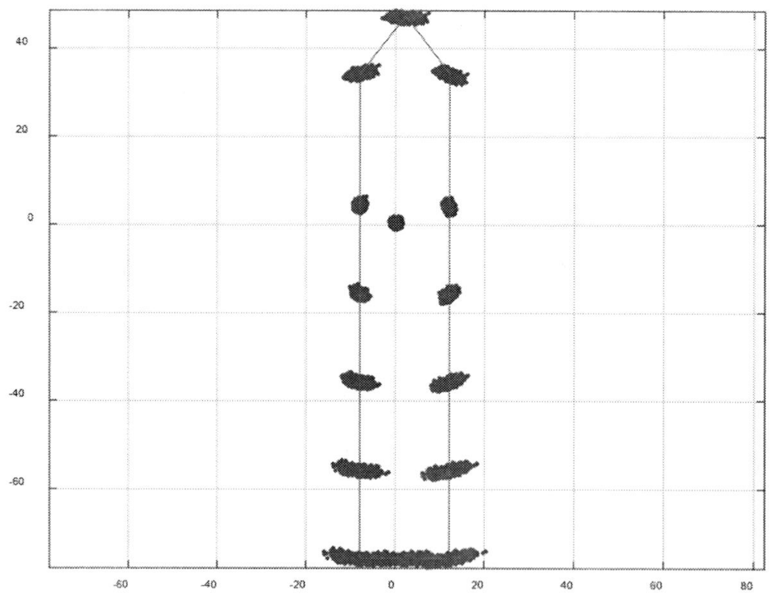

Figure 7. *Computed position accuracy as points for different locations around the hull (MS Nordnorge). Both axes in meters. Notice, the heading standard deviation is exaggerated to illustrate the effect of varying position accuracy around the hull. Axes in meters.*

The GNSS receiver is located as position (0,0) and the analysis is based on the aft port point (-76, -8). All coastal ferries are docking alongside with port side to the quay.

3. Closed-loop analysis of automatic docking

To propose new requirements for PVT used for autodocking, a simple mathematical model describing the closed-loop dynamics can be used.

In low speed operation, surge, sway and yaw motion of a ship can be modelled as 3 individual linear mass-damper-spring systems, see Figure 8. Surge motion are coupled with heave and pitch motion, but for the sake of simplicity, the heave and pitch motion are not considered for the analysis. Sway and yaw are coupled dynamics, but in this simple analysis we further assume that the motions are decoupled. Detailed information on mathematical modelling of ship's motion can be found in Fossen [4].

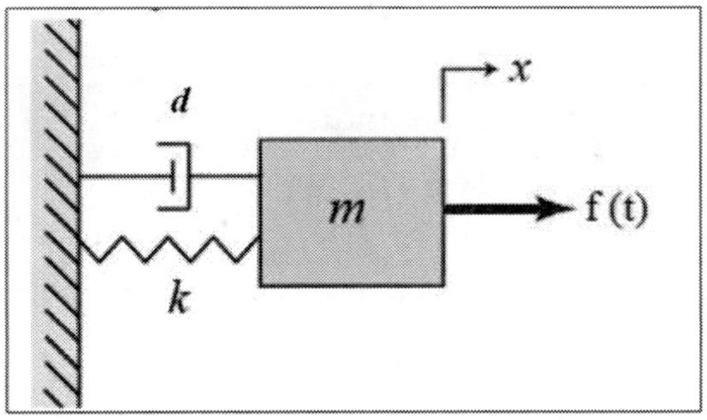

Figure 8. *Linear mass-damper-spring system.*

Introducing a simple PD control law by:

$$f = -K_d\dot{x} - K_p x$$

yields the closed-loop dynamics described by the following mass-damper-spring model. K_p and K_d are controller gains for the PD controller. Notice, that for surge, sway and yaw motion there is no natural restoring/spring force (k=0).

$$m\ddot{x} + (d + K_d)\dot{x} + K_p x = 0$$

where

$$2\zeta\omega_n = \frac{d + K_d}{m}$$

and

$$\omega_n = \sqrt{\frac{K_p}{m}}$$

In closed-loop, Kp and Kd are both positive constants and the natural period $T_n = \frac{2\pi}{\omega_n}$ satisfies

$$2\zeta\omega_n = 2\zeta\frac{2\pi}{T_n} = \frac{d + K_d}{m} = \frac{1}{T} + \frac{K_d}{m}$$

where d is the hydrodynamic damping coefficient, ω_n is the natural frequency (undamped oscillator when d=0) and ζ is the relative damping ratio. In principle, the feedback from velocity will increase the damping, whilst feedback from position will define the closed-loop bandwidth. Choice of controller gains must be based on feasibly values due to desired response needed and effect of the "noisy" control signals to the propulsion system.

To obtain a **physical motion** accuracy of 0.1 meter as recommended by IMO, we have changed the noise model to fit the closed-loop requirement of 0.1 meter sideways and justified control gains to fit the desired closed-loop bandwidth requirement. Justified requirements are then:

$$p = 0.3\ m\ (95\%)$$
$$v = 0.04\ m/s\ (95\%)$$
$$\psi = 0.2\ \text{deg}\ (95\%)$$

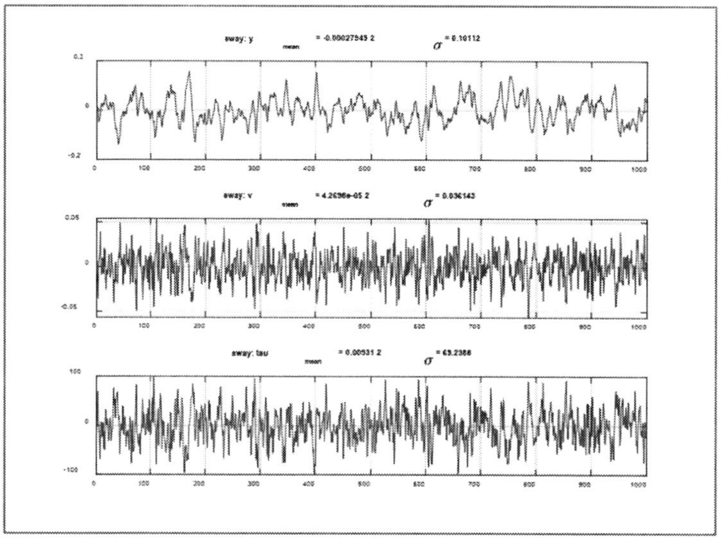

Figure 9. Closed-loop response in sway at selected point. Position in [m], velocity in [m/s]and force in [tonnes].

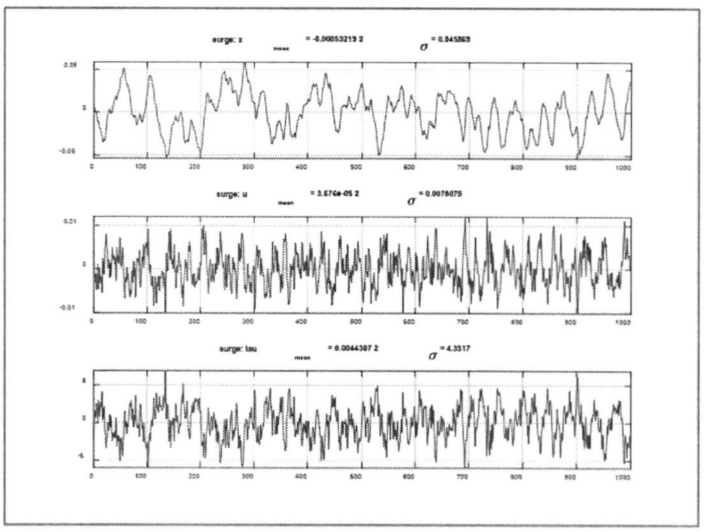

***Figure 10**. Closed-loop response in surge at selected point. Position in [m], velocity in [m/s]and force in [tonnes].*

Based on the simulation, the closed-loop control system will then have a sideways motion with an accuracy of 0.1 m (95%), whilst longship accuracy is 0.05 m (95%). The thrust needed for sideways control are relatively much higher compared to surge control because since heading noise will contribute "more" in longship direction than the sideways direction. The accuracy requirement for the heading sensor can be tightened to obtain less thrust usage of course. This study illustrate that the sensor configuration and sensor accuracies will be a trade-off of complexity, performance and cost for each vessel and selected operations. Shortly, **the proposed methodology can be applied for different vessel and different type of operations as well. Due to scaling effect, a similar smaller or larger vessel can be scaled accordingly using Froude scaling.**

4. Integrity, continuity and availability requirements

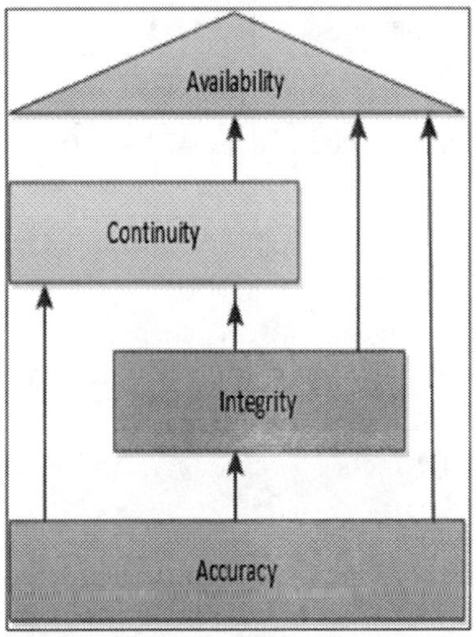

Figure 11*. Dependency and hierarchy of service performance parameters*

In addition to accuracy requirement, the integrity, continuity and availability requirements are also important with regards of safety. **Integrity risk** is the probability of providing a signal that is out of tolerance without warning the user in a given period of time. It defines the maximum probability with which a receiver is allowed to provide position failures not detected by the integrity monitoring system. The **continuity** of a system is the ability of the total system (comprising all elements necessary to maintain vessel's position within the defined area) to perform its function without interruption during the intended operation. The **availability** of a navigation system is the percentage of time that the services of the system are usable by the navigator. The dependencies of the requirements are shown in Figure 11.

Absolute safety cannot be achieved in any human or automatic control activity.

To justify the integrity requirements, which is highly connected to passenger's safety, we propose to justify the integrity requirements according to the effect of failure(s) during autonomous navigation. For example, the time to alarm (TTA) and alarm limit (AL) for example will depend on the actual requirement for the closed-loop control and the desired trajectory for the docking case. For the specific docking case, it is possible to define the desired trajectory to ensure a **safe speed** towards the dock if the control system fails. A separate fail-safe mechanism independent on the PVT solution can be implemented to avoid risk for the vessel or passengers.

For passenger safety, the probability concept can be used to define different safety levels as maximum horizontal acceleration values.

4.1. Probability concept

Based on the International Code on Safety for High-Speed Craft [5], we propose to use the probability concept to define **maximum horizontal acceleration levels.** If the actual velocity is below a safe limit, failure in position measurement will not cause any severe damage to vessel and thus reduce safety for passengers.

The probabilities quoted shall be on an hourly or per-journey basis, depending on which is more appropriate to the assessment in question.

Probability	Explanation	Probability value
Frequent	**Likely** to occur often during the operational life of a particular vessel	10^{-3}
Reasonably probable	**Unlikely** to occur often, but may occur **several times** during the total operational life of a particular vessel.	10^{-3} to 10^{-5}
Remote	**Unlikely** to occur to every vessel, but may occur to a **few vessel** of a type over the total operational life of a **number of vessel of the same type**.	10^{-5} to 10^{-7}
Extremely remote	**Unlikely** to occur when considering the total operational life of a number of vessels of the type, but nevertheless shall be considered as being possible.	10^{-7} to 10^{-9}
Extremely improbable	is one which is so extremely remote that it shall not be considered as possible to occur.	Less than 10^{-9}

Table 4. *Probability, explanation with values, [5].*

4.2. Safety levels

Safety level is a numerical value characterizing the relationship between vessel performance represented as horizontal single-amplitude acceleration (g) and the severity of acceleration-load effects on standing and sitting humans. The safety levels and the corresponding severity of effects on passengers and safety criteria for vessel performance is defined in Figure 12.

EFFECT	CRITERIA NOT TO BE EXCEEDED		COMMENTS
	Type of load	Value[2]	
LEVEL 1 MINOR EFFECT Moderate degradation of safety	Maximum acceleration measured horizontally[1]	0.20g	0.08g: Elderly person will keep balance when holding. 0.15g: Mean person will keep balance when holding. 0.15g: Sitting person will start holding.
LEVEL 2 MAJOR EFFECT Significant degradation of safety	Maximum acceleration measured horizontally[1]	0.35g	0.25g: Maximum load for mean person keeping balance when holding. 0.45g: Mean person falls out of seat when not wearing seat belts.
LEVEL 3 HAZARDOUS EFFECT Major degradation of safety	Collision design condition (calculated) Maximum structural design load, based on vertical acceleration at centre of gravity	Ref. 4.3.3 Ref. 4.3.1	Risk of injury to passengers; safe emergency operation after collision. 1.0g: Degradation of passenger safety
LEVEL 4 CATASTROPHIC EFFECT			Loss of craft or/and fatalities.

Figure 12. *Safety level with acceleration limits, [5].*

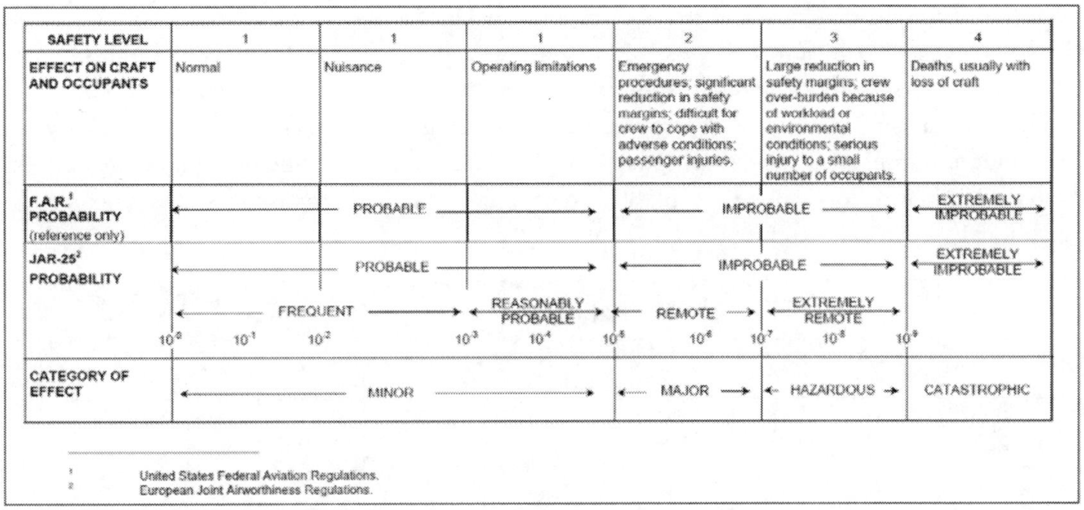

Figure 13. *Safety level with probability, [5].*

Based on the integrity risk probability requirement of 10^{-5}, which is the limit of safety level 1 (minor), and safety level 2 (major), the maximum peak horizontal acceleration level should be less than 0.35g = 3.5 m/s^2.

Notice, if applying Froude scaling, the maximum acceleration level will be the same for different vessel types and operations.

4.3. Safety analysis

For the automatic docking case, we can assume that the vessel will be able to stop within 1 seconds, which is rather conservative. If we assume a continuity requirement of 5 sigmas for the velocity and uses the approach speed of 1 m/s for the distance 50-75 meter from the quay, the **maximum safe speed will be around 2 m/s** and corresponding g-force of 0.2 g horizontally. By using safety level 3 (hazardous and extremely remote) to define the acceptable integrity risk of 10^{-8} during docking phase (approximately 5 minute) it is possible to increase the alarm limit to 50 meter in this case. Time to alarm depends on the ships dynamics and must be chosen to small enough to give necessary response to reduce the speed below the safe speed limit. In our case, the time constant for this vessel is around 30 seconds, so alarm time less than 10 seconds will be sufficient time to make necessary correction in case of warning or alarm. Finally, we can propose the integrity requirements for this vessel's docking operation as:

- Alarm limit <= 50 meter
- Time to alarm <= 10 seconds
- Integrity risk 10^{-8} per 5 minutes
- Continuity 99.97% per 5 minutes

5. Conclusions

As seen from the study, the requirements for a feasible PVT sensor solution for autonomous operations depends highly on selected vessel and operation. The recommended performance values proposed by IMO for different operations must be adapted for autonomous operations and must include requirements for also heading and velocities. By using feedback control, it is necessary to include heading accuracy as additional requirements to calculate the vessel uncertainty zone describing the probability of the outer hull boundary. Different uncertainties covering both sensor noise, mounting error and error in the geometry model must be considered when computing the real physical location of selected points around the vessel. In addition, the feedback control system needs both position and velocity measurements to obtain the necessary damping and desired response when approaching the dock. To reduce risk and effect of unwanted collision, it is possible to introduce the safe speed concept, which is maximum velocity the vessel can dock without significant degradation of safety. Based on the AIS analysis, it is possible to establish feasible safe trajectories for the docking operation and adjust the integrity risk and alarm limit accordingly. If the vessel speed is within safe limit during the docking phase, any failures due to navigational errors will be acceptable. To conclude, the final proposed requirements for the PVT navigational system are then proposed in Table 5.

	System level parameters						Service level parameters			
	Accuracy (95%)			Integrity						
Autonomous Operations	Horizontal (m)	Heading (deg)	Linear velocity (m/s)	Alarm limit (m)	Time to alarm (sec)	Integrity risk	Availability % per 30 days	Continuity	Coverage	Fixed interval (sec)
Coastal	10	1.0	-	25	<10	10^{-5} 3 hours	99.8%	99.97% 3 hours	Global	2
Autodocking	0.3	0.2	0.04	50	<10	10^{-8} 5 min	99.8%	99.97% 5 min	Local	<1

Table 5*. Justified PVT requirements for coastal and autodocking for the coastal express vessel.*

References

[1] IMO/Res.A.915/22. (2001). *REVISED MARITIME POLICY AND REQUIREMENTS FOR A FUTURE GLOBAL NAVIGATION SATELLITE SYSTEM (GNSS). November 2001*

[2] Göge, Dennis & Engler, Evelin & Wrede, Carl. (2018). From E-Navigation to Resilience Engineering.

[3] H2020, EU project Hull-to hull, https://www.sintef.no/projectweb/hull-to-hull/

[4] Fossen, T. I, (2011). Handbook of Marine Craft Hydrodynamics and Motion Control, https://doi.org/10.1002/9781119994138

[5] International Code on Safety for High-Speed Craft (1995), IMO.

[6] NAVISP-1, ESA project, Artificial Intelligence/Machine Learning Sensor Fusion for Autonomous Vessel Navigation (ARTISAN), https://navisp.esa.int/project/details/86/show

Acknowledgements

This paper is based on several reports submitted as project deliveries to the European GNSS Agency (GSA) and is based on initial concept definition, user requirements and gap analysis of current state-of-the-art technologies and standards related to autonomous navigation. The project reports are based on a cooperation between Kongsberg Seatex AS (NO), SINTEF Ocean AS (NO), SINTEF Digital (NO), Mampaey Offshore Industries (NL) and KU Leuven (BE). The H2H project has its own project website [3]. The H2H project has received funding from the European GNSS Agency under the European Union's Horizon 2020 research and innovation programme grant agreement No 775998. In addition, the proposed methodology for establishing justified PVT requirements is a part of the project ARTISAN, funded by European Space Agency (ESA), a cooperation between SINTEF Ocean AS, SINTEF Digital and Kongsberg Seatex AS, see website [6].

AUTHOR INDEX

Institute of Physics
Dirac House, Temple Back
Bristol BS1 6BE UK

ISSN: 1757-8981
ISBN 978-1-7138-3192-1